Christian Kirches

Fast Numerical Methods for Mixed-Integer Nonlinear Model-Predictive Control

Christian Kirches

# Fast Numerical Methods for Mixed-Integer Nonlinear Model-Predictive Control

VIEWEG+TEUBNER RESEARCH

Bibliographic information published by the Deutsche Nationalbibliothek
The Deutsche Nationalbibliothek lists this publication in the Deutsche Nationalbibliografie; detailed bibliographic data are available in the Internet at http://dnb.d-nb.de.

Dissertation Heidelberg University, 2010

1st Edition 2011



Editorial Office: Ute Wrasmann | Anita Wilke

Vieweg+Teubner Verlag is a brand of Springer Fachmedien.
Springer Fachmedien is part of Springer Science+Business Media.
www.viewegteubner.de

Cover design: KünkelLopka Medienentwicklung, Heidelberg
Printed on acid-free paper

ISBN 978-3-8348-1572-9

# Acknowledgments

*„Aber in der Beschäftigung selbst Vergnügen finden*
*— dies ist das Geheimnis des Glücklichen!"*

Sophie Mereau

This work was presented to the combined faculty for mathematics and natural sciences of Heidelberg University as a doctoral thesis in mathematics on July 23rd, 2010, and was successfully defended on November 2nd.

My deeply felt gratitude for their excellent support goes to my teachers and mentors *Professor Dr. Dr. h.c. Hans Georg Bock* and *Professor Dr. Gerhard Reinelt*, as well as *Dr. Sebastian Sager* and *Dr. Johannes P. Schlöder*. Their extensive knowledge laid the essential foundation for this thesis. With their warm and open–minded spirit and the friendly and cooperative atmosphere they maintain in their research groups, they made my work in the past three years a pleasure.

Many members of the work group *Simulation and Optimization* as well as the junior research group *Mathematical and Computational Optimization* have made significant contributions to the creation and success of this work. Among them, it is my desire to say a special thank you to *Jan Albersmeyer, Dörte Beigel, Chris Hoffmann, Andreas Potschka* and *Leo Wirsching*. Over countless cups of coffee they gave many and much needed inspirations and contributed to the progress of this work during many discussions. Without you, it wouldn't have been even half the fun, but certainly twice the time. Moreover, I would like to point out *Alexander Buchner* and *Florian Kehrle*, who both made most valuable contributions during their diploma theses.

For their support in proof–reading of this thesis I owe thanks to *Sebastian Sager* and *Johannes Schlöder*. For any remaining errors I assume the sole reponsibility. I cordially thank *Margret Rothfuss* and *Thomas Klöpfer* for their valuable support with all organisational and technical issues.

Financial support by *Heidelberg University*, the *Steinbeis–Transferzentrum 582 „Simulation und Optimierung"*, the DFG Graduate School 220 *"Heidelberg Graduate School of Mathematical and Computational Methods for the Sciences"*, and the *7. Framework Programme of the European Union under contract n$^o$ FP7-ICT-2009-4 248940* is gratefully acknowledged.

To my parents *Claus* and *Ulrike*, my brother *Michael* and my sister *Anja* I wish to say a big thank you for your love and support throughout.

My final lines go to my girlfriend *Simone Evke*. Thank you for your love, for your patience and support whenever my mind was entangled in upper and lower triangles, and for every plan for the future we made up together.

Christian Kirches

# Abstract

This work aims at the investigation and development of fast numerical methods for nonlinear mixed–integer optimal control and model–predictive control problems. A new algorithm is developed based on the direct multiple shooting method for optimal control and on the idea of real–time iterations, and using a convex reformulation and relaxation of dynamics and constraints of the original predictive control problem. This algorithm relies on theoretical results and is based on a nonconvex Sequential Quadratic Programming method and a new active set method for nonconvex parametric quadratic programming. It achieves real–time capable control feedback though block structured linear algebra for which we develop new matrix updates techniques. The applicability of the developed methods is demonstrated on several applications.

This thesis presents novel results and advances over previously established techniques in a number of areas as follows:

- We develop a new algorithm for mixed–integer nonlinear model–predictive control. It consists of a combination of Bock's direct multiple shooting method, a reformulation based on partial outer convexification and relaxation of the integer controls, a rounding scheme, and a real–time iteration scheme.
- For this new algorithm we establish an interpretation in the framework of inexact Newton–type methods and give a proof of local contractivity assuming an upper bound on the sampling time, implying nominal stability of this new algorithm.
- We propose a convexification of path constraints directly depending on integer controls that guarantees feasibility after rounding, and investigate the properties of the obtained nonlinear programs. We show that these programs can be treated favorably as Mathematical Program with Vanishing Constraints, a young and challenging class of nonconvex problems.
- We describe a Sequential Quadratic Programming method and develop a new parametric active set method for the arising nonconvex

quadratic subproblems. This method is based on strong stationarity conditions for Mathematical Program with Vanishing Constraints under certain regularity assumptions. We further present a heuristic for improving stationary points of the nonconvex quadratic subproblems to global optimality.

- The mixed–integer control feedback delay is determined by the computational demand of our active set method. We describe a block structured factorization that is tailored to Bock's direct multiple shooting method. It has favorable run time complexity for problems with long horizons or many controls unknowns, as is the case for mixed–integer optimal control problems after outer convexification.
- We develop new matrix update techniques for this factorization that reduce the run time complexity of all but the first active set iteration by one order.
- All developed algorithms are implemented in a software package that allows for the generic, efficient solution of nonlinear mixed–integer optimal control and model–predictive control problems using the developed methods.

# Contents

# List of Figures

# List of Tables

# List of Acronyms

| | |
|---|---|
| ACQ | Abadie Constraint Qualification |
| BDF | Backward Differentiation Formula |
| BFGS | Broyden–Fletcher–Goldfarb–Shanno |
| BVP | Boundary Value Problem |
| CQ | Constraint Qualification |
| DAE | Differential Algebraic Equation |
| DFP | Davidon–Fletcher–Powell |
| END | External Numerical Differentiation |
| EQP | Equality Constrained Quadratic Program |
| FLOP | Floating–Point Operation |
| GCQ | Guignard Constraint Qualification |
| HPSC | Hessian Projection Schur Complement |
| IND | Internal Numerical Differentiation |
| IVP | Initial Value Problem |
| KKT | Karush–Kuhn–Tucker |
| LICQ | Linear Independence Constraint Qualification |
| LLSCC | Lower Level Strict Complementarity Condition |
| LMPC | Linear Model Predictive Control |
| LP | Linear Program |
| MFCQ | Mangasarian–Fromovitz Constraint Qualification |
| MILP | Mixed–Integer Linear Program |
| MINLP | Mixed–Integer Nonlinear Program |
| MIOC | Mixed–Integer Optimal Control |
| MIOCP | Mixed–Integer Optimal Control Problem |
| MIQP | Mixed–Integer Quadratic Program |
| MPBVP | Multi–Point Boundary Value Problem |
| MPC | Model Predictive Control |
| MPCC | Mathematical Program with Complementarity Constraints |
| MPEC | Mathematical Program with Equilibrium Constraints |
| MPVC | Mathematical Program with Vanishing Constraints |
| NLP | Nonlinear Program |
| NMPC | Nonlinear Model Predictive Control |
| OCP | Optimal Control Problem |

| | |
|---|---|
| ODE | Ordinary Differential Equation |
| PQP | Parametric Quadratic Program |
| QP | Quadratic Program |
| QPVC | Quadratic Program with Vanishing Constraints |
| RTI | Real–Time Iteration |
| SOS | Special Ordered Set |
| SQP | Sequential Quadratic Programming |

# 0 Introduction

A dynamic process in the spirit of this work is a naturally occurring or specifically designed phenomenon whose properties, varying in time, can be observed, measured, and affected by external manipulation. It is an old and natural question to then ask for a description of the special way in which such a process should be affected in order to serve an intended purpose. The development and availability of mathematical methods for the simulation and optimization of dynamic processes has had an enormous impact on our lives in the past that until today continues to rise. An ever growing number of increasingly complex dynamic processes from various scientific disciplines such as biology, chemistry, economy, engineering, and physics can be simulated and optimized for various criteria.

Certain features of a dynamic process however make this optimization task harder to conduct from a mathematical and computational point of view. One such feature is the presence of controls that may attain one of an only finite selection of different values. One may think here of a simple switch that either enables or disables a certain part or property of the dynamic process. As long as this process is running, the decision on whether to turn this switch on or off can obviously be made afresh at every instant. The question we ask here is this: How to operate this switch in a way that allows the dynamic process to achieve a certain prescribed goal, without violating certain other prescribed constraints? Certainly, even if we limit the number of times the switch may be operated, there are a great many of different possibilities to consider and it is in general all but clear what an optimal answer might look like. Moreover, we may find ourselves in a hurry to decide on a switching strategy as the process keeps running while we ponder on this question. A late decision may prevent the process from reaching the desired goal. Worse yet, the process may violate critical constraints relevant to our safety.

This doctoral thesis in applied mathematics is meant to be understood as one new step towards real–time optimal control of dynamic processes that can be affected by both continuous and discrete controls.

### Optimal Control

The optimization of dynamic processes that are in our case described by systems of Ordinary Differential Equations (ODEs) or Differential Algebraic Equations (DAEs) is referred to as the discipline of optimal control. For the numerical solution of optimal control problems, direct methods and in particular simultaneous or all–at–once methods have emerged as the methods of choice for most practical applications [27, 112]. Amongst them, direct collocation methods [10, 26, 101] and direct multiple shooting methods [36, 134, 167] are the most prominent ones. Using a discretization of the infinite–dimensional control space by a finite number of control parameters, these methods transform the optimal control problem into a large and possibly structured Nonlinear Program (NLP). Efficient numerical methods are based on the exploitation of this structure, see chapter 7 of this thesis. We refer to e.g. [134] for Sequential Quadratic Programming (SQP) type active set methods [100, 170] and e.g. [27] for interior point methods [91, 113, 155].

### Model–Predictive Control

The idea of Model Predictive Control (MPC) is to determine an optimal control at time instant $t_0$ by solving an optimal control problem on a prediction horizon $[t_0, t_0 + h]$. The solution of this problem depends on an observation or estimate of the actual mode of operation of the physical process. It yields optimal values for the process controls that are fed back to the physical process for a short time $\delta t$. Meanwhile, a new optimal control problem is solved for the next prediction horizon $[t_0 + \delta t, t_0 + \delta t + h]$ that moved forward in time by $\delta t$. Continuous repetition of this scheme amounts to solving under real–time conditions a sequence of optimal control problems on a moving horizon, based on varying process observations. This opens up the possibility of reacting to disturbances and unforeseen changes in the behavior of the physical process.

Linear MPC has over the past decades matured to widespread applicability in a large number of industrial scenarios. See for example the reviews [77], [171] and the overviews found in [7], [46], [177] and [217]. It is frequently the case that nonlinear models, derived from first principles such as fundamental physical, mechanical, or chemical laws, lead to more accurate models of the actual process under consideration. Overviews over theoretical investigation of Nonlinear Model Predictive Control (NMPC) can be found in [8, 152, 178] and overview over nonlinear applications in [147, 171]. The computational speed and reliability of algorithms for NMPC has seen considerable improvement by major algorithmic developments found e.g. in [51, 52] for SQP type methods, later transferred to interior point methods in e.g. [224].

### Mixed–Integer Optimal Control

Process controls with a finite number of admissible values arise naturally in a large number of relevant applications. Immediately obvious examples are valves in chemical plants that can either be open or closed, gear shifts in vehicles that select between several available transmission ratios, or design and operational alternatives, e.g. choosing a vessel or tray to fill or deplete. Here, a large potential for optimization is found as the combinatorial nature of the problem frequently eludes engineering intuition, and the vast number of possible modes of operation is hard to explore in an exhaustive way. The discipline of Mixed–Integer Optimal Control (MIOC), also referred to as mixed–logic dynamic optimization or hybrid optimal control by different authors, addresses optimal control problems of this structure. A first application, namely the optimal choice of discrete acceleration in a subway train, has already been considered in the early eighties [35]. Applications in chemical operations research are ample, see [40, 94, 161, 183, 185, 195], and for vehicle control problems we refer to e.g. [80, 103, 122, 209]. Mixed–integer control problems in biology can be found e.g. in [130, 183, 196].

### Mixed–Integer Programming

Algorithmic approaches differentiate between state dependent i.e., implicitly defined discrete decisions, also referred to as implicit switches, and explicitly controllable switches which in the following are referred to as binary or integer controls. Algorithms and applications for state dependent switches can be found e.g. in [40, 60, 118, 156, 208].

Several authors are concerned with Mathematical Program with Complementarity Constraints (MPCC) or Mathematical Program with Equilibrium Constraints (MPEC) reformulations of time–invariant discrete decisions in optimization and optimal control problems. We refer to e.g.

[16, 173, 174, 193] for problem formulations, numerical methods, and applications.

We will in the following be exclusively concerned with explicitly controllable switches, i.e., binary or integer valued control functions that may vary over time. One way of approaching Mixed–Integer Optimal Control Problems (MIOCPs) is to apply a direct, simultaneous method in order to obtain a discretized counterpart, a Mixed–Integer Nonlinear Program (MINLP). The class of MINLPs has been proven $\mathcal{NP}$–hard [78]. Hence, assuming $\mathcal{P} \neq \mathcal{NP}$, there exist MINLP instances that cannot be solved on a deterministic machine in polynomial time. In addition, the high accuracy required for many problems may potentially require a fine control discretization, leading to a large number of binary or integer variables in the resulting MINLP, thus adding to its difficulty and increasing the computational effort required to find a solution. Several authors have solved MIOC problems by branching techniques [80], dynamic programming [42, 103], relaxation and penalization [179], or optimization of switching times [81]. Progress towards the efficient solution of MIOCPs has been made recently by a convexification and relaxation scheme [183, 189] with guaranteed lower bounds. Despite the high complexity of MIOC problems from a theoretical point of view, it allows to solve many problem instances of practical relevance without exponential computational effort. This scheme has found successful application to a number of problems e.g. in biology [130, 131], chemistry [141, 185], and automotive control [122, 187].

### Mixed–Integer Model–Predictive Control

As we have seen, model predictive control is a well established technique in the linear case and has become computationally tractable in real–time in the nonlinear case in the last decade. At the same time, process controls with a finite number of admissible values arise naturally in a large number of applications where predictive control would be of high relevance to current industrial practice, while it is a challenging task to apply MINLP techniques for their solution. Consequentially, real–time capable mixed–integer model–predictive control techniques are scarcely considered in the literature, certainly due to the inherent difficulty of the problem class and the apparent lack of fundamental results. As an example, in the representative proceedings collection [148] not a single contribution on mixed–integer algorithms or applications in MPC can be found. In [7] several approaches to mixed–integer linear MPC are reported that rely on Mixed–Integer Quadratic Program (MIQP) solvers. Related techniques have recently been applied by [21] to a wireless sensor feedback problem. MPC for hybrid systems is addressed in [46] by MIQP techniques or piecewise affine models. A mixed–integer formulation is used in [142] to prioritize two competing objectives in a linear vehicle MPC application. An MPC problem with optimal back-off robustification is considered by [199] who solve the resulting bi-level optimization problem using a reformulation of the lower-level problem's optimality conditions that involves complementarity constraints.

## Aims and Contributions of this Thesis

The aim of this thesis is to develop an efficient numerical algorithm, underlying theory, and an actual implementation of a numerical method for real–time capable model predictive control of nonlinear dynamic processes with both continuous and integer–valued controls. To this end, this thesis presents novel results and advances over previously established techniques in a number of areas. They are described in more detail in the following.

### Mixed–Integer Nonlinear Model Predictive Control

In this thesis we develop a new algorithm for mixed–integer nonlinear model–predictive control. It combines Bock's direct multiple shooting method [36, 167] and the real–time iteration scheme [51, 52] with a reformulation of the integer part of the predictive control problem based on outer convexification [183] and relaxation of both the dynamics and the constraints with respect to the discrete controls. Based on the contractivity statement for classical real–time iterations, we show that the mixed–integer real–time iteration scheme can be interpreted as an inexact Newton–type method. The inexactness conveys the application of the rounding scheme taking place after each Newton–type iteration. Using this idea we give a formal proof of local convergence of real–time iterations in the presence of rounding schemes that relies on contractivity conditions for inexact Newton–type methods. The established contractivity condition allows to derive a bound on the sampling time of the mixed–integer model–predictive control scheme that ensures nominal stability of this new algorithm.

### Switch Costs

An important property of dynamic process models with discrete controls is the cost incurred by operating a discrete switch. Here we may be interested in limiting or penalizing the number of switch operations, or in approximating fast transient behavior of the dynamic process by state discontinuities. This question has e.g. been considered in [42] for a dynamic programming framework, and is often included in the modeling of hybrid systems with implicit, i.e., state dependent switches [40, 60, 118, 156, 208]. Limitations of these approaches often apply to changes of the number and relative order in time of the switch events. The inclusion of switch costs in direct methods for optimal control is seldom considered. We propose an Mixed–Integer Linear Program (MILP) switch cost formulation that determines an integer feasible solution from a relaxed but fractional one obtained using the outer convexification reformulation. We develop a convexification for this MILP formulation that can be readily included in a direct multiple shooting discretization of the Optimal Control Problem (OCP) and hence allows to immediately obtain relaxed solutions that satisfy a switch cost constraint. Our formulation in addition avoids attracting fractional solutions of the relaxed problem.

### Convexification and Relaxation

The outer convexification and relaxation method [183] did not previously consider path constraints directly depending on a binary or integer control function. We propose a new reformulation of point constraints and discretized path constraints directly depending on integer controls. This reformulation guarantees feasibility after rounding of the relaxed optimal solution obtained for the discretized OCP. The properties of the obtained NLPs are investigated for the first time, and explanations for frequently observed ill–behavedness of SQP and Quadratic Program (QP) methods on these programs are given. We identify sources of lack of constraint qualification, ill–conditioning, infeasible steps, and cycling of active set method. Addressing these issues we show that the arising NLPs can instead be treated favorably as Mathematical Programs with Vanishing Constraints (MPVCs), a young and challenging class of nonconvex nonlinear problems that commonly arise in truss optimization and only recently attracted increased research interest on its own [3, 105, 109]. The connection to mixed–integer convexification and relaxation approaches however is, to the best of our knowledge, a new contribution. We show that certain strong regularity assumptions for MPVCs are satisfied for the NLPs arising from outer convexification of path constraints. This regularity allows to retain the concept of Karush–Kuhn–Tucker (KKT) based local optimality.

### A Nonconvex Parametric SQP Method

We describe a concrete realization of a nonconvex SQP framework due to [193] and develop a new nonconvex parametric active set method derived from the convex case presented in [25, 67] and is method is based on strong stationarity conditions for MPVC. This method is applied to the nonconvex quadratic subproblems, called Quadratic Programs with Vanishing Constraints (QPVCs) in this thesis, that arise from outer convexification of path constraints. Parametric programming techniques are used to enable highly efficient warm starts of this method required to iterate in the nonconvex feasible set. We further present a heuristic for improving strongly stationary points of the nonconvex quadratic subproblems to global optimality that is based on a sufficient condition for global optimality.

### Block Structured Linear Algebra

The mixed–integer control feedback delay is determined by the computational demand of our QPVC active set method. The QP and QPVC subproblems for discretized, convexified, and relaxed MIOCPs divert from quadratic subproblems typically encountered in optimal control in that the vector of unknowns frequently comprises many more control parameters than differential states. This has a significant impact on the block structure of the problems. We show that classical condensing methods [36, 167] are inappropriate for structure exploitation in this situation.

Addressing this issue, we present a block structured factorization derived from [202, 204] that is tailored to the block structure of the QP's KKT matrices induced by Bock's direct multiple shooting method. Our factorization can be applied equally well to time discrete systems with separable constraints. Its run time complexity of $\mathcal{O}(mn^3)$ and in particular of $\mathcal{O}(mn^{\mathrm{q}2})$ is favorable for predictive control problems with large horizon lengths $m$ or many controls unknowns $n^{\mathrm{q}}$. This is the particular case arising for mixed–integer optimal control problems after reformulation using the outer convexification approach.

**Matrix Update Techniques**

We develop and prove new block structured matrix update techniques for the block structured factorization that reduce the run time complexity of all but the first active set iteration by one order to $\mathcal{O}(mn^2)$, making our method attractive also for dynamic systems with more than a few differential states. We stress the novelty of these update techniques for the case of block structured KKT matrices by quoting [158] who state that *"In some areas of application, the KKT matrix [...] contains special structure. For instance, the QPs that arise in optimal control and MPC have banded matrices [...] When active–set methods are applied to this problem, however, the advantages of bandedness and sparsity are lost after just a few updates of the basis."* For the case of KKT matrices with block diagonal Hessian and block bi–diagonal constraints, we improve upon the described situation by providing basis updates that fully maintain the block structure. Our techniques are based on a combination of existing QR, Schur complement, and Cholesky updates and generate no fill–in.

**Analysis of Computational Demand**

Targetting embedded systems, it is of vital importance to be able to specify the computational demand of all employed algorithms. To this end, an effort has been made in this thesis to analyze the number of floating–point operations required by the core QP algorithms to compute the control feedback, and to relate this number to the properties of the MIOCP under investigation.

**Software Package**

All developed algorithms are implemented in two software packages, named `MuShROOM` and `qpHPSC`, which together allow for the generic and fast solution of nonlinear mixed–integer optimal control and model–predictive control problems.

**Case Studies**

The main theoretical contributions of this thesis, being the mixed–integer real–time iteration scheme and its sampling time bound ensuring local contractivity, the outer convexification of path constraints and the treatment of the resulting NLPs as MPVCs, the nonconvex SQP and active set QP algorithms, and the block structured factorization with matrix updates, are demonstrated at the example of several mixed–integer optimal control problems and – where possible – compared to existing algorithmic approaches.

**Realtime Predictive Cruise Control**

A challenging predictive cruise control problem with high relevance to current and future research efforts in the automotive industry is investigated. Increasing fuel prices, scarcity of unclaimed natural resources, and growing awareness for environmental concerns leads to consistently increasing efforts in energy saving. One particular concern is the comparably high fuel consumption of heavy–duty trucks for long–distance transportations [209]. It can be observed that significant savings can be realized by appropriate adaptation of the truck's mode of operation to road and traffic conditions [103, 209]. In particular, an intelligent choice of velocity and gear shift points based on in–advance knowledge of road conditions bears a significant potential for reducing operational costs.

The underlying mathematical problem however turns out to be hard to approach under real–time conditions. A predictive optimization problem with both discrete and continuous unknowns needs to be solved on–board the truck. As the optimal gear is decided upon not only for the instant, but for a prediction of the future that can be as distant as 4000 meters or around 200 seconds at full traveling speed, the combinatorial explosion of the number of possible gear shift patterns is a major difficulty and obstacle. Drastically reducing the maximum allowed number of gear shifts would sacrifice much of the potential for optimization, though.

The optimal control feedback has to be provided in real–time in order to be able to adapt to ever changing exterior conditions such as changes of the road's slope and curvature, but also to react to more critical conditions such as suddenly arising traffic conditions. Hence, the delay between observations of the truck's mode of operation and various exterior conditions on the one hand, and the availability of the computed control feedback on the other hand shall be as small as possible. Industrial demands here are in the range of some tens of milliseconds. The computational power of the equipment available for this task on–board the truck remains limited, however.

Previous research has considered this and closely related problems without taking multiple discrete gear shifts into account [209], or by applying exhaustive search algorithms of the dynamic programming type [42, 80, 103] that do either require precomputation of the optimal solution or do not meet the computational resource constraints.

In this thesis, this predictive cruise control problem is solved for the first time under demanding real–time constraints. The resulting algorithms are presently considered for application in an ongoing industrial cooperation.

## Thesis Overview

This thesis is laid out in nine chapters and one appendix as follows.

In chapter 1 we introduce Optimal Control Problems (OCPs) for dynamic processes described by systems of Ordinary Differential Equations (ODEs). We survey numerical methods for the algorithmic solution of such problems, and settle on multiple shooting methods. We present in great detail Bock's direct multiple shooting method for optimal control, and give pointers to new developments in this thesis affecting various components of this method.

Chapter 2 is concerned with Mixed–Integer Optimal Control Problems, a class of optimal control problems with both continuous and discrete controls. After a brief survey of possible algorithmic approaches towards solving problems of this class, we describe the outer convexification and relaxation approach. This approach allows to obtain a local solution of a Mixed–Integer Optimal Control Problem (MIOCP) or an approximation thereof by solving a single, continuous, but possibly larger OCP without observing exponential run time in practice. We present bounds on the quality of the approximation of the optimal control and ODE state trajectories. An extension of the MIOCP problem class including the notion of switch costs is presented. A Mixed–Integer Linear Program (MILP) formulation is developed that computes an integer feasible control trajectory satisfying a switch cost constraint from a relaxed optimal but possibly fractional one. In order to include the switch cost constraint in the OCP itself, a convexification of this formulation is developed. It avoids attracting fractional solutions and maintains separability, thus allowing for integration into the direct multiple shooting method.

In chapter 3 we cover the theory of Nonlinear Programming (NLP) and present Sequential Quadratic Programming (SQP) algorithms for the solution of the discretized, convexified, and relaxed MIOCP. Numerical methods for the solution of Initial Value Problems (IVPs) required to evaluate certain direct multiple shooting NLP constraints and for derivative and sensitivity generation are briefly presented.

In chapter 4 we describe algorithmic extensions of the direct multiple shooting method to Nonlinear Model Predictive Control (NMPC) problems. We introduce the concepts of real–time iterations and the online active–set strategy and mention major theoretical results such as conditions for local contractivity and a sketch of the proof due to [51]. We extend the real–time iteration scheme to mixed–integer NMPC problems for the first time. Two rounding schemes and warm start strategies are presented. We give a formal proof of local contractivity of the developed mixed–integer real–time iteration scheme that is based on the interpretation of our algorithm as an inexact Newton–type method. A bound on the mixed–integer NMPC sampling time ensuring local contractivity is derived based on this proof.

Previous work on outer convexification and relaxation schemes for MIOCPs has not considered constraints directly depending on the discrete controls. In chapter 5 we investigate for the first time the structure of NLPs obtained by direct discretization of a MIOCP with constraints treated by the outer convexification and relaxation approach of chapter 2. We show that important Constraint Qualifications (CQs) are violated and identify reasons for various convergence problems observed for descent based methods such as SQP. The obtained NLPs are identified as Mathematical Programs with Vanishing Constraints (MPVCs), a challenging class of problems that has only recently attracted increased research interest. We present a Nonlinear Programming framework for MPVCs. For MIOCPs with constraints treated by outer convexification, we show that a replacement CQ holds under reasonable assumptions. This allows us to retain the concept of Karush–Kuhn–Tucker (KKT) based local optimality.

Chapter 6 presents a new SQP method for MPVCs that carries the nonconvexity of the problem over to the local subproblems. A new parametric active set strategy for the solution of a sequence of those local subproblems, which are referred to as Quadratic Programs with Vanishing Constraints (QPVCs). Strong MPVC stationarity conditions are used to derive active set exchange rules that allow to find a strongly stationary point on the nonconvex feasible set of a QPVC. Even though QPVCs must be considered as nonconvex problems, global optimality can be verified locally, and a heuristic is presented that exploits this knowledge to improve strongly stationary points up to global optimality.

In chapter 7 the focus is put on sparse and block structured linear algebra for QPVCs with many control parameters due to outer convexification of MIOCP constraints. We survey existing approaches for exploiting the arising block structures and evaluate their applicability. We continue by presenting a factorization of the QPVC's KKT system that is tailored to the case of many control parameters and is free of any fill–in, called the Hessian Projection Schur Complement (HPSC) factorization in this thesis. It has a favorable runtime complexity for problems with long prediction horizons and for MIOCPs reformulated by outer convexification. Properties, applicability, and extensions of this factorization are investigated in detail, and the run time and storage space requirements are examined.

Matrix update for factorizations are of vital importance for the efficiency of any active set method. The rate at which QPVC solutions can be computed effectively determines the control feedback delay of the mixed–integer NMPC controller. Chapter 8 is concerned with detailed proofs of new matrix updates for the HPSC factorization that reduce the runtime complexity in terms of the number of unknowns by one order.

Chapter 9 presents numerical results for all new algorithms and makes comparisons to existing methods at the example of several mixed–integer optimal control and mixed–integer NMPC problems. In an extensive industrial case study, a challenging mixed–integer nonlinear model predictive control problem is solved under demanding real–time constraints. We show that by using the developed numerical methods, this problem could even be solved on an embedded system with limited computational resources.

This thesis is closed by appendix A, containing details on the implementation of the developed algorithms within the software packages `MuShROOM` and `qpHPSC`.

### Computing Environment

All computational results and run times presented in this thesis have been obtained on a 64–bit *Ubuntu*© *Linux*™ *9.10* system powered by an *Intel*© *Core*™ *i7 920* CPU at 2.67 GHz, with 6 GB main memory available. A single core of the available four physical cores of the CPU has been used. All source code is written in *ANSI C99* and compiled using version 4.4.1 of the *GNU C/C++ compiler collection*, with applicable machine–specific optimization flags enabled.

# 1 The Direct Multiple Shooting Method for Optimal Control

In this chapter we consider a class of continuous optimal control problems. We survey different numerical solution approaches that introduce computationally tractable variants of this problem and discuss their usefulness for the purpose of this thesis. A method particularly suitable is the direct multiple shooting method of section 1.2.5, the foundation for all further algorithms presented in this thesis.

## 1.1 Problem Formulations

**Definition 1.1 (Continuous Optimal Control Problem)**
A continuous optimal control problem is a constrained infinite–dimensional optimization problem of the form

$$\begin{aligned} \min_{x(\cdot),u(\cdot)} \quad & \varphi(x(\cdot),u(\cdot)) && && (1.1)\\ \text{s.t.} \quad & \dot{x}(t) = f(t,x(t),u(t)), && t\in\mathcal{T}, \\ & 0 \leqslant c(t,x(t),u(t)), && t\in\mathcal{T}, \\ & 0 \leqq r(\{x(t_i)\}), && \{t_i\}\subset\mathcal{T}, \end{aligned}$$

in which we determine a dynamic process $x : \mathcal{T} \to \mathbb{R}^{n^x}$ on a time horizon $\mathcal{T} \stackrel{\text{def}}{=} [t_0, t_f] \subset \mathbb{R}$, described by a system of Ordinary Differential Equations (ODEs) with right hand side $f : \mathcal{T} \times \mathbb{R}^{n^x} \times \mathbb{R}^{n^u} \to \mathbb{R}^{n^x}$, affected by a control $u : \mathcal{T} \to \mathbb{R}^{n^u}$, such that we minimize a performance index $\varphi : \mathcal{X} \times \mathcal{U} \to \mathbb{R}$ and satisfy path constraints $c : \mathcal{T} \times \mathbb{R}^{n^x} \times \mathbb{R}^{n^u} \to \mathbb{R}^{n^c}$ and point constraints $r : (\mathbb{R}^{n^x})^{m+1} \to \mathbb{R}^{n^r}$ on a finite number $m+1$ of grid points $\{t_i\} \subset \mathcal{T}$, $0 \leqslant i \leqslant m$. △

In definition 1.1 we have modeled a time dependent process $x : \mathcal{T} \to \mathbb{R}^{n^x}$, $t \mapsto x(t)$ on the time horizon $\mathcal{T} \stackrel{\text{def}}{=} [t_0, t_f] \subset \mathbb{R}$ by a system of ODEs

$$\dot{x}(t) = f(t, x(t), u(t)), \quad t \in \mathcal{T}. \tag{1.2}$$

This dynamic process can be affected by a control input $u(t)$ at any time. We assume the function $u : \mathcal{T} \to \mathbb{R}^{n^u}$ to be measurable and define $\mathcal{U} \stackrel{\text{def}}{=} \{u :$

$\mathcal{T} \to \mathbb{R}^{n^u} \mid \boldsymbol{u}$ measurable$\}$ to be the set of all such control functions. The variable $\boldsymbol{x}(t)$ describes the system state of this process at any time $t \in \mathcal{T}$, and we define $\mathcal{X} \stackrel{\text{def}}{=} \{\boldsymbol{x} : \mathcal{T} \to \mathbb{R}^{n^x}\}$ to be the set of all state trajectories. To ensure existence and uniqueness of the ODE system's solution, we assume $\boldsymbol{f} : \mathcal{T} \times \mathbb{R}^{n^x} \times \mathbb{R}^{n^u} \to \mathbb{R}^{n^x}$ to be piecewise Lipschitz continuous. The constraint function $\boldsymbol{c} : \mathcal{T} \times \mathbb{R}^{n^x} \times \mathbb{R}^{n^u} \to \mathbb{R}^{n^c}$ restricts the set of admissible state and control trajectories $\boldsymbol{x}(\cdot)$ and $\boldsymbol{u}(\cdot)$. It may contain mixed path and control constraints, restrict the set of initial values $\boldsymbol{x}(t_0)$, and contain boundary conditions for the trajectories. Finally, the point constraint function $\boldsymbol{r} : (\mathbb{R}^{n^x})^{m+1} \to \mathbb{R}^{n^r}$ imposes point–wise constraints on the states in a finite number $m+1$ of grid points $\{t_i\} \subset \mathcal{T}$, $0 \leqslant i \leqslant m$ that may be coupled in time. Possible uses are the specification of boundary conditions such as initial and terminal states or periodicity constraints. The presented Optimal Control Problem (OCP) clearly is an infinite–dimensional optimization problem, the unknowns to be determined being the control trajectory $\boldsymbol{u}(\cdot)$ and the resulting state trajectory $\boldsymbol{x}(\cdot)$ of the process.

Problem (1.1) can be extended and specialized to include a large number of additional characteristics. The following concretizations of the above problem class can easily be incorporated by reformulations:

### Objective Functions of Bolza and Least–Squares Type

The performance index $\varphi(\boldsymbol{x}(\cdot), \boldsymbol{u}(\cdot))$ evaluated on the time horizon $\mathcal{T}$ usually is a general objective function that consists of an integral contribution, the Lagrange type objective with integrand $l(t, \boldsymbol{x}(t), \boldsymbol{u}(t))$, and an end–point contribution, the Mayer type objective $m(t_f, \boldsymbol{x}(t_f))$,

$$\varphi(\boldsymbol{x}(\cdot), \boldsymbol{u}(\cdot)) = \int_{t_0}^{t_f} l(t, \boldsymbol{x}(t), \boldsymbol{u}(t)) \, \mathrm{d}t + m(t_f, \boldsymbol{x}(t_f)). \tag{1.3}$$

Least–squares objective functions are of particular interest in tracking problems where they minimize a weighted measure of deviation from a desired path in state space, and they can be employed for numerical regularization of the control trajectory $\boldsymbol{u}(t)$. The general form is

$$\varphi(\boldsymbol{x}(\cdot), \boldsymbol{u}(\cdot)) = \int_{t_0}^{t_f} \|\boldsymbol{r}(t, \boldsymbol{x}(t), \boldsymbol{u}(t))\|_2^2 \, \mathrm{d}t + \left\|\boldsymbol{m}(t_f, \boldsymbol{x}(t_f))\right\|_2^2. \tag{1.4}$$

The Gauß–Newton approximation that exploits the particular form of this objective function is presented in section 3.2.3.

#### Constraint Types

We distinguish several different types of constraints by the structure of the constraint function $\boldsymbol{c}(\cdot)$. Decoupled constraints do not couple different time points of the state or control trajectories. They may be imposed on the entire horizon

$$\mathbf{0} \leqslant \boldsymbol{c}(\boldsymbol{x}(t), \boldsymbol{u}(t)), \qquad t \in \mathcal{T},$$

or on grid points $\{t_i\}_{0 \leqslant i \leqslant m} \subset \mathcal{T}$ only,

$$\mathbf{0} \leqslant \boldsymbol{c}(\boldsymbol{x}(t_i)), \qquad 0 \leqslant i \leqslant m.$$

Exploiting their property of separability is crucial in the design of efficient numerical methods as presented in chapter 7. Coupled constraints

$$0 \leqslant \boldsymbol{c}(\boldsymbol{x}(t_0), \dots, \boldsymbol{x}(t_m))$$

couple the process states in finitely many different points $t_i \in \mathcal{T}$, $0 \leqslant i \leqslant m$ of the horizon. We refer to $\{t_i\}$ as a constraint grid in this case. Boundary constraints impose restrictions on the trajectories only in the initial and the final point of the horizon,

$$\mathbf{0} \leqslant \boldsymbol{c}(\boldsymbol{x}(t_0), \boldsymbol{x}(t_\mathrm{f})),$$

and a special instance are periodicity constraints

$$\mathbf{0} = \boldsymbol{\pi}_0(\boldsymbol{x}(t_0)) - \boldsymbol{\pi}_\mathrm{f}(\boldsymbol{x}(t_\mathrm{f})),$$

wherein $\boldsymbol{\pi}_0$ and $\boldsymbol{\pi}_\mathrm{f}$ may e.g. contain permutations of the state vector's components. Both types of constraints can be reformulated as decoupled constraints e.g. by introducing an augmented state vector $\hat{\boldsymbol{x}} = [\, \boldsymbol{x} \; \boldsymbol{r} \,]$ holding the constraint residuals,

$$\hat{\boldsymbol{f}}(t, \hat{\boldsymbol{x}}(t), \boldsymbol{u}(t)) = \begin{bmatrix} \boldsymbol{f}(t, \boldsymbol{x}(t), \boldsymbol{u}(t)) \\ \mathbf{0} \end{bmatrix}, \tag{1.5}$$

$$\boldsymbol{r}_k(t_k) = \boldsymbol{c}_k(\boldsymbol{x}(t_k)).$$

An appropriately chosen decoupled terminal constraint can then be imposed on the augmented state vector,

$$\mathbf{0} \leqslant \mathbf{c}(\mathbf{r}_0(t_\mathrm{f}),\dots,\mathbf{r}_m(t_\mathrm{f})). \tag{1.6}$$

In some Nonlinear Model Predictive Control (NMPC) applications, zero terminal constraints

$$\mathbf{r}(\mathbf{x}(t_\mathrm{f})) = \mathbf{x}(t_\mathrm{f}) - \mathbf{x}_\mathrm{steady} = \mathbf{0},$$

where $\mathbf{x}_\mathrm{steady}$ is the differential steady state of the process, help to guarantee nominal stability [153]. Initial value constraints

$$\mathbf{x}(t_0) = \mathbf{x}_0$$

play a special role in the real–time iteration scheme presented in chapter 4 and the parametric Sequential Quadratic Programming (SQP) and Quadratic Programming (QP) algorithms of chapter 6.

### Variable Time Horizons

Problem (1.1) is given on a fixed time horizon $\mathcal{T} = [t_0, t_\mathrm{f}] \subset \mathbb{R}$ for simplicity. Free initial times or free end times, and thus time horizons of variable length, are easily realized by the time transformation

$$t(\tau) \overset{\mathrm{def}}{=} t_0 + h\tau, \quad h \overset{\mathrm{def}}{=} t_\mathrm{f} - t_0. \tag{1.7}$$

that allows to restate the OCP on the normalized horizon $\tau \in [0,1] \subset \mathbb{R}$,

$$\begin{aligned} \min_{\mathbf{x}(\cdot),\mathbf{u}(\cdot),h} \quad & \varphi(\mathbf{x}(\cdot),\mathbf{u}(\cdot),h) \\ \text{s. t.} \quad & \dot{\mathbf{x}}(\tau) = h \cdot \mathbf{f}\big(t(\tau),\mathbf{x}(t(\tau)),\mathbf{u}(t(\tau))\big), \quad \tau \in [0,1], \\ & \mathbf{0} \leqslant \mathbf{c}\big(\mathbf{x}(t(\cdot)),\mathbf{u}(t(\cdot)),h\big). \end{aligned} \tag{1.8}$$

### Global Parameters

A vector of model parameters $\mathbf{p} \in \mathbb{R}^{n^\mathrm{p}}$ describing global properties of the dynamic process and its environment may enter the objective function, the ODE system, and the constraints,

$$\min_{x(\cdot),u(\cdot),p} \quad \varphi(x(\cdot),u(\cdot),p) \tag{1.9}$$
$$\text{s.t.} \quad \dot{x}(t) = f(t,x(t),u(t),p), \quad t \in \mathcal{T},$$
$$0 \leqslant c(x(\cdot),u(\cdot),p).$$

Model parameters can either be regarded as special control functions that are constant on the whole of $\mathcal{T}$, or can be replaced by introducing an augmented state vector

$$\hat{x} = \begin{bmatrix} x \\ p \end{bmatrix}$$

with augmented ODE right hand side function

$$\hat{f}(t,\hat{x}(t),u(t)) = \begin{bmatrix} f(t,x(t),u(t),p) \\ 0 \end{bmatrix}.$$

Depending on the choice of the method in a numerical code, either formulation may be more efficient. Note however that only those model parameters that actually attain different values under practically relevant process conditions should be included in the problem formulation in this way. Parameters that attain one constant value under all thinkable conditions should not be exposed to the problem formulation, but rather be considered part of the model equations.

### Linear Semi–Implicit Differential Algebraic Equations

The dynamic process in problem (1.1) is described by a system of ODEs. We may extend this class to that of semi–implicit index one Differential Algebraic Equation (DAE) systems with differential states $x(\cdot) \in \mathbb{R}^{n^x}$ and algebraic states $z(\cdot) \in \mathbb{R}^{n^z}$ satisfying

$$A(t,x(t),z(t),u(t),p)\,\dot{x}(t) = f(t,x(t),z(t),u(t),p), \quad t \in \mathcal{T}, \tag{1.10}$$
$$0 = g(t,x(t),z(t),u(t),p), \quad t \in \mathcal{T}.$$

Here the left hand side matrix $A(\cdot)$ and the Jacobian $g_z(\cdot)$ are assumed to be regular. We may then regard the algebraic state trajectory $z(\cdot)$ as an implicit function $z(t) = g^{-1}(t,x(t),u(t),p)$ of the differential state and control trajectories.

We refer to [4, 5, 15, 61, 166] for the numerical solution of semi–implicit index one DAE systems. In [133, 191] partial reduction techniques for the algebraic states are presented in an SQP framework.

## 1.2 Solution Methods for Optimal Control Problems

For the solution of optimal control problems of type (1.1) several different methods have been developed. They differ in the choice of the optimization space, the applied type of discretization, the precision of the obtained optimal solutions, the computational demand of the method that limits the size of the problem instances that can be treated in a given amount of time, and finally in the ease of implementation as a numerical code. We mention *indirect methods* based on Pontryagin's Maximum Principle, and review *dynamic programming*, based on Bellman's Principle of Optimality. *Direct methods* are presented in greater detail, and we distinguish *collocation* methods from *shooting* methods here. Finally, the *direct multiple shooting* method is presented as a hybrid approach combining the advantages of both direct methods. We go into more detail here, as all numerical algorithms presented in this thesis are based on and focused on direct multiple shooting.

### 1.2.1 Indirect Methods

*Indirect methods* are based on Pontryagin's *Maximum Principle* and optimize in an infinite–dimensional function space. Necessary conditions of optimality are used to transform the optimal control problem into a Multi–Point Boundary Value Problem (MPBVP), which can then be solved by appropriate numerical methods. For very special small cases, an analytical solution is also possible. Indirect methods are therefore suitable for a theoretical analysis of a problem's solution structure. They yield highly accurate solutions for the optimal control profile, as the infinite–dimensional problem is solved and no approximation of the control functions took place. This is in sharp contrast to direct methods, which are discussed below. As all control degrees of freedom vanish in the MPBVP to be solved, indirect methods appear attractive for optimal control problems with a large number of controls, such a convexified Mixed–Integer Optimal Control Problems (MIOCPs) discussed in chapter 2.

The disadvantages of indirect methods are quite obvious, however. The necessary conditions of optimality have to be derived analytically for every problem instance, and require several different special cases to be distinguished. In particular, general path and control constraints $\boldsymbol{c}(\cdot)$ frequently lead to an a priori unknown structure of the optimal solution. This laborious derivation possibly has to be repeated for even small changes of initial conditions, parameters, or for small changes of the problem's structure, e.g. the introduction of an additional constraint, as the optimal solution's structure often is very sensitive to changes of the problem data. In addition, this derivation is difficult if not practically impossible for problem instances with more than a few states and controls. Finally, the solution of the MPBVP typically requires initial guesses to be available for all variables, which are often difficult to obtain especially for the adjoint variables associated with the constraints. For the numerical solution of the MPBVP, a Nonlinear Programming method may be applied. For its convergence, it is crucial that these initial guesses even come to lie inside the domain of local convergence. Consequentially, the numerical solution of an optimal control problem using indirect methods cannot be fully automated. It remains an interactive process, requiring insight into the problem and paperwork for the specific instance to be solved. Due to these inconveniences, indirect methods have not emerged as the tool of choice for fast or even online numerical solution of optimal control problems.

### 1.2.2 Dynamic Programming

*Dynamic Programming* is based on Bellman's principle of optimality, which states that *"an optimal policy has the property that whatever the initial state and initial decision are, the remaining decisions must constitute an optimal policy with regard to the state resulting from the first decision"* [18].

**Theorem 1.1 (Principle of Optimality)**
Let $(\boldsymbol{x}^\star(\cdot), \boldsymbol{u}^\star(\cdot))$ be an optimal solution on the interval $\mathcal{T} = [t_0, t_\mathrm{f}] \subset \mathbb{R}$, and let $\hat{t} \in \mathcal{T}$ be a point on that interval. Then $(\boldsymbol{x}^\star(\cdot), \boldsymbol{u}^\star(\cdot))$ is an optimal solution on $[\hat{t}, t_\mathrm{f}] \subseteq \mathcal{T}$ for the initial value $\hat{\boldsymbol{x}} = \boldsymbol{x}^\star(\hat{t})$. △

Put in brief words, Bellman's principle states that any subarc of an optimal solution is optimal. The converse is not true in general, i.e., the concatenation of optimal solutions on a partition of the interval $\mathcal{T}$ is not necessarily an optimal solution on the whole interval $\mathcal{T}$. We refer to [23] for an extensive treatment of the applications of the Dynamic Programming approach to optimal control and only briefly describe the algorithm here. Based on the principle of optimality, we define the *cost–to–go* function

**Definition 1.2 (Continuous Cost–to–go Function)**
On an interval $[\hat{t}, t_\mathrm{f}] \subset \mathcal{T} \subset \mathbb{R}$ the *cost–to–go* function $\varphi$ for problem (1.1) with a Bolza type objective is defined as

$$\varphi(\hat{t}, \hat{\boldsymbol{x}}) \stackrel{\text{def}}{=} \min \left\{ \int_{\hat{t}}^{t_\mathrm{f}} l(\boldsymbol{x}(t), \boldsymbol{u}(t))\, \mathrm{d}t + m(\boldsymbol{x}(t_\mathrm{f})) \;\middle|\; \boldsymbol{x}(\hat{t}) = \hat{\boldsymbol{x}},\ \boldsymbol{x}(\cdot), \boldsymbol{u}(\cdot) \text{ feasible} \right\}. \quad \triangle$$

Imposing a time grid on the horizon $\mathcal{T}$,

$$t_0 < t_1 < \ldots < t_{m-1} < t_m = t_\mathrm{f}, \tag{1.11}$$

a recursive variant of the cost–to–go function can be defined as follows.

**Definition 1.3 (Recursive Cost–to–go Function)**
In the point $t_k$, $k \in \{0, \ldots, m-1\}$ the recursive *cost–to–go* function $\varphi$ for problem (1.1) with a Bolza type objective is defined as

$$\begin{aligned} \varphi(t_k, \boldsymbol{x}_k) \stackrel{\text{def}}{=} \min \Bigg\{ & \int_{t_k}^{t_{k+1}} l(\boldsymbol{x}(t), \boldsymbol{u}(t))\, \mathrm{d}t + \varphi(t_{k+1}, \boldsymbol{x}_{k+1}) \\ & \Bigg|\; \boldsymbol{x}(t_k) = \boldsymbol{x}_k,\ \boldsymbol{x}(\cdot), \boldsymbol{u}(\cdot) \text{ feasible} \Bigg\}. \end{aligned} \tag{1.12}$$

The recursion ends in $k = m$, $t_m = t_\mathrm{f}$ with $\varphi(t_m, \boldsymbol{x}_m) \stackrel{\text{def}}{=} m(\boldsymbol{x}_m)$. $\triangle$

Using this recursive definition, the optimal objective function values together with the values of the optimal state trajectory $\boldsymbol{x}(\cdot)$ and control trajectory $\boldsymbol{u}(\cdot)$ can be computed in the grid points $t_k$ by backward recursion, starting with $k = m$ and proceeding to $k = 0$. The minimization problem (1.12) has to be solved for each time interval $[t_k, t_{k+1}]$ and for all feasible initial values $\boldsymbol{x}_k$. The resulting cost–to–go function value together with the corresponding optimal control $\boldsymbol{u}_k$ have to be tabulated in state space for use in the next backward recursion step, which requires the choice of discretizations for the sets of feasible values of $\boldsymbol{x}_k$ and $\boldsymbol{u}_k$.

Dynamic Programming has been used in some Model Predictive Control (MPC) applications, cf. [103, 42], and has the advantage of searching the entire state space, thus finding a globally optimal solution. The technique suffers from the "curse of dimensionality", though, and delivers sufficiently fast run times for tiny problem instances only. Control algorithms based on Dynamic Programming also lack extendability because a slight increase of the problem's dimensions, e.g. by using a more sophisticated model of the process, frequently induces an unacceptable growth of the required run time.

### 1.2.3 Direct Single Shooting

In contrast to indirect approaches, direct methods are based on a discretization of the infinite–dimensional OCP (1.1) into a finite–dimensional nonlinear optimization problem.

In direct single shooting one chooses to regard the state trajectory as a dependent value of the controls $\boldsymbol{u}(\cdot)$, and to solve an Initial Value Problem (IVP) using an ODE or DAE solver to compute it. To this end, a finite–dimensional discretization

$$\boldsymbol{q} = \left(\boldsymbol{q}_0, \boldsymbol{q}_1, \ldots, \boldsymbol{q}_{m-1}\right) \in \mathbb{R}^{m \times n^{\mathrm{q}}},$$

of the control trajectory $\boldsymbol{u}(\cdot)$ on $\mathcal{T} = [t_0, t_m]$ is chosen for the solution of the IVP

$$\begin{aligned} \dot{\boldsymbol{x}}(t) &= \boldsymbol{f}(t, \boldsymbol{x}(t), \boldsymbol{u}(t)), \qquad t \in \mathcal{T}, \\ \boldsymbol{x}(t_0) &= \boldsymbol{s}_0, \end{aligned} \tag{1.13}$$

e.g. by using piecewise constant functions ($n^{\mathrm{q}} = n^{\mathrm{u}}$)

$$\boldsymbol{u}(t) = \boldsymbol{q}_i, \qquad t \in [t_i, t_{i+1}] \subset \mathcal{T}, \qquad 0 \leqslant i \leqslant m. \tag{1.14}$$

From this approach an Nonlinear Program (NLP) in the $n_{\mathrm{x}} + m n_{\mathrm{q}}$ unknowns

$$\boldsymbol{v} \stackrel{\text{def}}{=} \left(\boldsymbol{s}_0, \boldsymbol{q}_0, \boldsymbol{q}_1, \ldots, \boldsymbol{q}_{m-1}\right)$$

is obtained which can be solved using e.g. the SQP techniques presented in chapter 3. Control and path constraints frequently are discretized and enforced on the control discretization grid only, or may be included in the objective using penalty terms.

The direct single shooting method suffers from a number of drawbacks. As only the initial state and the control variables enter the NLP, initialization of the state trajectory using prior knowledge about the process is not possible. Further, for a chosen initialization of $\boldsymbol{x}_0$ and the controls $\boldsymbol{u}$, the

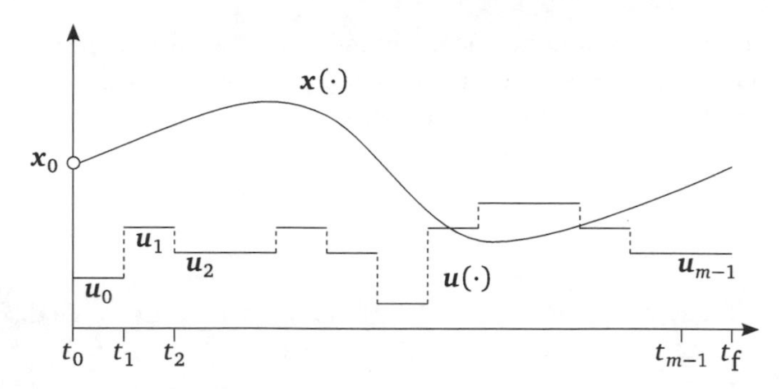


Figure 1.1: Illustration of the direct single shooting discretization applied to the optimal control problem.

IVP's solution need not even exist, e.g. due to a singularity in time. Even if it exists, it possibly cannot be computed numerically due to propagation of errors over the course of the integration. Typically, to guarantee existence of a numerical solution of highly nonlinear or unstable ODEs, an initial guess of these variables is required that is already very close to the true optimal solution. Such a guess may of course be hard to obtain. The convergence speed of the NLP solver is effectively governed by the amount of nonlinearity present in the ODE system, and single shooting methods cannot improve upon this situation. Finally, even well–conditioned IVPs may induce unstable Boundary Value Problems (BVPs), and a small step in the initial value $\boldsymbol{x}_0$ may result in a large step in the ODE's solution $\boldsymbol{x}(\cdot)$ or induce large violations of constraints. This behavior is a challenge for derivative–based NLP methods.

Still, direct single shooting is often used in practice as the idea is easy to grasp, the implementation is straightforward, and the resulting NLP has a small number of unknowns only.

### 1.2.4 Direct Collocation

Collocation methods for BVPs go back to the works [182, 211] and one of the first works on direct collocation for OCPs is [210]. Collocation methods discretize both the states and the controls on a fine grid of $m$ intervals with $k$ collocation points per interval. To this end, the ODE system

$$\dot{\boldsymbol{x}}(t) = \boldsymbol{f}(t, \boldsymbol{x}(t), \boldsymbol{u}(t)), \qquad t \in \mathcal{T},$$

is replaced by a $k$-point collocation scheme that introduces a discretization using $k$ state vectors of dimension $n_\mathrm{x}$ per interval $[t_i, t_{i+1}]$, $0 \leqslant i \leqslant m-1$, coupled by nonlinear equality constraints. Algebraic variables $\boldsymbol{z}(t)$ in DAE systems can be treated by enforcing the discretized condition $\mathbf{0} = \boldsymbol{g}(t, \boldsymbol{x}(t), \boldsymbol{z}(t), \boldsymbol{u}(t))$ on the collocation grid points. Path and point constraints are included in a similar way. Summing up, we obtain from direct collocation methods a large but sparse NLP in the $(km+1)n_\mathrm{x} + kmn_\mathrm{u}$ unknowns

$$\boldsymbol{v} \stackrel{\text{def}}{=} \left(\boldsymbol{x}_0, \boldsymbol{u}_0, \boldsymbol{x}_1, \boldsymbol{u}_1, \ldots, \boldsymbol{x}_{km-1}, \boldsymbol{u}_{km-1}, \boldsymbol{x}_{km}\right),$$

which can be solved with NLP methods such as sparse interior point methods or tailored Sequential Quadratic Programming methods. Refinements and applications of direct collocation can be found in e.g. [10, 26, 101, 194, 206, 225].

In contrast to direct single shooting, collocation allows to initialize the state trajectory variables, thereby allowing more a–priori knowledge about the process and its optimal solution to enter the problem. Small violations of the matching conditions over the course of the NLP solution help dampen the spreading of perturbations. Although the resulting NLP is very large, efficient solution methods exist, e.g. [215].

A drawback of collocation methods is the difficulty to include *adaptivity* of the ODE or DAE solution process. This concept is of particular importance for the treatment of highly nonlinear or stiff systems, which require small step sizes in a priori unknown regions of the time horizon, and thus high values of $m$ in the collocation approach. In addition, certain collocation schemes may exhibit highly oscillatory behavior on *singular arcs* of the optimal control, cf. [112], which may degrade the NLP solver's performance, and which can be overcome by elaborate regularization approaches only. Direct multiple shooting methods, to be presented in the next section, overcome these limitations by making efficient use of state–of–the–art adaptive ODE and DAE solvers, cf. [5, 4, 15, 61, 166], thereby decoupling state and control discretization. Finally, the large–scale sparse NLP obtained from the collocation discretization of the OCP is most efficiently solved by interior–point methods that can easily apply sparse linear algebra, cf. [28, 215]. For most model predictive control tasks, though, active set methods are preferred as detailed in chapter 4 due to their better warm starting abilities [13, 138]. The sparsity patterns introduced by collocation however are not easily exploited in active set methods.

### 1.2.5 Direct Multiple Shooting

First descriptions of multiple shooting methods for the solution of BVPs can be found in [29, 30, 31, 43, 162]. The direct multiple shooting method for OCP goes back to the works of [36, 167]. Direct multiple shooting methods are hybrids of direct collocation and direct single shooting approaches, as they discretize the state trajectory, but still rely on solving IVPs. They typically have an advantage over direct collocation methods in that they easily allow to use of highly efficient adaptive solvers for the IVPs, e.g. [5, 4, 15, 61, 166]. Contrary to single shooting, the initialization of the state trajectory variables is easily possible, also permitting infeasible initializations. Stability of both the IVP and the BVP solution process is improved. In addition, computational experience shows that nonlinearity present in the BVP is diminished, thus improving the speed of convergence [6].

The remainder of this chapter is exclusively concerned with the direct multiple shooting method for optimal control, which serves as the foundation for all algorithms to be presented in this thesis.

## 1.3 The Direct Multiple Shooting Method for Optimal Control

The direct multiple shooting method for optimal control has its origins in the diploma thesis [167], supervised by Hans Georg Bock, and was first published in [36]. Extensions can be found in e.g. [4, 34, 51, 133, 134, 191]. The direct multiple shooting code `MUSCOD-II` is described in detail in [132], and the direct multiple shooting code `MuShROOM` developed as part of this thesis is described in appendix A.

We consider the numerical solution of the following class of OCPs that have been introduced in the previous sections together with various possible extensions:

$$\min_{\boldsymbol{x}(\cdot),\boldsymbol{u}(\cdot)} \; m(t_{\mathrm{f}},\boldsymbol{x}(t_{\mathrm{f}})) + \int_{t_0}^{t_{\mathrm{f}}} l(t,\boldsymbol{x}(t),\boldsymbol{u}(t))\,\mathrm{d}t \tag{1.15a}$$

$$\text{s.t.}\quad \dot{\boldsymbol{x}}(t) = \boldsymbol{f}(t,\boldsymbol{x}(t),\boldsymbol{u}(t)), \qquad t\in\mathcal{T}, \tag{1.15b}$$

$$\boldsymbol{0} = \boldsymbol{r}_i^{\mathrm{eq}}(t_i,\boldsymbol{x}(t_i)), \qquad \{t_i\}_{0\leqslant i\leqslant n^{\mathrm{r}}}\subset\mathcal{T}, \tag{1.15c}$$

$$\boldsymbol{0} \leqslant \boldsymbol{r}_i^{\mathrm{in}}(t_i,\boldsymbol{x}(t_i)), \qquad \{t_i\}_{0\leqslant i\leqslant n^{\mathrm{r}}}\subset\mathcal{T}, \tag{1.15d}$$

$$\boldsymbol{0} \leqslant \boldsymbol{c}(t,\boldsymbol{x}(t),\boldsymbol{u}(t)) \qquad t\in\mathcal{T}, \tag{1.15e}$$

in which we minimize a Bolza type objective function (1.15a) of a dynamic process $x(\cdot)$ defined on the horizon $\mathcal{T} \stackrel{\text{def}}{=} [t_0, t_f] \subset \mathbb{R}$ in terms of an ODE system with right hand side function $f$ (1.15b). The process is controlled by a control trajectory $u(\cdot)$ subject to optimization. It shall satisfy certain inequality path constraints $c(\cdot)$ (1.15e) as well as equality and inequality point constraints $r_i(\cdot)$ (1.15c, 1.15d) imposed on a constraint grid $\{t_i\} \subset \mathcal{T}$ of $n^r$ points on $\mathcal{T}$. We expect all functions to be twice continuously differentiable with respect to the unknowns of problem (1.15).

### 1.3.1 Control Discretization

In order to obtain a computationally tractable representation of the infinite-dimensional control trajectory $u(\cdot)$ we introduce a discretization of $u(\cdot)$ on the control horizon $\mathcal{T}$ by partitioning $\mathcal{T}$ into $m$ not necessarily equidistant intervals

$$t_0 < t_1 < \ldots < t_{m-1} < t_m = t_f \tag{1.16}$$

that define the shooting grid $\{t_i\}$. For simplicity of exposition, we assume it to coincide with the constraint grid of the point constraints used in the introduction of this chapter and in (1.15). Note however that the presentation to follow can be extended to differing grids for the controls and point constraints.

On each interval $[t_i, t_{i+1}]$, $0 \leqslant i \leqslant m-1$, we choose base functions $b_{ij} : \mathcal{T} \times \mathbb{R}^{n^q_{ij}} \to \mathbb{R}$ for each component $1 \leqslant j \leqslant n^u$ of the control trajectory. To ensure separability of the discretization the $b_{ij}$ shall have local support, and are parameterized by a vector of finitely many control parameters $q_{ij} \in \mathbb{R}^{n^q_{ij}}$. Popular choices for the base functions include piecewise constant controls ($n^q_{ij} = 1$)

$$b_{ij} : [t_i, t_{i+1}] \times \mathbb{R}^{n^q_{ij}} \longrightarrow \mathbb{R}, \quad (t, q_{ij}) \mapsto q_{ij},$$

piecewise linear controls ($n^{ij}_q = 2$)

$$b_{ij} : [t_i, t_{i+1}] \times \mathbb{R}^{n^q_{ij}} \longrightarrow \mathbb{R}, \quad (t, \boldsymbol{q}_{ij}) \mapsto \frac{t_{i+1} - t}{t_{i+1} - t_i} q_{ij1} + \frac{t - t_i}{t_{i+1} - t_i} q_{ij2},$$

or piecewise cubic spline controls ($n^{ij}_q = 4$)

$$b_{ij} : [t_i, t_{i+1}] \times \mathbb{R}^{n^{\mathrm{q}}_{ij}} \longrightarrow \mathbb{R}, \quad (t, \boldsymbol{q}_{ij}) \mapsto \sum_{k=1}^{4} q_{ij_k} \beta_k \left( \frac{t - t_i}{t_{i+1} - t_i} \right)^{k-1}$$

with appropriately chosen spline function coefficients $\boldsymbol{\beta}$. Evidently, the type of discretization may be chosen differently for each of the $n^{\mathrm{u}}$ components of the control trajectory.

For the ease of notation we introduce an additional discretized control $\boldsymbol{b}_m(t_m, \boldsymbol{q}_m)$ in the last point $t_m$ of the shooting grid $\{t_i\}$, which shall be implicitly fixed to the final control value of the previous shooting interval,

$$\boldsymbol{b}_m(t_m, \boldsymbol{q}_m) \stackrel{\text{def}}{=} \boldsymbol{b}_{m-1}(t_m, \boldsymbol{q}_{m-1}). \tag{1.17}$$

For certain choices of the control discretization, e.g. for piecewise linear controls, continuity of the discretized control trajectory may be desired. This can be enforced by imposing additional control continuity conditions for the trajectory $u_j(t)$ in all points of the control discretization grid $\{t_i\}$,

$$0 = b_{ij}(t_{i+1}, \boldsymbol{q}_i) - b_{i+1,j}(t_{i+1}, \boldsymbol{q}_{i+1}), \quad 0 \leqslant i \leqslant m-1. \tag{1.18}$$

The choice of the control discretization obviously affects the quality of the discretized OCP's solution approximation that of the infinite–dimensional one, see e.g. [122].

### 1.3.2 State Parameterization

In addition to the control discretization, and notably different from single shooting described in section 1.2.3, we also introduce a parameterization of the state trajectory $\boldsymbol{x}(\cdot)$ on the shooting grid $\{t_i\}$ that yields $m$ IVPs with initial values $\boldsymbol{s}_i \in \mathbb{R}^{n^{\mathrm{x}}}$ on the intervals $[t_i, t_{i+1}]$ of the horizon $\mathcal{T}$,

$$\dot{\boldsymbol{x}}_i(t) = \boldsymbol{f}(t, \boldsymbol{x}_i(t), \boldsymbol{b}_i(t, \boldsymbol{q}_i)), \qquad t \in [t_i, t_{i+1}],\ 0 \leqslant i \leqslant m-1, \tag{1.19a}$$

$$\boldsymbol{x}_i(t) = \boldsymbol{s}_i. \tag{1.19b}$$

In order to ensure continuity of the obtained trajectory $\boldsymbol{x}(\cdot)$ on the whole of $\mathcal{T}$, we introduce $m-1$ additional matching conditions

$$\boldsymbol{0} = \boldsymbol{x}_i(t_{i+1};\ t_i, \boldsymbol{s}_i, \boldsymbol{q}_i) - \boldsymbol{s}_{i+1}, \quad 0 \leqslant i \leqslant m-1. \tag{1.20}$$

Here, the expression $\boldsymbol{x}_i(t_{i+1};\ t_i, \boldsymbol{s}_i, \boldsymbol{q}_i)$ denotes the final value $\boldsymbol{x}(t_{i+1})$ obtained as the solution of the IVP (1.19) on the interval $[t_i, t_{i+1}]$ when starting in the initial value $\boldsymbol{x}(t_i) = \boldsymbol{s}_i$ and applying the control trajectory $\boldsymbol{u}(t) = \boldsymbol{b}_i(t, \boldsymbol{q}_i)$ on $[t_i, t_{i+1}]$. The evaluation of the residual of constraint (1.20) thus requires the solution of an IVP by an appropriate numerical method, see [5, 4, 15, 61] and chapter 3.

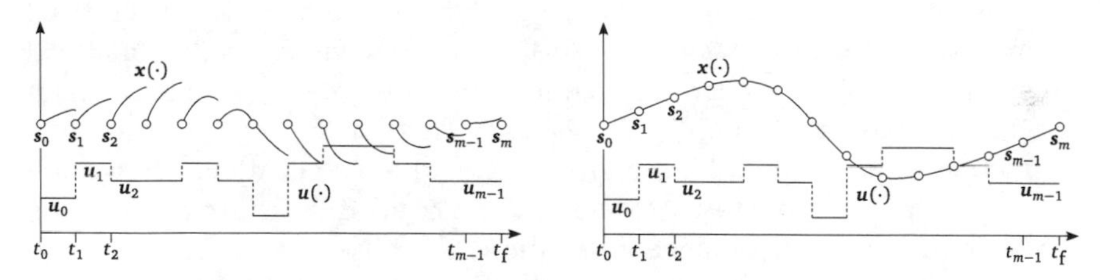


Figure 1.2: Illustration of the direct multiple shooting discretization applied to the optimal control problem. On the left, all shooting nodes were initialized identically and the solution of the $m$ IVPs violates the matching conditions. On the right, the matching conditions are fulfilled after convergence of the NLP solver.

More generally, the IVPs (1.19) together with the matching conditions (1.20) may be replaced by a sufficiently smooth but otherwise arbitrarily chosen mapping

$$\boldsymbol{0} = \boldsymbol{j}_i(t_i, t_{i+1}, \boldsymbol{s}_i, \boldsymbol{s}_{i+1}, \boldsymbol{q}_i, \boldsymbol{q}_{i+1}), \quad 0 \leqslant i \leqslant m-1. \tag{1.21}$$

In particular, the Jacobians of the mappings $\boldsymbol{j}_i$ with respect to the subsequent state $\boldsymbol{s}_{i+1}$ or the control $\boldsymbol{q}_{i+1}$ need not necessarily be regular. This effectively increases the number of degrees of freedom of the discretized optimal control problem. In case of singularity of $\boldsymbol{j}_i$ with respect to $\boldsymbol{s}_{i+1}$, we allow for discontinuities of the state trajectory $\boldsymbol{x}(\cdot)$, while singularity of $\boldsymbol{j}_i$ with respect to $\boldsymbol{q}_{i+1}$ is just a generalization of choosing a discontinuous control discretization.

The resulting vector of $(m+1)n^{\mathrm{x}} + mn^{\mathrm{q}}$ unknowns, assuming $n^{\mathrm{q}}_{ij}$ to be identical for all nodes and controls, is denoted by

$$\boldsymbol{v} \stackrel{\text{def}}{=} \begin{bmatrix} \boldsymbol{s}_0 & \boldsymbol{q}_0 & \ldots & \boldsymbol{s}_{m-1} & \boldsymbol{q}_{m-1} & \boldsymbol{s}_m \end{bmatrix}. \tag{1.22}$$

### 1.3.3 Constraint Discretization

The inequality path constraint $\boldsymbol{c}(\cdot)$ is discretized by enforcing it in the points $\{t_i\}$ of the shooting grid only,

$$\mathbf{0} \leqslant \boldsymbol{c}_i(t_i, \boldsymbol{s}_i, \boldsymbol{b}_i(t_i, \boldsymbol{q}_i)), \quad 0 \leqslant i \leqslant m. \tag{1.23}$$

This discretization will in general enlarge the feasible set of the discretized optimal control problem compared to that of the continuous one, and will thus affect the obtained optimal solution. For most real-world problems, it can be observed that an optimal trajectory $(\boldsymbol{x}^\star(\cdot), \boldsymbol{u}^\star(\cdot))$ obtained as solution of the discretized problem shows only negligible violations of the path constraints $\boldsymbol{c}(\cdot)$ in the interior of shooting intervals if those are enforced on the shooting nodes. If significant violations are observed or strict feasibility on the whole of $\mathcal{T}$ is of importance, remaining violations can sometimes be treated successfully by choosing an adapted, possibly finer shooting grid $\{t_i\}$. Alternatively, a semi-infinite programming algorithm for tracking of constraint violations in the interior of shooting intervals is discussed in [168, 169].

### 1.3.4 The Nonlinear Problem

The discretized optimal control problem resulting from application of the direct multiple shooting discretization to problem (1.15) reads

$$\begin{aligned}
\min_{\boldsymbol{s},\boldsymbol{q}} \quad & \sum_{i=0}^{m} l_i(t_i, \boldsymbol{s}_i, \boldsymbol{q}_i) && \qquad (1.24)\\
\text{s. t.} \quad & \mathbf{0} = \boldsymbol{x}_i(t_{i+1};\ t_i, \boldsymbol{s}_i, \boldsymbol{q}_i) - \boldsymbol{s}_{i+1}, && 0 \leqslant i \leqslant m-1,\\
& \mathbf{0} = \boldsymbol{r}_i^{\text{eq}}(t_i, \boldsymbol{s}_i, \boldsymbol{b}_i(t_i, \boldsymbol{q}_i)), && 0 \leqslant i \leqslant m,\\
& \mathbf{0} \leqslant \boldsymbol{r}_i^{\text{in}}(t_i, \boldsymbol{s}_i, \boldsymbol{b}_i(t_i, \boldsymbol{q}_i)), && 0 \leqslant i \leqslant m,\\
& \mathbf{0} \leqslant \boldsymbol{c}_i(t_i, \boldsymbol{s}_i, \boldsymbol{b}_i(t_i, \boldsymbol{q}_i)), && 0 \leqslant i \leqslant m.
\end{aligned}$$

Here we have written the Mayer term $m(t_m, \boldsymbol{s}_m)$ as final term $l_m(t_m, \boldsymbol{s}_m, \boldsymbol{q}_m)$ of the objective.

Constrained nonlinear programming theory and the SQP method for the solution of problem (1.24) are presented in chapter 3. In chapter 4, we present the concept of real-time iterations that improves the efficiency of SQP methods applied to solve NMPC problems. In chapter 6 we present a nonconvex SQP method for a special class of NLPs obtained from the discretization of convexified MIOCPs. The quadratic subproblems arising in this SQP method are studied and an active set algorithm for ttheir solution is presented.

### 1.3.5 Separability

We refer to a function of problem (1.24) as *separable* if it coupling of unknowns $(s_i, q_i)$ and $(s_j, q_j)$ on different nodes $0 \leqslant i, j \leqslant m$, $i \neq j$ by this function is either absent or linear. In this case, the Jacobians of that function with respect to $(s, q)$ show block structure. The property of separability is crucial for the design of efficient linear algebra for the solution of problem (1.24) by derivative–based algorithms. Chapters 7 and 8 are concerned with techniques and contributions of this thesis to the efficient exploitation of structures in problem (1.24).

By choice of the control discretization and the state parameterization, the objective function and the constraints $c_i$ and $r_i$ are separable with respect to the unknowns $(s_i, q_i)$ on the shooting grid nodes. Unknowns on adjacent nodes of the shooting grid are coupled by the matching conditions (1.20) exclusively.

## 1.4 Summary

In this chapter we have defined a class of optimal control problems for dynamic processes modeled by systems of ODE or semi–implicit index one DAE systems and have described various extensions of this problem class. A survey of numerical approaches for the solution of problems of this class considered direct and indirect as well as sequential and simultaneous approaches. We presented in more detail the direct multiple shooting method for optimal control, a hybrid approach that can be located between single shooting and collocation methods. It is known for its excellent convergence properties, yields a highly structured NLP, but allows at the same time the use of state–of–the–art adaptive solvers for the IVPs. The resulting discretized OCPs are best treated by SQPs methods which will be investigated from different points of view in several places in this thesis.

# 2 Mixed–Integer Optimal Control

In this chapter we extend the problem class of continuous optimal control problems discussed in chapter 1 to include control functions that may at each point in time attain only a finite number of values from a discrete set. We briefly survey different approaches for the solution of the discretized Mixed–Integer Optimal Control Problem (MIOCP), such as Dynamic Programming, branching techniques, and Mixed–Integer Nonlinear Programming methods. These approaches however turn out to be computationally very demanding already for off–line optimal control, and this fact becomes even more apparent in a real–time on–line context. The introduction of an outer convexification and relaxation approach for the MIOCP class allows to obtain an approximation of the MIOCP's solution by solving only one single but potentially much larger continuous optimal control problem using the techniques of chapter 1. We describe theoretical properties of this problem reformulation that provide bounds on the loss of optimality for the infinite dimensional MIOCP. Rounding schemes are presented for the discretized case that maintain these guarantees. We develop techniques for the introduction of switch costs into the class of MIOCPs and give reformulations that allow for a combination with direct multiple shooting and outer convexification.

## 2.1 Problem Formulations

We consider in this chapter the extension of the Optimal Control Problem (OCP) class of chapter 1 to the following class of Mixed–Integer Optimal Control Problems (MIOCPs).

**Definition 2.1 (Mixed–Integer Optimal Control Problem)**
A Mixed–Integer Optimal Control Problem (MIOCP) is a continuous optimal control problem with integer feasibility requirement on a subset of the control trajectories of the following form:

$$\begin{aligned}
\min_{x(\cdot),u(\cdot),w(\cdot)} \quad & x(t_f, m(t_f)) + \int_{t_0}^{t_f} l(t, x(t), u(t), w(t))\,\mathrm{d}t && && (2.1)\\
\text{s.t.} \quad & \dot{x}(t) = f(t, x(t), u(t), w(t)), && t \in \mathcal{T}, &\\
& 0 = r_i^{\mathrm{eq}}(t_i, x(t_i)), && \{t_i\}_{0 \leqslant i \leqslant n^{\mathrm{r}}} \subset \mathcal{T}, &\\
& 0 \leqslant r_i^{\mathrm{in}}(t_i, x(t_i)), && \{t_i\}_{0 \leqslant i \leqslant n^{\mathrm{r}}} \subset \mathcal{T}, &\\
& 0 \leqslant c(t, x(t), u(t), w(t)), && t \in \mathcal{T}, &\\
& w(t) \in \Omega \subset \mathbb{R}^{n^{\mathrm{w}}}, && t \in \mathcal{T}, & |\Omega| = n^{\Omega} < \infty.
\end{aligned}$$

In addition to the conventions of definition 1.1, the dynamic process $x(\cdot)$ is affected by an additional vector–valued control function $w : \mathcal{T} \to \Omega \subset \mathbb{R}^{n^{\mathrm{w}}}$, $t \mapsto w(t)$ which only attains values from a finite discrete set $\Omega \overset{\mathrm{def}}{=} \{\omega^1, \dots, \omega^{n^{\Omega}}\} \subset \mathbb{R}^{n^{\mathrm{w}}}$ with cardinality $|\Omega| = n^{\Omega} < \infty$. △

This problem class differs from the class of OCPs treated in chapter 1 by the integer feasibility requirement imposed on the additionally introduced control function $w : \mathcal{T} \to \Omega$. Here $n^{\mathrm{w}}$ denotes the number of scalar integer control functions, and $n^{\Omega}$ denotes the finite number of discrete choices the vector valued integer control trajectory $w(\cdot)$ can attain at each point in time. For the discussion in this chapter, we restrict the problem class (2.1) to binary control functions $w(t) \in \{0,1\}^{n^{\mathrm{w}}}$, i.e., $n^{\Omega} = 2^{n^{\mathrm{w}}}$.

**Remark 2.1 (Restriction to Binary Control Functions)**
The restriction to binary control functions is not substantial as it is possible to reformulate any MIOCP involving integer control functions $v(t) \in \{v^1, \dots, v^{n^{\mathrm{v}}}\} \subset \mathbb{R}$ by transformation of the model functions to $\hat{v}(t) \in \{1, \dots, n^{\mathrm{v}}\} \subset \mathbb{N}$ and letting

$$\hat{v}(t) \overset{\mathrm{def}}{=} 1 + \sum_{i=1}^{\lceil \log_2 n^{\mathrm{v}} \rceil} 2^i w_i(t). \tag{2.2}$$

This reformulation requires $\lceil \log_2 n^{\mathrm{v}} \rceil$ binary control functions $w_i(t)$ to replace $v(t)$.

## 2.2 Mixed–Integer Nonlinear Programming

### 2.2.1 Discretization to a Mixed–Integer Nonlinear Program

A straightforward approach to solving problem (2.1) is to apply one of the discretization approaches for optimal control presented in chapter 1, i.e., collocation or direct shooting methods. The discretization variables introduced for a piecewise constant discretization of the integer control function

$w(\cdot)$ then inherit the integrality property. This approach yields a Mixed–Integer Nonlinear Program (MINLP) in place of the continuous NLP (1.24). The multiple shooting discretization of this problem is given in the following definition.

**Definition 2.2 (Multiple Shooting Discretized MIOCP)**
The multiple shooting discretized counterpart of the mixed–integer optimal control problem (2.1) with binary control functions is

$$\begin{aligned}
\min_{s,q,w} \quad & \sum_{i=0}^{m} l_i(t_i, s_i, q_i, w_i) && && (2.3)\\
\text{s.t.} \quad & 0 = x_i(t_{i+1};\ t_i, s_i, b_i(t_i, q_i), w_i) - s_{i+1}, && 0 \leqslant i \leqslant m-1, \\
& 0 = r_i^{\text{eq}}(t_i, s_i, b_i(t_i, q_i), w_i), && 0 \leqslant i \leqslant m, \\
& 0 \leqslant r_i^{\text{in}}(t_i, s_i, b_i(t_i, q_i), w_i), && 0 \leqslant i \leqslant m, \\
& 0 \leqslant c_i(t_i, s_i, b_i(t_i, q_i), w_i), && 0 \leqslant i \leqslant m, \\
& w_i \in \{0,1\}^{n^{\text{w}}}, && 0 \leqslant i \leqslant m.
\end{aligned}$$

△

This multiple shooting discretized MIOCP can be cast as a MINLP as follows.

**Definition 2.3 (Mixed–Integer Nonlinear Program)**
A Mixed–Integer Nonlinear Program (MINLP) is a Nonlinear Program (NLP) in the unknowns $x \in \mathbb{R}^{n^x}$ and $w \in \mathbb{R}^{n^w}$ with binary feasibility requirement imposed on $w$,

$$\begin{aligned}
\min_{x,w} \quad & f(x, w) && (2.4)\\
\text{s.t.} \quad & 0 = g(x, w), \\
& 0 \leqslant h(x, w), \\
& w \in \{0,1\}^{n^{\text{w}}}.
\end{aligned}$$

△

Note that in this formulation $w$ contains the binary control variables $w_i$ of all shooting nodes $0 \leqslant i \leqslant m-1$, thus $n^{\text{w}}$ has grown by a factor $m$ compared to the last section.

The class of MINLP problems is $\mathcal{NP}$–hard [78]. Hence, there exist instances of problem (2.4) than cannot be solved in polynomial runtime on a deterministic machine. Moreover, this fact serves as a first indicator that computing exact solutions to problems of practical interest in real–time is likely to be a most challenging task. In this section we briefly mention various approaches targeted either at solving the MIOCP or the MINLPs to optimality, or at approximating a local or global solution.

### 2.2.2 Enumeration Techniques

#### Full Enumeration

A naïve approach to solving (2.4) is to fix the binary control variables for every possible choice of $w$ and solve a continuous optimal control problem. The solution process either indicates infeasibility of the problem given the chosen $w$, or yields an optimal objective function value $f(w)$. The MINLP's optimal solution is choice $w^\star$ that resulted in the smallest objective of the associated continuous problem. It is a global solution if all solvable continuous OCPs have been solved to global optimality.

The obvious drawback of this approach is the exponentially increasing number of $2^{n^w}$ continuous optimal control problems that have to be solved. Even for a scalar binary control function, the optimal solutions of $2^m$ multiple shooting discretized NLPs need to be computed. The computational effort very quickly becomes prohibitive.

#### Dynamic Programming

The principle and technique of dynamic programming has already been presented in section 1.2.2 for continuous optimal control problems. There we needed to choose a discretization of the continuous control space. For the integer controls $w(\cdot)$, this discretization is given in a natural way and from this point of view, mixed–integer optimal control problems are obvious candidates for treatment by dynamic programming approaches. Still, the curse of dimensionality prevents computational access to problems with more than a few system states.

Some examples of MIOCPs that have been solved using dynamic programming techniques with applications in the automotive control area can be found in [42, 103]. Fulfilling real–time constraints on the computation times has however been shown to be possible only by precomputation or by drastically reducing the search space.

### 2.2.3 Branching Techniques

#### Branch & Bound

Branch & Bound is a general framework for the solution of combinatorial problems that improves upon full enumeration of the search space. The fundamental idea is to perform a tree search in the space of binary or integer variables, and to solve a continuous linear or nonlinear problem in every node of the tree. The search tree's root node holds a complete relaxation of the (mixed–)integer problem. Descending the tree's branches to its leaves, more and more integer variables get fixed to one of the admissible choices. This amounts to recursively partitioning the original problem into multiple smaller subproblems, each giving a *lower* bound for the objective of all leaf problem on the branch. The minimum of the optimal solutions found for these subproblems is the optimal solution of the original problem.

The recursive partitioning process can be bounded whenever a subproblem is found to be infeasible or an integer solution is found for it. In the first case, further branching is unneccesary as all problems on the subtree will be infeasible as well. In the second case, an *upper* bound to the original problem's objective is found. Subproblems with higher objective – being a lower bound — need not be branched on as all subproblems on the subtree will be suboptimal as well.

In practice, good heuristics for the choice of the next subproblem and the selection of fractional variables to branch on are crucial to quickly obtain feasible solutions and tight upper bounds. This avoids visits to all possible subproblems, in which case Branch & Bound would degenerate to full enumeration again. Such heuristics frequently must be tailored to the specific problem under investigation, and details can be found in e.g. [58, 140].

The first Branch & Bound algorithm for integer linear programming is due to [49, 128]. Extensions to convex nonlinear mixed–integer problems can e.g. be found in [19, 39, 135, 136].

### Branch & Cut

The fixation of binary or integer variables to a selected admissible value in the above Branch & Bound algorithm can be regarded as the introduction of additional inequality constraints $x_i \leqslant \lfloor x_i^\star \rfloor$ and $x_i \geqslant \lceil x_i^\star \rceil$ respectively for a fractional relaxed optimal solution $\boldsymbol{x}^\star$. These constraints cut off a part of the feasible set of the preceding problem, and this concept is further generalized in the Branch & Cut method that adds cutting planes to the subproblems as the method descends the search tree's branches.

In the linear case, the optimal solution is known to coincide with a vertex of the convex hull of the feasible set. The algorithm then aims at adding cutting planes to the subproblems until the convex hull in a neighborhood of the optimal integer solution is approximated well enough for the relaxed solution to be integer.

Earliest sources for cutting planes methods can be found in [48, 90]. Again, the determination of strong cutting planes that retain all integer solutions and invalidate as much as possible of the relaxed feasible set is crucial and subject of intense research efforts, see e.g. [95, 110, 159, 207]. The application of Branch & Cut to nonlinear integer programs is subject to ongoing research, see [159, 207].

In the case of a MIOCP, every tree node would hold a separate continuous optimal control problem, and a full evaluation of the search tree would have to be carried out to find an optimal mixed–integer control feedback. Clearly, the achievable run times are not likely to be competitive. An application of a Branch & Bound technique to an offline time optimal MIOCP can be found in [80].

### 2.2.4 Outer Approximation

#### Outer Approximation

The Outer Approximation method due to [59] was developed for convex MINLPs explicitly. It aimed at avoiding the potentially large number of NLPs to be solved in a branching method by replacing them with more accessible Mixed–Integer Linear Programs (MILPs), for which advanced techniques and solvers have been readily available. To this end, an alternating sequence of MILPs, called *master* problems, and NLPs is solved. Therein, the MILP yields an integer solution and a *lower* bound to the original problem's solution while the NLP with fixed integer variables yields — if feasible — yields an *upper* bound and a new linearization point, improving the outer approximation of the convex feasible set by linearizations of the constraints.

For applications of Outer Approximation to MIOCPs with time–independent binary variables we refer to [93, 160].

#### Generalized Bender's Decomposition

Generalized Bender's decomposition, first described by [79], is older than outer approximation and a straightforward comparison reveals that it is identical except for a weaker formulation of the master program, e.g. [74].

### LP/NLP based Branch & Bound

The number of MILPs to be solved in an Outer Approximation method can possibly be further reduced by an LP/NLP based branch & Bound, cf. [38, 135, 172]. A single master MILP problem is solved by a Branch & Bound type method. An NLP is then solved for every feasible point giving a possibly tightened upper bound to the original problem's solution. Linearizations of the constraints in the obtained solution are added to the outer approximation of the feasible set, and pending subproblems in the Branch & Bound search tree are updated appropriately before continuing the descent along the tree's branches.

## 2.2.5 Reformulations

The difficulty of the problem class (2.1) comes from the integer restrictions on some of the control variables. The idea of replacing the integrality requirement by adding one or more constraints to the continuous formulation therefore is of obvious appeal. To this end, several propositions can be found in the literature.

### Relaxation and Homotopies

A first approach is to replace for $w_i \in \{0,1\}$ by the set of constraints

$$w_i(1-w_i) = 0, \quad w_i \in [0,1] \subset \mathbb{R}.$$

The feasible sets coincide, and the reformulation thus is exact. As the first constraint is nonconvex, the feasible set is disjoint, and Linear Independence Constraint Qualification (LICQ) is violated, descent based methods such as Sequential Quadratic Programming (SQP) show bad performance on this reformulation. This situation can be ameliorated to a certain extent by introducing the formulation

$$w_i(1-w_i) \leqslant \beta, \quad w_i \in [0,1] \subset \mathbb{R},$$

together with a homotopy $\beta \to 0^+$. This technique has found applications in the field of Mathematical Programs with Equilibrium Constraints (MPECs), cf. [16, 137, 144, 173, 174]. Geometrically motivated reformulations of less general applicability have been proposed by e.g. [173] for the feasible set

$\{(0,1),(1,0)\} \subset \mathbb{N}^2$ and homotopies for relaxation are applied here as well. In [137] the Fischer–Burmeister function

$$F^{\text{FB}}(w_1, w_2) = w_1 + w_2 - \sqrt{w_1^2 + w_2^2}$$

which is zero if $(w_1, w_2)$ is binary feasible, is used to extend an SQP method to treat MPECs. We'll revisit some of these reformulations in chapter 5.

The disadvantage of virtually all reformulation for Nonlinear Model Predictive Control (NMPC) is that they involve a homotopy parameter that needs to be driven to zero or infinity in order to approach the MIOCP's solution. This requires the solution of multiple optimal control problems at an increased computational effort. Moreover, the homotopy steps can in general not be determined a priori, such that bounds on the computation time for the solution of an MIOCP cannot be established.

**Convexification and Relaxation**

As we have seen, all presented methods require for the case of MIOCP the solution of a potentially large number of discretized optimal control subproblems. In view of the fast solution times required for the computation of the control feedback, such techniques are unlikely to be successful approaches for fast mixed–integer NMPC. We therefore continue our presentation with the introduction of a convexification and relaxation approach that serves to obtain a good approximation of a local solution of the discretized MIOCP by solving only a single but possibly larger discretized OCP.

Convexification and relaxation approaches aim at reformulating (2.4) as a purely continuous NLP for which computationally highly efficient solution methods exist and are presented in the sequel of this thesis. The crucial point here is that the reformulation should be as tight as possible in the sense that the feasible set is enlarged as little as possible compared to the original MINLP. The reader is referred to e.g. [125] for reformulations of a number of popular combinatorial problems. We mention here two approaches named inner convexification and outer convexification that have found application to MINLPs arising from the discretization of MIOCPs in e.g. [130, 131, 141, 183, 185], and our contributions [119, 120, 122, 187]. The term convexification in the following refers to a convex reformulation of the dynamics – and later also the constraints – of problem (2.4) with respect to the integer control. All other parts of the discretized optimal control problem remain untouched and hence possibly nonlinear and/or nonconvex. In this sense the techniques to be presented are *partial* convexifications of the problem with respect to a selected subset of the unknowns only. In the next section we study in detail the outer convexification and relaxation technique.

### Inner Convexification

A naïve idea for transforming problem (2.1) into a purely continuous NLP is to simply drop the integrality requirement on the unknown $\boldsymbol{w}$. This requires that all model functions, comprising the objective function, the Ordinary Differential Equation (ODE) system's right hand side function, and all constraint functions can be evaluated in fractional points of the space of unknowns, and yield meaningful results that do not prevent optimal solutions from being found. The locally optimal solution found for the NLP obtained by the inner convexification approach will in general not be an integer one. Guarantees for integer feasibility of a rounded solution in the neighborhood of the locally optimal one can in general not be derived, nor can bounds on the loss of optimality be given. This approach has been used for mixed–integer optimal control in e.g. [80] in combination with a branching algorithm to overcome the mentioned problems.

### Outer Convexification

Outer convexification due to [183] introduces a new binary variable $\omega_i \in \{0,1\}$ (note the subscript index) for each choice $\omega^i$ (indicated by a superscript index) contained in the set $\Omega$ of feasible choices for the discrete control. All model functions directly depending on the integer control function $\boldsymbol{w}(\cdot)$ are partially convexified, i.e., convexified with respect to this integer control only. The introduced binary variables $\omega_i$ act as convex multipliers. Replacing the binary variables $\omega_i$ by relaxed ones $\alpha_i \in [0,1] \subset \mathbb{R}$ is referred to as *relaxation*. The outer convexification reformulation ensures that even after relaxation all model functions are evaluated in integer feasible choices from the set $\Omega$ only. Bounds on the loss of optimality can be established as will be shown in the next section. In addition, we develop here and in chapter 5 the application of outer convexification to constraints directly depending on a binary or integer control function, and show how integer feasibility can be guaranteed after rounding of the possibly fractional solution obtained for the NLP. For certain optimal control problems that enjoy a bang–bang property, the convexified and relaxed OCP's locally optimal solution itself can be shown to be an optimal solution of the MIOCP.

**Example 2.1 (Inner and Outer Convexification)**
To stress the difference between the two proposed convexification approaches for nonlinear functions, we exemplarily study the function $f(x,w) = (x-3w)^2$ depending on the continuous variable $x \in \mathbb{R}$ and the integer variable $w \in \mathbb{Z}$. The graph $gr\, f^w(x)$ of this function for integer and relaxed choices of $w$ is shown in figure 2.1. Inner convexification of $f(x,w)$ with respect to the integer variable $w$ effectively means dropping the integer constraint on $w$. Outer convexification of $f(x,w)$, here for the choices $w \in \Omega = \{\omega^1, \omega^2, \omega^3\} = \{-1,0,1\}$, yields after relaxation

$$\begin{aligned} f^{\text{OC}}(x,\boldsymbol{\alpha}) &\stackrel{\text{def}}{=} \alpha_1 f(x,-1) + \alpha_2 f(x,0) + \alpha_3 f(x,1), \\ 1 &= \alpha_1 + \alpha_2 + \alpha_3, \qquad \boldsymbol{\alpha} \in [0,1]^3. \end{aligned}$$

If $f$ contributes to the objective function, inner convexification results for relaxed choices of $w$ in objective function values unattainable for integer choices. If the graph of $f$ denotes the border of the feasible set, inner convexification leads to constraint violations for relaxed choices of $w$. Outer convexification offers a remedy for both situations, as we will see in the next section.

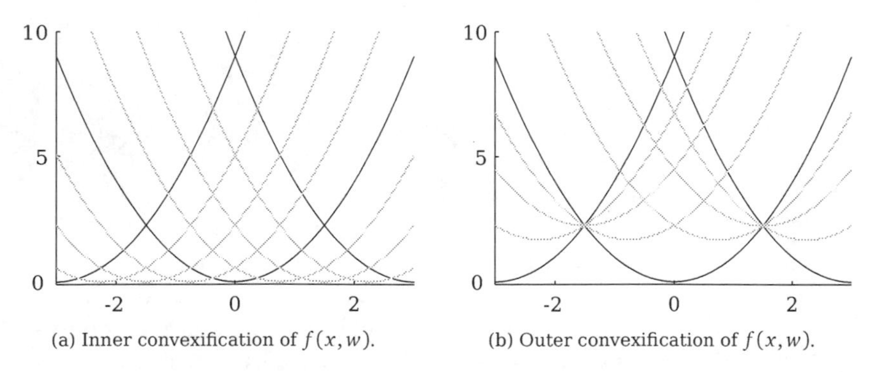


(a) Inner convexification of $f(x,w)$. (b) Outer convexification of $f(x,w)$.

Figure 2.1: Inner and outer convexification at the example $f(x,w) = (x-3w)^2$. (—) shows the function's graph for the integer choices $w \in \{-1,0,1\}$ and (⋯) shows the function's graph in the relaxed choices $w \in \left\{-\frac{3}{4}, -\frac{1}{2}, -\frac{1}{4}, \frac{1}{4}, \frac{1}{2}, \frac{3}{4}\right\}$ for inner convexification and corresponding choices $\boldsymbol{\alpha} \in \left\{(\frac{3}{4},\frac{1}{4},0), (\frac{1}{2},\frac{1}{2},0), (\frac{1}{4},\frac{3}{4},0), (0,\frac{3}{4},\frac{1}{4}), (0,\frac{1}{2},\frac{1}{2}), (0,\frac{1}{4},\frac{3}{4})\right\}$ for outer convexification.

## 2.3 Outer Convexification and Relaxation

In this section we investigate the approximation of solutions of MIOCPs by solving a convexified and relaxed counterpart problem. To this end it is crucial to derive maximal lower bounds on the attained objective value in order to judge on the quality of the obtained solution. For nonlinear or nonconvex problems, the relaxed solution will in general not be binary feasible and the optimal objective function value can in general not be attained by a binary feasible solution.

The results of this section were first presented in [183] and apply to the infinite–dimensional MIOCP prior to any discretization taking place. We present bounds on the objective function and the constraint residuals of the optimal solutions obtained when applying the outer convexification approach to the MIOCP. We also state the bang–bang principle which allows to deduce binary feasibility and optimality of relaxed solutions for certain linear optimal control problems.

### 2.3.1 Convexified and Relaxed Problems

For the presentation of theoretical results we restrict ourselves to the following more limited problem class of MIOCPs.

**Definition 2.4 (Binary Nonlinear Problem)**
The class of *binary nonlinear optimal control problems* is the following subclass of (2.1),

$$\begin{aligned} \min_{x(\cdot),u(\cdot),w(\cdot)} \quad & \varphi(x(\cdot),u(\cdot),w(\cdot)) && \text{(BN)} \\ \text{s.t.} \quad & \dot{x}(t)=f(x(t),u(t),w(t)), \quad t\in\mathcal{T}, \\ & x(t_0)=x_0, \\ & w(t)\in\{0,1\}^{n^{\mathrm{w}}}, \qquad\qquad t\in\mathcal{T}. \end{aligned}$$

△

Here we consider the minimization of an objective function computed from the trajectories of an unconstrained Initial Value Problem (IVP) controlled by continuous and binary control functions. The set of admissible choices for the binary control here is $\Omega=\{0,1\}^{n^{\mathrm{w}}}$ and the set members $\omega^i$ enumerate all $n^{\Omega}=2^{n^{\mathrm{w}}}$ possible assignments of binary values to the components of $w(t)$.

**Definition 2.5 (Relaxed Nonlinear Problem)**
The *relaxed nonlinear* variant of (BN) is the continuous optimal control problem

$$\begin{aligned}
\min_{x(\cdot),u(\cdot),w(\cdot)} \quad & \varphi(x(\cdot),u(\cdot),w(\cdot)) && && \text{(RN)}\\
\text{s.t.} \quad & \dot{x}(t)=f(x(t),u(t),w(t)), && t\in\mathcal{T},\\
& x(t_0)=x_0,\\
& w(t)\in[0,1]^{n^{\mathrm{w}}}, && t\in\mathcal{T}.
\end{aligned}$$

△

Clearly due to the relaxation that equals the inner convexification approach described above, the right hand side function $f(\cdot)$ will be evaluated in non-integer choices for the relaxed binary control $w(t)$.

We now introduce counterparts to the above two problems obtained by applying outer convexification to the set $\Omega = \{0,1\}^{n^{\mathrm{w}}}$ of discrete choices. For each choice $w(t)=\omega^i\in\Omega$ we introduce a new binary control function $\omega_i(t)$ indicating whether or not the choice $\omega^i$ was made at time $t\in\mathcal{T}$.

**Definition 2.6 (Binary Convexified Linear Problem)**
The *binary convexified linear problem* is the following convexified counterpart of problem (BN),

$$\min_{x(\cdot),u(\cdot),\omega(\cdot)} \sum_{i=1}^{2^{n^{\mathrm{w}}}} \varphi(x(\cdot),u(\cdot),\omega^i)\,\omega_i(\cdot) \tag{BC}$$

$$\begin{aligned}
\text{s.t.} \quad & \dot{x}(t)=\sum_{i=1}^{2^{n^{\mathrm{w}}}} f(x(t),u(t),\omega^i)\,\omega_i(t), && t\in\mathcal{T},\\
& x(t_0)=x_0,\\
& \omega(t)\in\{0,1\}^{2^{n^{\mathrm{w}}}}, && t\in\mathcal{T},\\
& 1=\sum_{i=1}^{2^{n^{\mathrm{w}}}}\omega_i(t), && t\in\mathcal{T}. && (2.5)
\end{aligned}$$

△

Here the constant parameters $\omega^i$ indicate evaluation of the right hand side $f(\cdot)$ in the admissible choice $w(t)=\omega^i$. The Special Ordered Set (SOS) type 1 property holds for the trajectory of convex multipliers $\omega(\cdot)$ and is enforced by the additionally imposed constraint (2.5).

**Definition 2.7 (Special Ordered Set Property)**
We say that the variables $\omega_1,\dots,\omega_n$ fulfill the *special ordered set type 1 property* if they satisfy

$$\sum_{i=1}^{n} \omega_i = 1, \qquad \omega_i \in \{0,1\}, \quad 1 \leqslant i \leqslant n. \tag{SOS1}$$

△

The relaxed counterpart problem to (BC) is given in the following definition.

**Definition 2.8 (Relaxed Convexified Linear Problem)**
The *relaxed convexified linear problem* is the convexified counterpart of (RN) and the relaxed counterpart of (BC), the continuous optimal control problem

$$\min_{x(\cdot),u(\cdot),\alpha(\cdot)} \sum_{i=1}^{2^{n^w}} \varphi(x(\cdot), u(\cdot), \omega^i)\, \alpha_i(\cdot) \tag{RC}$$

$$\begin{aligned} \text{s. t.} \quad \dot{x}(t) &= \sum_{i=1}^{2^{n^w}} f(x(t), u(t), \omega^i)\, \alpha_i(t), && t \in \mathcal{T}, \\ x(t_0) &= x_0, \\ \alpha(t) &\in [0,1]^{2^{n^w}}, && t \in \mathcal{T}, \\ 1 &= \sum_{i=1}^{2^{n^w}} \alpha_i(t), && t \in \mathcal{T}. \end{aligned}$$

△

**Remark 2.2 (Combinatorial Explosion)**
The convexified problems (BC) and (RC) involve a number of control functions $\omega(\cdot)$ resp. $\alpha(\cdot)$ that grows exponentially with the number $n^w$ of binary control trajectories $w(\cdot)$ in problem (BN). Still for most practical problems it is possible to eliminate many binary combinations for $w(\cdot)$ that are physically meaningless or can be logically excluded. The number of binary controls after convexification is e.g. linear in $n^w$ in the applications [119, 120, 122, 183, 185, 209]. A second consequence of this convexification is the need for tailored structure exploiting algorithms that can treat problems with a large number of control parameters efficiently. We revisit this question when we discuss such methods in chapter 7.

**Remark 2.3 (Scalar Binary Control Functions)**
For scalar binary control functions $w(\cdot)$ with $n^w = 1$ the nonlinear and the convexified formulation have the same number of control functions. If the problems (BN) and (RN) are affine linear in $w(\cdot)$, the nonlinear and the convexified formulations are identical.

**Remark 2.4 (Elimination using the SOS1 Constraint)**
The SOS1 constraint allows the elimination of one binary control function $\omega_i(\cdot)$,

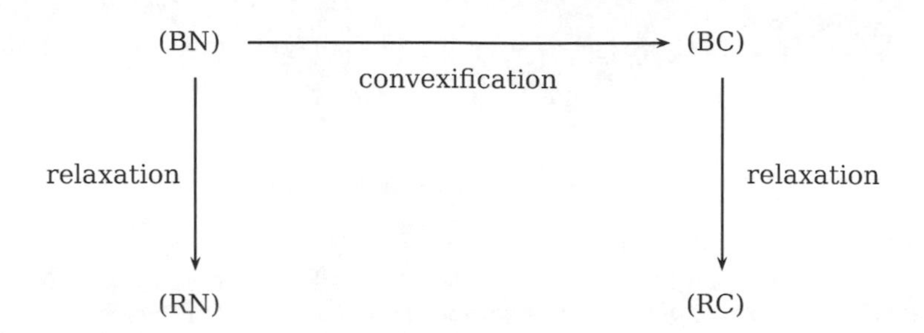


Figure 2.2: Relation of the four auxiliary problems.

which can be replaced by

$$\omega_i(t) = 1 - \sum_{\substack{j=1 \\ j \neq i}}^{2^{n^{\mathrm{w}}}} \omega_j(t), \qquad t \in \mathcal{T}. \tag{2.6}$$

The same is obviously true for the relaxed counterpart function $\boldsymbol{\alpha}(\cdot)$. In mixed-integer linear programming, this elimination is usually avoided as it destroys to a certain extent the sparsity structure present in the constraints matrices. For the direct multiple shooting discretization, this elimination has shown itself to be of advantage for some MIOCPs. Sparsity of the node constraint Jacobians is usually not exploited and the number of active set iterations spent in the Quadratic Program (QP) subproblem solution tends to become smaller.

### 2.3.2 The Bang-Bang Principle

The convexified relaxed control trajectories $\alpha_i(\cdot)$ may be optimal on the boundary or in the interior of $[0,1] \subset \mathbb{R}$. In this section we use the bang-bang principle to show that for a certain subclass of optimal control problems, the relaxed optimal solutions come to lie on the boundary of the unit hypercube and are thus binary feasible without any further effort. We consider linear control problems of the following structure.

**Definition 2.9 (Linear Control Problem)**
The problem

$$\begin{aligned} \dot{\boldsymbol{x}}(t) &= \boldsymbol{A}(t)\boldsymbol{x}(t) + \boldsymbol{B}(t)\boldsymbol{u}(t), \quad && t \in \mathcal{T}, \\ \boldsymbol{x}(t_0) &= \boldsymbol{x}_0, \\ \boldsymbol{u}^{\mathrm{lo}} \leqslant \boldsymbol{u}(t) &\leqslant \boldsymbol{u}^{\mathrm{up}}, && t \in \mathcal{T}, \end{aligned} \tag{2.7}$$

is referred to as a *linear control problem*. The function $\boldsymbol{u} : \mathcal{T} \to \mathbb{R}^{n^{\mathrm{u}}}$, $t \mapsto \boldsymbol{u}(t)$ shall be measurable and satisfy lower and upper bounds $\boldsymbol{u}^{\mathrm{lo}}$ and $\boldsymbol{u}^{\mathrm{up}}$. The matrix functions $\boldsymbol{A} : \mathcal{T} \to \mathbb{R}^{n^{\mathrm{x}} \times n^{\mathrm{x}}}$, $\boldsymbol{B} : \mathcal{T} \to \mathbb{R}^{n^{\mathrm{x}} \times n^{\mathrm{u}}}$ shall be continuous. △

We require the definition of the controllable set, and define further two classes $\mathcal{U}^{\mathrm{meas}}$ and $\mathcal{U}^{\mathrm{bb}}$ of control functions as follows.

**Definition 2.10 (Controllable Set)**
The *controllable set at time* $t \in \mathcal{T}$ is the set of all points $\boldsymbol{x}_0 \in \mathbb{R}^{n^{\mathrm{x}}}$ that can be steered back to the origin $\boldsymbol{x}(t) = \boldsymbol{0}$ in time $t$ by control functions from a given set $\mathcal{U}$,

$$\mathcal{C}(\mathcal{U}, t) \stackrel{\text{def}}{=} \left\{ \boldsymbol{x}_0 \in \mathbb{R}^{n^{\mathrm{x}}} \;\middle|\; \exists \boldsymbol{u}(\cdot) \in \mathcal{U} :\; \boldsymbol{x}(t; \boldsymbol{x}_0, \boldsymbol{u}(\cdot)) = \boldsymbol{0} \right\} \tag{2.8}$$

The *controllable set* is defined to be the union of all controllable sets at time $t$,

$$\mathcal{C}(\mathcal{U}) \stackrel{\text{def}}{=} \bigcup_{t>0} \mathcal{C}(\mathcal{U}, t) \tag{2.9}$$

△

**Definition 2.11 (Set of measurable control functions)**
The *set* $\mathcal{U}^{\mathrm{meas}}$ *of measurable control functions* is defined as

$$\mathcal{U}^{\mathrm{meas}} \stackrel{\text{def}}{=} \left\{ \boldsymbol{u} : \mathcal{T} \to \mathbb{R}^{n^{\mathrm{u}}} \;\middle|\; \boldsymbol{u}(\cdot) \text{ measurable} \right\}. \tag{2.10}$$

△

**Definition 2.12 (Set of bang–bang control functions)**
The *set* $\mathcal{U}^{\mathrm{bb}}$ *of bang–bang control functions* is defined as

$$\mathcal{U}^{\mathrm{bb}} \stackrel{\text{def}}{=} \left\{ \boldsymbol{u} : \mathcal{T} \to \mathbb{R}^{n^{\mathrm{u}}} \;\middle|\; t \in \mathcal{T}, 1 \leqslant i \leqslant n^{\mathrm{u}} :\; u_i(t) = u_i^{\mathrm{up}} \vee u_i(t) = u_i^{\mathrm{lo}} \right\}. \tag{2.11}$$

△

With these definitions, the bang–bang principle can be stated as follows.

**Theorem 2.1 (Bang–bang principle)**
For the class of linear control problems (2.7) it holds that

$$\mathcal{C}(\mathcal{U}^{\mathrm{meas}}, t) = \mathcal{C}(\mathcal{U}^{\mathrm{bb}}, t) \quad t > 0. \tag{2.12}$$

This set is compact, convex, and continuous in $t$. △

**Proof** Proofs can be found in [104] and [146]. □

An immediate consequence of theorem 2.1 is that binary feasible and optimal solutions to a linear time–optimal MIOCP coincide with optimal solutions of the OCP obtained using the outer convexification and relaxation reformulation, cf. [122, 183, 184, 185, 187] for applications. On a given discretization grid, this holds for the discretized OCP's solution except in the neighborhood of switching points of the convex multiplier trajectories $\boldsymbol{\alpha}(\cdot)$. If those do not fall onto grid points of the control discreization grid, fractional solutions may be obtained on the corresponding shooting interval, cf. [122, 183].

**Remark 2.5 (Bang–Bang Optimal Problems and Active Set Methods)**
Exploiting the bang–bang property also in the numerical methods used to solve the convexified and relaxed NLP is crucial to achieve maximum computational efficiency. Controls that enter linearly should be removed from a major part of the involved linear algebra by suitable active set methods as soon as they attain one of the two extremal values. We revisit this question in chapter 7.

### 2.3.3 Bounds on the Objective Function

For nonlinear or nonconvex problems, the bang–bang principle 2.1 does not hold, and the relaxed solution will in general not be binary feasible. In this setion, bounds on the objective function gap and the constraint residual gap are presented that correlate the auxiliary problems to each other.

**Theorem 2.2 (Objective functions of (BN) and (BC))**
The binary nonlinear problem (BN) has an optimal solution $(\boldsymbol{x}^\star, \boldsymbol{u}^\star, \boldsymbol{w}^\star)$ if and only if the binary convexified problem (BC) has an optimal solution $(\boldsymbol{x}^\star, \boldsymbol{u}^\star, \boldsymbol{\omega}^\star)$. Their objective function values $\varphi^{\text{BN}}$ and $\varphi^{\text{BC}}$ are identical. △

**Proof** A proof can be found in [183]. □

Theorem 2.2 holds for the pair of binary problems only and is in general not true for the pair of relaxed problems. This is evident as the feasible set of (RN) will in general be larger than that of (RC) as could already be seen in example 2.1. We can however relate the solutions of (BC) and (RC) to each other, which shows the advantage of the convexified formulation (RC) over the nonlinear formulation (RN).

**Theorem 2.3 (Objective functions of (BC) and (RC))**
Let $(\boldsymbol{x}^\star, \boldsymbol{u}^\star, \boldsymbol{\alpha}^\star)$ be an optimal solution of the relaxed convexified problem (RC) with objective function value $\varphi^{\text{RC}}$. Then for every $\varepsilon > 0$ there exists a binary feasible control function $\boldsymbol{\omega}_\varepsilon$ and a state trajectory $\boldsymbol{x}_\varepsilon$ such that $(\boldsymbol{x}_\varepsilon, \boldsymbol{u}^\star, \boldsymbol{\omega}_\varepsilon)$ is a feasible solution of (BC) with objective function value $\varphi^{\text{BC}}_\varepsilon$ and it holds that $\varphi^{\text{BC}}_\varepsilon \leqslant \varphi^{\text{RC}} + \varepsilon$. △

**Proof** A proof can be found in [183]. □

A synthesis of the interrelations of the optimal solutions of the four auxiliary problems can now be given in the form of the following theorem.

**Theorem 2.4 (Objective Functions of the Auxiliary Problems)**
Let $(\boldsymbol{x}^\star, \boldsymbol{u}^\star, \boldsymbol{\alpha}^\star)$ be an optimal solution of the relaxed convexified problem (RC) with objective function value $\varphi^{\mathrm{RC}}$. Then for every $\varepsilon > 0$

1. there exists a binary feasible $2^{n^{\mathrm{w}}}$–dimensional control function $\boldsymbol{\omega}_\varepsilon$ and a state trajectory $\boldsymbol{x}_\varepsilon$ such that $(\boldsymbol{x}_\varepsilon, \boldsymbol{u}^\star, \boldsymbol{\omega}_\varepsilon)$ is a feasible solution of (BC) with objective function value $\varphi_\varepsilon^{\mathrm{BC}}$,
2. there exists a binary feasible $n^{\mathrm{w}}$–dimensional control function $\boldsymbol{w}_\varepsilon$ such that $(\boldsymbol{x}_\varepsilon, \boldsymbol{u}^\star, \boldsymbol{w}_\varepsilon)$ is a feasible solution of (BN) with objective function value $\varphi_\varepsilon^{\mathrm{BN}}$,
3. it holds for the objective function values that

$$\varphi^{\mathrm{RN}} \leqslant \varphi^{\mathrm{RC}} \leqslant \varphi_\varepsilon^{\mathrm{BC}} = \varphi_\varepsilon^{\mathrm{BN}} \leqslant \varphi^{\mathrm{RC}} + \varepsilon.$$

△

**Proof** A proof can be found in [183]. □

The optimal solution of an OCP can thus be approximated arbitrarily close by a feasible solution on the boundary of the feasible region of the relaxed control functions.

### 2.3.4 Bounds on the Infeasibility

The four auxiliary problems formulation in section 2.3.1 depart from the more general class of MIOCPs in a single but critical way: We have so far assumed that no path constraints are present.

#### Binary Control Independent Path Constraints

Concerning the introduction of path constraints, we distinguish two substantially different cases. For inequality path constraints $\boldsymbol{c}(\boldsymbol{x}(\cdot), \boldsymbol{u}(\cdot), \boldsymbol{p})$ that do not explicitly depend on the binary control function $\boldsymbol{w}(\cdot)$, these constraints can be satisfied to any prescribed tolerance with bang–bang controls $\tilde{\boldsymbol{w}}(\cdot)$ by virtue of theorem 2.4 as the state trajectory $\boldsymbol{x}(\cdot)$ can be approximated arbitrarily close.

### General Path Constraints

If the binary control function $\boldsymbol{w}(\cdot)$ enters the path constraints $\boldsymbol{c}(\cdot)$ explicitly, the picture is different. The bang–bang control $\boldsymbol{\omega}(\cdot)$ may lead to violations of the path constraints that could be satisfied by the relaxed trajectory $\boldsymbol{\alpha}(\cdot)$ and moreover, the original problem may not even have a binary feasible solution as the following example due to [183, 186] shows.

**Example 2.2 (Path constraints depending on a binary control function)**
Consider for problem (BN) the one–dimensional control constraint

$$\mathbf{0} \leqslant \boldsymbol{c}(w(t)) = \begin{bmatrix} 1 - 10^{-n} - w(t) \\ w(t) - 10^{-n} \end{bmatrix}, \quad n \geqslant 1.$$

These constraints exclude all binary feasible solutions for $w(\cdot)$, while the relaxed problem (RC) may still have a feasible solution in the interior of $\mathcal{T} \times [0,1]$.

### Outer Convexification of Constraints

One remedy is to extend the outer convexification approach to those constraint functions that directly depend on the binary control. For the above example 2.2 this results in the following formulation with convex relaxed multiplier functions $\alpha_1(\cdot), \alpha_2(\cdot) \in [0,1]$ and four constraints instead of two,

$$\mathbf{0} \leqslant \boldsymbol{c}^{\text{OC}}(w(t)) = \begin{bmatrix} \alpha_1(t)(1 - 10^{-n}) \\ -\alpha_2(t)10^{-n} \\ \alpha_2(t)(1 - 10^{-n}) \\ -\alpha_1(t)10^{-n} \end{bmatrix}, \quad n \geqslant 1, \quad \alpha_1(t) + \alpha_2(t) = 1.$$

In this variant a feasible solution does not exist even for the convexified and relaxed problem (RC), which immediately allows to verify infeasibility of the original problem (BN). This reformulation however has a number of implications for the NLPs obtained from transferring (2.3) into its convexified and relaxed counterpart problem. Chapters 5 and 6 are dedicated to this issue.

## 2.4 Rounding Strategies

For practical computations we will usually be interested in solving the convexified relaxed counterpart problem (RC) by means of one of the optimal control problem solving algorithms of chapter 1, and construct an integer feasible solution from the relaxed optimal one. To this end we discuss in this section the construction and application of rounding schemes. We assume a relaxed optimal solution $\boldsymbol{\alpha}^\star(\cdot)$ of the convexified relaxed problem (RC) to be available. In general, $\boldsymbol{\alpha}^\star(\cdot)$ will not be binary feasible, and so we are interested in the reconstruction of binary feasible solutions. We present direct and sum–up rounding schemes due to [183] that respect the SOS1 property, and give a bound due to [186, 188] on the loss of optimality of the rounded solution.

### 2.4.1 The Linear Case

We first consider rounding strategies in the case of $n^{\mathrm{w}}$ binary control functions $\boldsymbol{w}(\cdot)$ entering linearly, in which case no convexification is required and relaxation can be applied immediately. We first define the two rounding strategies.

**Definition 2.13 (Direct Rounding)**
Let $\boldsymbol{\alpha}^\star(\cdot)$ be the solution of (RC). Then the *direct rounding* solution $\boldsymbol{\omega}(t) \stackrel{\text{def}}{=} \boldsymbol{p}_i \in \{0,1\}^{n^{\mathrm{w}}}$ for $t \in [t_i, t_{i+1}]$ on the grid $\{t_i\}$, $0 \leqslant i \leqslant m$ is defined by

$$p_{i,j} \stackrel{\text{def}}{=} \begin{cases} 1 & \text{if } \int_{t_i}^{t_{i+1}} \alpha_j^\star(t)\,\mathrm{d}t \geqslant \frac{1}{2}, \\ 0 & \text{otherwise.} \end{cases} \quad 1 \leqslant j \leqslant n^{\mathrm{w}}. \tag{2.13}$$

△

**Definition 2.14 (SUR–0.5 Rounding)**
Let $\boldsymbol{\alpha}^\star(\cdot)$ be the solution of (RC). Then the *sum–up rounding* solution $\boldsymbol{\omega}(t) \stackrel{\text{def}}{=} \boldsymbol{p}_i \in \{0,1\}^{n^{\mathrm{w}}}$ for $t \in [t_i, t_{i+1}]$ on the grid $\{t_i\}$, $0 \leqslant i \leqslant m$ obtained by applying strategy SUR–0.5 is defined by

$$p_{i,j} \stackrel{\text{def}}{=} \begin{cases} 1 & \text{if } \int_{t_0}^{t_{i+1}} \alpha_j^\star(t)\,\mathrm{d}t - \sum_{k=0}^{i-1} p_{k,j} \geqslant \frac{1}{2}, \\ 0 & \text{otherwise.} \end{cases} \quad 1 \leqslant j \leqslant n^{\mathrm{w}}. \tag{2.14}$$

△

Direct rounding is immediately obvious, but does not exploit the fact that the $p_{\star j}$ belong to a control trajectory in time. Figure 2.3 illustrates the idea of sum–up rounding which strives to minimize the deviation of the control integrals at any point $t$ on the time horizon $\mathcal{T}$,

$$\left| \int_{t_0}^{t} \boldsymbol{\omega}(\tau) - \boldsymbol{\alpha}(\tau)\,\mathrm{d}\tau \right|.$$

This strategy is easily extended to non–equidistant discretizations of the control. The following theorem allows to estimate the deviation from the relaxed optimal control based on the granularity of the control discretization if the strategy SUR–0.5 is applied.

**Theorem 2.5 (Sum–up Rounding Approximation of the Control Integral)**
Let $\boldsymbol{\alpha}(t) : \mathcal{T} \times [0,1]^{n^{\mathrm{w}}}$ be a measurable function. Define on a given approximation grid $\{t_i\}$, $0 \leqslant i \leqslant m$, a binary control trajectory $\boldsymbol{\omega}(t) \stackrel{\mathrm{def}}{=} \boldsymbol{p}_i \in \{0,1\}^{n^{\mathrm{w}}}$ for $t \in [t_i, t_{i+1}]$

$$p_{i,j} \stackrel{\mathrm{def}}{=} \begin{cases} 1 & \text{if } \int_{t_0}^{t_{i+1}} \alpha_j(t)\,\mathrm{d}t - \sum_{k=0}^{i-1} p_{k,j}\delta t_k \geqslant \frac{1}{2}\delta t_i, \\ 0 & \text{otherwise.} \end{cases} \qquad 1 \leqslant j \leqslant n^{\mathrm{w}} \tag{2.15}$$

Then it holds for all $t \in \mathcal{T}$ that

$$\left\| \int_{t_0}^{t} \boldsymbol{\omega}(\tau) - \boldsymbol{\alpha}(\tau)\,\mathrm{d}\tau \right\|_{\infty} \leqslant \tfrac{1}{2} \max_i \delta t_i. \tag{2.16}$$

△

**Proof** A proof can be found in [186]. □

For control affine systems, the deviation of the state trajectory after rounding $\boldsymbol{x}(t;\boldsymbol{x}_0,\boldsymbol{\omega}(\cdot))$ from the relaxed optimal one $\boldsymbol{x}(t;\boldsymbol{x}_0,\boldsymbol{\alpha}(\cdot))$ can be bounded using the following important result based on theorem 2.5 and Gronwall's lemma.

**Theorem 2.6 (Sum–up Rounding of the State Trajectory)**
Let two initial value problems be given on the time horizon $\mathcal{T} \stackrel{\mathrm{def}}{=} [t_0, t_{\mathrm{f}}]$,

$$\begin{aligned} \dot{\boldsymbol{x}}(t) &= \boldsymbol{A}(t,\boldsymbol{x}(t))\boldsymbol{\alpha}(t), \quad t \in \mathcal{T}, \qquad \boldsymbol{x}(t_0) = \boldsymbol{x}_0, \\ \dot{\boldsymbol{y}}(t) &= \boldsymbol{A}(t,\boldsymbol{y}(t))\boldsymbol{\omega}(t), \quad t \in \mathcal{T}, \qquad \boldsymbol{y}(t_0) = \boldsymbol{y}_0, \end{aligned}$$

for given measurable functions $\boldsymbol{\alpha}(t), \boldsymbol{\omega}(t) : \mathcal{T} \to [0,1]^{n^{\mathrm{w}}}$ and for $\boldsymbol{A} : \mathcal{T} \times \mathbb{R}^{n^{\mathrm{x}}} \to \mathbb{R}^{n^{\mathrm{x}}} \times \mathbb{R}^{n^{\mathrm{w}}}$ differentiable. Assume that there exist constants $C, L \in \mathbb{R}^+$ such that it holds for $t \in \mathcal{T}$ almost everywhere that

$$\begin{aligned} \left\| \frac{\mathrm{d}}{\mathrm{d}t} \boldsymbol{A}(t,\boldsymbol{x}(t)) \right\| &\leqslant C, \\ \left\| \boldsymbol{A}(t,\boldsymbol{x}(t)) - \boldsymbol{A}(t,\boldsymbol{y}(t)) \right\| &\leqslant L \left\| \boldsymbol{x}(t) - \boldsymbol{y}(t) \right\|, \end{aligned}$$

and assume that $\boldsymbol{A}(\cdot,\boldsymbol{x}(\cdot))$ is essentially bounded on $\mathcal{T} \times \mathbb{R}^{n^{\mathrm{x}}}$ by a constant $M \in \mathbb{R}^+$. Assume further that there exists a constant $\varepsilon > 0$ such that it holds for all $t \in \mathcal{T}$ that

$$\left\| \int_0^t \boldsymbol{\omega}(s) - \boldsymbol{\alpha}(s)\,\mathrm{d}s \right\| \leqslant \varepsilon.$$

Then it holds for all $t \in \mathcal{T}$ that

$$\left\|y(t) - x(t)\right\| \leqslant \left(\left\|y_0 - x_0\right\| + (M + C(t - t_0))\varepsilon\right) \exp(L(t - t_0)). \tag{2.17}$$

△

**Proof** A proof can be found in [186]. □

### 2.4.2 The Nonlinear Case

In the presence of SOS1 constraints that arise from outer convexification, the above rounding strategies are not directly applicable as the rounded solution $\boldsymbol{\omega}(\cdot)$ may violate the SOS1 property. More specifically, the SOS1 property e.g. is satisfied after rounding if there exists for all $t \in \mathcal{T}$ a component $\alpha_j^\star(t) \geqslant \frac{1}{2}$ of the relaxed optimal trajectory $\boldsymbol{\alpha}^\star(t)$. The following modifications of the above two schemes are proposed in [183, 186].

**Definition 2.15 (Direct SOS1 Rounding)**
Let $\boldsymbol{\alpha}^\star(\cdot)$ be the solution of (RC). Then the *direct SOS1 rounding* solution $\boldsymbol{\omega}(t) \stackrel{\text{def}}{=} \boldsymbol{p}_i \in \{0,1\}^{n^{\mathrm{w}}}$ for $t \in [t_i, t_{i+1}] \subset \mathcal{T}$ on the grid $\{t_i\}$, $0 \leqslant i \leqslant m$ is defined for $1 \leqslant j \leqslant n^{\mathrm{w}}$ by

$$p_{i,j} \stackrel{\text{def}}{=} \begin{cases} 1 & \text{if } (\forall k : \int_{t_i}^{t_{i+1}} \tilde{\alpha}_{ij}^\star(t)\,\mathrm{d}t \geqslant \int_{t_i}^{t_{i+1}} \tilde{\alpha}_{ik}^\star(t)\,\mathrm{d}t) \\ & \wedge (\forall k, \int_{t_i}^{t_{i+1}} \alpha_{ij}^\star(t)\,\mathrm{d}t = \int_{t_i}^{t_{i+1}} \tilde{\alpha}_{ik}^\star(t)\,\mathrm{d}t : \ j < k), \\ 0 & \text{otherwise.} \end{cases} \tag{2.18}$$

△

This rounding strategy chooses for each shooting interval the largest relaxed optimal choice $\tilde{q}_{i,j}^\star$ amongst all choices $1 \leqslant j \leqslant n^{\mathrm{w}}$. If the index $j$ of the largest choice is not unique, smaller indices are arbitrarily preferred. The equivalent for sum–up rounding is given in the following definition.

**Definition 2.16 (SOS1–SUR Rounding)**
Let $\boldsymbol{\alpha}^\star(\cdot)$ be the solution of (RC). Let further control differences $\hat{p}_{ij}$ for the $j$-th step of sum–up rounding be defined as

$$\hat{p}_{i,j} \stackrel{\text{def}}{=} \int_{t_0}^{t_i} \alpha_j^\star(t)\,\mathrm{d}t - \sum_{k=0}^{i-1} p_{k,j}. \tag{2.19}$$

Then the *sum-up rounding* solution $\boldsymbol{\omega}(t) \stackrel{\text{def}}{=} \boldsymbol{p}_i$ for $t \in [t_i, t_{i+1}] \subset \mathcal{T}$ on the grid $\{t_i\}$, $0 \leqslant i \leqslant m$ obtained by applying strategy SOS1–SUR–0.5 is defined for $1 \leqslant j \leqslant n^{\mathrm{w}}$ by

$$p_{i,j} \stackrel{\text{def}}{=} \begin{cases} 1 & \text{if } (\forall k : \ \hat{p}_{i,j} \geqslant \hat{p}_{i,k}) \wedge (\forall k, \ \hat{p}_{i,j} = \hat{p}_{i,k} : \ j < k), \\ 0 & \text{otherwise.} \end{cases} \tag{2.20}$$

△

The following variant of theorem 2.5 holds for the control approximation obtained from application of the SOS1–SUR–0.5 rounding scheme to the relaxed optimal control trajectory.

**Theorem 2.7 (SOS1–SUR–0.5 Approximation of the Control Integral)**
Let $\boldsymbol{\alpha}(t) : \mathcal{T} \times [0,1]^{n^{\mathrm{w}}}$ be a measurable function. Define on a given approximation grid $\{t_i\}$, $0 \leqslant i \leqslant m$, a binary control trajectory $\boldsymbol{\omega}(t) \stackrel{\mathrm{def}}{=} \boldsymbol{p}_i \in \{0,1\}^{n^{\mathrm{w}}}$ for $t \in [t_i, t_{i+1}]$ by

$$p_{i,j} \stackrel{\mathrm{def}}{=} \begin{cases} 1 & \text{if } (\forall k:\ \hat{p}_{i,j} \geqslant \hat{p}_{i,k}) \wedge (\forall k,\ \hat{p}_{i,j} = \hat{p}_{i,k}:\ j < k), \\ 0 & \text{otherwise.} \end{cases} \tag{2.21}$$

where $1 \leqslant j \leqslant n^{\mathrm{w}}$. Then it holds for all $t \in \mathcal{T}$ that

$$\left\| \int_{t_0}^{t} \boldsymbol{\omega}(\tau) - \boldsymbol{\alpha}(\tau)\, \mathrm{d}\tau \right\|_{\infty} \leqslant (n^{\mathrm{w}} - 1) \max_i \delta t_i. \tag{2.22}$$

△

**Proof** A proof is given in [186]. □

**Remark 2.6**
It is conjectured that the factor $n^{\mathrm{w}} - 1$ in (2.22) is not tight.

### 2.4.3 The Discretized Case

In contrast to the discussion of the previous sections, the convexified relaxed problem is solved e.g. using a direct discretization of the control trajectory $\boldsymbol{\omega}(\cdot)$. The optimization space is thus restricted to a finite number $m$ of degrees of freedom for the relaxed control trajectory $\boldsymbol{\omega}(\cdot)$, parameterized by control parameters $\boldsymbol{q}_i \in \Omega \subset \mathbb{R}^{n^{\mathrm{w}}}$ on a control grid $\{t_i\} \subset \mathcal{T}$ with $m+1$ grid points,

$$\boldsymbol{\omega}_i(t) = \boldsymbol{q}_i \quad t \in [t_i, t_{i+1}) \subset \mathcal{T}, \quad 0 \leqslant i \leqslant m-1. \tag{2.23}$$

The integer control parameters $\boldsymbol{q}_i$ are convexified and relaxed, leading to the introduction of $n^{\Omega}$ control trajectories $\boldsymbol{\alpha}(\cdot)$ parameterized by control parameters $\boldsymbol{q}_i \in \{0,1\}^{n^{\Omega}}$ on each node of the grid. In general, the solution $\boldsymbol{q}^\star$ obtained will not be a vertex of the binary hypercube. A special property that sets the situation apart from more generic MINLPs is then exploited by any of the sum–up rounding strategies. The control parameters $q_{ij} \in [0,1]$, $0 \leqslant i \leqslant m$ are for any given $j$ discretizations of the same discretized binary control trajectory $\omega_j(\cdot)$ in different time points $t_i \in \mathcal{T}$, and the rounding strategies take this knowledge into account.

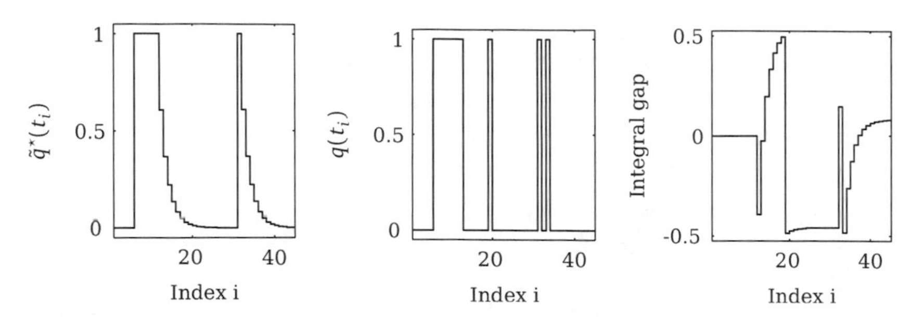


Figure 2.3: Sum–up rounding minimizes the deviation of the control integrals. On this equidistant grid, the integral gap never exceeds one half.

## 2.5 Switch Costs

In this section we are concerned with the introduction of switch costs into the problem class (2.1) in order to avoid solutions that show frequent switching of the binary control trajectory.

### 2.5.1 Frequent Switching

Theorem 2.4 is a theoretical result for the infinite–dimensional OCP. The optimal solution may have to switch infinitely often between the two extremal values on the finite horizon $\mathcal{T}$ in order to attain the relaxed fractional solution's objective and state trajectory $\boldsymbol{x}(\cdot)$. Hence, frequent switching of the binary control may be part of the actual optimal solution of the infinite–dimensional MIOCP, in which case it is referred to as chattering or Zeno's phenomenon, referring to the ancient Greek philosopher Zeno of Elea (Ζήνων ὁ Ἐλεάτης) who was the first to give examples of obviously paradoxical statements drawing attention to the interpretative difficulties in this situation.

**Remark 2.7 (Zeno's Phenomenon)**
Theorem 2.4 holds for the infinite–dimensional optimization problems prior to any discretization. If a relaxed optimal control trajectory $\boldsymbol{w}(\cdot)$ has singular or path constrained arcs, i.e., parts of the optimal trajectory attain values in the interior of the relaxed feasible set $[0,1]^{n^w}$, the bang–bang optimal trajectory approximating it may have to switch between the extremal points infinitely often on a finite time horizon.

The first example of an optimal control problem with a solution that shows Zeno's phenomenon was given in [76]. Chattering of an integer control may be undesirable from a point of view of the process under control, and the inclusion of switch costs in the MIOCPs then becomes a modeling issue. For example, the change of a discrete decisions might take some process time to complete, incur some operational cost that is to be kept as low as possible, or may be discouraged due to mechanical wear–off in the physical device realizing the switch, e.g. in a valve or a gearbox [42, 103]. Closely related is the problem of seeking for sparse controls that switch only a limited number of times and remain on the lower bound for most of the time, e.g. when optimizing the application of drugs over time during medical treatment [63].

We summarize under the term switch costs three different but related settings. First, the number of switches permitted on the time horizon may be bounded from above by a number smaller than the number $m$ of available control discretization intervals. Second, it may be desirable to penalize the occurrence of a switch by a positive contribution to the problem's objective, striving for a pareto–optimal solution that constitutes a compromise between maximum performance and minimum operation of the switch. Third, operating a switch may trigger fast transients in the dynamic process states. One possible way of representing these transients is by approximation using state discontinuities. For the case of implicitly defined, i.e., state dependent switches this is discussed e.g. in [156, 40, 118]. For the case of externally operated switches, i.e., here discretized time–varying control functions with a finite number of admissible values, such state jumps occur in the discretization points.

It is immediately evident that any inclusion of positive costs of switches of a binary or integer control, either by penalization of the number of switches in the objective function or by limiting the number of switches through an additionally imposed constraint, prevents Zeno's phenomenon from occurring as the resulting solution would show infinite cost in the first case and be infeasible in the second case.

### 2.5.2 Switch Costs in a MILP Formulation

This observation leads us to discuss the inclusion of switch costs on a discretization of the binary convexified linear problem (BC). Here, the optimal solution can switch only a finite number of times, at most once in each of the control discretization grid points. Still, frequent switching of the binary control trajectory may occur after sum–up rounding of the discretized solution on singular or path constrained arcs of the convexified relaxed optimal solution. For fine discretizations, the inclusion of switch costs may have a significant impact on the shape of optimal solutions.

Instead of applying a sum–up rounding strategy to the relaxed OCP solution we proposed in [190] to solve the MILP

$$\begin{aligned}
\min_{p} \quad & \max_{1\leqslant k\leqslant n^{\Omega}} \max_{0\leqslant i\leqslant m-1} \left| \sum_{j=0}^{i} \left( \tilde{q}^{\star}_{jk} - p_{jk} \right) \delta t_j \right| && (2.24)\\
\text{s.t.} \quad & 1 = \sum_{k=1}^{n^{\Omega}} p_{ik}, && 0 \leqslant i \leqslant m-1,\\
& \sigma_k \geqslant \sum_{i=0}^{m-1} \left| p_{ik} - p_{i+1,k} \right|, && 0 \leqslant k \leqslant n^{\Omega},\\
& \boldsymbol{p}_i \in \{0,1\}^{n^{\Omega}}. && 0 \leqslant i \leqslant m-1
\end{aligned}$$

that computes a binary feasible solution $\boldsymbol{\omega}(\cdot)$ from a relaxed optimal one $\boldsymbol{\alpha}^{\star}(\cdot)$ subject to givens limits $\sigma_j > 0$, $1 \leqslant j \leqslant n^{\mathrm{w}}$ for the number $\sigma_j$ of switches of the binary control trajectory $\omega_j(\cdot)$ after rounding. Here $\boldsymbol{q}^{\star}_i \in [0,1]^{n^{\Omega}}$, $0 \leqslant i \leqslant m-1$ denote the control parameters of a relaxed optimal solution of a direct discretization of (BC), and let $\boldsymbol{p}_i \in \{0,1\}^{n^{\Omega}}$ denote the rounded solution respecting switch costs.

The relaxed solution serves as an excellent initial guess that is very cheap to obtain. A combinatorial Branch & Bound algorithm is developed for this problem in [190] that exploits this information to gain significant computational speed–ups.

In view of the inherent complexity of this MILP and the fast computation times required for mixed–integer NMPC we focus in the following on an approximate switch cost formulation that can be readily included in the convexified and relaxed OCP.

### 2.5.3 Switch Costs for Outer Convexification

In this section we develop a switch cost formulation that is suitable for inclusion into the convexified relaxed problem (RC) and fits into a direct multiple shooting discretization. The key issues here are maintaining differentiability and separability of the problem formulation.

We introduce the following discretized variant of the auxiliary problem (BC) with switch costs as shown in definition 2.17. Herein, a discretization on a fixed grid $\{t_i\}$ with $m$ nodes,

$$\boldsymbol{\omega}(t) \stackrel{\text{def}}{=} \boldsymbol{q}_i \in \Omega \quad t \in [t_i, t_{i+1}), \quad 0 \leqslant i \leqslant m-1, \tag{2.25}$$

is assumed for the binary convexified control trajectory $\boldsymbol{\omega}(\cdot)$, and we remind the reader of the convenience definition $\boldsymbol{\omega}(t_m) \stackrel{\text{def}}{=} \boldsymbol{q}_m = \boldsymbol{q}_{m-1}$. The additionally introduced variable $\boldsymbol{\sigma}$ counts the number of changes in each component of the discretized binary control trajectory $\boldsymbol{\omega}(\cdot)$ over time. The number of switches may be constrained by an upper limit $\boldsymbol{\sigma}_{\max}$ or included in the objective function using a weighting penalization factor $\boldsymbol{\pi}$. We further denote by (RCS) the relaxed counterpart problem of (BCS) which only differs in

$$\boldsymbol{q}_i \in [0,1]^{n^{\Omega}}, \qquad 0 \leqslant i \leqslant m-1. \tag{RCS}$$

**Definition 2.17 (Discretized Problem (BC) with Switch Costs)**
The multiple shooting discretization of the *binary convexified linear problem with switch costs* is the following extension of problem (BC),

$$\min_{\boldsymbol{s},\boldsymbol{q}} \quad \sum_{i=0}^{m} \sum_{j=1}^{n^{\Omega}} l_i(t_i, \boldsymbol{s}_i, \boldsymbol{\omega}^j)\, q_{ij} + \sum_{j=1}^{n^{\Omega}} \pi_j \sigma_j \tag{BCS}$$

$$\begin{aligned}
\text{s.t.} \quad & \boldsymbol{0} = \sum_{j=1}^{n^{\Omega}} \boldsymbol{x}_i(t_{i+1};\ t_i, \boldsymbol{s}_i, \boldsymbol{\omega}^j)\, q_{ij} - \boldsymbol{s}_{i+1}, && 0 \leqslant i \leqslant m-1,\\
& \boldsymbol{s}_0 = \boldsymbol{x}_0, &&\\
& \boldsymbol{q}_i \in \{0,1\}^{n^{\Omega}}, && 0 \leqslant i \leqslant m-1,\\
& 1 = \sum_{j=1}^{n^{\Omega}} q_{ij}, && 0 \leqslant i \leqslant m-1,\\
& \sigma_j = \sum_{i=0}^{m-1} \left| q_{i+1,j} - q_{i,j} \right|, && 1 \leqslant j \leqslant n^{\Omega},\\
& \sigma_j \leqslant \sigma_{j,\max}, && 1 \leqslant j \leqslant n^{\Omega}.
\end{aligned}$$

△

### 2.5.4 Reformulations

In the following, we address several issues with the problem formulation (BCS). The defining constraint for $\boldsymbol{\sigma}$ is nondifferentiable with respect to $\boldsymbol{q}$, and we present two reformulations that overcome this. Second, the behavior of (RCS) after relaxation of the binary requirement is studied. Finally, the above formulation obviously comprises a coupled constraint connecting control parameters from adjacent shooting intervals. This impairs the separability property of the NLP's Lagrangian and we present a separable reformulation in order to maintain computational efficiency.

### Differentiable Reformulation

Addressing the issue of nondifferentiability, we use a reformulation introducing slack variables for the constraint defining $\sigma_k$, $1 \leqslant k \leqslant n^{\Omega}$ as in

$$\begin{aligned} \sigma_k &= \tfrac{1}{2} \sum_{i=0}^{m-1} \delta_{i,j}, \\ \delta_{i,j} &\geqslant q_{i+1,j} - q_{i,j}, \\ \delta_{i,j} &\geqslant q_{i,j} - q_{i+1,j}. \end{aligned} \tag{2.26}$$

Here, positivity of the slacks $\delta_{i,j}$ is ensured by the two additionally introduced constraints. This moves the nondifferentiability to the active set method solving the NLP. At least one of the two constraints on the positive differences $\delta_{i,j}$ will be active at any time.

Note that the introduction of switch costs by this formulations introduces for each of the $m \cdot n^{\Omega}$ convex multipliers one additional control parameter into the NLP. Linear algebra techniques for solution of the discretized MIOCP that are tailored to problems with many control parameters are presented in chapter 7.

After relaxation of the convex multipliers $q_{i,j}$ emerging from the outer convexification approach, the above differential reformulation has the drawback of attracting fractional solutions. As an example, the relaxed optimal solution $q_{ij} = \frac{1}{2}$ for all intervals $0 \leqslant i \leqslant m-1$ and $j = 1, 2$ should be recognized as a "switching" solution, as the sum–up rounding strategy would yield a rounded control alternating between zero and one. The differences of adjacent relaxed optimal control parameters are zero however, which yields $\boldsymbol{\sigma} = \mathbf{0}$.

**Convex Reformulation**

In order to address this issue, we develop a new second approach making use of a convex reformulation of the nondifferentiability. For two binary control parameters $q_{i,j}$ and $q_{i+1,j}$ adjacent in the time discretization of the control trajectory, the switch cost $\sigma_{i,j}$ is given by the expression

$$\sigma_{i,j} \stackrel{\text{def}}{=} \alpha_{i,j}(q_{i,j} + q_{i+1,j}) + \beta_{i,j}(2 - q_{i,j} - q_{i+1,j}), \quad \alpha_{i,j} + \beta_{i,j} = 1, \tag{2.27}$$

in which $\alpha_{i,j}$ and $\beta_{i,j}$ are binary convex multipliers introduced as additional degrees of freedom into the optimization problem. Note that the SOS1 constraint can again be used to eliminate the multiplier $\beta_{i,j}$,

$$\sigma_{i,j} = (2\alpha_{i,j} - 1)(q_{i,j} + q_{i+1,j} - 1) + 1. \tag{2.28}$$

Under minimization of the switch cost, this expression yields the solutions listed in table 2.1a. Figure 2.4a depicts the evaluation points and values.

For the relaxed problem (RCS), this convex reformulation ensures that fractional solutions are assigned a nonzero cost, in particular any fractional solution is more expensive than the nonswitching binary ones, and that the switching binary solutions are assigned the highest cost. Table 2.1b shows the switch costs under minimization for fractional values of the relaxed control parameters. The convexification envelope is depicted in figure 2.4b.

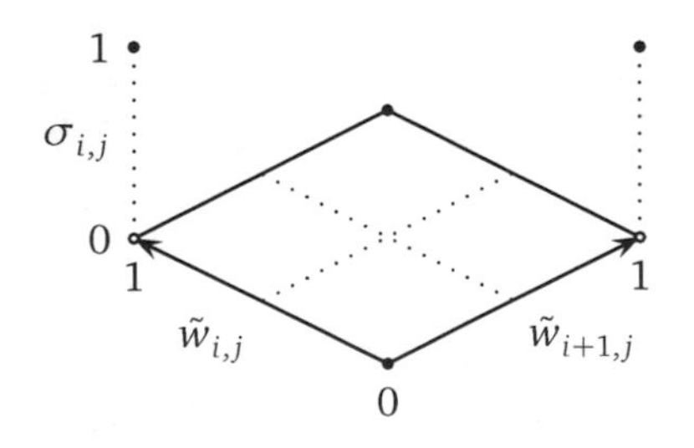


(a) Switch costs for binary controls.

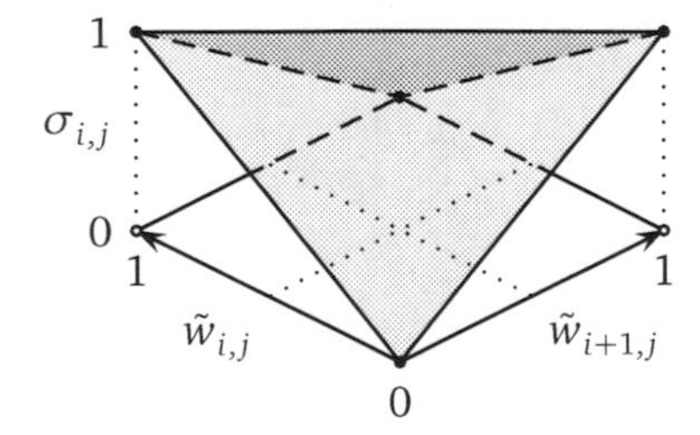


(b) Switch costs for relaxed controls.

Figure 2.4: Convex reformulation of the switch cost constraint for two binary controls adjacent in the time discretization.

Numerical results obtained from this strategy look promising as detailed in chapter 9. The connection to the MILP (2.24) yet remains to be clarified.

| $q_{i,j}$ | $q_{i+1,j}$ | $\alpha_{i,j}$ | $\beta_{i,j}$ | $\frac{1}{2}\sigma_{i,j}$ |
|---|---|---|---|---|
| 0 | 0 | 1 | 0 | 0 |
| 0 | 1 | free | free | 1 |
| 1 | 0 | free | free | 1 |
| 1 | 1 | 0 | 1 | 0 |

(a) Switch costs for binary controls.

| $q_{i,j} + q_{i+1,j}$ | $\alpha_{i,j}$ | $\beta_{i,j}$ | $\frac{1}{2}\sigma_{i,j}$ |
|---|---|---|---|
| $<1$ | 1 | 0 | $<1$ |
| $=1$ | free | free | 1 |
| $>1$ | 0 | 1 | $<1$ |

(b) Switch costs for relaxed controls.

Table 2.1: Binary and relaxed optimal solutions for the convex switch cost reformulation.

### Separable Reformulation for Direct Multiple Shooting

Separability of the objective function and all constraint functions with respect to the discretization in time is a crucial property of the direct multiple shooting discretization. It introduces a block structure into the discretized OCP that can be exploited very efficiently as detailed in chapter 7. The only coupling between adjacent shooting intervals allowed so far has been the consistency condition imposed on the IVP solutions in the shooting nodes.

Separability of the above differential reformulation can be recovered by formally introducing an augmented vector of differential states $\hat{\boldsymbol{x}} = [\,\boldsymbol{x}\;\boldsymbol{d}\,]$ together with the augmented matching condition

$$\begin{bmatrix} \sum_{j=1}^{n^\Omega} \boldsymbol{x}_i(t_{i+1};\, t_i, \boldsymbol{s}_i, \boldsymbol{\omega}^j)\, q_{ij} - \boldsymbol{s}_{i+1} \\ \boldsymbol{q}_i - \boldsymbol{q}_{i+1} - \boldsymbol{d} \end{bmatrix} = \boldsymbol{0}. \tag{2.29}$$

This matching condition deviates from the classical direct multiple shooting method [36, 167] in that it depends on the control parameter vector of the subsequent shooting interval. In chapter 7 we investigate structured linear algebra techniques for SQP subproblems with many control parameters that support this generalized type of matching condition. The separable linear reformulation reads

$$\sigma = \frac{1}{2} \sum_{i=0}^{m-1} \sum_{j=1}^{n^\Omega} \delta_{ij}, \qquad \boldsymbol{\delta}_i \geqslant \boldsymbol{d}_i \geqslant -\boldsymbol{\delta}_i, \quad 0 \leqslant i \leqslant m-1. \tag{2.30}$$

Separability of the convex reformulation can be recovered in a similar way.

## 2.6 Summary

In this chapter we have been concerned with reformulations of MIOCPs that allow for the efficient numerical solution or approximation. Focus has been put on a convex reformulation of the MIOCP with respect to the binary or integer control. After relaxation, this reformulation allows to obtain an approximation to a solution of an MIOCP by solving only a single continuous but possibly larger OCP. This is due to the fact that the convexified OCP's optimal solution can be approximated with arbitrary quality by a control trajectory on the boundary of the feasible set. For a discretization of this reformulated OCP, bounds on the loss of optimality and on the infeasibility of those constraints independent of the integer control can be shown. Constraints directly depending on the integer control are considered in chapter 5. Rounding schemes were presented that in the case of fractional optimal solutions of the relaxed OCP allow to reconstruct an integer feasible solution with a known bound on the loss of optimality. We have considered an MILP formulation replacing sum–up rounding in the case of upper limits constraining the permitted number of switches. A convexification and relaxation has been developed for this formulation that does not attract fractional solutions of the convexified relaxed OCP. It further maintains separability of the objective and constraints functions, and can thus be included in a direct multiple shooting discretization of the OCP. The presented switch cost formulations double the number of control parameters emerging out of the convex reformulation. Linear algebra techniques for the solution of the discretized OCP with many control parameters are considered in chapter 7.

# 3 Constrained Nonlinear Programming

We have already briefly touched nonlinear programming in the last two chapters, in which we introduced the multiple shooting discretized optimal control problem, a Nonlinear Program (NLP), and presented the outer convexification and relaxation approach that allows to compute approximations to local Mixed–Integer Optimal Control Problem (MIOCP) solutions by solving a reformulated and discretized but possibly much larger Optimal Control Problem (OCP). This chapter is concerned with theory and numerical methods for the solution of NLPs and equips the reader with definitions and algorithms required for the following chapters of this thesis. We present optimality conditions characterizing locally optimal solutions and introduce Sequential Quadratic Programming (SQP) methods for the iterative solution of NLPs. The evaluation of the matching condition constraints of the discretized optimal control problem requires the solution of Initial Value Problems (IVPs). To this end, we present one step methods for non–stiff Ordinary Differential Equations (ODEs) and discuss the efficient and consistent computation of sensitivities of IVPs solutions. The familiar reader may wish to continue with chapter 4 rightaway, and refer back to this chapter should the need arise.

## 3.1 Constrained Nonlinear Programming

In this section we prepare our investigation of SQP methods for the solution of the discretized optimal control problem by repeating definitions of a number of terms commonly arising in constrained nonlinear programming for further reference. We give an overview over major results characterizing optimal solutions of NLPs. These can be found in any standard textbook on nonlinear programming, e.g. [24, 72, 158].

### 3.1.1 Definitions

This section is concerned with the solution of constrained nonlinear programs of the general form given in definition 3.1.

**Definition 3.1 (Nonlinear Program)**
An optimization problem of the general form

$$\begin{aligned} \min_{x \in \mathbb{R}^n} \quad & f(x) \\ \text{s.t.} \quad & g(x) = 0, \\ & h(x) \geqslant 0, \end{aligned} \tag{3.1}$$

with objective function $f : \mathbb{R}^n \to \mathbb{R}$, equality constraints $g : \mathbb{R}^n \to \mathbb{R}^{n^g}$, and inequality constraints $h : \mathbb{R}^n \to \mathbb{R}^{n^h}$ is called a *Nonlinear Program*. △

The functions $f$, $g$, and $h$ are assumed to be twice continuously differentiable with respect to $x$. We are interested in finding a point $x^\star \in \mathbb{R}^n$ that satisfies all constraints and minimizes, in a neighborhood, the objective function. To this end, we define the subset of feasible points of problem (3.1).

**Definition 3.2 (Feasible Point, Feasible Set)**
A point $\overline{x} \in \mathbb{R}^n$ is called a *feasible point* of problem (3.1) if it satisfies the constraints

$$\begin{aligned} & g(\overline{x}) = 0, \\ & h(\overline{x}) \geqslant 0. \end{aligned}$$

The set of all feasible points of problem (3.1) is denoted by

$$\mathcal{F} \stackrel{\text{def}}{=} \{x \in \mathbb{R}^n \mid g(x) = 0,\ h(x) \geqslant 0\} \subseteq \mathbb{R}^n. \tag{3.2}$$

△

Applying this definition, we may restate problem (3.1) as a pure minimization problem over a set $\mathcal{F}$ with possibly complicated shape,

$$\min_{x \in \mathcal{F}} f(x).$$

For any point of the feasible set, the active set denotes the subset of inequality constraints that are satisfied to equality.

**Definition 3.3 (Active Constraint, Active Set)**
Let $\overline{x} \in \mathbb{R}^n$ be a feasible point of problem (3.1). An inequality constraint $h_i$, $i \in \{1, \dots, n^h\} \subset \mathbb{N}$, is called *active* if $h_i(\overline{x}) = 0$ holds. It is called *inactive* otherwise. The set of indices of all active constraints

$$\mathcal{A}(\overline{x}) \stackrel{\text{def}}{=} \{i \mid h_i(\overline{x}) = 0\} \subseteq \{1, \dots, n^h\} \subset \mathbb{N} \tag{3.3}$$

is called the *active set* associated with $\overline{x}$. △

The restriction of the inequality constraint function $\boldsymbol{h}$ onto the active inequality constraints is denoted by $\boldsymbol{h}_{\mathcal{A}} : \mathbb{R}^n \to \mathbb{R}^{|\mathcal{A}|}$, $\boldsymbol{x} \mapsto \boldsymbol{h}_{\mathcal{A}}(\boldsymbol{x})$.

Constraint Qualifications (CQs) ensure a certain well–behavedness of the feasible set in a neighborhood of a feasible point $\overline{\boldsymbol{x}} \in \mathcal{F}$. We often require the set of active constraints to be linear independent.

**Definition 3.4 (Linear Independence Constraint Qualification, Regular Point)**
We say that Linear Independence Constraint Qualification (LICQ) holds for problem (3.1) in $\overline{\boldsymbol{x}} \in \mathbb{R}^n$ if it holds that

$$\operatorname{rank} \begin{bmatrix} \boldsymbol{g}_x(\overline{\boldsymbol{x}}) \\ (\boldsymbol{h}_{\mathcal{A}})_x(\overline{\boldsymbol{x}}) \end{bmatrix} = n^{\text{g}} + n^{\text{h}}_{\mathcal{A}}. \tag{3.4}$$

Then, $\overline{\boldsymbol{x}}$ is referred to as a *regular point* of problem (3.1). △

In irregular points, constraints in the active set $\mathcal{A}$ are linearly dependent. Numerical methods then frequently require that a linear independent subset $\mathcal{W} \subset \mathcal{A}$ is chosen, referred to as the *working set*.

Finally, we have the following formal definition of a local solution to problem (3.1). In the next section, alternative conditions will be given that can be used to devise numerical algorithms for finding candidate points for local solutions.

**Definition 3.5 (Locally Optimal Point)**
A point $\boldsymbol{x}^\star \in \mathcal{F} \subseteq \mathbb{R}^n$ is called a *locally optimal* point of problem (3.1) if there exists an open ball $\mathcal{B}_\varepsilon(\boldsymbol{x}^\star)$ with $\varepsilon > 0$ such that

$$\forall \boldsymbol{x} \in \mathcal{B}_\varepsilon(\boldsymbol{x}^\star) \cap \mathcal{F} : \; f(\boldsymbol{x}) \geqslant f(\boldsymbol{x}^\star).$$

If in addition it holds that

$$\forall \boldsymbol{x} \in \mathcal{B}_\varepsilon(\boldsymbol{x}^\star) \cap \mathcal{F}, \boldsymbol{x} \neq \boldsymbol{x}^\star : \; f(\boldsymbol{x}) > f(\boldsymbol{x}^\star)$$

then $\boldsymbol{x}^\star$ is called a *strict local optimum* of problem (3.1). △

### 3.1.2 First Order Necessary Optimality Conditions

In order to state necessary optimality conditions for a candidate point $\boldsymbol{x}$ to be a locally optimal solution of (3.1), we require the following definitions.

**Definition 3.6 (Lagrangian Function)**
The function $L : \mathbb{R}^{n^{\text{x}}} \times \mathbb{R}^{n^{\text{g}}} \times \mathbb{R}^{n^{\text{h}}} \to \mathbb{R}$,

$$L(\boldsymbol{x}, \boldsymbol{\lambda}, \boldsymbol{\mu}) \stackrel{\text{def}}{=} f(\boldsymbol{x}) - \boldsymbol{\lambda}^T \boldsymbol{g}(\boldsymbol{x}) - \boldsymbol{\mu}^T \boldsymbol{h}(\boldsymbol{x}) \tag{3.5}$$

with *Lagrange multipliers* $\boldsymbol{\lambda} \in \mathbb{R}^{n^g}$ and $\boldsymbol{\mu} \in \mathbb{R}^{n^h}$ is called the *Lagrangian (function)* of problem (3.1). △

The following famous theorem independently found by [114, 126] specifies necessary conditions for local optimality of a regular point that are based on definition 3.6. We will return to this theorem in chapter 6 and consider it again under weaker assumptions.

**Theorem 3.1 (Karush–Kuhn–Tucker Conditions)**
Let $\boldsymbol{x}^\star \in \mathbb{R}^n$ be a locally optimal point of problem (3.1), and assume that LICQ holds in $\boldsymbol{x}^\star$. Then there exist Langrange multipliers $\boldsymbol{\lambda}^\star \in \mathbb{R}^{n^g}$ and $\boldsymbol{\mu}^\star \in \mathbb{R}^{n^h}$ such that the following conditions are satisfied:

$$\begin{aligned}
\mathbf{0} &= L_x(\boldsymbol{x}^\star, \boldsymbol{\lambda}^\star, \boldsymbol{\mu}^\star),\\
\mathbf{0} &= \boldsymbol{g}(\boldsymbol{x}^\star),\\
\mathbf{0} &\leqslant \boldsymbol{h}(\boldsymbol{x}^\star),\\
\mathbf{0} &\leqslant \boldsymbol{\mu}^\star,\\
0 &= \boldsymbol{\mu}^{\star T} \boldsymbol{h}(\boldsymbol{x}^\star).
\end{aligned}$$

The point $(\boldsymbol{x}^\star, \boldsymbol{\lambda}^\star, \boldsymbol{\mu}^\star)$ is then referred to as a Karush–Kuhn–Tucker *(KKT) point.* △

**Proof** Proofs under LICQ can be found in any standard textbook on nonlinear programming, such as [72, 158]. □

The condition $\boldsymbol{\mu}^{\star T} \boldsymbol{h}(\boldsymbol{x}^\star) = 0$ in theorem 3.1 is referred to as complementarity condition. It specifies that Lagrange multipliers for inactive inequality constraints shall be zero. This condition may be sharpened as follows.

**Definition 3.7 (Strict Complementarity)**
Let $\boldsymbol{x}^\star \in \mathbb{R}^n$ be a locally optimal point of problem (3.1) and let $\boldsymbol{\lambda}^\star \in \mathbb{R}^{n_g}$, $\boldsymbol{\mu}^\star \in \mathbb{R}^{n_h}$ be Lagrange multipliers such that the conditions of theorem 3.1 are satisfied. We say that *strict complementarity* holds in $(\boldsymbol{x}^\star, \boldsymbol{\lambda}^\star, \boldsymbol{\mu}^\star)$ if $\mu_i^\star > 0$ for all active constraints $h_i$. △

Active constraints in violation of strict complementarity are called weakly active.

**Definition 3.8 (Weakly Active Constraint)**
Let $\boldsymbol{x}^\star \in \mathbb{R}^n$ be a locally optimal point of problem (3.1) and let $\boldsymbol{\lambda}^\star \in \mathbb{R}^{n_g}$, $\boldsymbol{\mu}^\star \in \mathbb{R}^{n_h}$ be some Lagrange multipliers such that the conditions of theorem 3.1 are satisfied. Then a constraint $h_i(\boldsymbol{x}^\star) \geqslant 0$ is called *stongly active* or *binding* if $h_i(\boldsymbol{x}^\star) = 0$ and $\mu_i^\star > 0$. The constraint $h_i$ is called *weakly active* if $h_i(\boldsymbol{x}^\star) = 0$ and $\mu_i = 0$ for all $\boldsymbol{\mu}$ satisfying the KKT conditions. △

For a given point KKT point $\boldsymbol{x}^\star$ there may be many choices of $\boldsymbol{\lambda}$ and $\boldsymbol{\mu}$ that satisfy the KKT condition of theorem 3.1. Under LICQ and strict complementarity, uniqueness of the Lagrange multipliers in a KKT point can be shown. For the design of most derivative based algorithms for constrained nonlinear optimization, LICQ thus is the CQ assumed most often.

**Theorem 3.2 (Uniqueness of Lagrange Multipliers)**
Let $\boldsymbol{x}^\star$ be a locally optimal point of problem (3.1), and let $\boldsymbol{\lambda}^\star \in \mathbb{R}^{n_g}$, $\boldsymbol{\mu}^\star \in \mathbb{R}^{n_h}$ be Lagrange multipliers such that the conditions of theorem 3.1 are satisfied. Let LICQ hold in $\boldsymbol{x}^\star$ and let strict complementarity hold in $(\boldsymbol{x}^\star, \boldsymbol{\lambda}^\star, \boldsymbol{\mu}^\star)$. Then it holds that the values $\boldsymbol{\lambda}^\star$ and $\boldsymbol{\mu}^\star$ are unique. △

**Proof** For the equality constraints and the subset of active inequality constraints, this is evident from linear independence of the gradients $\boldsymbol{g}_x(\boldsymbol{x}^\star)$ and $(\boldsymbol{h}_{\mathcal{A}})_x(\boldsymbol{x}^\star)$. For the inactive inequality constraints, $\mu_i^\star = 0$ for $i \notin \mathcal{A}(\boldsymbol{x}^\star)$ is enforced by strict complementarity. □

### 3.1.3 Second Order Conditions

A necessary and a sufficient condition for local optimality that both use second order derivative information are given in this section. We require the definition of the reduced or null–space Hessian.

**Definition 3.9 (Reduced Hessian)**
Let $(\boldsymbol{x}, \boldsymbol{\lambda}, \boldsymbol{\mu}) \in \mathbb{R}^{n^x} \times \mathbb{R}^{n^g} \times \mathbb{R}^{n^h}$ be a primal–dual point and let $\boldsymbol{Z}(\boldsymbol{x}) \in \mathbb{R}^{n^x \times n^z}$ be a column basis of the null space of the active constraints in $\boldsymbol{x}$. The projection of the Hessian $L_{xx}(\boldsymbol{x}, \boldsymbol{\lambda}, \boldsymbol{\mu})$ of the Lagrangian onto the null space of the active constraints,

$$\boldsymbol{H}^{\text{red}}(\boldsymbol{x}, \boldsymbol{\lambda}, \boldsymbol{\mu}) \stackrel{\text{def}}{=} \boldsymbol{Z}^T(\boldsymbol{x})\, L_{xx}(\boldsymbol{x}, \boldsymbol{\lambda}, \boldsymbol{\mu})\, \boldsymbol{Z}(\boldsymbol{x}), \tag{3.6}$$

is called the *reduced Hessian*. △

**Theorem 3.3 (Second Order Necessary Conditions)**
Let $\boldsymbol{x}^\star \in \mathbb{R}^{n^x}$ be a locally optimal point of problem (3.1). Let LICQ hold in $\boldsymbol{x}^\star$ and let $\boldsymbol{\lambda}^\star \in \mathbb{R}^{n^g}$, $\boldsymbol{\mu}^\star \in \mathbb{R}^{n^h}$ be the unique Lagrange multipliers such that the KKT conditions are satisfied. Then it holds that the reduced Hessian is positive semidefinite,

$$\forall \boldsymbol{d} \in \{\boldsymbol{Z}(\boldsymbol{x}^\star)\boldsymbol{d}_z \mid \boldsymbol{d}_z \in \mathbb{R}^{n^z}\} : \ \boldsymbol{d}^T L_{xx}(\boldsymbol{x}^\star, \boldsymbol{\lambda}^\star, \boldsymbol{\mu}^\star)\, \boldsymbol{d} \geqslant 0.$$

△

**Proof** A proof can be found in [158]. □

**Theorem 3.4 (Strong Second Order Sufficient Conditions)**
Let $\boldsymbol{x}^\star \in \mathbb{R}^n$ be a feasible point of problem (3.1) and let $\boldsymbol{\lambda} \in \mathbb{R}^{n_g}$, $\boldsymbol{\mu} \in \mathbb{R}^{n_h}$ be Lagrange multipliers such that the KKT conditions are satisfied. Further, let the reduced Hessian be positive definite,

$$\forall \boldsymbol{d} \in \{Z(\boldsymbol{x}^\star)\boldsymbol{d}_z \mid \boldsymbol{d}_z \in \mathbb{R}^{n^z}\} \setminus \{\boldsymbol{0}\} : \ \boldsymbol{d}^T L_{xx}(\boldsymbol{x}^\star, \boldsymbol{\lambda}^\star, \boldsymbol{\mu}^\star)\, \boldsymbol{d} > 0.$$

Then $\boldsymbol{x}^\star$ is a locally strictly optimal point of problem (3.1). △

**Proof** A proof can be found in [158]. □

### 3.1.4 Stability

We are concerned with the stability of a KKT point under small perturbations of the problem data in (3.1). Under strict complementarity, the active set can be shown to remain invariant for small perturbations of the solution. To this end we consider the perturbed problem

$$\begin{aligned} \min_{\boldsymbol{x} \in \mathbb{R}^n} \quad & f(\boldsymbol{x}, \varepsilon) \\ \text{s.t.} \quad & \boldsymbol{g}(\boldsymbol{x}, \varepsilon) = \boldsymbol{0}, \\ & \boldsymbol{h}(\boldsymbol{x}, \varepsilon) \geqslant \boldsymbol{0} \end{aligned} \tag{3.7}$$

with perturbation parameter $\varepsilon > 0$ for the objective function and the constraint functions. Further, let the problem functions $f$, $\boldsymbol{g}$, and $\boldsymbol{h}$ be continuously differentiable with respect to the disturbation parameter $\varepsilon$.

**Theorem 3.5 (Stability)**
Consider the perturbed problem (3.7), and assume that for $\varepsilon = 0$ this problem coincides with problem (3.1). Let $(\boldsymbol{x}^\star, \boldsymbol{\lambda}^\star, \boldsymbol{\mu}^\star)$ be a regular point satisfying both strict complementarity and the assumptions of theorem 3.4 in $\varepsilon = 0$. Then there exists an open interval $\mathcal{V}(0) \subset \mathbb{R}$, an open ball $\mathcal{W}(\boldsymbol{x}^\star, \boldsymbol{\lambda}^\star, \boldsymbol{\mu}^\star) \subset \mathbb{R}^{n^x} \times \mathbb{R}^{n^g} \times \mathbb{R}^{n^h}$, and a continuously differentiable mapping $\varphi : \mathcal{V} \to \mathcal{W}$ such that $\boldsymbol{x}^\star(\varepsilon)$ is a strict local minimum of (3.7) and $\varphi(\varepsilon) = (\boldsymbol{x}^\star(\varepsilon), \boldsymbol{\lambda}^\star(\varepsilon), \boldsymbol{\mu}^\star(\varepsilon))$ is the unique KKT point of (3.7) in $\mathcal{W}$. The set $\mathcal{A}(\boldsymbol{x}^\star(0))$ of active inequality constraints remains unchanged in $\boldsymbol{x}^\star(\varepsilon)$. △

**Proof** A proof can be found in [34]. □

## 3.2 Sequential Quadratic Programming

In this section we introduce two SQP methods, the full–step exact Hessian SQP method and the constrained Gauß–Newton method, and mention their local convergence properties. They may be used to solve OCPs of the class (1.1) after a direct multiple shooting discretization.

SQP methods for constrained nonlinear programming were first proposed by [221]. The first practically successful implementations are due to [100, 170]. For a more extensive treatment of the underlying theory and possible algorithmic variants we refer to the textbooks [72, 158]. Descriptions of an SQP method tailored to the direct multiple shooting method can be found e.g. in [132, 134]. Further extensions for large scale systems are described e.g. in [6, 191].

### 3.2.1 Basic Algorithm

SQP methods are iterative descent-based methods for finding a KKT point of problem (3.1) by computing a sequence of iterates $\{(\boldsymbol{x}^k, \boldsymbol{\lambda}^k, \boldsymbol{\mu}^k)\}$ starting in an initial guess $(\boldsymbol{x}^0, \boldsymbol{\lambda}^0, \boldsymbol{\mu}^0)$. The steps are found from the minimization of local quadratic models of the Lagrangian function on a linearization of the feasible set in the most recent iterate.

**Definition 3.10 (Local Quadratic Subproblem)**
The *local quadratic subproblem* in $(\boldsymbol{x}^k, \boldsymbol{\lambda}^k, \boldsymbol{\mu}^k) \in \mathbb{R}^{n^x} \times \mathbb{R}^{n^g} \times \mathbb{R}^{n^h}$ for problem (3.1) is given by

$$\begin{aligned} \min_{\delta \boldsymbol{x}^k \in \mathcal{D}^k} \quad & \tfrac{1}{2} \delta \boldsymbol{x}^{k^T} \boldsymbol{B}^k \delta \boldsymbol{x}^k + \delta \boldsymbol{x}^{k^T} \boldsymbol{b}^k \qquad (3.8) \\ \text{s.t.} \quad & \boldsymbol{g}(\boldsymbol{x}^k) + \boldsymbol{g}_x(\boldsymbol{x}^k) \delta \boldsymbol{x}^k = \boldsymbol{0}, \\ & \boldsymbol{h}(\boldsymbol{x}^k) + \boldsymbol{h}_x(\boldsymbol{x}^k) \delta \boldsymbol{x}^k \geqslant \boldsymbol{0}. \end{aligned}$$

The matrix $\boldsymbol{B}^k$ denotes the Hessian $L_{xx}(\boldsymbol{x}^k, \boldsymbol{\lambda}^k, \boldsymbol{\mu}^k)$ of the Lagrangian or a suitable approximation thereof, and the vector $\boldsymbol{b}^k$ denotes the gradient $f_x(\boldsymbol{x}^k)$ of the objective function. The set $\mathcal{D}^k \subseteq \mathbb{R}^n$ may be used to further restrict the set of permitted steps. △

The solution $\delta \boldsymbol{x}^k$ of (3.8) is used as a step direction to obtain the next iterate $\boldsymbol{x}^{k+1}$,

$$\boldsymbol{x}^{k+1} \stackrel{\text{def}}{=} \boldsymbol{x}^k + \alpha^k \delta \boldsymbol{x}^k. \qquad (3.9)$$

SQP methods differ in the choice of the Hessian $\boldsymbol{B}^k$ or its approximation, in the choice of the set $\mathcal{D}^k$ of candidate steps, and in the way the length $\alpha^k \in (0,1] \subset \mathbb{R}$ of the step is determined. We refer the reader to [72, 158] for details. By construction of $\boldsymbol{b}^k$ the duals $\hat{\boldsymbol{\lambda}}^k$ and $\hat{\boldsymbol{\mu}}^k$ obtained from the solution of (3.8) are the new SQP dual iterates after a full step $\alpha^k = 1$, thus we have

$$\begin{aligned} \boldsymbol{\lambda}^{k+1} &\stackrel{\text{def}}{=} (1-\alpha^k)\boldsymbol{\lambda}^k + \alpha^k\hat{\boldsymbol{\lambda}}^k, \\ \boldsymbol{\mu}^{k+1} &\stackrel{\text{def}}{=} (1-\alpha^k)\boldsymbol{\mu}^k + \alpha^k\hat{\boldsymbol{\mu}}^k. \end{aligned} \tag{3.10}$$

For a proof we refer to e.g. [132]. The sequence of SQP iterates can be shown to converge to a KKT point of (3.1) under certain conditions as detailed below. In practice, it is terminated once a prescribed convergence criterion is satisfied.

**Algorithm 3.1:** A basic SQP algorithm.

**input** : $\boldsymbol{x}^0$, $\boldsymbol{\lambda}^0$, $\boldsymbol{\mu}^0$
**output**: $\boldsymbol{x}^\star$, $\boldsymbol{\lambda}^\star$, $\boldsymbol{\mu}^\star$
**while** $\neg$**terminate**$(\boldsymbol{x}^k, \boldsymbol{\lambda}^k, \boldsymbol{\mu}^k)$ **do**
  Evaluate $\boldsymbol{b}^k$, $\boldsymbol{g}$, $\boldsymbol{g}_x$, $\boldsymbol{h}$, $\boldsymbol{h}_x$ in $\boldsymbol{x}^k$;
  Evaluate $\boldsymbol{B}^k$ in $(\boldsymbol{x}^k, \boldsymbol{\lambda}^k, \boldsymbol{\mu}^k)$;
  Determine $\mathcal{D}^k \subseteq \mathbb{R}^n$;
  Solve subproblem (3.8) for $(\boldsymbol{\delta x}^k, \hat{\boldsymbol{\lambda}}^k, \hat{\boldsymbol{\mu}}^k)$;
  Determine step length $\alpha^k \in (0,1]$;
  $\boldsymbol{x}^{k+1} = \boldsymbol{x}^k + \alpha^k \boldsymbol{\delta x}^k$;
  $\boldsymbol{\lambda}^{k+1} = (1-\alpha^k)\boldsymbol{\lambda}^k + \alpha^k\hat{\boldsymbol{\lambda}}^k$;
  $\boldsymbol{\mu}^{k+1} = (1-\alpha^k)\boldsymbol{\mu}^k + \alpha^k\hat{\boldsymbol{\mu}}^k$;
  $k = k+1$;
**end**

### 3.2.2 The Full Step Exact Hessian SQP Method

The full step exact Hessian SQP method is named for its choice

$$\begin{aligned} \boldsymbol{B}^k &= L_{xx}(\boldsymbol{x}^k, \boldsymbol{\lambda}^k, \boldsymbol{\mu}^k), \\ \mathcal{D}^k &= \mathbb{R}^n, \\ \alpha^k &= 1 \end{aligned} \tag{3.11}$$

of the algorithmic parameters described in section 3.2.1. It can be shown to be equivalent to Newton–Raphson iterations applied to the KKT conditions.

**Theorem 3.6 (Equivalence of SQP and Newton's method)**
Let $\boldsymbol{B}^k(\boldsymbol{x}^k, \boldsymbol{\lambda}^k, \boldsymbol{\mu}^k) = L_{xx}(\boldsymbol{x}^k, \boldsymbol{\lambda}^k, \boldsymbol{\mu}^k)$, $\mathcal{D}^k = \mathbb{R}^n$, and $\alpha^k = 1$ for all $k \geqslant 1$. Then the SQP method is equivalent to Newton's method. △

For equality constrained problems, this can be seen from the KKT conditions of the local quadratic subproblem (3.8),

$$\begin{aligned} \boldsymbol{B}^k \delta \boldsymbol{x}^k + \boldsymbol{b}^k - \hat{\boldsymbol{\lambda}}^{k^T} \boldsymbol{g}_x(\boldsymbol{x}^k) &= \boldsymbol{0}, \\ \boldsymbol{g}(\boldsymbol{x}^k) + \boldsymbol{g}_x(\boldsymbol{x}^k)\delta \boldsymbol{x}^k &= \boldsymbol{0}. \end{aligned} \tag{3.12}$$

After substitution $\hat{\boldsymbol{\lambda}}^k = \boldsymbol{\lambda}^k + \delta \boldsymbol{\lambda}^k$ (3.12) may be written as

$$\begin{aligned} \boldsymbol{B}^k \delta \boldsymbol{x}^k + L_x(\boldsymbol{x}^k, \boldsymbol{\lambda}^k, \boldsymbol{\mu}^k) - \delta \boldsymbol{\lambda}^{k^T} \boldsymbol{g}_x(\boldsymbol{x}^k) &= \boldsymbol{0}, \\ \boldsymbol{g}(\boldsymbol{x}^k) + \boldsymbol{g}_x(\boldsymbol{x}^k)\delta \boldsymbol{x}^k &= \boldsymbol{0}. \end{aligned} \tag{3.13}$$

This is the Newton–Raphson iteration for $(\delta \boldsymbol{x}^k, \delta \boldsymbol{\lambda}^k)$ on the KKT system of (3.8),

$$\begin{bmatrix} L_x(\boldsymbol{x}^k, \boldsymbol{\lambda}^k, \boldsymbol{\mu}^k) \\ \boldsymbol{g}(\boldsymbol{x}^k) \end{bmatrix} + \frac{\mathrm{d}}{\mathrm{d}(\boldsymbol{x}, \boldsymbol{\lambda})} \begin{bmatrix} L_x(\boldsymbol{x}^k, \boldsymbol{\lambda}^k, \boldsymbol{\mu}^k) \\ \boldsymbol{g}(\boldsymbol{x}^k) \end{bmatrix} \begin{bmatrix} \delta \boldsymbol{x}^k \\ \delta \boldsymbol{\lambda}^k \end{bmatrix} = \boldsymbol{0}. \tag{3.14}$$

The local convergence rate of the full step exact Hessian SQP method in the neighborhood of a KKT point thus is quadratic. Good initial guesses for $\boldsymbol{x}^0$ are required, though. The following theorem shows that such guesses are not required for the Lagrange multipliers $\boldsymbol{\lambda}^0$.

**Theorem 3.7 (Convergence of the full step exact Hessian SQP Method)**
Let $(\boldsymbol{x}^\star, \boldsymbol{\lambda}^\star)$ satisfy theorem 3.4. Let $(\boldsymbol{x}^0, \boldsymbol{\lambda}^0)$ be chosen such that $\boldsymbol{x}^0$ is sufficiently close to $\boldsymbol{x}^\star$ and that the KKT matrix of (3.8) is regular. Then the sequence of iterates generated by the full step exact Hessian SQP method shows local q-quadratic convergence to $(\boldsymbol{x}^\star, \boldsymbol{\lambda}^\star)$. △

**Proof** A proof can be found in [72]. □

In the presence of inequality constraints, if theorem 3.4 and strict complementarity hold, then the active set does not change in a neighborhood of the KKT point. Indeed, under these assumptions even if $\boldsymbol{B}^k$ only is some positive definite approximation of the exact Hessian, both the optimal active set and the optimal Lagrange multipliers $\boldsymbol{\lambda}^\star$, $\boldsymbol{\mu}^\star$ can be identified knowing the primal optimal solution $\boldsymbol{x}^\star$ only [51, 180].

### 3.2.3 The Gauß-Newton Approximation

The Gauß–Newton approximation of the Hessian is applicable to NLPs with objective function of the structure

$$f(\boldsymbol{x}) = \tfrac{1}{2}\,||\boldsymbol{r}(\boldsymbol{x})||_2^2 = \tfrac{1}{2}\sum_{i=1}^{n^r} r_i^2(\boldsymbol{x}), \tag{3.15}$$

where $\boldsymbol{r} : \mathbb{R}^{n^x} \to \mathbb{R}^{n^r}$, $\boldsymbol{x} \mapsto \boldsymbol{r}(\boldsymbol{x})$ is a vector valued residual function. The gradient $f_{\boldsymbol{x}}$ is

$$f_{\boldsymbol{x}}(\boldsymbol{x}) = \sum_{i=1}^{n^r} r_i(\boldsymbol{x}) r_{i\boldsymbol{x}}(\boldsymbol{x}) = \boldsymbol{r}_{\boldsymbol{x}}^T(\boldsymbol{x})\boldsymbol{r}(\boldsymbol{x}), \tag{3.16}$$

and the Hessian of this objective is

$$\begin{aligned} f_{\boldsymbol{xx}}(\boldsymbol{x}) &= \sum_{i=1}^{n^r} \left( r_{i\boldsymbol{x}}^T(\boldsymbol{x}) r_{i\boldsymbol{x}}(\boldsymbol{x}) + r_i(\boldsymbol{x}) r_{i\boldsymbol{xx}}(\boldsymbol{x}) \right) \\ &= \boldsymbol{r}_{\boldsymbol{x}}^T(\boldsymbol{x})\boldsymbol{r}_{\boldsymbol{x}}(\boldsymbol{x}) + \sum_{i=1}^{n^r} r_i(\boldsymbol{x}) r_{i\boldsymbol{xx}}(\boldsymbol{x}). \end{aligned} \tag{3.17}$$

The first part of the Hessian $f_{\boldsymbol{xx}}$ can be calculated only from gradient information and in addition often dominates the second order contribution in the case of small residuals $\boldsymbol{r}(\boldsymbol{x})$ or because the model is almost linear close to the solution. This gives rise to the approximation

$$\boldsymbol{B}_{\mathrm{GN}}(\boldsymbol{x}) \stackrel{\text{def}}{=} \boldsymbol{r}_{\boldsymbol{x}}^T(\boldsymbol{x})\boldsymbol{r}_{\boldsymbol{x}}(\boldsymbol{x}) \tag{3.18}$$

which is independent of the Lagrangian multipliers belonging to $\boldsymbol{x}$. The error in this approximation can be shown to be of order $\mathcal{O}(||\boldsymbol{r}(\boldsymbol{x}^\star)||)$, which leads us to expect the Gauß–Newton method to perform well for small residuals $\boldsymbol{r}(\boldsymbol{x})$. For more details on the constrained Gauß–Newton method as an important special case of SQP methods applicable to NLPs with the objective function (3.15) we refer to [34].

Note that the Gauß–Newton approximation can be applied separately to the Hessian blocks $\boldsymbol{B}_i^k$, $0 \leqslant i \leqslant m$ of the direct multiple shooting system.

### 3.2.4 BFGS Hessian Approximation

The BFGS (Broyden–Fletcher–Goldfarb–Shanno) approximation belongs to a larger family of quasi–Newton update formulas, of which it is presently considered the most effective one. One possible way to derive it is the following. Starting with an existing symmetric and positive definite Hessian approximation $\boldsymbol{B}^k$, that need not necessarily be a BFGS one, we compute the new approximation $\boldsymbol{B}^{k+1}$ as

$$\boldsymbol{B}^{k+1} \stackrel{\text{def}}{=} \boldsymbol{B}^k - \frac{\boldsymbol{B}^k \boldsymbol{\delta x}^k \boldsymbol{\delta x}^{k^T} \boldsymbol{B}^{k^T}}{\boldsymbol{\delta x}^{k^T} \boldsymbol{B}^k \boldsymbol{\delta x}^k} + \frac{\boldsymbol{\delta f}_x^k \boldsymbol{\delta f}_x^{k^T}}{\boldsymbol{\delta f}_x^{k^T} \boldsymbol{\delta x}^k} \tag{3.19}$$

wherein $\boldsymbol{\delta x}^k$ is the step from $\boldsymbol{x}^k$ to $\boldsymbol{x}^{k+1}$ and $\boldsymbol{\delta f}_x^k$ is the associated gradient change $f_x(\boldsymbol{x}^{k+1}) - f_x(\boldsymbol{x}^k)$. Different derivations of this update formula can be found in the original papers [41, 71, 87, 198] as well as in the textbooks [72, 158].

For larger dimensions $n^{\text{x}}$, and thus larger and possibly dense Hessian approximations $\boldsymbol{B}^k$, the limited memory variant L-BFGS of this approximation is attractive. Instead of storing the matrix $\boldsymbol{B}^k$, a limited number $l$ of historical vector pairs $(\boldsymbol{\delta x}^i, \boldsymbol{\delta f}_x^i)$, $k - l + 1 \leqslant i \leqslant k$, is stored from which the approximation $\boldsymbol{B}^k$ is rebuilt. In each SQP iteration the oldest vector pair is replaced by the current step and gradient step. While $\boldsymbol{B}^k$ could in principle be computed by repeated application of (3.19), this requires $\mathcal{O}(l^2 n^{\text{x}})$ multiplications and we refer to e.g. [158] for compact representations of the L-BFGS update that reduce the computational effort to only $\mathcal{O}(l n^{\text{x}})$ multiplications.

This approximation scheme has also proved itself successful for ill–conditioned problems and for initial guesses far away from the optimal solution, in which case outdated secant information unrelated to the locally optimal solution would accumulate in a conventional BFGS approximation.

It is of vital importance to note that the rank two update (3.19) can be applied independently to each Hessian block $\boldsymbol{B}_i^k$, $0 \leqslant i \leqslant m$ of the direct multiple shooting system. This leads to a rank $2m$ update for the NLP that significantly improves the convergence speed of the SQP method as first noted in [36].

### 3.2.5 Local Convergence

We are interested in sufficient conditions for local convergence for general SQP methods that produce a series of iterates

$$y^{k+1} = y^k + \delta y^k = y^k - M(y^k) r(y^k) \tag{3.20}$$

towards a root $y^\star$ of the function $r : \mathcal{D} \to \mathbb{R}^{n^r}$, $\mathcal{D} \subseteq \mathbb{R}^n$, with $r(y^\star) = \mathbf{0}$. Here the matrices $M(y^k)$ denote approximations of the inverse of the Jacobians $J(y^k)$ of the residuals $r(y^k)$. In the case of SQP methods, $r$ is the gradient of the Lagrangian and $M$ a suitable approximation of the inverse of the Hessian of the Lagrangian.

We define the set of Newton pairs of the iteration (3.20) as follows.

**Definition 3.11 (Set of Newton Pairs)**
The set $\mathcal{N}$ of Newton pairs is defined as

$$\mathcal{N} \stackrel{\text{def}}{=} \{(y_1, y_2) \in \mathcal{D} \times \mathcal{D} \mid y_2 = y_1 - M(y_1) r(y_1)\}. \tag{3.21}$$

△

We require two definitions concerned with the quality of the approximations $M(y^k)$ to $J^{-1}(y^k)$.

**Definition 3.12 ($\omega$–Condition)**
The approximation $M$ satisfies the *$\omega$–condition* in $\mathcal{D}$ if there exists $\omega < \infty$ such that for all $t \in [0,1] \subset \mathbb{R}$ and all $(y_1, y_2) \in \mathcal{N}$ it holds that

$$\left|\left| M(y_2)(J(y_1 + t(y_2 - y_1)) - J(y_1))(y_1 - y_2) \right|\right| \leqslant \omega t \left|\left| y_1 - y_2 \right|\right|^2. \tag{3.22}$$

△

**Definition 3.13 (Compatibility or $\kappa$–Condition)**
The approximation $M$ satisfies the *$\kappa$–condition* in $\mathcal{D}$ if there exists $\kappa < 1$ such that for all $(y_1, y_2) \in \mathcal{N}$ it holds that

$$\left|\left| M(y_2)(I - J(y_1) M(y_1)) r(y_1) \right|\right| \leqslant \kappa \left|\left| y_1 - y_2 \right|\right|. \tag{3.23}$$

△

With this, we define with $\delta y^k \stackrel{\text{def}}{=} -M(y^k) r(y^k)$ the constant

$$\delta^k \stackrel{\text{def}}{=} \kappa + \tfrac{\omega}{2} \left|\left| \delta y^k \right|\right| \tag{3.24}$$

and the closed ball

$$\mathcal{D}^0(\boldsymbol{y}^0) \stackrel{\text{def}}{=} \overline{\mathcal{B}}_\varepsilon(\boldsymbol{y}^0), \quad \varepsilon \stackrel{\text{def}}{=} \left|\left|\delta \boldsymbol{y}^0\right|\right| / (1 - \delta^0). \tag{3.25}$$

We are now prepared to state a local contraction theorem for the sequence $\{\boldsymbol{y}^k\}$ of SQP iterates.

**Theorem 3.8 (Local Contraction Theorem)**
Let $\boldsymbol{M}$ satisfy the $\omega$– and $\kappa$–conditions and let $\boldsymbol{y}^0 \in \mathcal{D}$. If $\delta^0 < 1$ and $\mathcal{D}^0(\boldsymbol{y}^0) \subset \mathcal{D}$, then $\boldsymbol{y}^k \in \mathcal{D}^0(\boldsymbol{y}^0)$ and $\{\boldsymbol{y}^k\} \to \boldsymbol{y}^\star \in \mathcal{D}^0(\boldsymbol{y}^0)$ with convergence rate

$$\left|\left|\delta \boldsymbol{y}^{k+1}\right|\right| \leqslant \delta^k \left|\left|\delta \boldsymbol{y}^k\right|\right| = \kappa \left|\left|\delta \boldsymbol{y}^k\right|\right| + \tfrac{\omega}{2} \left|\left|\delta \boldsymbol{y}^k\right|\right|^2. \tag{3.26}$$

Furthermore, the following a–priori estimates hold for $j \geqslant 1$,

$$\left|\left|\boldsymbol{y}^{k+j} - \boldsymbol{y}^\star\right|\right| \leqslant \tfrac{\delta_k^j}{1-\delta^k} \left|\left|\delta \boldsymbol{y}^k\right|\right| \leqslant \tfrac{\delta_0^{k+j}}{1-\delta^0} \left|\left|\delta \boldsymbol{y}^0\right|\right|. \tag{3.27}$$

If $\boldsymbol{M}(\boldsymbol{y})$ is continuous and regular in $\boldsymbol{y}^\star$ then $\boldsymbol{r}(\boldsymbol{y}^\star) = \boldsymbol{0}$. △

**Proof** The proof can be found in [34]. □

We will make use of this local contraction theorem in chapter 4 to proof convergence of the real–time iteration scheme and its mixed–integer extension.

### 3.2.6 Termination Criterion

In algorithm 3.1 we have left open the issue of finding a suitable termination criterion. Certainly, due to approximation errors in the derivatives and due to finite precision and conditioning issues we will in general be able to identify a primal–dual point $(\boldsymbol{x}^k, \boldsymbol{\lambda}^k, \boldsymbol{\mu}^k)$ that is satisfies the optimality conditions only to a certain prescribed tolerance. In [133] it has been proposed to use the KKT tolerance

$$\mathbf{kkttol}(\boldsymbol{x}, \boldsymbol{\lambda}, \boldsymbol{\mu}) \stackrel{\text{def}}{=} \left|\left|f_x(\boldsymbol{x}, \boldsymbol{\lambda}, \boldsymbol{\mu}^T \delta \boldsymbol{x})\right|\right| + \sum_{i=1}^{n^g} |\lambda_i g_i(\boldsymbol{x})| + \sum_{i=1}^{n^g} |\mu_i h_i(\boldsymbol{x})|. \tag{3.28}$$

We terminate the SQP iterations once a prescribed threshold value kktacc, e.g. kktacc $= 10^{-8}$ is satisfied for some iteration $k$,

$$\mathbf{kkttol}(\boldsymbol{x}^k, \boldsymbol{\lambda}^k, \boldsymbol{\mu}^k) < \texttt{kktacc}. \tag{3.29}$$

### 3.2.7 Scaling

The step direction, step length, and the termination criterion are susceptible to scaling of the unknowns and of the objective function and the constraints functions. Most NLP algorithm therefore either determine appropriate scale factors automatically, or provide a means for the user to specify suitable positive scale factors $\sigma_{\mathrm{x}} \in \mathbb{R}^{n^{\mathrm{x}}}$, $\sigma_{\mathrm{f}} \in \mathbb{R}$, $\boldsymbol{\sigma}_{\mathrm{g}} \in \mathbb{R}^{n^{\mathrm{g}}}$, $\boldsymbol{\sigma}_{\mathrm{h}} \in \mathbb{R}^{n^{\mathrm{h}}}$. The equivalent NLP

$$\begin{aligned}
\min_{\boldsymbol{x}\in\mathbb{R}^n} \quad & \sigma_{\mathrm{g}} \tilde{f}(\Sigma_{\mathrm{x}} \boldsymbol{x}) && \qquad (3.30) \\
\text{s.t.} \quad & \sigma_{\mathrm{g},i} \tilde{g}_i(\Sigma_{\mathrm{x}} \boldsymbol{x}) = 0, && 1 \leqslant i \leqslant n^{\mathrm{g}}, \\
& \sigma_{\mathrm{h},i} \tilde{h}_i(\Sigma_{\mathrm{x}} \boldsymbol{x}) \geqslant 0, && 1 \leqslant i \leqslant n^{\mathrm{h}},
\end{aligned}$$

is then solved in place of the original one (3.1). Here $\Sigma_{\mathrm{x}} = \operatorname{diag} \sigma_{\mathrm{x}}$ and $\tilde{f}(\cdot)$, $\tilde{g}(\cdot)$, and $\tilde{h}(\cdot)$ are appropriately rescaled counterparts of the original problem functions. In order to avoid rounding and cutoff errors to be introduced due to limited machine precision, the scale factors should be chosen as positive powers of two which can be represented exactly and allow for fast and lossless multiplication and division.

## 3.3 Derivative Generation

The numerical optimization algorithms presented in this thesis strongly rely on the availability of derivative information for the various functions modeling the controlled process. This includes gradients, directional derivatives, full Jacobian matrices, and second order derivative information in the form of exact or approximate Hessians. In this section we survey several numerical and algorithmic strategies to obtain such derivatives and study their respective precisions and efficiencies. For further details, we refer to e.g. [4] and the textbook [92].

### 3.3.1 Analytical Derivatives

Using analytical derivatives of model functions assumes that a symbolic expression of the derivative is available, e.g. in the form of an expression tree. Elementary differentiation rules known from basic calculus are applied to the operators and elementary functions at the tree nodes in order to derive a symbolic expression for the desired derivative. While this process can easily be carried out manually, it is cumbersome for larger model functions and has been automated in computer algebra systems like Maple V [150] and Mathematica [222]. These tools also readily provide facilities to translate the symbolic derivative expression into e.g. Fortran or C source code.

Analytical derivatives of model functions obviously are exact up to machine precision. A symbolic expression may however not always be available or its transferral to a computer algebra system may be difficult to accomplish. Second, the obtained symbolic expression for the derivative may not be optimally efficient in terms of evaluation time, as demonstrated by the classical example

$$f(x_1,\dots,x_n) = \prod_{i=1}^{n} x_i$$

due to [200]. Here, the gradient's entries $(f_{\boldsymbol{x}})_i$ consist of all partial products that omit one factor $x_i$, while the Hessian's entries $(f_{\boldsymbol{xx}})_{ij}$ are the partial products that omit any two factors $x_i$, $x_j$. The efficient reuse of common subexpressions is up to the compiler that translates the possibly inefficient symbolic expression into machine code.

### 3.3.2 Finite Difference Approximations

Directional derivatives $\boldsymbol{f_x d}$ of a function $\boldsymbol{f} : \mathbb{R}^n \to \mathbb{R}^m$, $\boldsymbol{x} \mapsto \boldsymbol{f}(\boldsymbol{x})$ into a direction $\boldsymbol{d} \in \mathbb{R}^n$ may be approximated in a point $\boldsymbol{x}_0 \in \mathbb{R}^n$ based on the Taylor series expansion of $\boldsymbol{f}$ in $\boldsymbol{x}_0$,

$$\boldsymbol{f}(\boldsymbol{x}_0 + h\boldsymbol{d}) = \boldsymbol{f}(\boldsymbol{x}_0) + h\boldsymbol{f_x}(\boldsymbol{x}_0)\boldsymbol{d} + \mathcal{O}(h^2), \tag{3.31}$$

which yields the one-sided finite difference scheme

$$\boldsymbol{f_x d} = \frac{\boldsymbol{f}(\boldsymbol{x}_0 + h\boldsymbol{d}) - \boldsymbol{f}(\boldsymbol{x}_0)}{h} + \mathcal{O}(h), \tag{3.32}$$

Combination with a second Taylor series expansion with negative increment $h$ yields the improved central difference scheme

$$\boldsymbol{f_x d} = \frac{\boldsymbol{f}(\boldsymbol{x}_0 + h\boldsymbol{d}) - \boldsymbol{f}(\boldsymbol{x}_0 - h\boldsymbol{d})}{2h} + \mathcal{O}(h^2). \tag{3.33}$$

Finite difference schemes are generally easy to implement as they only rely on repeated evaluations of the model function $f$ in perturbed evaluation points $x_0 \pm hd$. This does not require knowledge about the internals of the representation of $f$ other than the assurance of sufficient differentiability. The computation of a single directional derivative $f_x(x_0)d \in \mathbb{R}^m$ for a direction $d \in \mathbb{R}^n$ comes at the cost of only two function evaluations. The full Jacobian matrix $f_x(x_0) \in \mathbb{R}^{m\times n}$ is available at $2n$ function evaluations.

The precision of finite difference approximations crucially depends on the magnitude $h||d||$ of the increments. For larger increments, the truncation error incurred by neglecting the higher–order terms of the Taylor series expansion become predominant, while for tiny increments cancellation errors due to limited machine precision become predominant. Though dependent on the actual function $f$, the recommended perturbations are $h||d|| = \mathsf{eps}^{\frac{1}{2}}$ for one–sided and $h||d|| = \mathsf{eps}^{\frac{1}{3}}$ for central finite difference schemes. The number of significant digits of the obtained derivative approximation is at most one half of that of the function values for one–sided schemes, and at most two thirds for central schemes.

### 3.3.3 Complex Step Approximation

Cancellation errors introduced into finite difference approximations of derivatives can be avoided by using complex arithmetic. As first noted by [145], we can evaluate the function $\tilde{f} : \mathbb{C}^n \to \mathbb{C}^m$ obtained from $f$ and apply perturbations $\imath hd$ to the imaginary part of $x_0 + 0\imath \in \mathbb{C}$. The directional derivative may be obtained as

$$f_x(x_0)d = \Im\left(\frac{f(x_0 + \imath hd)}{h}\right) + \mathcal{O}(h^2). \tag{3.34}$$

As the perturbation is applied to the imaginary part this approach does not suffer from cancellation errors. The perturbation magnitude $h||d||$ should be chosen small but representable, e.g. $h||d|| = 10^{-100}$.

For the complex step approximation to be applicable, the model function $f$ has to be analytic and the symbolic representation has to support evaluation in the complex domain. The use of linear algebra subroutines may prevent this. The computational efficiency of the complex step approximation depends on the availability of native hardware support for complex arithmetic on the computational platform in use.

### 3.3.4 Automatic Differentiation

Like symbolic differentiation to obtain analytical expressions for the derivatives, automatic differentiation is based on the idea of decomposing the function $f$ into a concatenation of certain elemental functions. The derivative is obtained by systematic application of the chain rule. Unlike in symbolic differentiation, this procedure is not applied to the symbolic expression tree, but instead takes place while evaluating the function $f$ itself in a given point $\boldsymbol{x}_0$. Pioneering works in the field of automatic differentiation are [115, 218], and for a comprehensive reference we refer to [92].

#### Principle

The idea of representing the function $f$ as a concatenation of elemental functions is sharpened by the following definition.

**Definition 3.14 (Factorable Function)**
Let $\mathcal{L}$ be a finite set of real–valued elemental functions $\varphi_i : \mathbb{R}^n \to \mathbb{R}$, $\boldsymbol{x} \mapsto \varphi(\boldsymbol{x})$. A function $\boldsymbol{f} : \mathbb{R}^n \to \mathbb{R}^m$, $\boldsymbol{x} \mapsto \boldsymbol{f}(\boldsymbol{x})$ is called a *factorable function* iff there exists a finite sequence $\{\varphi_{1-n}, \dots, \varphi_k\}$, $k \geqslant m$, such that it holds:

1. For $1 \leqslant i \leqslant n$ the function $\varphi_{i-n} = \pi_i^n$ is the projection on the $i$-th component of the evaluation point $\boldsymbol{x} \in \mathbb{R}^n$,
2. For $1 \leqslant i \leqslant m$ the function $\varphi_{k-m+i} = \pi_i^m$ is the projection on the $i$-th component of the evaluation result $\boldsymbol{y} = f(\boldsymbol{x}) \in \mathbb{R}^m$,
3. For $1 \leqslant i \leqslant k - m$ the function $\varphi_i$ is constant or a concatenation of one or more functions $\varphi_j$ with $1 - n \leqslant j \leqslant i - 1$, i.e., of functions preceding $\varphi_i$ in the concatenation sequence.

$\triangle$

For simplicity, we assumed scalar valued elemental functions $\varphi_i$ here, though these can in principle be vector valued functions as well if it is of computational advantage to treat for example linear algebra operations on vectors or matrices as elemental. Using a representation of a factorable function $f$ by a sequence $\{\varphi_{1-n}, \dots, \varphi_k\}$, the function $f$ can be evaluated by algorithm 3.2.

**Algorithm 3.2:** Zero order forward sweep of automatic differentiation.

**input** : $\varphi_{1-n}, \ldots, \varphi_k, \boldsymbol{x}_0$
**output**: $y = f(\boldsymbol{x}_0)$
$v_{[1-n:0]} = x_{[1:n]}$;
**for** $i = 1 : k$ **do**
  $v_i = \varphi_i(v_{j \prec i})$;
**end**
$y_{[1:m]} = v_{[k-m+1:k]}$;

### Forward Mode of Automatic Differentiation

The forward mode of automatic differentiation is used to compute a (forward) directional derivative $\dot{y} = f_x(\boldsymbol{x}_0)\dot{\boldsymbol{x}}$. The required procedure is algorithm 3.3.

**Algorithm 3.3:** First order forward sweep of automatic differentiation.

**input** : $\varphi_{1-n}, \ldots, \varphi_k, \boldsymbol{x}_0, \dot{\boldsymbol{x}}$
**output**: $y = f(\boldsymbol{x}_0)$, $\dot{y} = f_x(\boldsymbol{x}_0)\dot{\boldsymbol{x}}$
$v_{[1-n:0]} = x_{[1:n]}$;
$\dot{v}_{[1-n:0]} = \dot{x}_{[1:n]}$;
**for** $i = 1 : k$ **do**
  $v_i = \varphi_i(v_{j \prec i})$;
  $\dot{v}_i = \sum_{j \prec i} (\varphi_i)_{v_j}(v_{j \prec i}) \cdot \dot{v}_j$;
**end**
$y_{[1:m]} = v_{[k-m+1:k]}$;
$\dot{y}_{[1:m]} = \dot{v}_{[k-m+1:k]}$;

Comparing this algorithm to the zero order forward sweep, we see that the directional derivative information is initialized with the direction $\dot{\boldsymbol{x}} \in \mathbb{R}^n$. Along with the nominal evaluation of the function sequence $\varphi$, derivative information is accumulated in $\dot{v}$ by application of the chain rule to the elemental functions $\varphi_i$. Algorithm 3.3 can easily be extended to compute multiple directional derivatives in one sweep. The overall computational effort is bounded by $1 + \frac{3}{2}n^{\mathrm{d}}$ function evaluations [92], where $n^{\mathrm{d}}$ is the number of forward directions. The computed derivative is exact within machine precision, while slightly more expensive than finite difference approximations.

**Reverse Mode of Automatic Differentiation**

The reverse mode of automatic differentiation is used to compute an adjoint directional derivative $\bar{\boldsymbol{x}} = \bar{\boldsymbol{y}}^T \boldsymbol{f}_{\boldsymbol{x}}(\boldsymbol{x}_0)$. This is achieved by applying the chain rule to the function's evaluation procedure executed in reverse order, i.e., in a backward sweep. The required procedure is algorithm 3.4.

---

**Algorithm 3.4:** First order backward sweep of automatic differentiation.

---

**input** : $\varphi_{1-n}, \ldots, \varphi_k, \boldsymbol{v}, \boldsymbol{x}_0, \bar{\boldsymbol{y}}$
**output**: $\boldsymbol{y} = \boldsymbol{f}(\boldsymbol{x}_0)$, $\bar{\boldsymbol{x}} = \bar{\boldsymbol{y}}^T \boldsymbol{f}_{\boldsymbol{x}}(\boldsymbol{x}_0)$
$\bar{v}_{[1-n:k-m]} = 0$;
$\bar{v}_{[k-m+1:k]} = \bar{y}_{[1:m]}$;
**foreach** $j \prec i$ **do** $\bar{v}_j \mathrel{+}= \bar{v}_i \cdot (\varphi_i)_{v_j}(v_{j \prec i})$;
$\bar{x}_{[1:n]} = \bar{v}_{[0:n]}$;

---

In the backward sweep procedure, the directional derivative information is initialized with the adjoint direction $\boldsymbol{\lambda} \in \mathbb{R}^m$. Each intermediate derivative value $\bar{v}_j$ accumulates derivative information from all intermediate values $v_i$ to which $v_j$ contributes during a function evaluation, i.e., during a zero order forward sweep. This procedure can easily be extended to simultaneous computation of multiple adjoint derivatives as well. The results $v_j$, $j = 1-n, \ldots, k$ of the intermediate function evaluations need to be known in advance before algorithm 3.4 can be executed. Thus, a backward sweep is usually preceded by a zero order forward sweep evaluating the function $\boldsymbol{f}$. The computational effort is bounded by $\frac{3}{2} + \frac{5}{2} n^d$ function evaluations [92], where $n^d$ is the number of adjoint directions. As this bound does not depend on the number $n$ of independent variables, the computation of adjoint derivatives is especially cheap if $m \ll n$.

### 3.3.5 Second Order Derivatives

Occasionally we will require second order derivatives of model functions. These can be computed efficiently by combining two sweeps of automatic differentiation on the function $\boldsymbol{f}$ as follows. The forward directional derivative $\dot{\boldsymbol{x}}$ obtained from a first order forward sweep can be viewed as a function $\dot{\boldsymbol{f}}$ of the forward direction $\dot{\boldsymbol{x}}$,

$$\dot{\boldsymbol{f}}(\boldsymbol{x}_0, \dot{\boldsymbol{x}}) = \boldsymbol{f}_{\boldsymbol{x}}(\boldsymbol{x}_0)\dot{\boldsymbol{x}}. \tag{3.35}$$

Application of the reverse sweep of automatic differentiation to $\dot{f}$ in the point $(\boldsymbol{x}_0, \dot{\boldsymbol{x}})$ for the adjoint direction $(\bar{y}, \bar{\dot{x}})$ yields

$$\bar{\dot{f}}(\boldsymbol{x}_0, \dot{\boldsymbol{x}}) = \bar{y}^T \left(f_x(\boldsymbol{x}_0)\dot{\boldsymbol{x}}\right)_x = \bar{y}^T f_{xx}(\boldsymbol{x}_0)\dot{\boldsymbol{x}} + \bar{\dot{x}}^T f_x(\boldsymbol{x}_0), \tag{3.36}$$

an automatic differentiation scheme that can be used to compute the second–order directional derivative $\bar{y}^T f_{xx}\dot{\boldsymbol{x}}$ of $f$ in $\boldsymbol{x}_0$.

Computing second–order derivatives using finite difference approximations leads to precisions of only about $\mathsf{eps}^{\frac{1}{4}} \approx 10^{-4}$, and is considerably more expensive. Using the complex step method is possible only for first order derivatives, but its combination with finite differences for the second order reduces the approximation error to about $\mathsf{eps}^{\frac{1}{2}} \approx 10^{-8}$. In [92], Taylor coefficient propagation schemes are described that also allow for the efficient computation of higher order derivatives.

## 3.4 Initial Value Problems and Sensitivity Generation

The evaluation of the dynamic process $\boldsymbol{x}(t)$ on the time horizon $\mathcal{T}$ requires the solution of IVPs on the shooting intervals. In this section, we present Runge–Kutta methods as popular representatives of one–step methods for non–stiff ODEs.

### 3.4.1 Runge–Kutta Methods for ODE IVPs

We start by recalling the definition of a parameter dependent IVP on $\mathcal{T} = [t_0, t_\mathrm{f}] \subset \mathbb{R}$ with initial value $\boldsymbol{x}_0 \in \mathbb{R}^n$, a time–independent parameter vector $\boldsymbol{p} \in \mathbb{R}^{n^\mathrm{p}}$, and ODE right hand side function $\boldsymbol{f} : \mathcal{T} \times \mathbb{R}^{n^\mathrm{x}} \times \mathbb{R}^{n^\mathrm{p}} \to \mathbb{R}^{n^\mathrm{x}}, (t, \boldsymbol{x}(t), \boldsymbol{p}) \mapsto \dot{x}(t)$,

$$\begin{aligned} \dot{\boldsymbol{x}}(t) &= \boldsymbol{f}\big(t, \boldsymbol{x}(t), \boldsymbol{p}\big), \quad t \in \mathcal{T}, \\ \boldsymbol{x}(t_0) &= \boldsymbol{x}_0. \end{aligned} \tag{3.37}$$

We assume $\boldsymbol{f}$ to be Lipschitz continuous on $\mathcal{T} \times \mathbb{R}^{n^\mathrm{x}}$, which ensures existence, uniqueness, and continuous differentiability of the IVP's solution $\boldsymbol{x}(t)$ on the whole of $\mathcal{T}$. For methods applicable to piecewise Lipschitz continuous functions with implicitly defined discontinuities we refer to e.g. [156, 40, 118]. One–step methods for the approximate solution of this class of IVPs can be defined as follows.

**Definition 3.15 (One–Step Method)**
Let $(t_0, \boldsymbol{x}_0)$ be an initial time and value for IVPs (3.37) and let a discretization grid $\{t^k\}$, $0 \leqslant k \leqslant N$ be given. A *one–step method* for the solution of (3.37) is given by a function $\boldsymbol{\Phi}(t,h,\boldsymbol{x})$ which defines a sequence of approximations $\{\boldsymbol{\eta}^k\}$ to the exact solutions $\{\boldsymbol{x}(t^k)\}$ of (3.37) on the discretization grid $\{t^k\}$ by starting in $(t^0, \boldsymbol{\eta}^0) \stackrel{\text{def}}{=} (t_0, \boldsymbol{x}_0)$ and applying the iteration scheme

$$\boldsymbol{\eta}^{k+1} \stackrel{\text{def}}{=} \boldsymbol{\eta}^k + h^k\, \boldsymbol{\Phi}\left(t^k, h^k, \boldsymbol{\eta}^k\right), \qquad h^k \stackrel{\text{def}}{=} t^{k+1} - t^k, \qquad 0 \leqslant k \leqslant N-1. \tag{3.38}$$

△

Within this framework, Runge–Kutta methods, named for the authors of [127, 181], are one–step methods with a generating function $\boldsymbol{\Phi}$ of the following special structure.

**Definition 3.16 (Runge–Kutta Method)**
A Runge–Kutta *method* with $s \in \mathbb{N}$ stages is a one–step method with the generating function

$$\boldsymbol{\Phi}(t,h,\boldsymbol{x}) \stackrel{\text{def}}{=} \sum_{i=1}^{s} c_i \boldsymbol{k}_i, \quad \boldsymbol{k}_i \stackrel{\text{def}}{=} \boldsymbol{f}\Big(t + \alpha_i h, \boldsymbol{x} + h \sum_{j=1}^{s} B_{ij} \boldsymbol{k}_j\Big), \tag{3.39}$$

where $\boldsymbol{c} \in \mathbb{R}^s$, $\boldsymbol{\alpha} \in \mathbb{R}^s$, and $\boldsymbol{B} \in \mathbb{R}^{s \times s}$ are suitably chosen coefficients. △

Appropriately chosen coefficients and number of stages yield a consistent and stable Runge–Kutta method of convergence order $p \geqslant 1$ if $\boldsymbol{f}$ is sufficiently often continuously differentiable, as stated by the following proposition.

**Definition 3.17 (Consistency, Stability, Convergence)**
A one–step method $\boldsymbol{\Phi}$ is called a *consistent* method if the consistency error or local discretization error $\boldsymbol{\tau}$ satisfies

$$\lim_{h \to 0} \sup_{t \in \mathcal{T}} ||\boldsymbol{\tau}(t,h,\boldsymbol{x})|| = 0, \qquad \boldsymbol{\tau}(t,h,\boldsymbol{x}) \stackrel{\text{def}}{=} \frac{\boldsymbol{x}(t+h) - \boldsymbol{x}(t)}{h} - \boldsymbol{\Phi}(t,h,\boldsymbol{x}). \tag{3.40}$$

It is said to have *consistency order* $p$ if $||\boldsymbol{\tau}(h)|| \in \mathcal{O}(h^p)$.

A one–step method $\boldsymbol{\Phi}$ is called *convergent* if

$$\lim_{h \to 0} \sup_{t \in \mathcal{T}} ||\boldsymbol{\varepsilon}(t,h,\boldsymbol{x})|| = 0, \qquad \boldsymbol{\varepsilon}(t,h,\boldsymbol{x}) \stackrel{\text{def}}{=} \boldsymbol{\eta} - \boldsymbol{x}(t+h). \tag{3.41}$$

It is said to have *convergence order* $p$ if $||\boldsymbol{\varepsilon}(h)|| \in \mathcal{O}(h^p)$.

A one–step method $\boldsymbol{\Phi}$ is called *stable* if there exists $\kappa < \infty$ such that $||\boldsymbol{\varepsilon}(h)|| \leqslant \kappa\, ||\boldsymbol{\tau}(h)||$.

△

**Proposition 3.1 (Consistency, Stability, Convergence of RK methods)**
A Runge–Kutta method is consistent if $\sum_{i=1}^{s} c_i = 1$. It is stable if the ODE's right hand side function $f$ is Lipschitz continuous. A consistent and stable method is convergent with order $p$ if it is consistent with order $p$. △

**Proof** Proofs can be found in any standard textbook on numerical analysis, e.g. [205]. □

For general coefficient matrices $\boldsymbol{B}$, (3.39) is an implicit definition of the vectors $\boldsymbol{k}_i$ that requires the solution of a system of nonlinear equations. A Runge–Kutta method is said to be of the *explicit* type if the rows of $\boldsymbol{B}$ can be reordered to form a lower triangular matrix of rank $s-1$. This allows for iterative computation of the values $\boldsymbol{k}_i$. Algorithm 3.5 implements a basic explicit Runge–Kutta method on a given discretization grid $\{t^k\}$ with $N+1$ grid points.

**Algorithm 3.5:** A basic explicit Runge–Kutta method.

**input** : $\boldsymbol{f}, \boldsymbol{x}_0, \boldsymbol{p}, \{t^k\}_{k=0:N}$
**output**: $\{\boldsymbol{\eta}^k\}$
$\boldsymbol{\eta}^0 = \boldsymbol{x}_0$;
**for** $k = 0 : N-1$ **do**
    $h^k = t^{k+1} - t^k$;
    **for** $i = 1 : s$ **do** $\boldsymbol{k}_i = \boldsymbol{f}\left(t^k + \alpha_i h^k, \boldsymbol{\eta}^k + h^k \sum_{j=1}^{s-1} B_{ij} \boldsymbol{k}_j, \boldsymbol{p}\right)$ ;
    $\boldsymbol{\eta}^{k+1} = \boldsymbol{\eta}^k + h^k \sum_{i=1}^{s} c_i \boldsymbol{k}_i$;
**end**

The method's coefficients are commonly given in the form of a so–called Butcher tableau [45], and the tableau of the classical explicit 4$^{\text{th}}$ order method is shown in figure 3.1. Other well–known explicit methods include Euler's method, Heun's method, and Simpson's rule.

$$\begin{array}{c|cccc} 0 & & & & \\ \frac{1}{2} & \frac{1}{2} & & & \\ \frac{1}{2} & 0 & \frac{1}{2} & & \\ 1 & 0 & 0 & 1 & \\ \hline & \frac{1}{6} & \frac{1}{3} & \frac{1}{3} & \frac{1}{6} \end{array}$$

Figure 3.1: Butcher tableau of the classical 4$^{\text{th}}$ order explicit RK method.

The choice of the discretization grid $\{t^k\}$ is critical for the accuracy of the obtained approximation to $\boldsymbol{x}(t^N)$. A number of authors have extended these classical methods to provide various error estimates, see e.g. [55, 65, 66, 213]. A continuous representation of the discrete solution $\{\boldsymbol{\eta}^k\}$ on the grid may be desirable in order to evaluate functions of the trajectory $\boldsymbol{x}(\cdot)$ independent of the discretization grid $\{t^k\}$. To address this issue, interpolation polynomials of a certain smoothness and with various error estimates have been described by e.g. [64, 163, 197]. For further details, we refer to these works and the references found therein.

### 3.4.2 Sensitivities of Initial Value Problems

The computation of the Lagrangian gradient and Hessian of (1.24) obtained from the direct multiple shooting discretization requires the computation of first- and second–order sensitivities of the IVP's solution with respect to the unknowns. In this section, we discuss several numerical approaches for the computation of forward and adjoint directional sensitivities. Emphasis is put on the discussion of methods that satisfy the principle of Internal Numerical Differentiation (IND) [32, 33].

#### Perturbed Trajectories

A straightforward approach to obtain sensitivities of the IVP's solution $\boldsymbol{x}(t_{\mathrm{f}})$ with respect to the initial values $\boldsymbol{x}_0$ and parameters $\boldsymbol{p}$ is to apply one of the finite difference procedures of section 3.3.2 to the IVP solution method. For one–sided finite differences in directions $\boldsymbol{d}^{\mathrm{x}} \in \mathbb{R}^{n^{\mathrm{x}}}$ and $\boldsymbol{d}^{\mathrm{p}} \in \mathbb{R}^{n^{\mathrm{p}}}$, we obtain

$$\begin{aligned} \boldsymbol{x}_{\mathrm{d}}(t_{\mathrm{f}};\ t_0,\boldsymbol{x}_0,\boldsymbol{p}) = {} & \frac{\boldsymbol{\eta}(t_{\mathrm{f}};\ t_0,\boldsymbol{x}_0+h\boldsymbol{d}^{\mathrm{x}},\boldsymbol{p}+h\boldsymbol{d}^{\mathrm{p}}) - \boldsymbol{\eta}(t_{\mathrm{f}};\ t_0,\boldsymbol{x}_0,\boldsymbol{p})}{h} \\ & + \mathcal{O}(\mathtt{tol}/h) + \mathcal{O}(h) \end{aligned} \tag{3.42}$$

where $\boldsymbol{\eta}(t_{\mathrm{f}};\ t,\boldsymbol{x})$ denotes the approximation of the IVP's solution $\boldsymbol{x}(t_{\mathrm{f}})$ obtained when starting at time $t$ with initial values $\boldsymbol{x}$, and $\mathtt{tol}$ is the local error bound, i.e., the integration tolerance. From (3.42) it can be seen that the optimal perturbation is $h = \mathtt{tol}^{\frac{1}{2}}$, which reveals that very tight integration tolerances, i.e., very fine discretization grids $\{t^k\}$, are required to obtain sufficiently precise approximations of the IVP sensitivities.

This approach is referred to as External Numerical Differentiation (END), as it treats the IVP solution method as a "black box" to which finite differences are applied *externally*. Indeed in (3.42) we have implicitly assumed the numerical method to be a sufficiently smooth mapping $\eta : \mathcal{T} \times \mathbb{R}^{n^x} \to \mathbb{R}^{n^x}, (t_0, \boldsymbol{x}_0) \mapsto \eta(t_f)$. For most numerical codes actually available, this assumption however is not satisfied. For the unperturbed initial values $(\boldsymbol{x}_0, \boldsymbol{p})$ and the perturbed ones $(\boldsymbol{x}_0 + h\boldsymbol{d}^x, \boldsymbol{p} + h\boldsymbol{d}^p)$, nondifferentiabilities and even discontinuities of the mapping $\eta$ are introduced by adaptive and possibly different choices of the discretization grids $\{t^k\}$, the use of pivoting in linear algebra subroutines, and the use of iterative or inexact solvers.

## The Principle of Internal Numerical Differentiation

The principle of Internal Numerical Differentiation (IND) was introduced in [32] and is based on the idea of differentiating the discretization scheme used to compute an approximation of the nominal solution. This encompasses the need to fix for the perturbed evaluations all adaptive components of the mapping $\eta$ to those settings that were used for the unperturbed evaluation. As this selective freeze of adaptive components requires access to the numerical code, the differentiation procedure actually becomes an *internal* component of that numerical code. The approximation (3.42) is in this case identical to the simultaneous solution of

$$\begin{aligned} \dot{\boldsymbol{x}}_d(t) &= \frac{\boldsymbol{f}(t, \boldsymbol{x}(t) + h\boldsymbol{x}_d, \boldsymbol{p} + h\boldsymbol{d}^p) - \boldsymbol{f}(t, \boldsymbol{x}(t), \boldsymbol{p})}{h}, \qquad t \in \mathcal{T}, \\ \boldsymbol{x}_d(t_0) &= \boldsymbol{d}. \end{aligned} \tag{3.43}$$

together with the IVP (3.37). The expected local approximation error for $\boldsymbol{x}_d$ is reduced to $\mathcal{O}(\texttt{tol}) + \mathcal{O}(\texttt{eps}^{\frac{1}{2}})$ if $h$ is chosen such that $h\|\boldsymbol{x}_d\| \in \mathcal{O}(\texttt{eps}^{\frac{1}{2}})$.

Thus IND results in a considerable *increase in precision* of the derivatives for low tolerances `tol` and delivers derivatives that are *consistent* with the discretization scheme. The possibility of using lower tolerances results in considerable *speedups* of the sensitivity generation. For applications of the IND principle to extrapolation schemes and linear multistep methods we refer to [34]. A discussion of arbitrary–order forward and adjoint sensitivity generation according to the IND principle in the variable order variable step size Backward Differentiation Formula (BDF) method `DAESOL-II` is found in [4].

### Variational Differential Equations

A second possibility of computing IVP sensitivities is to solve the variational differential equations along with the ODE. The sensitivity IVP into directions $\boldsymbol{d}^{\mathrm{x}} \in \mathbb{R}^{n^{\mathrm{x}}}$ and $\boldsymbol{d}^{\mathrm{p}} \in \mathbb{R}^{n^{\mathrm{p}}}$ reads

$$\begin{aligned}\dot{\boldsymbol{x}}_{\mathrm{d}}(t) &= \boldsymbol{f}_{\boldsymbol{x}}(t,\boldsymbol{x}(t),\boldsymbol{p})\cdot\boldsymbol{x}_{\mathrm{d}}(t)+\boldsymbol{f}_{\boldsymbol{p}}(t,\boldsymbol{x}(t),\boldsymbol{p})\cdot\boldsymbol{d}^{\mathrm{p}}, \qquad t\in\mathcal{T}, \qquad (3.44)\\ \boldsymbol{x}_{\mathrm{d}}(t_0) &= \boldsymbol{d}^{\mathrm{x}},\end{aligned}$$

assuming the initial value $\boldsymbol{x}_0$ to be independent of $\boldsymbol{p}$. Here again the IND principle is fulfilled only if both the IVP and the variational system (3.44) are solved simultaneously using the same adaptively chosen discretization scheme. For all linear methods it can be shown that this approach is the limit case of the finite difference approach for $h \to 0$, cf. [34]. The expected local approximation error for $\boldsymbol{x}_{\mathrm{d}}$ is further reduced to $\mathcal{O}(\mathtt{tol})$. This approach also allows to exploit sparsity patterns in the Jacobians $\boldsymbol{f}_{\boldsymbol{x}}$ and $\boldsymbol{f}_{\boldsymbol{p}}$ of the ODE system's right hand side.

### Adjoint Sensitivities by Automatic Differentiation

Adjoint sensitivities $\bar{\boldsymbol{x}}_{\mathrm{d}} \in \mathbb{R}^{n^{\mathrm{x}}}$, $\bar{\boldsymbol{p}}_{\mathrm{d}} \in \mathbb{R}^{n^{\mathrm{p}}}$ of the discretization scheme for IVP (3.37) into a direction $\bar{\boldsymbol{d}} \in \mathbb{R}^{n^{\mathrm{x}}}$ can be computed by applying the reverse mode of automatic differentiation, cf. section 3.3.4. For an explicit Runge–Kutta method (3.39), this yields for $N \geqslant k \geqslant 1$ the adjoint iteration schemes

$$\bar{\boldsymbol{x}}_{\mathrm{d}}^{N} \stackrel{\mathrm{def}}{=} \bar{\boldsymbol{d}}, \qquad \bar{\boldsymbol{x}}_{\mathrm{d}}^{k-1} \stackrel{\mathrm{def}}{=} \bar{\boldsymbol{x}}_{\mathrm{d}}^{k} - h^{k-1}\bar{\boldsymbol{\Phi}}^{\mathrm{x}}(t^{k-1},h^{k-1},\boldsymbol{x}^{k-1}), \qquad (3.45\mathrm{a})$$

$$\bar{\boldsymbol{p}}_{\mathrm{d}}^{N} \stackrel{\mathrm{def}}{=} \boldsymbol{0}, \qquad \bar{\boldsymbol{p}}_{\mathrm{d}}^{k-1} \stackrel{\mathrm{def}}{=} \bar{\boldsymbol{p}}_{\mathrm{d}}^{k} + h^{k-1}\bar{\boldsymbol{\Phi}}^{\mathrm{p}}(t^{k-1},h^{k-1},\boldsymbol{x}^{k-1}), \qquad (3.45\mathrm{b})$$

with the adjoint one–step method's generating functions $\bar{\boldsymbol{\Phi}}^{\mathrm{x}}$ and $\bar{\boldsymbol{\Phi}}^{\mathrm{p}}$ being

$$\bar{\boldsymbol{\Phi}}^{\mathrm{x}}(t,h,\boldsymbol{x}) \stackrel{\mathrm{def}}{=} \sum_{i=1}^{s} \bar{\boldsymbol{k}}_i^{\mathrm{x}}, \qquad \bar{\boldsymbol{k}}_i^{\mathrm{x}} \stackrel{\mathrm{def}}{=} -\bar{\boldsymbol{d}}_i^T \boldsymbol{f}_{\boldsymbol{x}}(t_i,\boldsymbol{x}_i,\boldsymbol{p}), \qquad (3.46\mathrm{a})$$

$$\bar{\boldsymbol{\Phi}}^{\mathrm{p}}(t,h,\boldsymbol{x}) \stackrel{\mathrm{def}}{=} \sum_{i=1}^{s} \bar{\boldsymbol{k}}_i^{\mathrm{p}}, \qquad \bar{\boldsymbol{k}}_i^{\mathrm{p}} \stackrel{\mathrm{def}}{=} \bar{\boldsymbol{d}}_i^T \boldsymbol{f}_{\boldsymbol{p}}(t_i,\boldsymbol{x}_i,\boldsymbol{p}). \qquad (3.46\mathrm{b})$$

Therein, we have used the shortcut notation $\bar{\boldsymbol{d}}_i$ to denote the adjoint direction of the $i$-th stage,

$$\bar{\boldsymbol{d}}_i \stackrel{\text{def}}{=} c_i \bar{\boldsymbol{x}}_{\text{d}}(t) - h \sum_{j=i+1}^{s} B_{ji} \bar{\boldsymbol{k}}_j^{\text{x}}. \tag{3.47}$$

A detailed derivation can be found in [34]. The evaluation points $(t_i, \boldsymbol{x}_i, \boldsymbol{p})$ of $\boldsymbol{f_x}$ are chosen as in the nominal scheme (3.39), which requires these point to be stored during the nominal solution for later reuse during the computation of adjoint directional sensitivities (*taping*). Adjoint directional derivatives $\bar{\boldsymbol{d}}_i^T \boldsymbol{f}_{\boldsymbol{x},\boldsymbol{p}}$ of the ODE system's right hand side can be computed by backward mode of automatic differentiation, section 3.3.4, by multiplication with a sparse Jacobian $\boldsymbol{f_x}$ if this can be done efficiently, or by directional finite difference approximations.

#### Adjoint Variational Differential Equations

The described approach can be interpreted as the integration of the adjoint system

$$\dot{\boldsymbol{x}}_{\text{d}}(t) = -\boldsymbol{f_x}^T(t, \boldsymbol{x}(t), \boldsymbol{p}) \cdot \boldsymbol{x}_{\text{d}}(t) \qquad t \in \mathcal{T}, \tag{3.48a}$$
$$\boldsymbol{x}_{\text{d}}(t_{\text{f}}) = \bar{\boldsymbol{d}},$$
$$\dot{\boldsymbol{p}}_{\text{d}}(t) = -\boldsymbol{f_p}^T(t, \boldsymbol{x}(t), \boldsymbol{p}) \cdot \bar{\boldsymbol{d}} \qquad t \in \mathcal{T}, \tag{3.48b}$$
$$\boldsymbol{p}_{\text{d}}(t_{\text{f}}) = \boldsymbol{0},$$

backwards in time using a special Runge–Kutta method. Certain methods such as the classical 4$^{\text{th}}$ order method are *self-adjoint* in the sense that $\bar{\boldsymbol{\Phi}}^{\text{x}}$ in (3.46a) actually is the generating function $\boldsymbol{\Phi}$ in (3.39), cf. [34]. The principle of IND is satisfied in this case. Note however that if system (3.48a) is solved forwards in time, i.e., simultaneously with the nominal IVP, then the obtained adjoint sensitivity in general is only an approximation of the inverse of the discretization scheme's adjoint and does not not satisfy the principle of IND, cf. again [34].

### 3.4.3 Second Order Sensitivities

The automatic differentiation approach of section 3.3.5 can be applied to Runge–Kutta methods to compute a directional second order sensitivity

$$\bar{\boldsymbol{d}}^T \frac{\mathrm{d}\boldsymbol{x}}{\mathrm{d}\boldsymbol{x}_0, \boldsymbol{p}}(t_\mathrm{f};\ t_0, \boldsymbol{x}_0, \boldsymbol{p})\, \boldsymbol{d} \tag{3.49}$$

into directions $\boldsymbol{d} \in \mathbb{R}^{n^\mathrm{x}+n^\mathrm{p}}$, $\bar{\boldsymbol{d}} \in \mathbb{R}^{n^\mathrm{x}}$ of the IVP solution. During the forward sweep the sensitivities $\boldsymbol{x}_\mathrm{d}(t_i^k)$ have to be stored in addition to the evaluation points $(t_i^k, \boldsymbol{x}_i^k)$, $1 \leqslant i \leqslant s$, $0 \leqslant k \leqslant N$. In the backward sweep, these are used to evaluate directional derivatives of $\boldsymbol{f}_{xx}$, $\boldsymbol{f}_{xp}$, and $\boldsymbol{f}_{pp}$.

The computation of a second order adjoint directional derivative as required for the computation of the exact Hessian of the Lagrangian requires a forward sweep through the Runge–Kutta scheme with $n^\mathrm{x} + n^\mathrm{q}$ canonical directions $\boldsymbol{d}$, and a single backward sweep with the Lagrangian multipliers $\bar{\boldsymbol{d}} = \boldsymbol{\lambda}_i^\mathrm{m}$ of the matching conditions.

## 3.5 Summary

In this chapter we have briefly reviewed the theory of nonlinear programming and have introduced SQP algorithms for the solution of NLPs in an active–set framework. Several different approximation schemes for the Hessian of the Lagrangian have been presented that avoid the costly computation of second order derivatives. The evaluation of the multiple shooting discretized OCP's matching condition constraints requires the solution of IVPs. To this end, we have briefly presented Runge–Kutta methods for non–stiff ODEs. Derivative and sensitivity generation for these methods according to the principle of IND have been discussed.

# 4 Mixed–Integer Real–Time Iterations

In this chapter we present an algorithmic framework for mixed–integer Nonlinear Model Predictive Control (NMPC) that builds on the MIOCP and NLP techniques and algorithms of the previous chapters. We present the real–time iteration scheme for moving horizons, cf. [51, 53] that allows to deal with challenges every real–time on–line optimal control algorithm has to face. Conditions for local contractivity of this algorithm are given. As one central part of this thesis we develop a new real–time iteration scheme for mixed–integer NMPC problems treated by the outer convexification reformulation of the previous chapter. Relying on local contractivity of the classical real–time iteration scheme, we give a proof of local contractivity for the mixed–integer case in the presence of an arbitrary rounding scheme. A sufficient condition coupling the contractivity statement to the sampling time is derived and an upper bound on the allowable sampling time is given that depends on Lipschitz and boundedness properties of the problem.

## 4.1 Real–Time Optimal Control

In an online optimal control application, we aim at solving not only one single instance of the optimal control problem, but rather a sequence of such problems. At every time instant $t$, an optimal control problem as described in chapter 1 with an initial value $\boldsymbol{x}_0(t)$ representing the differential state of the process has to be solved instantaneously. The resulting optimal control $\boldsymbol{u}^\star(t)$ is fed back to the process at time $t$. Hence, the trajectory $\boldsymbol{u}^\star(\cdot, \boldsymbol{x}_0(\cdot))$ constitutes an optimal feedback control.

### 4.1.1 Conventional NMPC Approach

Practical implementations obviously deviate from this idealized description in several points. First, it is not the optimal control problem that is solved at every time instant, but rather a finite dimensional approximation obtained by one of the approaches presented in chapter 1. Second, the solution $\boldsymbol{u}^\star(t)$

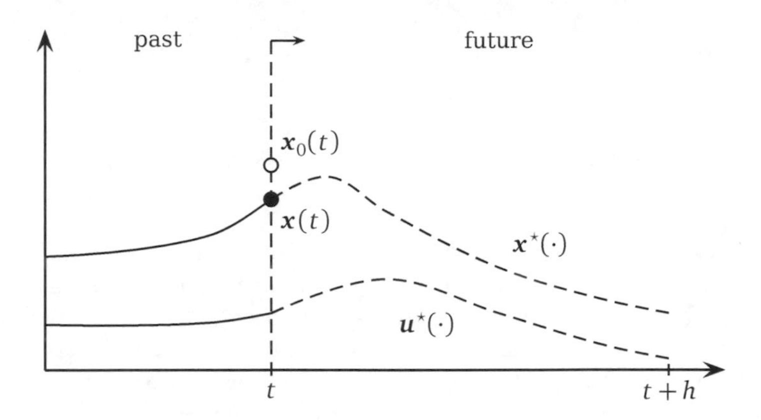


Figure 4.1: Idealized real–time optimal control. For the observed process state $x_0(t)$ the infinite–dimensional predictive control problem is solved instantaneously and the optimal control $u^\star(t)$ is fed back to the process at time $t$, steering it to an optimal state trajectory $x^\star(\cdot)$.

cannot be obtained instantaneously, but requires some computation time to pass until it is available for feedback to the process under control. The optimal control problems thus are solved only at discrete sampling times $t_0$, $t_1$, ... with durations $\delta t_0$, $\delta t_1$, ... at least as long as the solution of the next optimal control problem takes.

As a consequence, the optimal control is available for feedback with a delay only. If the sampling intervals are not short enough, this delay may impact the performance of the process controller to an unacceptable degree. On the other hand, the computational effort of solving an optimal control problem determines an unavoidable minimal delay. This delay in addition is not known in advance and theoretically may not even be bounded at all if iterative numerical methods are employed. Practical implementations therefore must use a prescribed stopping criterion.

The described conventional NMPC scheme may exhibit bad performance as sudden disturbances of the process state $x(t)$ can be accounted for in the optimal control problem only after a worst case delay of $\delta t$. In the meantime, a rather arbitrary and certainly suboptimal control unrelated to the actual process state is fed back to the process.

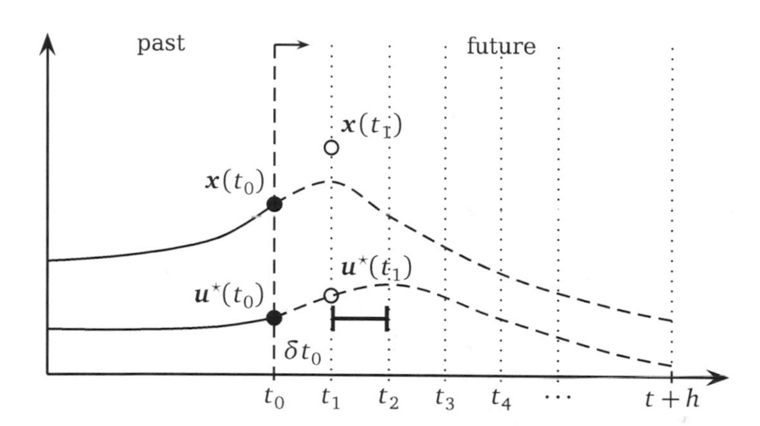


Figure 4.2: The conventional NMPC approach. The optimal feedback control $u^\star(t_0)$ for the process state $x(t_0)$ is available only after a computational delay of $\delta t_0$ and is fed back to the process for the time interval $[t_1, t_2]$. In the meantime, a rather arbitrary and suboptimal control is applied. The feedback control $u^\star(t_0)$ therefore is unrelated to the process state $x(t_1)$ that has since deviated from the optimal trajectory.

### 4.1.2 The Idea of Real–Time Iterations

It is the fundamental idea of the real–time iteration scheme, cf. [51, 53], to use a predictor $x^{\text{pred}}(t+\delta t)$ of the process state $x(\cdot)$ at the time instant $t+\delta t$ when the computed control $u^\star(t)$ is anticipated to be ready for feedback to the process. First, a careful initialization of the SQP method used to solve the discretized optimal control problem allows to reduce the computational effort to only one SQP iteration for each problem. Second, a separation of the computations necessary to obtain the feedback control into three phases is made, two of which can be completed in advance without knowledge of the actual process state. This approach reduces the feedback delay $\delta t$ by orders of magnitude.

Figure 4.3 depicts the real–time iteration scheme setting. A tangential prediction of the process state $x(t_1)$ based on the old one $x(t_0)$ provides an improved initializer for the computation of the control feedback $u(t_1)$. The deviation of this feedback from the actually optimal one $u^\star(t_1)$ is diminished as the feedback delay $\delta t_0$ is drastically reduced by the three–phase computation.

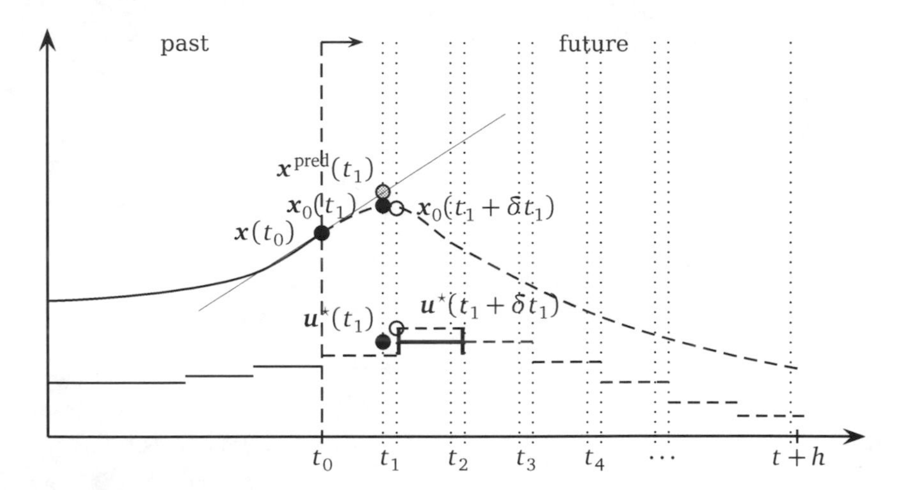


Figure 4.3: The real–time iteration scheme for a piecewise constant control discretization of the predictive control problem. A first order predictor of the process state in $t_1$ is used as an initialization for the computation of the feedback control $u^{\star}(t_1)$. The feedback delay $\delta t_1$ and thus the feedback duration for the unrelated and suboptimal control is drastically reduced by the three–phase computation. Consequentially, the actually optimal feedback control $u^{\star}(t_1+\delta t_1)$ does not deviate much from the computed one.

## 4.2 The Real–Time Iteration Scheme

Based on these ideas we now derive the real–time iteration scheme as a special SQP method solving a sequence of NLPs differing only in a homotopy parameter that enters the problems via a linear embedding.

### 4.2.1 Parametric Sequential Quadratic Programming

We start by considering the family of NLPs parameterized by a homotopy parameter $\tau \in \mathbb{R}$,

$$\begin{aligned} \min_{x,\tau} \quad & f(x,\tau) \\ \text{s.t.} \quad & 0 = g(x,\tau), \\ & 0 \leqslant h(x,\tau), \end{aligned} \tag{4.1}$$

under the assumptions of chapter 3. The primal–dual solution points $(\boldsymbol{x}^\star(\tau), \boldsymbol{\lambda}^\star(\tau), \boldsymbol{\mu}^\star(\tau))$ form a continuous and piecewise differentiable curve if they satisfy strong second order conditions for all $\tau$ and if certain technical assumptions hold in addition as shown by theorem 4.1.

**Theorem 4.1 (Piecewise Differentiability)**
Let $(\boldsymbol{x}^\star(0), \boldsymbol{\lambda}^\star(0), \boldsymbol{\mu}^\star(0))$ be a KKT point of (4.1) such that the strong second order conditions of theorem 3.4 are satisfied. Assume further that the step $\boldsymbol{\delta} \stackrel{\text{def}}{=} (\boldsymbol{\delta x}^\star, \boldsymbol{\delta \lambda}^\star, \boldsymbol{\mu}^\star)$ obtained from the solution of the Quadratic Program (QP)

$$\begin{aligned} \min_{\boldsymbol{\delta x}} \quad & \tfrac{1}{2}\boldsymbol{\delta x}^T \boldsymbol{B} \boldsymbol{\delta x} + \boldsymbol{\delta x}^T \boldsymbol{b} \\ \text{s.t.} \quad & \boldsymbol{0} = (\boldsymbol{g}_x \boldsymbol{\delta x})_t + (\boldsymbol{g}_x)_x^T \boldsymbol{\delta x}, \\ & \boldsymbol{0} = (\boldsymbol{h}_x^{\text{strong}})_t + (\boldsymbol{h}_x^{\text{strong}})_x^T \boldsymbol{\delta x}, \\ & \boldsymbol{0} \leqslant (\boldsymbol{h}_x^{\text{weak}})_t + (\boldsymbol{h}_x^{\text{weak}})_x^T \boldsymbol{\delta x}, \end{aligned} \tag{4.2}$$

satisfies strict complementarity. Herein $\boldsymbol{B} \stackrel{\text{def}}{=} L_{xx}$ and $\boldsymbol{b} \stackrel{\text{def}}{=} (L_x)_t$ are both evaluated in the KKT point $(\boldsymbol{x}^\star(0), \boldsymbol{\lambda}^\star(0), \boldsymbol{\mu}^\star(0))$ and $\boldsymbol{h}^{\text{strong}}$ and $\boldsymbol{h}^{\text{weak}}$ denote the restrictions of $\boldsymbol{h}$ onto the strongly resp. weakly active constraints.

Then there exists $\varepsilon > 0$ and a differentiable curve

$$\gamma : [0, \varepsilon] \to \mathbb{R}^{n^x} \times \mathbb{R}^{n^y} \times \mathbb{R}^{n^h}, \; \tau \mapsto (\boldsymbol{x}^\star(\tau), \boldsymbol{\lambda}^\star(\tau), \boldsymbol{\mu}^\star(\tau))$$

of KKT points for problem (4.1) in $\tau \in [0, \varepsilon]$. The one–sided derivative $\gamma_\tau(0)$ of the curve $\gamma$ in $\tau = 0^+$ is given by the QP step $\boldsymbol{\delta}$. △

**Proof** Proofs can be found in [51, 96]. □

### Tangential Predictor

We now consider the family (4.1) of NLPs including the additional constraint

$$\tau - \hat{\tau} = 0 \tag{4.3}$$

that fixes the free homotopy parameter to a prescribed value $\hat{\tau} \in \mathbb{R}$. The addition of this constraint to problem (4.1) allows for a transition between two subsequent NLPs of the parametric family (4.1) for different values of the homotopy parameter $\tau$. Derivatives with respect to $\tau$ are computed for the setup of the QP subproblem, such that a first order approximation of the solution manifold is provided already by the first SQP iterate.

**Theorem 4.2 (Exact Hessian SQP First Order Predictor)**
Let $(\boldsymbol{x}^\star(0), \boldsymbol{\lambda}^\star(0), \boldsymbol{\mu}^\star(0))$ be a KKT point of problem (4.3) in $\hat{\tau} = 0$ that satisfies theorem 3.4. Then the first step towards the solution of (4.3) in $\hat{\tau} > 0$ sufficiently small, computed by a full step exact Hessian SQP method starting in the KKT point, equals $\hat{\tau}\gamma_\tau(0)$. △

**Proof** A proof can be found in [51]. □

Due to its linearity, the additional embedding constraint (4.3) will already be satisfied after the first SQP iteration. The additional Lagrange multiplier introduced in the QP for this constraint does not play a role as the Hessian of the Lagrangian is unaffected.

### Parametric Quadratic Programming

The change in $\hat{\tau}$ is assumed to be sufficiently small in theorem 4.2 in order to ensure that the active set of the first SQP subproblem is identical to that of the solution point $\boldsymbol{x}^\star(0)$. Under this assumption theorem 4.2 holds even for points $\boldsymbol{x}^\star(0)$ in which the active set changes. In practice, active set changes will occur anywhere on the homotopy path from one homotopy parameter $\tau_0$ to another $\tau_1$, and this situation is best addressed by parametric QP methods. The approach here in to compute the first order predictor $(\delta\boldsymbol{x}^\star(\tau_1), \delta\boldsymbol{\lambda}^\star(\tau_1), \delta\boldsymbol{\mu}^\star(\tau_1))$ by solving the parametric QP on $\tau \in [0,1] \subset \mathbb{R}$ for its solution in $\tau = 1$,

$$\begin{aligned} \min_{\delta x} \quad & \tfrac{1}{2}\delta\boldsymbol{x}^T \boldsymbol{B}(\tau)\delta\boldsymbol{x} + \delta\boldsymbol{x}^T \boldsymbol{b}(\tau) \\ \text{s.t.} \quad & \boldsymbol{0} = \boldsymbol{g}_x(\tau)\delta\boldsymbol{x} + \boldsymbol{g}(\tau), \\ & \boldsymbol{0} \leqslant \boldsymbol{h}_x(\tau)\delta\boldsymbol{x} + \boldsymbol{h}(\tau), \end{aligned} \tag{4.4}$$

and initializing the solution process with the known solution

$$(\delta\boldsymbol{x}^\star(\tau_0), \delta\boldsymbol{\lambda}^\star(\tau_0), \delta\boldsymbol{\mu}^\star(\tau_0)) = (\boldsymbol{0}, \boldsymbol{0}, \boldsymbol{0})$$

in $\tau = 0$. The predictor of theorem 4.2 can then be understood as the initial piece of a piecewise affine linear homotopy path that potentially crosses multiple active set changes. This approach has been investigated for Linear Model Predictive Control (LMPC) in e.g. [67, 69]. We consider properties of Parametric Quadratic Programs (PQPs) and an active set method for their efficient solution in chapter 6 after the implications of our mixed–integer convexification and relaxation approach for the structure of these PQPs has been investigated in chapter 5.

### 4.2.2 Initial Value Embedding

We have seen from theorem 4.2 that the first iterate of the exact Hessian SQP method consistutes a first order tangential predictor of the solution of an NLP given an initializer in the neighborhood. The augmented problem formulation (4.3) was used for this purpose. In our setting of real–time optimal control, the parametric variable is the initial process state $\boldsymbol{s}_k$ that is fixed to the measured or estimated actual process state $\boldsymbol{x}_0(t^k)$ by a trivial linear equality constraint. For the direct multiple shooting discretization (1.24) this initial value embedding for problem $P(t^0)$ reads

$$\begin{aligned}
\min_{\boldsymbol{s},\boldsymbol{q}} \quad & \sum_{i=0}^{m} l_i(t_i, \boldsymbol{s}_i, \boldsymbol{q}_i) && && (4.5)\\
\text{s.t.} \quad & \boldsymbol{0} = \boldsymbol{x}_i(t_{i+1};\ t_i, \boldsymbol{s}_i, \boldsymbol{q}_i) - \boldsymbol{s}_{i+1}, && 0 \leqslant i \leqslant m-1, \\
& \boldsymbol{0} = \boldsymbol{r}_i^{\text{eq}}(t_i, \boldsymbol{s}_i, \boldsymbol{b}_i(t_i, \boldsymbol{q}_i)), && 0 \leqslant i \leqslant m, \\
& \boldsymbol{0} \leqslant \boldsymbol{r}_i^{\text{in}}(t_i, \boldsymbol{s}_i, \boldsymbol{b}_i(t_i, \boldsymbol{q}_i)), && 0 \leqslant i \leqslant m, \\
& \boldsymbol{0} \leqslant \boldsymbol{c}_i(t_i, \boldsymbol{s}_i, \boldsymbol{b}_i(t_i, \boldsymbol{q}_i)), && 0 \leqslant i \leqslant m, \\
& \boldsymbol{0} = \boldsymbol{s}_0 - \boldsymbol{x}_0(t^0).
\end{aligned}$$

Given an optimal solution $(\boldsymbol{s}^\star(\boldsymbol{x}_0(t^k)), \boldsymbol{q}^\star(\boldsymbol{x}_0(t^k)))$ to the discretized optimal control problem $P(t^k)$ for the process state $\boldsymbol{x}_0(t^k)$, the first full step computed by the exact Hessian SQP method for the neighboring problem with new embedded initial value $\boldsymbol{x}_0(t^{k+1})$ is a first order predictor of that NLP's solution.

### 4.2.3 Moving Horizons

So far we have assumed that an initializer $(\boldsymbol{s}_0, \boldsymbol{q}_0, \ldots, \boldsymbol{s}_{m-1}, \boldsymbol{q}_{m-1}, \boldsymbol{s}_m)$ required for the first iteration of the SQP algorithm on the multiple shooting discretization of the current problem is available. Depending on the characteristics of the problem instance under investigation, different strategies for obtaining an initializer from the previous real–time iteration's solution can be designed. In the context of real–time optimal control we are usually interested in moving prediction horizons that aim to approximate an infinite prediction horizon in a computationally tractable way. Consequentially, the optimal control problems of the sequence can be assumed to all have the same horizon length $m$ and differ only in the embedded initial value $\boldsymbol{x}_0(t)$ at sampling time $t$. Strategies for this case and also the case of shrinking horizons can be found in [51].

### Shift Strategy

The principle of optimality can be assumed to hold approximately also for the finite horizon if that horizon is chosen long enough such that the remaining costs can be safely neglected, e.g. because the controlled process has reached its desired state already inside the finite horizon. This is the motivation for the following shift strategy which uses the primal iterate

$$\boldsymbol{v}^k = (\boldsymbol{s}_0, \boldsymbol{q}_0, \ldots, \boldsymbol{s}_{m-1}, \boldsymbol{q}_{m-1}, \boldsymbol{s}_m),$$

the outcome of the first SQP iteration for the NLP with embedded initial value $\boldsymbol{x}_0(t^k)$ or equivalently the outcome of the $k$-th iteration of the real–time iteration scheme, to initialize the next iteration with $\boldsymbol{v}^{k+1}$ as follows,

$$\boldsymbol{v}^{k+1} \stackrel{\text{def}}{=} (\boldsymbol{s}_1, \boldsymbol{q}_1, \ldots, \boldsymbol{s}_{m-1}, \boldsymbol{q}_{m-1}, \boldsymbol{s}_m, \boldsymbol{q}_{m-1}^{\text{new}}, \boldsymbol{s}_m^{\text{new}}).$$

The new state and control values $\boldsymbol{q}_{m-1}^{\text{new}}$ and $\boldsymbol{s}_m^{\text{new}}$ can be chosen according to different strategies.

1. An obvious choice is $\boldsymbol{q}_{m-1}^{\text{new}} = \boldsymbol{q}^{\text{m}-1}$ and $\boldsymbol{s}_m^{\text{new}} = \boldsymbol{s}_m$. The only infeasibility introduced for iteration $k+1$ into a feasible trajectory from iteration $k$ is the potentially violated matching condition

$$\boldsymbol{x}(t_m; t_{m-1}, \boldsymbol{s}_{m-1}, \boldsymbol{q}_{m-1}) = \boldsymbol{s}_m.$$

2. A new state value $\boldsymbol{s}_m^{\text{new}}$ that satisfies the matching condition can be determined at the expense of solving an IVP on the new last shooting interval

$$\begin{aligned} \dot{\boldsymbol{x}}_{m-1}(t) &= \boldsymbol{f}(t, \boldsymbol{x}(t), \boldsymbol{b}(t, \boldsymbol{q}_{m-1}^{\text{new}}), \boldsymbol{p}), \qquad t \in [t_{m-1}, t_m] \subset \mathbb{R}, \\ \boldsymbol{x}_{m-1}(t_{m-1}) &= \boldsymbol{s}_{m-1}. \end{aligned}$$

employing the control parameter value $\boldsymbol{q}_{m-1}^{\text{new}}$. Path and terminal constraints may still be violated by this initialization.

Note however that constraint violations can be treated in a natural way in a direct multiple shooting framework and do not introduce additional difficulties.

The shift in the primal variables carries over to that of the dual variables, of the computed node function values and linearizations, and depending on the approximation scheme also to the Hessian.

### Warm Start Strategy

For short horizons and problem instances whose solution characteristics are dominated by terminal constraints or the Mayer–term objective that possibly are introduced to cover the neglegted cost of an infinite prediction horizon, the solutions of the subsequent problems may show similar features. In this case a better initialization of the NLP variables may be provided by $v^{k+1} = v^k$. The initial value embedding then introduces an infeasibility and the obtained first SQP iterate is exactly the first order predictor of theorem 4.2 if an exact Hessian method is used. In this warm start strategy, the sampling times also decouple from the shooting grid discretization, i.e., the discretization can be chosen coarser than the control feedback.

## 4.2.4 Local Feedback Laws

The real–time iteration scheme with warm start strategy can be seen as a generator of linear local feedback laws. In this sense, it's sampling rate can be fully decoupled from that of the control feedback. The linear feedback controller evaluates the most recently generated feedback law, a computationally very cheap operation, until the real–time iteration scheme has computed a more recent feedback law based on new function evaluations and linearizations [51]. This idea can be extended to adaptive reevaluation and relinearization of different parts of the discretized OCP, and has been realized in a so–called *multi–level iteration scheme*, see e.g. [37, 123].

Parts of the real–time iteration scheme can in priciple be transferred to interior–point methods, see e.g. [224, 225], although the excellent warm starting capabilities of active set methods are not maintained and the computed tangential predictors are inferior across active set changes [54].

### 4.2.5 Immediate Feedback

The quadratic subproblem solved for the step $\boldsymbol{\delta v} = (\boldsymbol{\delta s}, \boldsymbol{\delta q})$ in each real–time iteration on a moving horizon reads

$$\begin{aligned} \min_{\boldsymbol{\delta v}} \quad & \tfrac{1}{2}\boldsymbol{\delta v}^T \boldsymbol{B}(\boldsymbol{v})\boldsymbol{\delta v} + \boldsymbol{\delta v}^T \boldsymbol{b}(\boldsymbol{v}) \\ \text{s.t.} \quad & \boldsymbol{0} = \boldsymbol{g}_v(\boldsymbol{v})\boldsymbol{\delta v} + \boldsymbol{g}(\boldsymbol{v}), \\ & \boldsymbol{0} \leqslant \boldsymbol{h}_v(\boldsymbol{v})\boldsymbol{\delta v} + \boldsymbol{h}(\boldsymbol{v}), \\ & \boldsymbol{0} = \boldsymbol{s}_0 + \boldsymbol{\delta s}_0 - \boldsymbol{x}_0(t^k). \end{aligned} \tag{4.6}$$

It is noteworthy that problem (4.6) depends on the actual or estimated process state $\boldsymbol{x}_0(t^k)$ only in the initial value embedding constraint. This suggests that the overwhelming part of problem (4.5) can be set up without knowledge of $\boldsymbol{x}_0(t^k)$. In particular, the evaluation and linearization of all functions — except the linear embedding constraint — as well as the computation or approximation of the Hessian can be carried out *before* knowledge of $\boldsymbol{x}_0(t^k)$ is required. This includes the solution of the IVPs and the generation of IVPs sensitivities with respect to the unknowns $\boldsymbol{s}$ and $\boldsymbol{q}$, a task that usually contributes significantly to the control feedback delay in conventional NMPC. The follwing three–phase setup for a real–time iteration realizes this advantage:

1. *Preparation*: Prepare the solution of (4.6) as far as possible without knowledge of $\boldsymbol{x}_0(t^k)$.

2. *Feedback*: Solve problem (4.6) for the control feedback $\boldsymbol{\delta q}_0$ on the first multiple shooting interval and feed it back to the process.

3. *Transition*: Complete the solution of (4.6) for the full vector $\boldsymbol{\delta v}$ of unknowns and make the transition to the new QP subproblem for $t^{k+1}$.

In addition, any exploitation of the direct multiple shooting structure hidden in problem (4.6), such as by condensing methods [36, 51, 132, 167] or by block structured linear algebra [202] can take place at this stage of the algorithm. We will see in chapter 7 how the immediate feedback property can be extended to the QP solver's block structured linear algebra to further reduce the feedback delay.

## 4.3 Contractivity of Real–Time Iterations

Nominal stability of the closed–loop NMPC system when using the real–time iteration scheme to compute control feedback is of vital importance in order to ensure that the NMPC controller does not accidentally drive the controlled process away from a desired point of operation. This section sketches a proof of contractivity of the real–time iteration scheme due to [51]. In the next section we rely on this proof to derive conditions for local contractivity of our new mixed–integer extension to the real–time iteration scheme.

### 4.3.1 Fixed Control Formulation

In the previous section we have derived the real–time iteration scheme from a special SQP method for a family of parametric NLPs, performing one SQP iteration for each problem. We now consider the real–time iteration scheme first on a shrinking horizon of initial length $m$ and assume sampling times $\delta t^k$ that sum up to the horizon length $h$ such that the predictive control problem has been solved after the $k$-th feedback and the horizon has shrunk to zero length.

We assume throughout this section that the real–time iterations be started sufficiently close to a KKT point of the problems, such that the active set is known. Assuming strict complementarity to hold, we may restrict our presentation to equality–constrained problems. We denote by $P(t_0)$ the problem for the first control feedback on the full horizon of length $m$,

$$\begin{aligned} \min_{s,q} \quad & f(s,q) && (4.7) \\ \text{s.t.} \quad & 0 = g(s,q), \\ & 0 = s_0 - x_0(t_0), \end{aligned}$$

satisfying the standard assumptions of chapter 3. We further assume regularity of $g_s$ which allows us to regard the vector $s$ as dependent variables on the vector of unknowns $q$. After the first SQP iteration, the control feedback $u_0 = q_0 + \delta q_0$ is known and the next problem $P(t_1)$ is obtained from $P(t_0)$ by fixing the unknown $q_0$ to the realized value. Problem $P(t_1)$ thus comprises the following constraints:

$$\begin{aligned}
\mathbf{0} &= \boldsymbol{s}_0 - \boldsymbol{x}_0(t_0), && \text{(initial value embedding)} \\
\mathbf{0} &= \boldsymbol{x}_0(t_1;\; t_0, \boldsymbol{s}_0, \boldsymbol{q}_0) - \boldsymbol{s}_1, && \text{(matching condition)} \\
\mathbf{0} &= \boldsymbol{q}_0 - \boldsymbol{u}_0. && \text{(control fixation)}
\end{aligned} \tag{4.8}$$

Due to linearity, the correct values for $\boldsymbol{s}_0$ and $\boldsymbol{q}_0$ will be found already after the first SQP iteration on problem $P(t_1)$. This shows that fixing $\boldsymbol{q}_0$ is completely equivalent to solving $P(t_1)$ on a shrinked horizon of length $m-1$. For the case of Differential Algebraic Equation (DAE) dynamics see [51].

Based on this conclusion we define the sequence $\{P(t^k)\}_k$ of problems for $0 \leqslant k \leqslant m$ as

$$\begin{aligned}
\min_{\boldsymbol{s},\boldsymbol{q}} \quad & f(\boldsymbol{s},\boldsymbol{q}) \\
\text{s. t.} \quad & \mathbf{0} = \boldsymbol{g}(\boldsymbol{s},\boldsymbol{q}), \\
& \mathbf{0} = \boldsymbol{s}_k - \boldsymbol{x}_0(t^k), \\
& \mathbf{0} = \boldsymbol{q}_i - \boldsymbol{u}_i, \qquad 0 \leqslant i \leqslant k-1.
\end{aligned} \tag{4.9}$$

For $k \geqslant m$ no degrees of freedom remain in the problems $P(t^k)$ of this sequence, as the horizon length has shrunk to zero.

### 4.3.2 Contraction Constants

We are interested in the tractability of the problems $P(t^k)$ by a Newton–type method. In particular, the problems $P(t^k)$ should not become more and more difficult to solve as we keep adding constraints that fix more and more degrees of freedom. To this end, the contraction constants $\omega$ and $\kappa$ of section 3.2.5 need to be investigated for the problems $P(t^k)$. We first consider contractivity of the off–line problem $P(t_0)$ for the Newton–type iterations

$$\boldsymbol{y}^{k+1} \stackrel{\text{def}}{=} \boldsymbol{y}^k + \delta \boldsymbol{y}^k = \boldsymbol{y}^k - \boldsymbol{M}(\boldsymbol{y}^k)\boldsymbol{r}(\boldsymbol{y}^k) \tag{4.10}$$

wherein $\boldsymbol{r}(\boldsymbol{y}^k)$ denotes the Lagrangian gradient of problem (4.7) in the primal–dual point $\boldsymbol{y}^k = (\boldsymbol{x}^k, \boldsymbol{\lambda}^k)$ and $\boldsymbol{M}(\boldsymbol{y}^k)$ denotes the inverse of the KKT matrix $\boldsymbol{J}(\boldsymbol{y}^k)$, an approximation to the inverse of the exact Hessian KKT matrix in $\boldsymbol{y}^k$.

**Theorem 4.3 (Off–Line Convergence)**
Assume that the inverse $\boldsymbol{M}(\boldsymbol{y})$ of the KKT matrix $\boldsymbol{J}(\boldsymbol{y})$ exists and that $\boldsymbol{M}(\boldsymbol{y})$ be bounded by $\beta < \infty$ on the whole of a domain $\mathcal{D} \subseteq \mathbb{R}^{n^x} \times \mathbb{R}^{n^g}$,

$$\forall y \in \mathcal{D}: \; \|M(y)\| \leqslant \beta. \tag{4.11}$$

Let $J$ be Lipschitz on $\mathcal{D}$ with constant $\omega/\beta < \infty$,

$$\forall y_1, y_2 \in \mathcal{D}: \; \beta\|J(y_1) - J(y_2)\| \leqslant \omega\|y_1 - y_2\|. \tag{4.12}$$

Assume further that the deviation of the Hessian approximation $B(y)$ from the exact Hessian $L_{xx}(y)$ is bounded by $\kappa/\beta < 1/\beta$ on $\mathcal{D}$,

$$\forall y \in \mathcal{D}: \; \beta\|B(y) - L_{xx}(y)\| \leqslant \kappa < 1. \tag{4.13}$$

Finally, let the assumptions of the local contraction theorem 3.8 hold; let in particular the first step $\delta y^0$ starting in an initial guess $y^0$ be sufficiently small such that

$$\delta_0 \stackrel{\text{def}}{=} \kappa + \tfrac{\omega}{2} \left\|\delta y^0\right\| < 1. \tag{4.14}$$

Then the sequence of Newton–type iterates $\{y^k\}$ converges to a KKT point $y^\star$ whose primal part $x^\star$ is a strict local minimum of problem (4.7). △

**Proof** A proof can be found in [51]. □

The following theorem shows that this property carries over to all problems of the sequence.

**Theorem 4.4 (Contraction Constants of the On–Line Problems)**
Let the sufficient local convergence conditions (4.3) hold for problem $P(t_0)$ (4.7). Then the inverse Jacobian approximations $M$ for the problems $P(t^k)$ satisfy the $\kappa$– and $\omega$–conditions of definition 3.12 and 3.13 with the same values of $\kappa$ and $\omega$. △

**Proof** A proof can be found in [51]. □

### 4.3.3 Contractivity

We now consider contractivity of the sequence $\{y^k\}_k$ of iterates computed by the real–time iteration scheme as more and more control parameters are fixed in the sequence $\{P(t^k)\}_k$ of problems. Each problem $P(t^k)$ of this sequence is equivalent to finding the root $y^\star = (s^\star, q^\star, \lambda^\star)$ of the Lagrangian gradient function

$$r(y) \stackrel{\text{def}}{=} \begin{bmatrix} \mathfrak{P}_1^k(q - u^k) \\ \mathfrak{P}_2^k L_q(s, q, \lambda) \\ L_s(s, q, \lambda) \\ g(s, q) \end{bmatrix}. \tag{4.15}$$

Herein, the vector

$$\boldsymbol{u}^k = \begin{bmatrix} \boldsymbol{u}_0 & \dots & \boldsymbol{u}_{k-1} & \mathbf{0} & \dots & \mathbf{0} \end{bmatrix} \in \mathbb{R}^{m \cdot n^{\mathrm{q}}} \tag{4.16}$$

contains the determined feedback controls of the $k-1$ completed real–time iterations. The projectors $\mathfrak{P}_1^k$ and $\mathfrak{P}_2^k$ project onto the already determined fixed part $\boldsymbol{q}_0, \dots, \boldsymbol{q}_{k-1}$ and the unknown free part $\boldsymbol{q}_k, \dots, \boldsymbol{q}_{m-1}$ of the control parameters $\boldsymbol{q} \in \mathbb{R}^{m \cdot n^{\mathrm{q}}}$ respectively,

$$\begin{aligned} \mathfrak{P}_1^k \boldsymbol{q}^k &\stackrel{\text{def}}{=} \begin{bmatrix} \boldsymbol{q}_0^k & \cdots & \boldsymbol{q}_{k-1}^k & \mathbf{0} & \dots & \mathbf{0} \end{bmatrix}, \\ \mathfrak{P}_2^k \boldsymbol{q}^k &\stackrel{\text{def}}{=} \begin{bmatrix} \mathbf{0} & \dots & \mathbf{0} & \boldsymbol{q}_k^k & \cdots & \boldsymbol{q}_{m-1}^k \end{bmatrix}. \end{aligned} \tag{4.17}$$

**Theorem 4.5 (Contractivity of the Real–Time Iterations)**
Let the sufficient local contraction conditions of theorem 4.3 hold for the off–line problem $P(t_0)$. Then the sequence of real–time iterates contracts to a feasible point $\boldsymbol{y}^\star \in \mathcal{D}_0$. △

**Proof** Key to the proof of this theorem is the observation that the step

$$\delta \boldsymbol{y}^k \stackrel{\text{def}}{=} -\boldsymbol{M}^k(\boldsymbol{y}^k)\boldsymbol{r}^k(\boldsymbol{y}^k) \tag{4.18}$$

in the Newton–type iteration does not change if it is instead defined by

$$\delta \boldsymbol{y}^k \stackrel{\text{def}}{=} -\boldsymbol{M}^{k+1}(\boldsymbol{y}^k)\boldsymbol{r}^{k+1}(\boldsymbol{y}^k), \tag{4.19}$$

i.e., if the residual function and inverse Jacobian approximation of the next problem $P(t^{k+1})$ are used in which the feedback control parameter $\boldsymbol{q}_k$ to be determined in problem $P(t^k)$ under investigation *is already known*. To see this, we only need to show

$$\begin{aligned} \mathbf{0} &= \boldsymbol{r}^{k+1}(\boldsymbol{y}^k) + \boldsymbol{J}^{k+1}(\boldsymbol{y}^k)\delta \boldsymbol{y}^k \\ &= \begin{bmatrix} \mathfrak{P}_1^{k+1}(\boldsymbol{q}^k - \boldsymbol{q}^{k+1}) \\ \mathfrak{P}_2^{k+1} L_q(\boldsymbol{s}^k, \boldsymbol{q}^k) \\ L_s(\boldsymbol{s}^k, \boldsymbol{q}^k) \\ \boldsymbol{g}(\boldsymbol{s}^k, \boldsymbol{q}^k) \end{bmatrix} + \begin{bmatrix} \mathfrak{P}_1^{k+1} & \mathbf{0} & \mathbf{0} \\ \mathfrak{P}_2^{k+1} \boldsymbol{B}_{qq} & \mathfrak{P}_2^{k+1} \boldsymbol{B}_{qs}^T & \mathfrak{P}_2^{k+1} \boldsymbol{g}_q^T \\ \boldsymbol{B}_{qs} & \boldsymbol{B}_{ss} & \boldsymbol{g}_s^T \\ \boldsymbol{g}_q & \boldsymbol{g}_s & \mathbf{0} \end{bmatrix} \begin{bmatrix} \boldsymbol{q}^{k+1} - \boldsymbol{q}^k \\ \boldsymbol{s}^{k+1} - \boldsymbol{s}^k \\ -(\boldsymbol{\lambda}^{k+1} - \boldsymbol{\lambda}^k) \end{bmatrix}. \end{aligned} \tag{4.20}$$

Observe now that we have in particular for the already determined control parameters

$$\mathfrak{P}_1^{k+1}(\boldsymbol{q}^k - \boldsymbol{q}^{k+1}) + \mathfrak{P}_1^{k+1}(\boldsymbol{q}^{k+1} - \boldsymbol{q}^k) = \mathbf{0}. \tag{4.21}$$

Hence, any two subsequent real–time iterations can be treated as if they belonged *to the same problem* $P(t^{k+1})$. The contractivity property then directly follows from the local contraction theorem 3.8. The full proof can be found in [51].

□

## 4.4 Mixed–Integer Model Predictive Control

In this section we develop a new algorithm for mixed–integer model predictive control. It is established from a synthesis of the convexification and relaxation approach of chapter 2 and the real–time iteration scheme for NMPC of the previous sections. We transfer rounding schemes to real–time iterations in order to come up with a fast mixed–integer nonlinear model predictive control algorithm. We show for the first time that this algorithm can be interpreted as an inexact SQP method. Using this framework, conditions are derived under which contractivity of the mixed–integer real–time iteration scheme is guaranteed.

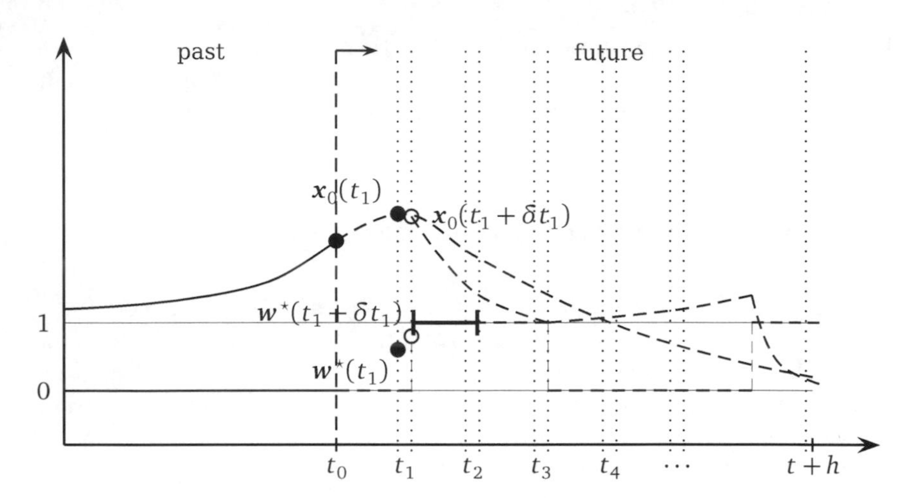


Figure 4.4: The mixed–integer real–time iteration scheme for a piecewise constant control discretization of a binary control function. The relaxed optimal feedback control $w^\star(t_1)$ is rounded to a binary feasible one according to a rounding scheme. The resulting deviation of the process state $x(t)$ from the computed optimal state trajectory can be bounded by a suitable choice of the discretization grid.

Figure 4.4 depicts the new mixed–integer variant of the real–time iteration scheme for the case of a single binary control. The computed integer feedback control $w(t_1)$ is available for feedback to the process after a very small feedback delay only, and thus does not deviate much from the theoretically optimal feedback $w^\star(t_1)$. It is rounded to a binary feasible control feedback that is applied to the process for the duration of the preparation phase of the next real–time iteration.

Several questions obviously arise. An appropriate rounding scheme should ensure binary or integer feasibility of the control while trying to approximate as closely as possible the relaxed optimal solution. Rounding the control results in a deviation of the state trajectory from the optimal one that is attainable only by a fractional control, and this deviation is coupled to the discretization grid granularity. Investigation of the contractivity properties of the resulting mixed–integer real–time iteration scheme is crucial in order to guarantee nominal stability. Ideally, the contractivity argument should yield an estimated upper bound on the allowable sampling times of the scheme.

### 4.4.1 Mixed–Integer Real–Time Iterations

For one modified Newton–type iteration of the mixed–integer real–time iteration scheme we introduce the notation

$$\begin{aligned} \boldsymbol{y}^{k+1} &\stackrel{\text{def}}{=} \boldsymbol{y}^k - \boldsymbol{M}^k(\boldsymbol{y}^k)\boldsymbol{r}^k(\boldsymbol{y}^k) + \boldsymbol{e}^k(\boldsymbol{y}^k), \\ \delta\tilde{\boldsymbol{y}}^k &\stackrel{\text{def}}{=} -\boldsymbol{M}^k(\boldsymbol{y}^k)\boldsymbol{r}^k(\boldsymbol{y}^k) + \boldsymbol{e}^k(\boldsymbol{y}^k). \end{aligned} \tag{4.22}$$

Herein, the vector $\boldsymbol{e}^k(\boldsymbol{y}^k)$ denotes the modification applied to the new iterate after the step $\boldsymbol{\delta y}^k$, required in order to obtain an integer feasible control part $\boldsymbol{q}_k^{k+1}$ of the new iterate $\boldsymbol{y}^{k+1}$. Assuming binary controls $\boldsymbol{q}_k$ and noting that after the Newton–type iteration for problem $P(t^k)$ only the $k$–th control is rounded, we have for the modification $\boldsymbol{e}^k(\boldsymbol{y}^k)$ the identity

$$\boldsymbol{e}^k(\boldsymbol{y}^k) = \begin{bmatrix} \boldsymbol{0} & \dots & \boldsymbol{0} & \boldsymbol{e}_k^k(\boldsymbol{y}_k^k) & \boldsymbol{0} & \dots & \boldsymbol{0} \end{bmatrix} \in \mathbb{R}^{m \cdot n^{\mathrm{q}}}, \tag{4.23}$$

wherein $\boldsymbol{e}_k^k(\boldsymbol{y}_k^k) \in \mathbb{R}^{n^{\mathrm{q}}}$ with bounds on $||\boldsymbol{e}_k^k(\boldsymbol{y}_k^k)||$ possibly depending on the actual rounding scheme applied. The introduced modification obviously affects the classical real–time iteration scheme argument in several ways as follows:

- The new fixed control $\boldsymbol{q}_k^{k+1} = \boldsymbol{q}_k^k + \delta\tilde{\boldsymbol{q}}_k^k$ to be determined after the Newton–type iteration for problem $P(t^k)$ can be found according to one of several possible rounding schemes.
- The fixation of $\boldsymbol{q}_k^{k+1}$ in problem $P(t^{k+1})$ to a possibly different value than that of the true next Newton–type iterate $\boldsymbol{q}^k + \boldsymbol{\delta q}^k$ introduces a deviation of the process state $\boldsymbol{x}(t^{k+1})$ from the predicted one $\boldsymbol{s}_{k+1}$.
- Note that in problem $P(t^{k+1})$ the correct values of $\boldsymbol{s}_k^k$ and $\boldsymbol{q}_k^k$ are still identified after the first Newton–type iteration for that problem as the two embedding constraints (4.8) remain linear.
- Nominal stability of this mixed–integer real–time iteration scheme is covered by the proofs due to [51], see theorem 4.5. Therein the solution of $P(t^{k+1})$ is identified with the solution of $P(t^k)$ assuming the controls $\boldsymbol{q}_k$ to already be fixed *to the outcome of the* Newton–*type step for* $P(t^k)$. Hence, the implications of modifying this Newton–type step in a rounding scheme need to be studied.

These topics are addressed one by one in the following sections.

### 4.4.2 Rounding Schemes

We are interested in obtaining after the $k$–th iteration from $\boldsymbol{q}_k^k$ and the Newton–type step $\boldsymbol{\delta q}_k^k$ a binary feasible new mixed–integer real–time iterate $\boldsymbol{q}_k^{k+1}$. Analogous to the rounding schemes of chapter 2 for off–line MIOCPs, two different on–line rounding schemes may be conceived that define a modified step $\boldsymbol{\delta\tilde{q}}_k^k$ to a new binary feasible feedback control $\boldsymbol{q}_k^{k+1}$.

#### Direct Rounding

The direct rounding scheme of definition 2.13 can immediately be transferred to the case of model–predictive control by defining the control part $\boldsymbol{\delta\tilde{q}}_k^k$ of the step $\boldsymbol{\delta\tilde{y}}^k$ to the new binary feasible mixed–integer real–time iterate $\boldsymbol{y}^{k+1}$ according to

$$\delta\tilde{q}_{kj}^k \stackrel{\text{def}}{=} \begin{cases} 1 - q_k^k & \text{if } q_{kj}^k + \delta q_{kj}^k > \frac{1}{2}, \\ -q_{kj}^k & \text{otherwise.} \end{cases} \qquad 1 \leqslant j \leqslant n^\Omega \tag{4.24}$$

Direct rounding does not take into account the accumulated past deviation from the relaxed optimal control trajectory incurred by rounding. This situation is improved by sum–up rounding as has been shown in chapter 2.

#### Sum–Up Rounding

The sum–up rounding strategy can be adapted to mixed–integer real–time iterations by introducing a memory of the past rounding decisions for $1 \leqslant j \leqslant n^\Omega$ according to

$$\delta\tilde{q}_{kj}^k \stackrel{\text{def}}{=} \begin{cases} 1 - q_{kj}^k & \text{if } \sum_{i=0}^{k} q_{ij}^i \delta t^i - \sum_{i=0}^{k-1} \tilde{q}_{ij}^i \delta t^i \geqslant \frac{1}{2}\delta t^k, \\ -q_{kj}^k & \text{otherwise.} \end{cases} \tag{4.25}$$

Note that this rounding strategy can be implemented without the need for a memory of all past mixed–integer real–time iterates. For the case of SOS1–constraints due to convexification of nonlinearly entering integer controls, these rounding strategies need to be adapted to maintain feasibility as shown in chapter 2.

### 4.4.3 Initial Value Embedding

The fixation of $q_k^{k+1}$ in problem $P(t^{k+1})$ to a possibly different value than that of the true next Newton–type iterate $q^k + \delta q^k$ introduces a deviation of the process state $x(t^{k+1})$ from the predicted one $s_{k+1}$. Two possibilities to address this can be thought of.

1. The deviation can be considered as an additional source of disturbances of the actual process under control. A worst–case estimate of the disturbance is provided by Gronwall's inequality using the rounded control $q_k^{k+1} = q_k^k + \delta\tilde{q}_k^k$,

$$\begin{aligned} &\left|\left|s_{k+1} - x_k(t^{k+1};\ t^k, s_k^k, q_k^{k+1}, p)\right|\right| \\ &\leqslant L \left|\left|\delta\tilde{q}_k^k - \delta q_k^k\right|\right| \exp(L(t^{k+1} - t^k)). \end{aligned} \tag{4.26}$$

   Here $L$ is the local Lipschitz constant of $f$. Hence for this approach to yield satisfactory performance of the mixed–integer real–time iteration scheme, the sampling time $\delta t^k$ and thus the granularity of the control discretization must be fine enough.

2. A new prediction of the state $s_{k+1}$ can be computed at the expense of one IVP solution on the $k$–th shooting interval using the rounded control $q_k^{k+1} = q_k^k + \delta\tilde{q}_k^k$

$$\begin{aligned} \dot{x}_k(t) &= f(t, x(t), q_k^{k+1}, p), \qquad t \in [t_k, t_{k+1}] \subset \mathbb{R}, \\ x_k(t_k) &= s_k^{k+1}. \end{aligned} \tag{4.27}$$

   This introduces an infeasibility of the matching condition constraint for node $s_{k+1}$ which can be naturally treated in the direct multiple shooting framework, as has already been noticed for the shift strategy.

### 4.4.4 Contractivity

We next investigate contractivity of the mixed–integer real–time iteration scheme. Nominal stability of this mixed–integer real–time iteration scheme is covered by the proofs due to [51] if the argument of identifying the solution of $P(t^{k+1})$ with the solution of $P(t^k)$ assuming the controls $q_k$ to be fixed *to the outcome of the* Newton–*type step for* $P(t^k)$ made in theorem 4.5 remains valid. In particular, if there exists an inexact inverse KKT matrix

$\tilde{M}^k(y^k)$ satisfying both the assumptions of the local contraction theorem 3.8 and the condition

$$\delta \tilde{y}^k = \delta y^k + e^k(y^k) = -\tilde{M}^k(y^k) r^k(y^k), \tag{4.28}$$

then the mixed–integer real–time iteration scheme can be written as Newton–type iterations

$$y^{k+1} = y^k - \tilde{M}^k(y^k) r^k(y^k) \tag{4.29}$$

and by the assumed properties of the inexact matrix $\tilde{M}^k$ local contractivity of the off–line problem, invariance of the contraction constants, and contractivity of the (then mixed–integer) real–time iterations follow from the presentation of the previous section.

**Theorem 4.6 (Local Contractivity of Mixed–Integer Real–Time Iterations)**
Assume that the modified Newton matrices $\tilde{M}^k(y^k)$ satisfying

$$\delta y^k + e^k(y^k) = -\tilde{M}^k(y^k) r^k(y^k) \tag{4.30}$$

exist, and let $\tilde{M}^k(y^k)$ satisfy the $\kappa$–condition with $\kappa < 1$. Then the mixed–integer real–time iteration scheme contracts to a feasible and integer feasible point $y^\star$. △

We first address the question of existence of the matrix $\tilde{M}^k(y^k)$ modifying the Newton–type step $\delta y^k$ such that it includes the rounding of the possibly fractional control component $q_k^{k+1}$ of the new iterate $y^{k+1}$. The following lemma shows that a construction of this modified matrix in $y^k$ is possible using symmetric updates of rank one or two.

**Lemma 4.1 (Construction of $\tilde{M}^k$)**
Let $y^k$ be a non–stationary point of problem $P(t^k)$ and let $M^k(y^k)$ denote the approximation to the inverse of the Jacobian of problem $P(t^k)$ in the point $y^k$, giving the Newton–type step

$$\delta y^k \stackrel{\text{def}}{=} -M^k(y^k) r^k(y^k).$$

Let $e^k(y^k)$ denote an arbitrary but fixed modification of the step $\delta y^k$. Then the modified matrix $\tilde{M}^k(y^k)$ satisfying

$$\delta \tilde{y}^k \stackrel{\text{def}}{=} \delta y^k + e^k(y^k) = -\tilde{M}^k(y^k) r^k(y^k)$$

exists. If $\boldsymbol{e}^k(\boldsymbol{y}^k)^T\boldsymbol{r}^k(\boldsymbol{y}^k) \neq 0$ it is given by the symmetric rank one update

$$\tilde{\boldsymbol{M}}^k(\boldsymbol{y}^k) \stackrel{\text{def}}{=} \boldsymbol{M}^k(\boldsymbol{y}^k) - \frac{\boldsymbol{e}^k(\boldsymbol{y}^k)\boldsymbol{e}^k(\boldsymbol{y}^k)^T}{\boldsymbol{e}^k(\boldsymbol{y}^k)^T\boldsymbol{r}^k(\boldsymbol{y}^k)}, \tag{4.31}$$

and it is otherwise given by the symmetric rank two (DFP) update

$$\tilde{\boldsymbol{M}}^k(\boldsymbol{y}^k) \stackrel{\text{def}}{=} \left(\boldsymbol{I} - \gamma^k\boldsymbol{\delta}\tilde{\boldsymbol{y}}^k\boldsymbol{r}^k(\boldsymbol{y}^k)^T\right)\boldsymbol{M}^k(\boldsymbol{y}^k)\left(\boldsymbol{I} - \gamma^k\boldsymbol{r}^k(\boldsymbol{y}^k)\boldsymbol{\delta}\tilde{\boldsymbol{y}}^{k^T}\right) - \gamma^k\boldsymbol{\delta}\tilde{\boldsymbol{y}}^k\boldsymbol{\delta}\tilde{\boldsymbol{y}}^{k^T}, \tag{4.32}$$

$$\gamma^k \stackrel{\text{def}}{=} 1/(\boldsymbol{\delta}\tilde{\boldsymbol{y}}^{k^T}\boldsymbol{r}^k(\boldsymbol{y}^k)).$$

△

**Proof** The proof is given by verifying the modified Newton–type step property. If $\boldsymbol{e}^k(\boldsymbol{y}^k)^T\boldsymbol{r}^k(\boldsymbol{y}^k) \neq 0$ we have

$$\begin{aligned}
&-\tilde{\boldsymbol{M}}^k(\boldsymbol{y}^k)\boldsymbol{r}^k(\boldsymbol{y}^k)\\
&= -\left(\boldsymbol{M}^k(\boldsymbol{y}^k) - \frac{\boldsymbol{e}^k(\boldsymbol{y}^k)\boldsymbol{e}^k(\boldsymbol{y}^k)^T}{\boldsymbol{e}^k(\boldsymbol{y}^k)^T\boldsymbol{r}^k(\boldsymbol{y}^k)}\right)\boldsymbol{r}^k(\boldsymbol{y}^k)\\
&= -\boldsymbol{M}^k(\boldsymbol{y}^k)\boldsymbol{r}^k(\boldsymbol{y}^k) + \boldsymbol{e}^k(\boldsymbol{y}^k)\frac{\boldsymbol{e}^k(\boldsymbol{y}^k)^T\boldsymbol{r}^k(\boldsymbol{y}^k)}{\boldsymbol{e}^k(\boldsymbol{y}^k)^T\boldsymbol{r}^k(\boldsymbol{y}^k)}\\
&= \boldsymbol{\delta y}^k + \boldsymbol{e}^k(\boldsymbol{y}^k)\\
&= \boldsymbol{\delta}\tilde{\boldsymbol{y}}^k.
\end{aligned}$$

Otherwise, if $\boldsymbol{\delta}\tilde{\boldsymbol{y}}^{k^T}\boldsymbol{r}^k(\boldsymbol{y}^k) \neq 0$ we have

$$\begin{aligned}
&-\tilde{\boldsymbol{M}}^k(\boldsymbol{y}^k)\boldsymbol{r}^k(\boldsymbol{y}^k)\\
&= \Big(\left(\boldsymbol{I} - \gamma^k\boldsymbol{\delta}\tilde{\boldsymbol{y}}^k\boldsymbol{r}^k(\boldsymbol{y}^k)^T\right)\boldsymbol{M}^k(\boldsymbol{y}^k)\left(\boldsymbol{I} - \gamma^k\boldsymbol{r}^k(\boldsymbol{y}^k)\boldsymbol{\delta}\tilde{\boldsymbol{y}}^{k^T}\right)\\
&\qquad -\gamma^k\boldsymbol{\delta}\tilde{\boldsymbol{y}}^k\boldsymbol{\delta}\tilde{\boldsymbol{y}}^{k^T}\Big)\boldsymbol{r}^k(\boldsymbol{y}^k)\\
&= \left(\boldsymbol{I} - \gamma^k\boldsymbol{\delta}\tilde{\boldsymbol{y}}^k\boldsymbol{r}^k(\boldsymbol{y}^k)^T\right)\boldsymbol{M}^k(\boldsymbol{y}^k)\left(\boldsymbol{r}^k(\boldsymbol{y}^k) - \boldsymbol{r}^k(\boldsymbol{y}^k)\right) + \boldsymbol{\delta}\tilde{\boldsymbol{y}}^k\\
&= \boldsymbol{\delta}\tilde{\boldsymbol{y}}^k.
\end{aligned}$$

If finally $\boldsymbol{\delta}\tilde{\boldsymbol{y}}^{k^T}\boldsymbol{r}^k(\boldsymbol{y}^k) = -\boldsymbol{\delta}\tilde{\boldsymbol{y}}^{k^T}\boldsymbol{J}^k(\boldsymbol{y}^k)\boldsymbol{\delta y} = 0$ then $\boldsymbol{\delta y} = \boldsymbol{0}$ by regularity of the exact KKT matrix $\boldsymbol{J}^k(\boldsymbol{y}^k)$ and $\boldsymbol{y}^k$ was a stationary point. □

For the next problem $P(t^{k+1})$, which is set up in the new iterate $\boldsymbol{y}^{k+1}$ with integer feasible control $\boldsymbol{q}_k^{k+1}$, we assume that control to be fixed by introduction of an additional constraint in the approximation $\tilde{\boldsymbol{J}}^{k+1}(\boldsymbol{y}^{k+1})$ and consequentially in its inverse $\tilde{\boldsymbol{M}}^{k+1}(\boldsymbol{y}^{k+1})$ as has been detailed in section 4.3.

Having shown existence of the modified matrix $\tilde{\boldsymbol{M}}^k$ producing the mixed–integer real–time iteration step, we are now interested in the contractivity properties of that matrix. To this end, we derive a condition under which the $\kappa$–condition of definition 3.13 is satisfied.

**Lemma 4.2 ($\kappa$–Condition for $\tilde{M}^k$)**
Let the the assumptions of theorem 4.3 be satisfied with constants $\beta$ and $\kappa_{\mathrm{M}}$. Assume further that there exists a constant $\varepsilon < 1 - \kappa_{\mathrm{M}}$ such that the exact Jacobian $\boldsymbol{J}^k(\boldsymbol{y}^k)$ satisfies

$$\beta||\boldsymbol{J}^k(\boldsymbol{y}^k)\boldsymbol{e}^k(\boldsymbol{y}^k)|| \leqslant \varepsilon||\delta\boldsymbol{y}^k + \boldsymbol{e}^k(\boldsymbol{y}^k)||. \tag{4.33}$$

Then the modified inverse $\tilde{\boldsymbol{M}}^k$ satisfies the $\kappa$–condition with $\kappa_{\tilde{\mathrm{M}}} = \kappa_{\mathrm{M}} + \varepsilon < 1$. △

**Proof** We evaluate the $\kappa$–condition (definition 3.13) for the modified matrix $\tilde{\boldsymbol{M}}^k$ assuming that $\boldsymbol{M}^k$ satisfies it with constant $\kappa_{\mathrm{M}}$,

$$\begin{aligned}
&||\tilde{\boldsymbol{M}}^k(\boldsymbol{y}^{k+1})\left(\boldsymbol{I} - \boldsymbol{J}^k(\boldsymbol{y}^k)\tilde{\boldsymbol{M}}^k(\boldsymbol{y}^k)\right)\boldsymbol{r}^k(\boldsymbol{y}^k)|| \\
&= ||\tilde{\boldsymbol{M}}^k(\boldsymbol{y}^{k+1})\left(\boldsymbol{r}^k(\boldsymbol{y}^k) - \boldsymbol{J}^k(\boldsymbol{y}^k)\tilde{\boldsymbol{M}}^k(\boldsymbol{y}^k)\boldsymbol{r}^k(\boldsymbol{y}^k)\right)|| \\
&= ||\tilde{\boldsymbol{M}}^k(\boldsymbol{y}^{k+1})\left(\boldsymbol{r}^k(\boldsymbol{y}^k) + \boldsymbol{J}^k(\boldsymbol{y}^k)\delta\tilde{\boldsymbol{y}}^k\right)|| \\
&= ||\tilde{\boldsymbol{M}}^k(\boldsymbol{y}^{k+1})\left(\boldsymbol{r}^k(\boldsymbol{y}^k) + \boldsymbol{J}^k(\boldsymbol{y}^k)\delta\boldsymbol{y}^k + \boldsymbol{J}^k(\boldsymbol{y}^k)\boldsymbol{e}^k(\boldsymbol{y}^k)\right)|| \\
&= ||\tilde{\boldsymbol{M}}^k(\boldsymbol{y}^{k+1})\left(\left(\boldsymbol{I} - \boldsymbol{J}^k(\boldsymbol{y}^k)\boldsymbol{M}^k(\boldsymbol{y}^k)\right)\boldsymbol{r}^k(\boldsymbol{y}^k) + \boldsymbol{J}^k(\boldsymbol{y}^k)\boldsymbol{e}^k(\boldsymbol{y}^k)\right)|| \\
&= ||\tilde{\boldsymbol{M}}^k(\boldsymbol{y}^{k+1})\left(\boldsymbol{I} - \boldsymbol{J}^k(\boldsymbol{y}^k)\boldsymbol{M}^k(\boldsymbol{y}^k)\right)\boldsymbol{r}^k(\boldsymbol{y}^k) + \tilde{\boldsymbol{M}}^k(\boldsymbol{y}^{k+1})\boldsymbol{J}^k(\boldsymbol{y}^k)\boldsymbol{e}^k(\boldsymbol{y}^k)|| \\
&\leqslant ||\tilde{\boldsymbol{M}}^k(\boldsymbol{y}^{k+1})\left(\boldsymbol{I} - \boldsymbol{J}^k(\boldsymbol{y}^k)\boldsymbol{M}^k(\boldsymbol{y}^k)\right)\boldsymbol{r}^k(\boldsymbol{y}^k)|| + ||\tilde{\boldsymbol{M}}^k(\boldsymbol{y}^{k+1})\boldsymbol{J}^k(\boldsymbol{y}^k)\boldsymbol{e}^k(\boldsymbol{y}^k)|| \\
&\leqslant \kappa_{\mathrm{M}}||\delta\tilde{\boldsymbol{y}}^k|| + ||\tilde{\boldsymbol{M}}^k(\boldsymbol{y}^{k+1})||\;||\boldsymbol{J}^k(\boldsymbol{y}^k)\boldsymbol{e}^k(\boldsymbol{y}^k)|| \\
&\leqslant \kappa_{\mathrm{M}}||\delta\tilde{\boldsymbol{y}}^k|| + \beta||\boldsymbol{J}^k(\boldsymbol{y}^k)\boldsymbol{e}^k(\boldsymbol{y}^k)||.
\end{aligned}$$

Under the assumptions made, this results in the estimate

$$||\tilde{\boldsymbol{M}}^k(\boldsymbol{y}^{k+1})\left(\boldsymbol{I} - \boldsymbol{J}^k(\boldsymbol{y}^k)\tilde{\boldsymbol{M}}^k(\boldsymbol{y}^k)\right)\boldsymbol{r}^k(\boldsymbol{y}^k)|| \leqslant (\kappa_{\mathrm{M}} + \varepsilon)||\delta\tilde{\boldsymbol{y}}^k||.$$

which is the $\kappa$–condition of definition 3.13 for $\tilde{\boldsymbol{M}}^k$ with a constant $\kappa_{\tilde{M}} = \kappa_M + \varepsilon$. □

It remains to be shown that the quantity $\boldsymbol{J}^k(\boldsymbol{y}^k)\boldsymbol{e}^k(\boldsymbol{y}^k)$ can indeed be made small enough such that $\varepsilon$ satisfies the bound $\kappa_{\mathrm{M}} + \varepsilon < 1$. The next lemma establishes a coupling of this quantity to the length $\delta t_k \stackrel{\text{def}}{=} t_{k+1} - t_k$ of the $k$–th multiple shooting interval $[t_k, t_{k+1}] \subset \mathbb{R}$ and shows that this quantity vanishes along with $\delta t$.

**Lemma 4.3 (Sampling Time $\kappa$–Condition for $\tilde{M}^k$)**
There exists a length $\delta \hat{t}_k > 0$ of the $k$–th multiple shooting interval such that the $\kappa$–condition of lemma 4.2 is satisfied for all $\delta t_k \in (0, \delta \hat{t}_k)$. △

**Proof** Observing the direct multiple shooting structure of the Jacobian $\boldsymbol{J}^k$ we compute $\boldsymbol{J}^k(\boldsymbol{y}^k)\boldsymbol{e}^k(\boldsymbol{y}^k)$ explicitly as follows in (4.34), see also chapter 7 for details on the block structure.

Observe now that after outer convexification the convex control multipliers $\boldsymbol{q}$ enter linearly in the objective, the dynamics, and the constraints. We thus have $\boldsymbol{B}^{\mathrm{qq}} = \boldsymbol{0}$. We further assume that bounds on the control parameter $\boldsymbol{q}$ to be rounded are inactive, as no rounding would occur otherwise. The SOS1 equality constraint is not affected by rounding. Finally, point and discretized path constraints treated by outer convexification are independent of $\boldsymbol{q}$. We thus have $\boldsymbol{R}^{\mathrm{q}}_k \boldsymbol{e}^{\mathrm{q}}_k = \boldsymbol{0}$.

$$\boldsymbol{J}^k(\boldsymbol{y}^k)\boldsymbol{e}^k(\boldsymbol{y}^k) = \begin{bmatrix} \ddots & & & & & & \\ & \boldsymbol{B}^{\mathrm{ss}}_k & \boldsymbol{B}^{\mathrm{sq}}_k & \boldsymbol{R}^{\mathrm{s}\,T}_k & \boldsymbol{G}^{\mathrm{s}\,T}_k & & \\ & \boldsymbol{B}^{\mathrm{qs}}_k & \boldsymbol{B}^{\mathrm{qq}}_k & \boldsymbol{R}^{\mathrm{q}\,T}_k & \boldsymbol{G}^{\mathrm{q}\,T}_k & & \\ & \boldsymbol{R}^{\mathrm{s}}_k & \boldsymbol{R}^{\mathrm{q}}_k & & & & \\ & \boldsymbol{G}^{\mathrm{s}}_k & \boldsymbol{G}^{\mathrm{q}}_k & & & -\boldsymbol{I} & \\ & & & & -\boldsymbol{I} & \boldsymbol{B}^{\mathrm{ss}}_{k+1} & \\ & & & & & & \ddots \end{bmatrix} \begin{bmatrix} \vdots \\ \boldsymbol{0} \\ \boldsymbol{e}^{\mathrm{q}}_k \\ \boldsymbol{0} \\ \boldsymbol{0} \\ \boldsymbol{0} \\ \vdots \end{bmatrix} = \begin{bmatrix} \vdots \\ \boldsymbol{B}^{\mathrm{sq}}_k \boldsymbol{e}^{\mathrm{q}}_k \\ \boldsymbol{B}^{\mathrm{qq}}_k \boldsymbol{e}^{\mathrm{q}}_k \\ \boldsymbol{R}^{\mathrm{q}}_k \boldsymbol{e}^{\mathrm{q}}_k \\ \boldsymbol{G}^{\mathrm{q}}_k \boldsymbol{e}^{\mathrm{q}}_k \\ \boldsymbol{0} \\ \vdots \end{bmatrix} \tag{4.34}$$

For the remaining two terms we derive estimates depending on $\delta t_k = t_{k+1} - t_k$. To this end, we denote by $\boldsymbol{x}_k(t_{k+1};\ t_k, \boldsymbol{s}_k, \boldsymbol{q}_k)$ the solution of the IVP

$$\begin{aligned} \dot{\boldsymbol{x}}(t) &= \sum_{i=1}^{n^{\mathrm{q}}} q_{k,i} \boldsymbol{f}(t, \boldsymbol{x}(t), \boldsymbol{\omega}^i), \qquad t \in [t_k, t_{k+1}], \\ \boldsymbol{x}(t_k) &= \boldsymbol{s}_k. \end{aligned}$$

and introduce integral forms for the linear time–variant IVPs for the sensitivities of the IVP solution with respect to $\boldsymbol{s}_k$

$$\frac{\mathrm{d}\boldsymbol{x}(t_{k+1})}{\mathrm{d}\boldsymbol{s}_k} = \boldsymbol{I} + \int_{t_k}^{t_{k+1}} \sum_{i=1}^{n^{\mathrm{q}}} q_{k,i} \boldsymbol{f}_{\boldsymbol{x}}(t, \boldsymbol{x}(t), \boldsymbol{\omega}^i) \frac{\mathrm{d}\boldsymbol{x}(t)}{\mathrm{d}\boldsymbol{s}_k}\, \mathrm{d}t,$$

and for the sensitivities of the IVP solution with respect to $\boldsymbol{q}_k$,

$$\frac{\mathrm{d}\boldsymbol{x}(t_{k+1})}{\mathrm{d}\boldsymbol{q}_k} = \int_{t_k}^{t_{k+1}} \sum_{i=1}^{n^{\mathrm{q}}} \left( q_{k,i} \boldsymbol{f}_{\boldsymbol{x}}(t, \boldsymbol{x}(t), \boldsymbol{\omega}^i) \frac{\mathrm{d}\boldsymbol{x}(t)}{\mathrm{d}\boldsymbol{q}_k} + \boldsymbol{f}(t, \boldsymbol{x}(t), \boldsymbol{\omega}^i) \boldsymbol{e}_i^T \right) \,\mathrm{d}t.$$

For their solutions we observe by Lipschitz continuity of $\boldsymbol{f}$ that for $t_{k+1} \to t_k$ it holds that

$$\frac{\mathrm{d}\boldsymbol{x}(t_{k+1})}{\mathrm{d}\boldsymbol{s}_k} \to \boldsymbol{I} \quad \text{and} \quad \frac{\mathrm{d}\boldsymbol{x}(t_{k+1})}{\mathrm{d}\boldsymbol{q}_k} \to \boldsymbol{0}.$$

We evaluate the two remaining nonzero terms in the Jacobian vector product. First, we have for $\boldsymbol{G}_k^{\mathrm{q}}$

$$\boldsymbol{G}_k^{\mathrm{q}} = \frac{\mathrm{d}}{\mathrm{d}\boldsymbol{q}} \left( \boldsymbol{x}_k(t_{k+1};\ t_k, \boldsymbol{s}_k, \boldsymbol{q}_k) - \boldsymbol{s}_{k+1} \right) = \frac{\mathrm{d}\boldsymbol{x}_k(t_{k+1})}{\mathrm{d}\boldsymbol{q}_k} \tag{4.35}$$

that $\boldsymbol{G}_k^{\mathrm{q}} \boldsymbol{e}_k^{\mathrm{q}} \to \boldsymbol{0} \boldsymbol{e}_k^{\mathrm{q}} = \boldsymbol{0}$ for $\delta_k \stackrel{\mathrm{def}}{=} t_{k+1} - t_k \to 0$. Second, we have for $\boldsymbol{B}_k^{\mathrm{sq}}$ that

$$\begin{aligned}
\boldsymbol{B}_k^{\mathrm{sq}} &= \frac{\mathrm{d}^2}{\mathrm{d}\boldsymbol{s}_k \mathrm{d}\boldsymbol{q}_k} \left( \left( \int_{t_k}^{t_{k+1}} l(t, \boldsymbol{x}(t), \boldsymbol{q}_k) \,\mathrm{d}t - \boldsymbol{\lambda}^T \left( \boldsymbol{x}_k(t_{k+1};\ t_k, \boldsymbol{s}_k, \boldsymbol{q}_k) - \boldsymbol{s}_{k+1} \right) \right) \right) \\
&= \int_{t_k}^{t_{k+1}} l_{\boldsymbol{x}\boldsymbol{q}}(t, \boldsymbol{x}(t), \boldsymbol{q}_k) \frac{\mathrm{d}\boldsymbol{x}(t)}{\mathrm{d}\boldsymbol{s}_k} \,\mathrm{d}t - \boldsymbol{\lambda}^T \frac{\mathrm{d}^2 \boldsymbol{x}_k(t_{k+1})}{\mathrm{d}\boldsymbol{s}_k \mathrm{d}\boldsymbol{q}_k}.
\end{aligned} \tag{4.36}$$

We further investigate the second term contributing to $\boldsymbol{B}_k^{\mathrm{sq}}$ which is again defined by a linear time–variant IVP,

$$\begin{aligned}
\frac{\mathrm{d}^2 \boldsymbol{x}_k(t_{k+1})}{\mathrm{d}\boldsymbol{s}_k \mathrm{d}\boldsymbol{q}_k} &= \frac{\mathrm{d}}{\mathrm{d}\boldsymbol{s}_k} \int_{t_k}^{t_{k+1}} \sum_{i=1}^{n^{\mathrm{q}}} \left( q_{k,i} \boldsymbol{f}_{\boldsymbol{x}}(t, \boldsymbol{x}(t), \boldsymbol{\omega}^i) \frac{\mathrm{d}\boldsymbol{x}(t)}{\mathrm{d}\boldsymbol{q}_k} + \boldsymbol{f}(t, \boldsymbol{x}(t), \boldsymbol{\omega}^i) \boldsymbol{e}_i^T \right) \,\mathrm{d}t \\
&= \int_{t_k}^{t_{k+1}} \sum_{i=1}^{n^{\mathrm{q}}} \Bigg( q_{k,i} \frac{\mathrm{d}\boldsymbol{x}(t)}{\mathrm{d}\boldsymbol{s}_k}^T \boldsymbol{f}_{\boldsymbol{x}\boldsymbol{x}}(t) \frac{\mathrm{d}\boldsymbol{x}(t)}{\mathrm{d}\boldsymbol{q}_k} \\
&\qquad + q_{k,i} \boldsymbol{f}_{\boldsymbol{x}}(t) \frac{\mathrm{d}^2 \boldsymbol{x}(t)}{\mathrm{d}\boldsymbol{s}_k \mathrm{d}\boldsymbol{q}_k} + \boldsymbol{f}_{\boldsymbol{x}}(t) \frac{\mathrm{d}\boldsymbol{x}(t)}{\mathrm{d}\boldsymbol{s}_k} \boldsymbol{e}_i^T \Bigg) \,\mathrm{d}t,
\end{aligned} \tag{4.37}$$

wherein we have omitted all but the leading argument of $\boldsymbol{f}$. The limit behavior for $t_{k+1} \to t_k$ is

$$\frac{\mathrm{d}^2 \boldsymbol{x}_k(t_{k+1})}{\mathrm{d}\boldsymbol{s}_k \mathrm{d}\boldsymbol{q}} \to \mathbf{0}.$$

Again by continuity of the integrand, we observe for $\delta_k \stackrel{\text{def}}{=} t_{k+1} - t_k \to 0$ that $\boldsymbol{B}_k^{\text{sq}} \boldsymbol{e}_k^{\text{q}} \to \mathbf{0}\boldsymbol{e}_k^{\text{q}} = \mathbf{0}$.

This shows $||\boldsymbol{J}^k(\boldsymbol{y}^k)\boldsymbol{e}^k|| \to 0$ for $\delta t \to 0$. By continuity in $t_{k+1}$ there exists a shooting interval length $\delta\hat{t}$ such that for all $\delta t < \delta\hat{t}$ it holds that $\beta \left|\left|\boldsymbol{J}^k(\boldsymbol{y}^k)\boldsymbol{e}^k\right|\right| < \varepsilon$ for any $\varepsilon > 0$. This completes the proof. □

Hence, by the proved lemmas, all assumptions of the contractivity theorem 4.6 for our mixed–integer real–time iteration scheme can actually be satisfied. In particular, for a shifting strategy we would need to chose a suitably fine choice of the control discretization, leading to a short sampling time during which rounding of the control causes a deviation of the process trajectory from the predictor. For a warm start strategy, the control discretization is independent of the sampling time. As long as the sampling time is short enough to satisfy lemma 4.3, the discretization may be chosen coarser.

### 4.4.5 Upper Bounds on the Sampling Time

We can extend the proof of the last theorem in order to derive a simple and accessible upper bound on the sampling time $\delta t$ to be chosen for the mixed–integer real–time iteration scheme. We only consider problems with vanishing cross–term $\boldsymbol{B}_k^{\text{sq}}$ of the Hessian of the Lagrangian here.

**Theorem 4.7 (Sampling Time for Contractivity of $\tilde{M}^k$)**
Let the approximate inverse of the KKT matrix in the new iterate be bounded by $\beta > 0$,

$$||\tilde{\boldsymbol{M}}^k(\boldsymbol{y})|| \leqslant \beta \quad \forall \boldsymbol{y} \in \mathcal{D},$$

and let the ODE system's right hand sides after outer convexification $\boldsymbol{f}(t, \boldsymbol{x}(t), \boldsymbol{\omega}^i)$ be bounded by $b_{\tilde{f}} > 0$,

$$||\boldsymbol{f}(t, \boldsymbol{x}, \boldsymbol{\omega}^i)|| \leqslant b_{\tilde{f}} \quad \forall t \in \mathcal{T},\ \boldsymbol{x} \in \mathcal{X} \subset \mathbb{R}^{n^x},$$

and satisfy a Lipschitz condition

$$||\boldsymbol{f}(t, \boldsymbol{x}_1, \boldsymbol{\omega}^i) - \boldsymbol{f}(t, \boldsymbol{x}_2, \boldsymbol{\omega}^i)|| \leqslant A_{\tilde{f}} ||\boldsymbol{x}_1 - \boldsymbol{x}_2|| \quad \forall t \in \mathcal{T},\ \boldsymbol{x}_1, \boldsymbol{x}_2 \in \mathcal{X} \subset \mathbb{R}^{n^x}.$$

Assume further that $\boldsymbol{B}_k^{\mathrm{sq}} = \boldsymbol{0}$. Then the term $\left|\left|\tilde{\boldsymbol{M}}^k(\boldsymbol{y}^{k+1})\boldsymbol{J}^k(\boldsymbol{y}^k)\boldsymbol{e}^k(\boldsymbol{y}^k)\right|\right|$ satisfies with $\delta t \stackrel{\text{def}}{=} t_{k+1} - t_k$

$$\left|\left|\tilde{\boldsymbol{M}}^k(\boldsymbol{y}^{k+1})\boldsymbol{J}^k(\boldsymbol{y}^k)\boldsymbol{e}^k(\boldsymbol{y}^k)\right|\right| \leqslant \beta b_{\tilde{\mathrm{f}}} \delta t \exp(A_{\tilde{\mathrm{f}}} \delta t). \tag{4.38}$$

△

**Proof** In order to consider estimates of the nonzero vector component $\boldsymbol{G}_k^{\mathrm{q}}\boldsymbol{e}_k^{\mathrm{q}}$ we first require the explicit solution to the sensitivity IVPs w.r.t. $\boldsymbol{q}_k$. Observe that the explicit solution to the linear time varying IVP in integral form

$$\boldsymbol{u}(t_1) = \boldsymbol{0} + \int_{t_0}^{t_1} \boldsymbol{A}(t)\boldsymbol{u}(t) + \boldsymbol{b}(t)\,\mathrm{d}t, \quad t_1 > t_0$$

is

$$\boldsymbol{u}(t_1) = \int_{t_0}^{t_1} \boldsymbol{b}(t) \exp\left(-\int_{t_0}^{t} \boldsymbol{A}(s)\,\mathrm{d}s\right)\mathrm{d}t \exp\left(\int_{t_0}^{t_1} \boldsymbol{A}(t)\,\mathrm{d}t\right), \quad t_1 > t_0.$$

By comparison of terms this gives for $\boldsymbol{G}_k^{\mathrm{q}}$ in (4.35) the explicit solution

$$\boldsymbol{G}_k^{\mathrm{q}} = \int_{t_k}^{t_{k+1}} \boldsymbol{f}(t)\boldsymbol{e}_i^T \exp\left(-\int_{t_k}^{t} q_{k,i}\boldsymbol{f}_{\boldsymbol{x}}(s)\,\mathrm{d}s\right)\mathrm{d}t \exp\left(\int_{t_k}^{t_{k+1}} \boldsymbol{f}_{\boldsymbol{x}}(t)\,\mathrm{d}t\right).$$

Using the estimate $\left|\left|\boldsymbol{u}(t_1)\right|\right| \leqslant \int_{t_0}^{t_1} ||\boldsymbol{A}(t)||\,||\boldsymbol{u}(t)|| + ||\boldsymbol{b}(t)||\,\mathrm{d}t$ for the general explicit solution, we have for the first component's norm $\left|\left|\boldsymbol{G}_k^{\mathrm{q}}\right|\right|$ that

$$\left|\left|\boldsymbol{G}_k^{\mathrm{q}}\boldsymbol{e}_k^{\mathrm{q}}\right|\right| \leqslant b_{\tilde{\mathrm{f}}} \delta t \exp(A_{\tilde{\mathrm{f}}} \delta t),$$

from $\left|\left|\boldsymbol{e}^k(\boldsymbol{y}^k)\right|\right| \leqslant 1$ guaranteed by rounding, and by using the assumed bound and Lipschitz constant on $\boldsymbol{f}(\cdot)$. □

We will see a practical application of this statement in chapter 9.

It is worth noting that the above contractivity statements and estimates crucially rely on the direct multiple shooting parameterization of the IVP that limits the effect of rounding $\boldsymbol{q}_k^k$ to the interval $[t^k, t^k + \delta t^k]$. A single shooting method would have to solve the IVP after rounding on the entire remaining horizon $[t^k, t_{\mathrm{f}}]$. Even if that solution existed, convergence of the Boundary Value Problem (BVP) might suffer from the potentially exponential deviation of the state trajectory on the long time horizon.

### 4.4.6 Mixed–Integer Real–Time Iteration Algorithm

We can summarize the developed mixed–integer real–time iteration scheme in the following algorithm:

1. Prepare the solution of problem $P(t^k)$ as far as possible without knowledge of the actual observed or estimated process state $\boldsymbol{x}_0(t^k)$.
2. As soon as an observation or estimate $\boldsymbol{x}_0(t^k)$ is available, compute the feedback control $\boldsymbol{q}_k$ by completing the prepared Newton–type iteration using the current iterate $\boldsymbol{y}^k$ as an initializer. This yields a first order corrector $\delta\boldsymbol{y}^k$.
3. Apply a rounding scheme to the potentially fractional feedback control parameters $\boldsymbol{q}_{k,0}$ on the first shooting interval and feed the rounded control back to the process.
4. Define the new primal–dual iterate $\boldsymbol{y}^{k+1}$ that will serve as an initializer for the next problem $P(t^{k+1})$ using a shift or warm start strategy.
5. Increase $k$ by one and start over in point 1.

As seen in section 4.2, step 2. of the above algorithm essentially involves the efficient solution of a QP or a parametric QP making use of the current iterate as an initializer. We need to make sure, though, that this solution completes as fast as required to maintain local contractivity and thus nominal stability of the scheme. Appropriate techniques are presented in chapters 7 and 8.

## 4.5 Summary

In this chapter we have presented the real–time iteration scheme for fast NMPC using active–set based NLP methods such as Sequential Quadratic Programming to repeatedly compute feedback controls based on a multiple shooting discretization of the predictive control problem. The underlying algorithmic ideas have been described and NLP theory has been studied as necessary to sketch a proof of local contractivity of the real–time iteration scheme. We have developed a new extension to the real–time iteration scheme that is applicable to mixed–integer NMPC problems. We have further derived a new sufficient condition for local contractivity of the mixed–integer real–time iterations. The proof relies on the existing contractivity statement for the continuous case. There, such a statement could be shown under the assumption that the repeatedly fixed controls are indeed fixed onto the exact outcome of the last Newton–type step. To make this framework accessible, we showed existence of a theoretical modification to the approximate inverse KKT matrix that provides the Newton–type step including the contribution due to a rounding scheme being applied afterwards. The proof of existence relies on a symmetric rank one or Broyden type rank two update. We derived a sufficient condition for the modified KKT matrix to satisfy the $\kappa$–condition for local contractivity, and showed that this condition can actually be satisfied by a sampling time chosen small enough. A dependency on the granularity of the binary or integer control discretization along the lines of chapter 2 has thus been established also for mixed–integer Model Predictive Control. Vice versa, we showed that, if appropriate bounds and Lipschitz constants on certain model functions are available, an upper bound on the sampling time can be derived.

# 5 Outer Convexification of Constraints

In this chapter, we take a closer look at combinatorial issues raised by applying outer convexification to constraints that directly depend on a discrete control. We describe several undesirable phenomena that arise when treating such problems with a Sequential Quadratic Programming (SQP) method, and give explanations based on the violation of Linear Independence Constraint Qualification (LICQ). We show that the Nonlinear Programs (NLPs) can instead be treated favorably as Mathematical Programs with Vanishing Constraints (MPVCs), a class of challenging problems with complementary constraints and nonconvex feasible set. We give a summary of a Lagrangian framework for MPVCs due to [3] and others, which allows for the design of derivative based descent methods for the efficient solution of MPVCs in the next chapter.

## 5.1 Constraints Depending on Integer Controls

We have seen in chapter 2 that for Mixed–Integer Optimal Control Problems (MIOCPs) with integer or binary controls that directly enter one or more of the constraints, rounding the relaxed optimal solution to a binary one is likely to violate these constraints.

### 5.1.1 The Standard Formulation after Outer Convexification

Having applied outer convexification to the binary vector of controls $\boldsymbol{w}(\cdot) \in \{0,1\}^{n^{\mathrm{w}}}$, a constraint $\boldsymbol{g}$ directly depending on $\boldsymbol{w}(t)$ can be written as constraining the convexified residual

$$\sum_{i=1}^{n^{\mathrm{w}}} \left( \tilde{w}_i(t) \cdot \boldsymbol{g}(t, \boldsymbol{x}(t), \boldsymbol{\omega}^i, \boldsymbol{p}) \right) \geqslant \boldsymbol{0}, \qquad t \in \mathcal{T}, \tag{5.1}$$

$$\sum_{i=1}^{n^{\mathrm{w}}} \tilde{w}_i(t) = 1, \qquad t \in \mathcal{T}.$$

This ensures that the constraint function $\boldsymbol{g}(\cdot)$ is evaluated in integer feasible controls $\boldsymbol{\omega}^i$ only. While in addition it is obvious that any rounded solution derived from the relaxed optimal one will be feasible, this formulation leads to relaxed optimal solutions that show *compensation effects*. Choices $\tilde{w}_i(t) > 0$ for some $1 \leqslant i \leqslant n^{\mathrm{w}}$ that are infeasible with respect to $\boldsymbol{g}(\boldsymbol{\omega}^i)$ can be compensated for by nonzero choices $\tilde{w}_j(t) > 0$ that are not part of the MIOCP's solution.

### 5.1.2 Outer Convexification of Constraints

In chapter 2 we already mentioned that this issue can be resolved by applying the outer convexification technique not only to the process dynamics but also to the affected constraints. This amounts to the substitution of

$$\begin{aligned} \boldsymbol{g}(t,\boldsymbol{x}(t),\boldsymbol{u}(t),\boldsymbol{w}(t),\boldsymbol{p}) \geqslant \boldsymbol{0}, \qquad & t \in \mathcal{T}, \\ \boldsymbol{w}(t) \in \{0,1\}^{n^{\mathrm{w}}}, \quad & t \in \mathcal{T}, \end{aligned} \tag{5.2}$$

by

$$\begin{aligned} \tilde{w}_i(t) \cdot \boldsymbol{g}(t,\boldsymbol{x}(t),\boldsymbol{\omega}^i,\boldsymbol{p}) \geqslant \boldsymbol{0}, \qquad & t \in \mathcal{T}, \quad 1 \leqslant i \leqslant n^{\mathrm{w}}, \\ \tilde{\boldsymbol{w}}(t) \in [0,1]^{n^{\mathrm{w}}}, \quad \sum_{i=1}^{n^{\mathrm{w}}} \tilde{w}_i(t) = 1, \qquad & t \in \mathcal{T}. \end{aligned} \tag{5.3}$$

It is immediately clear that an optimal solution satisfying the constraints $\boldsymbol{g}$ for some relaxed optimal control $\tilde{\boldsymbol{w}} \in [0,1]^{n^{\mathrm{w}}}$ will also satisfy these constraints for any rounded binary control. This convexification of the constraints $\boldsymbol{g}$ does not introduce any additional unknowns into the discretized MIOCP. The increased number of constraints is put in context by the observation that the majority of the convexified constraints can be considered inactive as the majority of the components of $\tilde{\boldsymbol{w}}$ is likely to be active at zero. Active set methods are an appropriate tool to exploit this property. As constraints of type (5.3) violate LICQ in $\tilde{w}_i(t) = 0$, we briefly survey further alternative approaches at maintaining feasibility of $\boldsymbol{g}$ after rounding, before we investigate the structure of (5.3) in more detail.

### 5.1.3 Relaxation and Homotopy

In the spirit of a *Big–M* method for feasibility determination [158], the constraints $\boldsymbol{g}$ can be relaxed using a relaxation parameter $M > 0$,

$$\boldsymbol{g}(t, \boldsymbol{x}(t), \boldsymbol{\omega}^i, \boldsymbol{p}) \geqslant -M(1 - \tilde{w}_i(t)), \qquad t \in \mathcal{T}, \quad 1 \leqslant i \leqslant n^{\mathrm{w}}, \tag{5.4}$$

$$\tilde{\boldsymbol{w}}(t) \in [0,1]^{n^{\mathrm{w}}}, \quad \sum_{i=1}^{n^{\mathrm{w}}} \tilde{w}_i(t) = 1, \qquad t \in \mathcal{T}.$$

Clearly for $\tilde{w}_i(t) = 1$ feasibility of the constraints $\boldsymbol{g}(\boldsymbol{\omega}^i)$ is enforced while for $\tilde{w}_i(t) = 0$ the constraint $\boldsymbol{g}(\boldsymbol{\omega}^i)$ does not affect the solution if $M$ is chosen large enough. This approach has been followed e.g. by [179]. The drawback of this formulation is the fact that the choice of the relaxation parameter $M$ is a priori unclear. In addition, the relaxation enlarges the feasible set of the Optimal Control Problem (OCP) unnecessarily, thereby possibly attracting fractional relaxed solutions of the OCP due to compensation effects.

### 5.1.4 Perspective Cuts

A recently emerging idea based on disjunctive programming [93] is to employ a reformulation based on perspective cuts for a constraint $w \cdot g(x) \geqslant 0$ depending on a continuous variable $x \in [0, x^{\mathrm{up}}] \subset \mathbb{R}$ and a binary variable $w \in \{0, 1\}$,

$$\begin{aligned} 0 &\leqslant \lambda g(x/\lambda), \\ 0 \leqslant x &\leqslant \lambda x^{\mathrm{up}}, \\ \lambda &\in [0,1] \subset \mathbb{R}. \end{aligned} \tag{5.5}$$

Clearly again for $\lambda = 1$ the original constraint on $x$ is enforced, while for $\lambda = 0$ is can be seen by Taylor's expansion of $\lambda g(x/\lambda)$ that the constraint is always satisfied. For simplicity of notation let us assume for a moment that $g(x(t), \boldsymbol{w}(t))$ be a scalar constraint function depending on a scalar state trajectory $x(t)$. The proposed reformulation then reads

$$
\begin{aligned}
\tilde{w}_i(t)\cdot g\left(\frac{\tilde{x}_i(t)}{\tilde{w}_i(t)},\boldsymbol{\omega}^i\right) &\geqslant 0, && t\in\mathcal{T}, && 1\leqslant i\leqslant n^{\mathrm{w}}, \\
\tilde{w}_i(t)\cdot x^{\mathrm{up}} &\geqslant \tilde{x}_i(t), && t\in\mathcal{T}, && 1\leqslant i\leqslant n^{\mathrm{w}}, \\
\tilde{x}_i(t)\in[0,x^{\mathrm{up}}], \quad \sum_{i=1}^{n^{\mathrm{w}}}\tilde{x}_i(t) &= x(t), && t\in\mathcal{T}, && 1\leqslant i\leqslant n^{\mathrm{w}}, \\
\tilde{\boldsymbol{w}}(t)\in[0,1]^{n^{\mathrm{w}}}, \quad \sum_{i=1}^{n^{\mathrm{w}}}\tilde{w}_i(t) &= 1, && t\in\mathcal{T}.
\end{aligned}
\tag{5.6}
$$

Here we have introduced slack variables $\tilde{x}_i(t)$, $1 \leqslant i \leqslant n^{\mathrm{w}}$ for the state $x(t)$. Clearly if $\tilde{w}_i(t) = 1$ the constraint $g(\boldsymbol{\omega}^i)$ will be satisfied by $\tilde{x}_i(t)$ and thus by $x(t)$, while if $\tilde{w}_i(t) = 0$ this constraint does not affect the solution. This reformulation is promising as LICQ holds for the resulting convexified NLPs on the whole of the feasible set if it held for the original one. As the number of unknowns increased further due to introduction of the slack variables $\tilde{x}_i(t)$, structure exploiting methods tailored to the case of many control parameters become even more crucial for the performance of the mixed–integer OCP algorithm. In addition, a connection between the convex combination $\tilde{\boldsymbol{x}}(t)$ of the state trajectory $x(t)$ and the convexification of the dynamics must be established.

## 5.2 Lack of Constraint Qualification

In this section we investigate numerical properties of an NLP with a constraint depending on integer or binary variables that has been treated by outer convexification as proposed in section 5.1.2. We show that certain constraint qualifications are violated which are commonly assumed to hold by numerical codes for the solution of NLPs. Linearization of constraints treated by outer convexification bears significant potential for severe ill–conditioning of the generated linearized subproblems, as has been briefly noted by [3] in the context of truss optimization, and we exemplarily investigate this situation. Finally, extensive numerical studies [116] have revealed that SQP methods attempting to solve MPVCs without special consideration of the inherent combinatorial structure of the feasible set are prone to very frequent cycling of the active set. We present an explanation of this phenomenon for the case of linearizations in NLP–infeasible iterates of the SQP method.

### 5.2.1 Constraint Qualifications

The local optimality conditions of theorem 3.1 were given under the assumption that LICQ holds in the candidate point $\boldsymbol{x}^\star$ in order to ensure a certain wellbehavedness of the feasible set in its neighborhood. The Karush–Kuhn–Tucker (KKT) theorem (3.1) can however be proven under less restrictive conditions, and various Constraint Qualifications (CQs) differing in strength and applicability have been devised by numerous authors. We present in the following four popular CQs in order of decreasing strength which will be of interest during the investigation of MPVCs in section 5.3.

We first require the definitions of the tangent and the linearized cone.

**Definition 5.1 (Tangent Cone, Linearized Cone)**
Let $\overline{\boldsymbol{x}} \in \mathbb{R}^n$ be a feasible point of problem (3.1). The set

$$\mathcal{T}(\overline{\boldsymbol{x}},\mathcal{F}) \overset{\text{def}}{=} \left\{\boldsymbol{d} \in \mathbb{R}^n \;\middle|\; \exists\{\boldsymbol{x}^k\} \subseteq \mathcal{F},\ \{t^k\} \to 0^+ :\ \boldsymbol{x}^k \to \overline{\boldsymbol{x}},\ \tfrac{1}{t^k}(\boldsymbol{x}^k - \overline{\boldsymbol{x}}) \to \boldsymbol{d}\right\} \quad (5.7)$$

is called the Bouligand *tangent cone* of the set $\mathcal{F}$ in the point $\overline{\boldsymbol{x}}$. The set

$$\mathcal{L}(\overline{\boldsymbol{x}}) \overset{\text{def}}{=} \left\{\boldsymbol{d} \in \mathbb{R}^n \;\middle|\; \boldsymbol{g}_x(\overline{\boldsymbol{x}})^T \boldsymbol{d} = \boldsymbol{0},\ (\boldsymbol{h}_{\mathcal{A}})_x(\overline{\boldsymbol{x}})^T \boldsymbol{d} \geqslant \boldsymbol{0}\right\} \quad (5.8)$$

is called the *linearized cone* of problem (3.1) in $\overline{\boldsymbol{x}}$. △

The dual $\mathcal{T}(\overline{\boldsymbol{x}},\mathcal{F})^\star$ of the tangent cone is also referred to as the *normal cone* of $\mathcal{F}$ in $\overline{\boldsymbol{x}}$. In the next section, these cones will play a role in characterizing the shape of the feasible set in a neighborhood of the point $\overline{\boldsymbol{x}}$.

**Definition 5.2 (Constraint Qualifications)**
Let $\overline{\boldsymbol{x}} \in \mathbb{R}^n$ be a feasible point of problem (3.1). We say that

- *Linear Independence Constraint Qualification (LICQ) [143]* holds in $\overline{\boldsymbol{x}}$ if the Jacobian

$$\begin{bmatrix} \boldsymbol{g}_x(\overline{\boldsymbol{x}}) \\ (\boldsymbol{h}_{\mathcal{A}})_x(\overline{\boldsymbol{x}}) \end{bmatrix} \quad \text{(LICQ)}$$

  has full row rank.

- Mangasarian–Fromovitz *Constraint Qualification (MFCQ) [149]* holds in $\overline{\boldsymbol{x}}$ if the Jacobian $\boldsymbol{g}_x(\overline{\boldsymbol{x}})$ has full row rank and there exists $\boldsymbol{d} \in \mathbb{R}^n$ such that

$$\begin{aligned} \boldsymbol{g}_x(\overline{\boldsymbol{x}})^T \boldsymbol{d} &= \boldsymbol{0}, \\ (\boldsymbol{h}_{\mathcal{A}})_x(\overline{\boldsymbol{x}})^T \boldsymbol{d} &> \boldsymbol{0}. \end{aligned} \quad \text{(MFCQ)}$$

- Abadie *Constraint Qualification (ACQ) [1]* holds in $\overline{\boldsymbol{x}}$ if

$$\mathcal{T}(\overline{\boldsymbol{x}},\mathcal{F})=\mathcal{L}(\overline{\boldsymbol{x}}). \tag{ACQ}$$

- Guignard *Constraint Qualification (GCQ) [97]* holds in $\overline{\boldsymbol{x}}$ if

$$\mathcal{T}(\overline{\boldsymbol{x}},\mathcal{F})^{\star}=\mathcal{L}(\overline{\boldsymbol{x}})^{\star}. \tag{GCQ}$$

△

In definition 5.2, each CQ implies the next weaker one,

$$\text{LICQ} \implies \text{MFCQ} \implies \text{ACQ} \implies \text{GCQ}$$

whereas the converse never holds. Proofs of theorem 3.1 assuming each of the above CQs as well as counterexamples for the converse cases can e.g. be found in [1, 165].

### 5.2.2 Checking Constraint Qualifications

We start by checking a minimal exemplary problem for constraint qualification according to definition 5.2.

**Example 5.1 (Violation of Constraint Qualifications)**
Consider the scalar constraint $x_1 \cdot g(x_2) \geqslant 0$ together with its associated simple bound $x_1 \geqslant 0$. Here, the possibly nonlinear constraint $g$ on $x_2$ vanishes if $x_1$ is chosen to be exactly zero. The Jacobian $\boldsymbol{C}$ found from a linearization of these two constraints with respect to the unknown $\boldsymbol{x}=(x_1,x_2)$ reads

$$\boldsymbol{C}=\begin{bmatrix} g(x_2) & x_1 \cdot g_x(x_2) \\ 1 & 0 \end{bmatrix}$$

We now observe the following:

1. In all points satisfying $x_1 = 0$ the Jacobian $\boldsymbol{C}$ becomes singular and LICQ is violated. All other points satisfy LICQ.
2. In $x_1 = 0$, $g(x_2) = 0$ we have

$$\boldsymbol{C}=\begin{bmatrix} 0 & 0 \\ 1 & 0 \end{bmatrix}.$$

   There is no vector $\boldsymbol{d} \in \mathbb{R}^2$ such that $\boldsymbol{C}^T\boldsymbol{d} > 0$, which shows that MFCQ is violated in these points. All other points satisfy MFCQ.

3. In $x_1 = 0$, $g(x_2) = 0$ we have the linearized cone

$$\mathcal{L}(\boldsymbol{x}) = \mathbb{R}_0^+ \times \mathbb{R}.$$

The Bouligand tangent cone is the union of two convex cones

$$\mathcal{T}(\boldsymbol{x}, \mathcal{F}) = (\mathbb{R}_0^+ \times \mathbb{R}_0^+) \cup (\{0\} \times \mathbb{R}_0^-),$$

which shows that ACQ is violated in this point. All other points satisfy ACQ by implication.

4. Observe that for the duals of the above cones it holds that

$$\mathcal{L}(\boldsymbol{x})^\star = \mathcal{T}(\boldsymbol{x}, \mathcal{F})^\star = \mathbb{R}_0^+ \times \{0\},$$

which finally shows that of the four introduced CQs, GCQ is the only one to hold on the entire feasible set.

These finding lead us to the following remark about CQs for NLPs with constraints treated by Outer Convexification.

**Remark 5.1 (Violation of CQs)**
For constraints treated by Outer Convexification, LICQ is violated if the binary control variable associated with a vanishing constraint is zero. If in addition the vanishing constraint is active at its lower or upper bound, MFCQ and ACQ are violated as well. As GCQ holds, locally optimal points are still KKT points.

### 5.2.3 Ill–Conditioning

We continue by investigating the numerical behavior of example 5.1 in a neighborhood of the area of violation of LICQ.

**Example 5.2 (Ill–conditioning)**
Consider again the constraints of example 5.1. As $x_1$ approaches zero, and thus the area of violation of LICQ, the Jacobian $\boldsymbol{C}$ approaches singularity and its condition number approaches infinity. Due to roundoff and truncation errors inevitably experienced by any numerical algorithm, a weakly active bound $x_1 \geqslant 0$ may be satisfied with a certain error only. If for example $x_1 = 10^{-12}$ instead of $x_1 = 0$, the condition number of $\boldsymbol{C}$ is $1/x_1 = 10^{12}$ if the vanishing constraint $g$ is active. If the constraint $g$ is inactive, the condition number even grows larger along with the constraint's residual.

The condition numbers of $\boldsymbol{C}$ obtained for this example if $g_x(x_2) = 1$ are shown in figure 5.1. We summarize our findings concerning ill–conditioning of the linearized subproblems with the following remark.

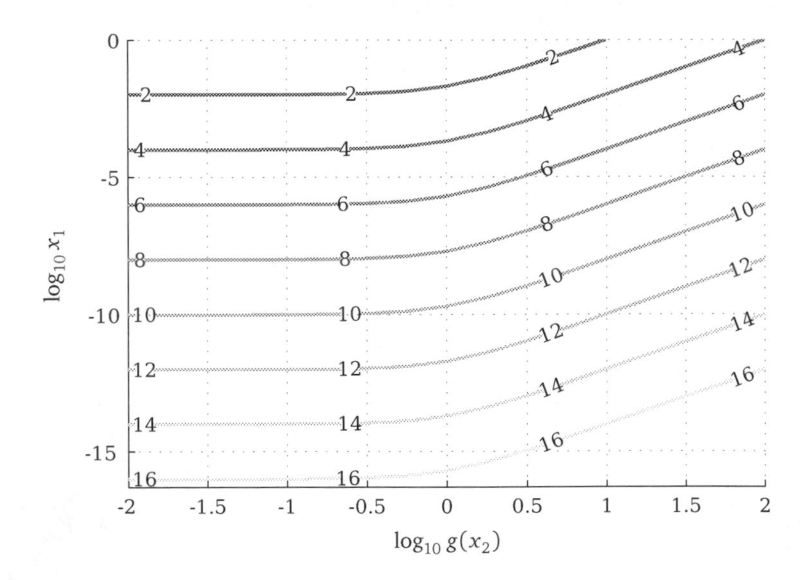


Figure 5.1: Logarithm of the Jacobian's condition number ($\log_{10} \operatorname{cond} \boldsymbol{C}$) for a vanishing constraint $x_1 \cdot g(x_2) \geqslant 0$ and a simple bound $x_1 \geqslant 0$, depending on the residuals $g(x_2)$ and $x_1$.

**Remark 5.2 (Ill–conditioning)**
In a neighborhood of the area of violation of LICQ, linearizations of constraints treated by Outer Convexification are ill–conditioned. For small perturbations $\varepsilon > 0$ of a binary control variable that is active at zero in the true solution, the condition number of the Jacobian obtained from linearization of the constraint is at least $1/\varepsilon$.

### 5.2.4 Infeasible and Suboptimal Steps

Due to the combinatorial structure of the feasible set, constraint linearizations may fail to approximate its structure. Figure 5.2 shows the mismatch of linearized feasible sets and the local neighborhood of the NLP's feasible set in a linearization point $\boldsymbol{x}$ that satisfies $g_i(\boldsymbol{x}) = 0$, $h_i(\boldsymbol{x}) = 0$. As clearly visible, both infeasible and suboptimal step decisions $(\boldsymbol{\delta g}, \boldsymbol{\delta h})$ may be taken by a descent based optimization method working on the linearized representation of the feasible set.

**Remark 5.3 (Infeasible and Suboptimal Steps)**
Linearizations of the feasible set of constraints treated by outer convexification may lead to infeasible NLP iterates if step decisions are made based on the linearizations.

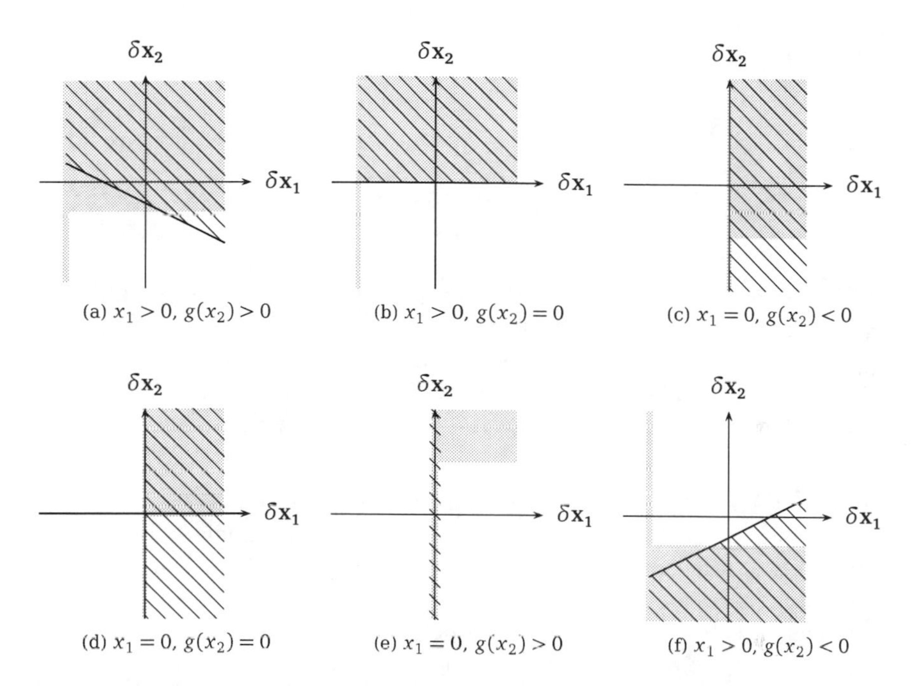


Figure 5.2: Linearized feasible sets (▧) entering the QP subproblems lead to infeasible and suboptimal steps of the SQP method. Actually globally feasible steps (░) for the NLP. In cases (a), (c), (d), and (f), the QP subproblem may determine an NLP–infeasible step. In cases (a), (b), (e), and (f), the determined step may be suboptimal.

### 5.2.5 Cycling of Active Set Methods

This third example uses a geometrical interpretation of the ill–conditioning observed numerically in example 5.2 in order to explain frequently observed cycling of the active set when solving problems with constraints treated by Outer Convexification with standard active set methods.

**Example 5.3 (Cycling of Active Set Methods)**
Consider again the pair of constraints

$$\begin{aligned} c(\boldsymbol{x}) = x_1 \cdot g(x_2) &\geqslant 0, \\ x_1 &\geqslant 0, \end{aligned}$$

and their linearization in $\overline{\boldsymbol{x}}$,

$$\begin{aligned} g(\overline{x}_2)\delta x_1 + \overline{x}_1 g_x(\overline{x}_2)\delta x_2 &\geqslant -\overline{x}_1 g(\overline{x}_2), \\ \delta x_1 &\geqslant -\overline{x}_1. \end{aligned}$$

For a linearization point $\overline{\boldsymbol{x}}$ that is not degenerate, i.e., $\overline{x}_1 > 0$, this may be written as

$$\delta x_1 + \beta \delta x_2 \begin{cases} \geqslant -\overline{x}_1 & \text{if } \boldsymbol{x} \text{ is feasible, i.e., } g(\overline{x}_2) > 0, \\ \leqslant -\overline{x}_1 & \text{if } \boldsymbol{x} \text{ is infeasible, i.e., } g(\overline{x}_2) < 0, \end{cases} \qquad \beta \overset{\text{def}}{=} \overline{x}_1 \frac{g_x(\overline{x}_2)}{g(\overline{x}_2)} \in \mathbb{R}.$$

Figure 5.3 shows the linearized feasible set of the quadratic subproblem for an SQP iterate that is feasible. With $\beta \to 0$, which means $\overline{x}_1 \to 0$ or $g(\overline{x}_2) \to \infty$, the angle between the two constraints approaches $\pi$ and the corner of the feasible set becomes degenerate. The resulting ill-conditioning of the linearization was also observed numerically in figure 5.1.

Figure 5.4 shows the linearized feasible set for an infeasible SQP iterate. In contrast to the previous case, with $\beta \to 0$ the angle between the constraints approaches 0 and the feasible set is spike-shaped. Besides ill-conditioning, this may potentially result in many tiny steps being made by an active set method if the stationary point is located near the spike's pin-point. This phenomenon is known as "zig–zagging".

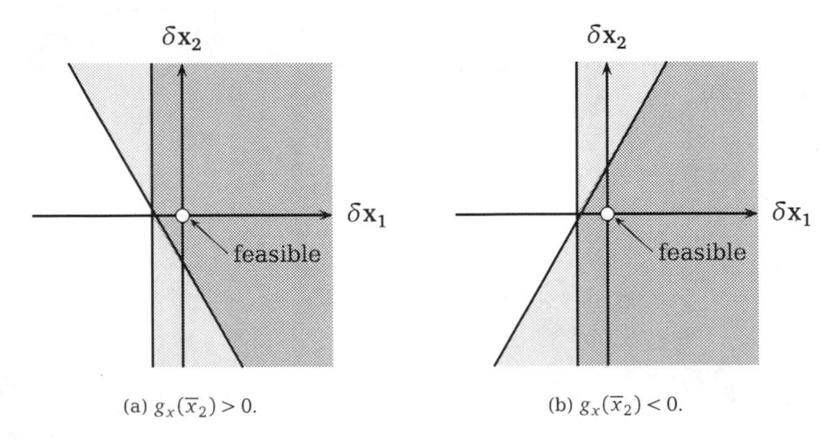


(a) $g_x(\overline{x}_2) > 0$. (b) $g_x(\overline{x}_2) < 0$.

Figure 5.3: Feasible sets of the subproblems obtained by linearization in a feasible iterate. The common feasible subset of both constraints is indicated by (▒) while parts feasible for one of the two constraints only are indicated by (░).

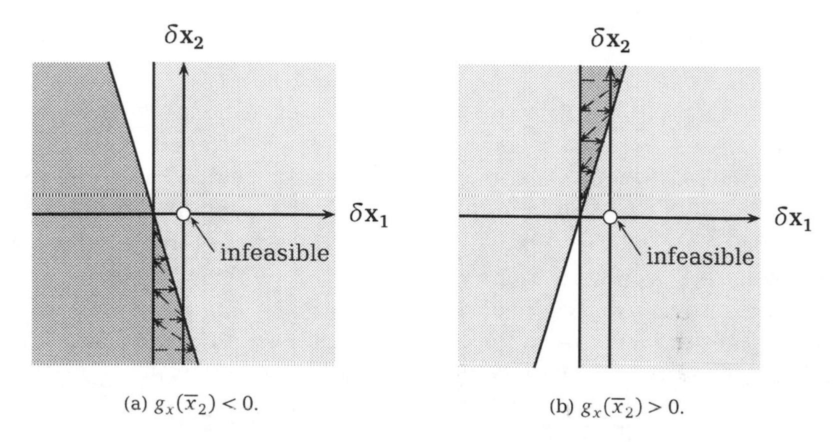


(a) $g_x(\overline{x}_2) < 0$.

(b) $g_x(\overline{x}_2) > 0$.

Figure 5.4: Feasible sets of the quadratic subproblems obtained by linearization in an iterate that is infeasible with respect to the vanishing constraint, i.e., $g(\overline{x}_2) < 0$.

**Remark 5.4 (Cycling of the Active Set)**
Linearizations of a constraint treated by Outer Convexification in SQP iterates with binary control variables close to zero or large residuals of the constraint will lead to degenerate corners of the feasible set polytope. If the linearization point is infeasible, an active set method solving the linearized subproblem is likely to suffer from zig–zagging.

**Remark 5.5 (Bang–Bang Optimal Controls)**
For optimal control problems that enjoy a bang–bang property (theorem 2.1), the described situation will actually be the one encountered most often. Close to the optimal solution, in each SOS1 set all but one relaxed control variable will be exactly zero for an active set method in exact arithmetic. In floating–point arithmetic, their values will be close to the machine precision `eps`, which results in severe ill–conditioning and spike shaped feasible sets.

## 5.3 Mathematical Programs with Vanishing Constraints

Theoretical and numerical challenges introduced into the NLP by outer convexification of constraints depending on binary or integer controls have been exemplarily studied in this section. Our findings clearly justify the need for deeper investigation of the combinatorial structure of such NLPs in order to take the structure into account in the design of a tailored SQP algorithm.

The findings of the first section lead us to investigate NLPs with constraints treated by outer convexification from a different point of view. In this section we introduce the problem class of Mathematical Programs with Vanishing Constraints (MPVCs), highly challenging nonconvex problems that have only recently attracted increased research interest. This problem class provides a framework for analysis of the shortcomings of the standard Lagrangian approach at solving MPVCs, and yields new stationarity conditions for a separated, i.e., non–multiplicative formulation of the vanishing constraints that will in general be much more well conditioned.

Pioneering work on the topic of MPVCs was brought forward only recently by [3, 106, 107]. In these works, MPVCs were shown to constitute a challenging class of problems, as standard constraint qualifications such as LICQ and MFCQ turn out to be violated for most problem instances. In the thesis [105] an exhaustive treatment of the theoretical properties of MPVCs can be found. The theoretical results presented in this section are a summary of those parts of [109] and the aforementioned works that are required for the numerical treatment of NLPs with constraints treated by outer convexification.

### 5.3.1 Problem Formulations

**Definition 5.3 (Nonlinear Program with Vanishing Constraints)**
The following NLP with $l \geqslant 1$ complementary inequality constraints

$$\begin{aligned} \min_{x \in \mathbb{R}^n} \quad & f(x) \\ \text{s.t.} \quad & g_i(x) \cdot h_i(x) \geqslant 0, \qquad 1 \leqslant i \leqslant l, \\ & \boldsymbol{h}(x) \geqslant \boldsymbol{0} \end{aligned} \tag{5.9}$$

is called a Mathematical Program with Vanishing Constraints (MPVC). △

The objective function $f : \mathbb{R}^n \to \mathbb{R}$ and the constraint functions $g_i, h_i : \mathbb{R} \to \mathbb{R}$, $1 \leqslant i \leqslant l$ are expected to be at least twice continuously differentiable. The domain of feasibility of problem (5.9) may be further restricted by a standard nonlinear constraint function $\boldsymbol{c}(x) \geqq \boldsymbol{0}$, including equality constraints. Such constraints do not affect the theory and results to be presented in this chapter, and we omit them for clarity of exposition.

### Constraint Logic Reformulation

It is easily observed that the constraint functions $g_i(\boldsymbol{x})$ do not impact the question of feasibility of an arbitrary point $\boldsymbol{x} \in \mathbb{R}^n$ iff $h_i(\boldsymbol{x}) = 0$ holds. The constraint $g_i$ is the said to have *vanished* in the point $\boldsymbol{x}$. This observation gives rise to the following constraint logic reformulation.

**Remark 5.6 (Constraint Logic Reformulation for MPVCs)**
Problem (5.9) can be equivalently cast as an NLP with $m$ logic constraints,

$$\begin{aligned} \min_{x\in\mathbb{R}^n} \quad & f(x) \\ \text{s.t.} \quad & h_i(x) > 0 \implies g_i(x) \geqslant 0, \qquad 1 \leqslant i \leqslant m, \\ & \boldsymbol{h}(x) \geqslant \boldsymbol{0}. \end{aligned} \tag{5.10}$$

The logical implication though appears to be unsuitable for derivative based numerical methods. We will however see in the sequel of this chapter that it can be realized in the framework of an active set method for Quadratic Programming.

### Equilibrium Constraint Reformulation

MPVCs can be related to the well studied class of Mathematical Programs with Equilibrium Constraints (MPECs) through the following reformulation.

**Remark 5.7 (MPEC Reformulation for MPVCs)**
Problem (5.9) can be equivalently cast as an MPEC by introduction of a vector $\xi \in \mathbb{R}^l$ of slack variables,

$$\begin{aligned} \min_{x\in\mathbb{R}^n} \quad & f(x) \\ \text{s.t.} \quad & \xi_i \cdot h_i(x) = 0, \qquad 1 \leqslant i \leqslant l, \\ & \boldsymbol{h}(x) \geqslant \boldsymbol{0}, \\ & \boldsymbol{\xi} \leqslant \boldsymbol{0}, \\ & \boldsymbol{g}(x) - \boldsymbol{\xi} \geqslant \boldsymbol{0}. \end{aligned} \tag{5.11}$$

Although this opens up the possibility of treating (5.9) using preexisting MPECs algorithms designed for interior point methods by e.g. [17, 139, 173] and for active set methods by e.g. [75, 111], this reformulation suffers from the fact that the additionally introduced vector $\boldsymbol{\xi}$ of slack variables is undefined if $h_i(\boldsymbol{x}^\star) = 0$ holds for some index $i$ in an optimal solution $\boldsymbol{x}^\star$.

In fact, MPECs are known to constitute an even more difficult class than MPVCs, as both LICQ and MFCQ are violated in *every* feasible point [47]. Consequentially, standard NLP sufficient conditions for optimality of $\boldsymbol{x}^\star$ as presented in section 3.1.3 do not hold. If on the other hand $h_i(\boldsymbol{x}^\star) \neq 0$ for all indices $1 \leqslant i \leqslant l$, we could as well have found the same solution $\boldsymbol{x}^\star$ by including all vanishing constraints $g_i$ as standard constraints, which allows to treat problem (5.9) as a plain NLP.

### Relaxation Method

It has frequently been proposed to solve problem (5.9) by embedding it into a family of perturbed problems. The vanishing constraint that causes violation of constraint qualifications is relaxed,

$$\begin{aligned} \min_{\boldsymbol{x}\in\mathbb{R}^n} \quad & f(\boldsymbol{x}) \\ \text{s.t.} \quad & g_i(\boldsymbol{x})\cdot h_i(\boldsymbol{x}) \geqslant -\tau, \qquad 1 \leqslant i \leqslant l, \\ & \boldsymbol{h}(\boldsymbol{x}) \geqslant \boldsymbol{0}, \end{aligned} \tag{5.12}$$

where the scalar $\tau \geqslant 0$ is the relaxation parameter. For $\tau > 0$, problem (5.12) is regular in its solution. The level of exactness of this relaxation is investigated e.g. in [109].

### Embedding Relaxation Method

A more general relaxation of the vanishing constraint $g_i(\boldsymbol{x}) \cdot h_i(\boldsymbol{x}) \geqslant 0$ is proposed in [105, 109] that works by embedding it into a function $\varphi$ satisfying

$$\varphi(a,b) = 0 \iff b \geqslant 0,\ a \cdot b \geqslant 0,$$

and introducing a relaxation parameter $\tau$ together with a family of functions $\varphi^\tau(a,b)$ for which it holds that $\varphi^\tau(a,b) \to \varphi(a,b)$ for $\tau \to 0$. It is argued that smooth functions $\varphi$ necessarily lead to lack of LICQ and MFCQ, and the use of a family of nonsmooth functions is proposed and convergence results are obtained similar in spirit to those of [175, 192] for MPECs.

Finally, two remarks are in order about the applicability of the surveyed reformulations.

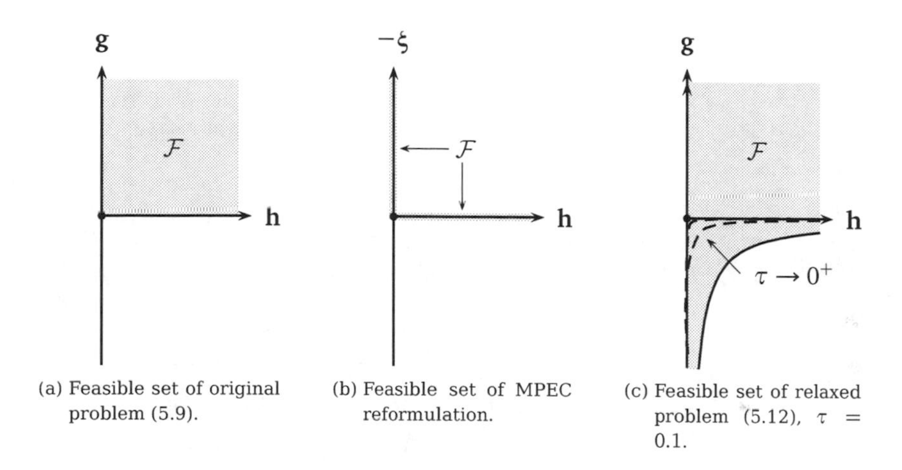


(a) Feasible set of original problem (5.9).

(b) Feasible set of MPEC reformulation.

(c) Feasible set of relaxed problem (5.12), $\tau = 0.1$.

Figure 5.5: Feasible sets of various reformulations of problem (5.9).

**Remark 5.8 (Ill–Conditioning of Reformulations)**
The ill–conditioning investigated in section 5.2.3, inherent to the multiplicative formulation of the vanishing constraint, is not resolved by either of the two relaxation methods.

**Remark 5.9 (Relaxation of Constraints)**
Relaxation methods require the problem's objective and constraint functions to be valid outside the feasible set originally described by problem (5.9). As we will see in chapter 9, this may lead to evaluations of model functions outside the physically meaningful domain. If relaxation methods are employed, it is important to allow for such evaluations in a way that does not prevent the true solution from being found.

## 5.4 An MPVC Lagrangian Framework

In the following we present an analysis of the standard Lagrangian approach presented in section 3.1, applied to problem (5.9) as derived by [3, 106] and related works. An alternative Lagrangian function that introduces separate multipliers for the two parts of a vanishing constraint is introduced that removes the major source of ill–conditioning. Stationarity and optimality conditions are presented.

### 5.4.1 Notation and Auxiliary Problems

We first require some notation to be defined. The notion of an active set for problem MPVC requires some extension as shown in the following definition of index sets.

**Definition 5.4 (Index Sets)**
Let $\overline{\boldsymbol{x}} \in \mathbb{R}^n$ be a feasible point. In a neighborhood $\mathcal{B}_\varepsilon(\overline{\boldsymbol{x}})$ of $\overline{\boldsymbol{x}}$, the set $\{1,\ldots,l\} \subset \mathbb{N}$ of indices of the $l$ vanishing constraints is partitioned into the following index sets:

$$\begin{aligned}
\mathcal{I}_+ &\overset{\text{def}}{=} \{i \mid h_i(\overline{\boldsymbol{x}}) > 0\}, & \mathcal{I}_{+0} &\overset{\text{def}}{=} \{i \mid h_i(\overline{\boldsymbol{x}}) > 0, g_i(\overline{\boldsymbol{x}}) = 0\}, \\
& & \mathcal{I}_{++} &\overset{\text{def}}{=} \{i \mid h_i(\overline{\boldsymbol{x}}) > 0, g_i(\overline{\boldsymbol{x}}) > 0\}, \\
\mathcal{I}_0 &\overset{\text{def}}{=} \{i \mid h_i(\overline{\boldsymbol{x}}) = 0\}, & \mathcal{I}_{0+} &\overset{\text{def}}{=} \{i \mid h_i(\overline{\boldsymbol{x}}) = 0, g_i(\overline{\boldsymbol{x}}) > 0\}, \\
& & \mathcal{I}_{00} &\overset{\text{def}}{=} \{i \mid h_i(\overline{\boldsymbol{x}}) = 0, g_i(\overline{\boldsymbol{x}}) = 0\}, \\
& & \mathcal{I}_{0-} &\overset{\text{def}}{=} \{i \mid h_i(\overline{\boldsymbol{x}}) = 0, g_i(\overline{\boldsymbol{x}}) < 0\}.
\end{aligned} \tag{5.13}$$

△

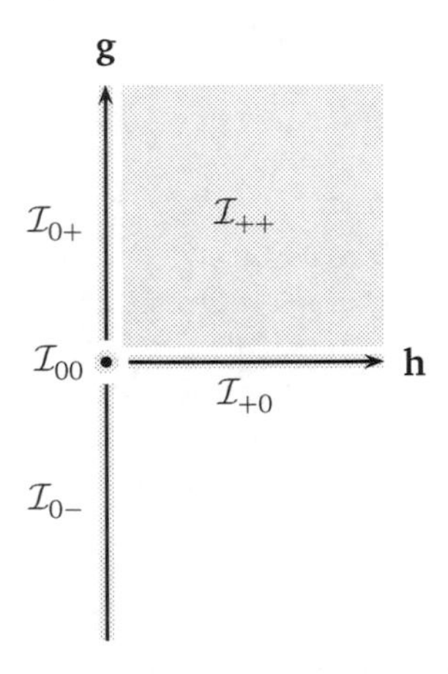


Figure 5.6: Index sets for a vanishing constraint. ( ) indicates the feasible set.

An interpretation of the index sets as parts of the feasible set is depicted in figure 5.6. In a neighborhood of $\overline{\boldsymbol{x}}$, the feasible set of problem (5.9) can then be expressed as

$$\begin{aligned}
\mathcal{F} = \big\{\boldsymbol{x} \in \mathbb{R}^n \;\big|\; & \boldsymbol{h}_{\mathcal{I}_{0+}}(\boldsymbol{x}) \geqslant \boldsymbol{0},\ \boldsymbol{h}_{\mathcal{I}_{00}}(\boldsymbol{x}) \geqslant \boldsymbol{0},\ \boldsymbol{h}_{\mathcal{I}_{0-}}(\boldsymbol{x}) = \boldsymbol{0}, \\
& \boldsymbol{g}_{\mathcal{I}_{+0}}(\boldsymbol{x}) \geqslant \boldsymbol{0},\ \forall i \in \mathcal{I}_{00} :\ g_i(\boldsymbol{x}) \cdot h_i(\boldsymbol{x}) \geqslant 0\big\}.
\end{aligned} \tag{5.14}$$

Clearly, the combinatorial structure of $\mathcal{F}$ is induced by exactly those constraints that are found in $\mathcal{I}_{00}$. If on the other hand $\mathcal{I}_{00} = \emptyset$, then $\mathcal{F}$ is the feasible set of a plain NLP and any combinatorial aspect disappears.

Based on (5.9) we further define two auxiliary NLPs, a tightened and a relaxed one, as follows.

**Definition 5.5 (Tightened Nonlinear Problem)**
Let $\overline{x} \in \mathbb{R}^n$ be a feasible point of (5.9). In a neighborhood $B_\varepsilon(\overline{x})$ of $\overline{x}$, the *tightened problem* (TNLP) is defined as

$$\begin{aligned} \min_{x \in \mathbb{R}^n} \quad & f(x) && \text{(TNLP)} \\ \text{s.t.} \quad & \boldsymbol{h}_{\mathcal{I}_{0+}}(x) \geqslant \boldsymbol{0}, \quad \boldsymbol{h}_{\mathcal{I}_{00}}(x) = \boldsymbol{0}, \quad \boldsymbol{h}_{\mathcal{I}_{0-}}(x) = \boldsymbol{0}, \\ & \boldsymbol{g}_{\mathcal{I}_{+0}}(x) \geqslant \boldsymbol{0}, \quad \boldsymbol{g}_{\mathcal{I}_{00}}(x) \geqslant \boldsymbol{0}. \end{aligned}$$

△

**Definition 5.6 (Relaxed Nonlinear Problem)**
Let $\overline{x} \in \mathbb{R}^n$ be a feasible point of (5.9). In a neighborhood $B_\varepsilon(\overline{x})$ of $\overline{x}$, the *relaxed problem* (RNLP) is defined as

$$\begin{aligned} \min_{x \in \mathbb{R}^n} \quad & f(x) && \text{(RNLP)} \\ \text{s.t.} \quad & \boldsymbol{h}_{\mathcal{I}_{0+}}(x) \geqslant \boldsymbol{0}, \quad \boldsymbol{h}_{\mathcal{I}_{00}}(x) \geqslant \boldsymbol{0}, \quad \boldsymbol{h}_{\mathcal{I}_{0-}}(x) = \boldsymbol{0}, \\ & \boldsymbol{g}_{\mathcal{I}_{+0}}(x) \geqslant \boldsymbol{0}. \end{aligned}$$

△

Both problems depart from the original problem (5.9) in those vanishing constraints imposed on $x$ that are contained in the critical index set $\mathcal{I}_{00}$ in $\overline{x}$. Given a partition $\{\mathcal{I}_1;\mathcal{I}_2\}$ of the critical set $\mathcal{I}_{00}$ into two subsets $\mathcal{I}_1$ and $\mathcal{I}_2$,

$$\mathcal{I}_1 \cup \mathcal{I}_2 = \mathcal{I}_{00}, \qquad \mathcal{I}_1 \cap \mathcal{I}_2 = \emptyset,$$

the *branch problem* for $\{\mathcal{I}_1;\mathcal{I}_2\}$, first used in [107, 109], requires the vanishing constraints in $\mathcal{I}_1$ to be feasible, $g_i(x) \geqslant 0$, and the constraints in $\mathcal{I}_2$ to have vanished, $h_i(x) = 0$.

**Definition 5.7 (Branch Problem)**
Let $\overline{x} \in \mathbb{R}^n$ be a feasible point of (5.9), and let $\{\mathcal{I}_1;\mathcal{I}_2\}$ be a partition of $\mathcal{I}_{00}$. In a neighborhood $B_\varepsilon(\overline{x})$ of $\overline{x}$, the *branch problem* (BNLP$\{\mathcal{I}_1,\mathcal{I}_2\}$) for $\{\mathcal{I}_1;\mathcal{I}_2\}$ is defined as

$$\begin{aligned} \min_{x \in \mathbb{R}^n} \quad & f(x) && \text{(BNLP}\{\mathcal{I}_1,\mathcal{I}_2\}\text{)} \\ \text{s.t.} \quad & \boldsymbol{h}_{\mathcal{I}_{0+}}(x) \geqslant \boldsymbol{0}, \quad \boldsymbol{h}_{\mathcal{I}_1}(x) \geqslant \boldsymbol{0}, \quad \boldsymbol{h}_{\mathcal{I}_2}(x) = \boldsymbol{0}, \quad \boldsymbol{h}_{\mathcal{I}_{0-}}(x) = \boldsymbol{0}, \\ & \boldsymbol{g}_{\mathcal{I}_{+0}}(x) \geqslant \boldsymbol{0}, \quad \boldsymbol{g}_{\mathcal{I}_1}(x) \geqslant \boldsymbol{0}. \end{aligned}$$

△

There obviously exist $|\mathcal{P}(\mathcal{I}_{00})| = 2^{|\mathcal{I}_{00}|} \leqslant 2^l$ different branch problems.

**Lemma 5.1 (Feasible Sets of the Auxiliary Problems)**
For the feasible sets of the auxiliary problems, the following relations hold:

$$\mathcal{F}_{\text{TNLP}} \overset{1.}{=} \bigcap_{\substack{\mathcal{J}_1 \cup \mathcal{J}_2 = \mathcal{I}_{00} \\ \mathcal{J}_1 \cap \mathcal{J}_2 = \emptyset}} \mathcal{F}_{\text{BNLP}\{\mathcal{J}_1;\mathcal{J}_2\}} \overset{2.}{\subset} \mathcal{F}_{\text{BNLP}\{\mathcal{I}_1;\mathcal{I}_2\}} \overset{3.}{\subset} \mathcal{F} \overset{4.}{=} \bigcup_{\substack{\mathcal{J}_1 \cup \mathcal{J}_2 = \mathcal{I}_{00} \\ \mathcal{J}_1 \cap \mathcal{J}_2 = \emptyset}} \mathcal{F}_{\text{BNLP}\{\mathcal{J}_1;\mathcal{J}_2\}} \overset{5.}{\subset} \mathcal{F}_{\text{RNLP}}. \quad (5.15)$$

△

**Proof** For 1., consider that for every constraint $i \in \{1, \dots, l\}$ there exists a partition with $i \in \mathcal{J}_1$ and one with $i \in \mathcal{J}_2$. Hence $g_i(\boldsymbol{x}) \geqslant 0$ and $h_i(\boldsymbol{x}) = 0$ hold for the intersection set, satisfying (TNLP). The reverse direction is obvious. Inclusion 2. is obvious as the intersection includes the case $\{\mathcal{I}_1; \mathcal{I}_2\} = \{\mathcal{J}_1; \mathcal{J}_2\}$. Inclusion 3. is obvious as any choice of $\mathcal{I}_2 \neq \emptyset$ is a proper restriction of $\mathcal{F}$. Forming the union of feasible sets of the branch problem lifts any restrictions imposed by choosing a specific $\mathcal{J}_2 \neq \emptyset$, and we get equivalence 4. Finally, problem (RNLP) allows for a violation of vanishing constraints contained in $\mathcal{I}_{00}$ in a neighborhood of $\overline{\boldsymbol{x}}$, which justifies 5. □

### 5.4.2 Lack of Constraint Qualification for MPVCs

The initial observation that the constraints found in $\mathcal{I}_{00}$ induce the combinatorial nature of MPVCs gives rise to the following constraint qualification for MPVC.

**Definition 5.8 (Lower Level Strict Complementarity Condition)**
Let $\overline{\boldsymbol{x}} \in \mathbb{R}^n$ be a feasible point of problem (5.9). If $\mathcal{I}_{00} = \emptyset$ in $\overline{\boldsymbol{x}}$, we say that *Lower Level Strict Complementarity* (LLSCC) holds for problem (5.9) in $\overline{\boldsymbol{x}}$. △

This complementarity condition is too restrictive, though, and will not be satisfied on the entire feasible set by most interesting MPVC problem instances. The same has been observed for strict complementarity conditions for MPECs as well, cf. [144].

If LLSCC is violated, the feasible set $\mathcal{F}$ has a combinatorial structure and the constraints of (5.9) inevitably are degenerate. Constraint qualifications commonly encountered in nonlinear programming are thus violated, as could already be seen in the first section of this chapter.

**Lemma 5.2 (Violation of LICQ)**
Let $\overline{\boldsymbol{x}} \in \mathbb{R}^n$ be a feasible point of (5.9), and let $\mathcal{I}_0 \neq \emptyset$. Then LICQ is violated in $\overline{\boldsymbol{x}}$. △

**Proof** For any $i \in \mathcal{I}_0$ we have $h_i(\overline{\boldsymbol{x}}) = 0$ and thus $(g_i \cdot h_i)_{\boldsymbol{x}}(\overline{\boldsymbol{x}}) = g_i(\overline{\boldsymbol{x}}) \cdot (h_i)_{\boldsymbol{x}}(\overline{\boldsymbol{x}})$, i.e., the gradient of $(g_i \cdot h_i)(\boldsymbol{x}) \geqslant 0$ is a multiple of that of $h_i(\boldsymbol{x}) \geqslant 0$. Since both constraints are active in $\overline{\boldsymbol{x}}$, LICQ is violated. □

**Lemma 5.3 (Violation of MFCQ)**
Let $\overline{\boldsymbol{x}} \in \mathbb{R}^n$ be a feasible point of (5.9), and let $\mathcal{I}_{00} \cup \mathcal{I}_{0-} \neq \emptyset$. Then MFCQ is violated in $\overline{\boldsymbol{x}}$. △

**Proof** For any $i \in \mathcal{I}_{00}$ we have $(g_i \cdot h_i)_{\boldsymbol{x}}(\overline{\boldsymbol{x}}) = \boldsymbol{0}$, so $(g_i \cdot h_i)_{\boldsymbol{x}}(\overline{\boldsymbol{x}})^T \boldsymbol{d} = \boldsymbol{0}$ for all vectors $\boldsymbol{d} \in \mathbb{R}^n$, hence MFCQ is violated. For any $i \in \mathcal{I}_{0-}$ and for any $\boldsymbol{d} \in \mathbb{R}^n$ which satisfies $(h_i)_{\boldsymbol{x}}(\overline{\boldsymbol{x}})^T \boldsymbol{d} \geqslant 0$ we find $(g_i \cdot h_i)_{\boldsymbol{x}}(\overline{\boldsymbol{x}})^T \boldsymbol{d} \leqslant 0$, hence MFCQ is violated in this case, too. □

In order to investigate the remaining CQs we require a representation of the tangent cone and the linearized cone of the feasible set of (5.9), which is given by the following lemma.

**Lemma 5.4 (Tangent and Linearized Cone of MPVC)**
For MPVCs, it holds that

$$\mathcal{T}_{\mathrm{MPVC}}(\overline{\boldsymbol{x}}, \mathcal{F}) = \bigcup_{\substack{\mathcal{J}_1 \cup \mathcal{J}_2 = \mathcal{I}_{00} \\ \mathcal{J}_1 \cap \mathcal{J}_2 = \emptyset}} \mathcal{T}_{\mathrm{BNLP}\{\mathcal{J}_1;\mathcal{J}_2\}}(\overline{\boldsymbol{x}}, \mathcal{F}) \subseteq \bigcup_{\substack{\mathcal{J}_1 \cup \mathcal{J}_2 = \mathcal{I}_{00} \\ \mathcal{J}_1 \cap \mathcal{J}_2 = \emptyset}} \mathcal{L}_{\mathrm{BNLP}\{\mathcal{J}_1;\mathcal{J}_2\}}(\overline{\boldsymbol{x}}) = \mathcal{L}_{\mathrm{MPVC}}(\overline{\boldsymbol{x}}). \quad \triangle$$

**Proof** A proof of this relation can be found in [105]. □

As can be seen, the tangent cone of (5.9) is the union of finitely many NLP tangent cones. Consequentially it is in general a non–polyhedral and non–convex set that does not coincide with the convex and polyhedral linearized cone, in which case ACQ is violated. Necessary conditions for ACQ to hold for problem (5.9) can be found in [105]. Under reasonable assumptions, only GCQ can be shown to hold for (5.9).

**Theorem 5.1 (Sufficient Condition for GCQ)**
Let $\overline{\boldsymbol{x}} \in \mathbb{R}^n$ be a feasible point of (5.9), and let LICQ for the auxiliary problem (TNLP) hold in $\overline{\boldsymbol{x}}$. Then GCQ holds in $\overline{\boldsymbol{x}}$ for problem (5.9). △

**Proof** A proof can be found in [105]. □

This gives rise to the introduction of the following constraint qualification.

**Definition 5.9 (MPVC Linear Independence Constraint Qualification)**
Let $\overline{x} \in \mathbb{R}^n$ be a feasible point of problem (5.9). We say that *MPVC–Linear Independence Constraint Qualification* holds in $\overline{x}$ if the Jacobian

$$\begin{bmatrix} \left(h_{\mathcal{I}_0}\right)_x (\overline{x}) \\ \left(g_{\mathcal{I}_{+0}}\right)_x (\overline{x}) \\ \left(g_{\mathcal{I}_{00}}\right)_x (\overline{x}) \end{bmatrix} \tag{5.16}$$

has full row rank. △

Using this definition we can restate theorem 5.1 by saying that for a feasible point of problem (5.9), MPVC–LICQ is a sufficient condition for GCQ to hold. Consequentially, under MPVC–LICQ a locally optimal point of problem (5.9) satisfies the KKT conditions of theorem 3.1.

**Remark 5.10 (Strength of MPVC–LICQ)**
The assumption of MPVC–LICQ is sometimes held for too strong, e.g. in the analogous case of MPECs, while on the other hand the classical KKT theorem cannot be shown to hold for MPVCs under less restrictive CQs.

For the case of constraints treated by outer convexification however, it is reasonable to assume GCQ and thus MPVC–LICQ to hold. Here, $h_{\mathcal{I}_0}$ holds active simple lower bounds on the convex multipliers, while $g_{\mathcal{I}_{00}}$ and $g_{\mathcal{I}_{+0}}$ hold active vanishing constraints independent of the convex multipliers. This allows us to retain the concept of iterating towards KKT based local optimality.

### 5.4.3 An MPVC Lagrangian Framework

The standard Lagrangian function according to definition 3.6 for problem 5.9 reads

$$L(x, \lambda) \stackrel{\text{def}}{=} f(x) - \lambda_{\text{h}}^T h(x) - \sum_{i=1}^{l} \lambda_{\text{gh},i} g_i(x) h_i(x), \tag{5.17}$$

wherein $\lambda_{\text{h}} \in \mathbb{R}^l$ and $\lambda_{\text{gh}} \in \mathbb{R}^l$ denote the Lagrange multipliers for the constraints $h(x) \geqslant 0$ and $g_i(x) \cdot h_i(x) \geqslant 0$ respectively. The multiplicative representation of the vanishing constraints has been shown to be the source of severe ill–conditioning in section 5.2. We introduce a modification of (5.17) that separates the vanishing constraints' parts $g$ and $h$.

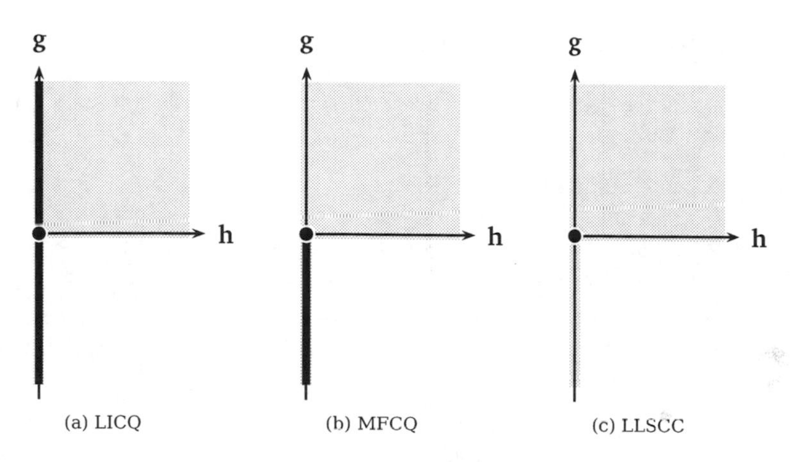


(a) LICQ (b) MFCQ (c) LLSCC

Figure 5.7: Areas of violation of constraint qualifications (■). The feasible set is indicated by ( ).

**Definition 5.10 (MPVC Lagrangian Function)**
The function $\Lambda : \mathbb{R}^{n^x} \times \mathbb{R}^l \times \mathbb{R}^l \to \mathbb{R}$,

$$\Lambda(x, \mu_g, \mu_h) \stackrel{\text{def}}{=} f(x) - \mu_h^T h(x) - \mu_g^T g(x) \tag{5.18}$$

with $x \in \mathbb{R}^n$ and $\mu_g, \mu_h \in \mathbb{R}^l$ is called the *MPVC Lagrangian (function)* of problem (5.9). △

Based on the KKT conditions for the auxiliary problem (RNLP) we define the notion of strong MPVC stationarity.

**Definition 5.11 (Strong Stationarity)**
Let $x^\star \in \mathbb{R}^n$ be a feasible point of (5.9). The point $x^\star$ is called a *strongly stationary* point of (5.9) if it is stationary for the relaxed auxiliary problem (RNLP) in the classical sense. There exist MPVC multipliers $\mu_g^\star \in \mathbb{R}^l$ and $\mu_h^\star \in \mathbb{R}^l$ such that

$$\begin{aligned} &\Lambda_x(x^\star, \mu_g^\star, \mu_h^\star) = 0, \\ &\qquad \mu^\star_{h,\mathcal{I}_{0+}} \geqslant 0, \quad \mu^\star_{h,\mathcal{I}_{00}} \geqslant 0, \quad \mu^\star_{h,\mathcal{I}_+} = 0, \\ &\qquad \mu^\star_{g,\mathcal{I}_{+0}} \geqslant 0, \quad \mu^\star_{g,\mathcal{I}_{++}} = 0, \quad \mu^\star_{g,\mathcal{I}_0} = 0. \end{aligned} \tag{5.19}$$

△

The strongly stationary points of an MPVC can be shown to coincide with the KKT points, as mentioned before.

**Theorem 5.2 (Strong Stationarity)**
Let $\boldsymbol{x}^\star \in \mathbb{R}^n$ be a feasible point of problem (5.9). Then it holds that $\boldsymbol{x}^\star$ is a KKT point of problem (5.9) satisfying theorem 3.1 if and only if $\boldsymbol{x}^\star$ is a strongly stationary point of problem (5.9). △

**Proof** A proof can be found in [3]. □

Hence, if the assumptions for theorem (3.1) are satisfied, strong stationarity is a necessary condition for local optimality.

**Theorem 5.3 (First Order Necessary Optimality Condition under GCQ)**
Let $\boldsymbol{x}^\star \in \mathbb{R}^n$ be a locally optimal point of (5.9) and let GCQ hold in $\boldsymbol{x}^\star$. Then $\boldsymbol{x}^\star$ is a strongly stationary point of (5.9). △

**Proof** Every locally optimal point satisfying GCQ is a KKT point by theorem 3.1. Every KKT point is strongly stationary by theorem 5.2 □

Theorem 5.3 can be recast using the newly introduced notion of MPVC–LICQ as follows. In addition, we have uniqueness of the MPVC multipliers analogous to the case of LICQ for NLPs.

**Theorem 5.4 (First Order Necessary Optimality Cond. under MPVC–LICQ)**
Let $\boldsymbol{x}^\star \in \mathbb{R}^n$ be a locally optimal point of (5.9) and let MPVC–LICQ hold in $\boldsymbol{x}^\star$. Then $\boldsymbol{x}^\star$ is a strongly stationary point of (5.9). The MPVC–multipliers $\boldsymbol{\mu}_{\mathrm{g}}$ and $\boldsymbol{\mu}_{\mathrm{h}}$ are unique. △

**Proof** MPVC–LICQ implies GCQ by definition. Uniqueness of the MPVC–multipliers follows from linear independence of the gradients. □

We have thus obtained a constrained nonlinear programming framework for MPVC under the strong assumption of MPVC–LICQ. We have however shown that this CQ may be assumed to hold for NLPs with constraints treated by outer convexification. The separation of the vanishing constraint functions $\boldsymbol{g}$ and $\boldsymbol{h}$ resolves ill–conditioning of the Jacobian and consequentially the undesirable phenomena examined in section 5.2.

Less restrictive stationarity conditions, such as the notions of Mordukhovich– or M–stationarity and Clarke– or C–stationarity, are derived in [105]. They do not require GCQ to hold for (5.9) but instead only assume GCQ for all branch problems, which is referred to as MPVC–GCQ. As strong stationarity cannot be expected to constitute a necessary condition for optimality under such CQs, M–stationary points are not necessarily KKT points of problem (5.9). Vice versa, it holds that every strongly stationary point is an M–stationary point. Again both notions of stationarity coincide if the critical index set $\mathcal{I}_{00}$ is empty.

### 5.4.4 Second Order Conditions

Second-order necessary and sufficient conditions for local optimality of a candidate point can be devised under MPVC–LICQ, similar to those of theorems 3.3 and 3.4 that hold for NLPs under LICQ. They are based on the MPVC reduced Hessian, defined analogously to the reduced Hessian of the Lagrangian in definition 3.9.

**Theorem 5.5 (Second Order Necessary Optimality Condition)**
Let $\boldsymbol{x}^\star$ be a locally optimal point of problem (5.9), and let MPVC–LICQ hold in $\boldsymbol{x}^\star$. Then it holds that

$$\forall \boldsymbol{d} \in \mathcal{L}_{\text{MPVC}}(\boldsymbol{x}^\star) : \ \boldsymbol{d}^T \Lambda_{xx}(\boldsymbol{x}^\star, \boldsymbol{\mu}_{\text{g}}, \boldsymbol{\mu}_{\text{h}})\, \boldsymbol{d} \geqslant 0,$$

where $\boldsymbol{\mu}_{\text{g}}$ and $\boldsymbol{\mu}_{\text{h}}$ are the MPVC–multipliers satisfying strong stationarity. △

**Proof** A proof can be found in [105]. □

**Theorem 5.6 (Second Order Sufficient Optimality Condition)**
Let $(\boldsymbol{x}^\star, \boldsymbol{\mu}_{\text{g}}, \boldsymbol{\mu}_{\text{h}})$ be a strongly stationary point of problem (5.9). Further, let

$$\forall \boldsymbol{d} \in \mathcal{L}_{\text{MPVC}}(\boldsymbol{x}^\star) \setminus \{\boldsymbol{0}\} : \ \boldsymbol{d}^T \Lambda_{xx}(\boldsymbol{x}^\star, \boldsymbol{\mu}_{\text{g}}, \boldsymbol{\mu}_{\text{h}})\, \boldsymbol{d} > 0,$$

hold. Then $\boldsymbol{x}^\star$ is a locally strictly optimal point of problem (5.9). △

**Proof** A proof can be found in [105]. □

## 5.5 Summary

In this chapter we have investigated the theoretical properties of NLPs with constraints that arise from the application of outer convexification to MIOCP constraints directly depending on a binary or integer control variable. CQs frequently assumed to hold in the design of descent based NLP algorithms, such as LICQ and MFCQ, have been shown to be violated by such NLPs. Consequences for the convergence behavior of such algorithms have been examined and new explanations for phenomena such as numerical ill–conditioning, infeasible steps, and cycling of the active set have been given. The NLPs under investigation have been identified as MPVCs for the first time, a challenging and very young problem class whose feasible set has a nonconvex combinatorial structure. An constrained nonlinear programming framework for MPVCs has been presented that allows to retain the KKT theorem and resolves a major cause of numerical ill–conditioning by introducing a Lagrangian function with separated constraint parts.

# 6 A Nonconvex Parametric SQP Method

The real–time iteration scheme described for Nonlinear Model Predictive Control (NMPC) in chapter 4 was based on the idea of repeatedly performing a single iteration of a Newton–type method to compute control feedback. We have seen that it can be combined with the convexification and relaxation method of chapter 2 to develop a new mixed–integer real–time iteration scheme. For Mixed–Integer Optimal Control Problems (MIOCPs) with constraints depending directly on an integer control, that method yields an Nonlinear Program (NLP) with vanishing constraints as seen in chapter 5. In this chapter we develop a new Sequential Quadratic Programming (SQP) method to solve the arising class of NLPs with vanishing constraints in an active–set framework. It generates a sequence of local quadratic subproblems inheriting the vanishing constraints property from the NLP. These local subproblems are referred to as Quadratic Programs with Vanishing Constraints (QPVCs) and are the subproblems to be solved in the mixed–integer real–time iteration scheme. For this purpose, we develop a new active set strategy that finds strongly stationary points of a QPVC with convex objective function but nonconvex feasible set. This active set strategy is based on the idea of using parametric quadratic programming techniques to efficiently move between convex subsets of the QPVC's nonconvex feasible set. Strongly stationary points of QPVCs are locally optimal, and we develop a heuristic that improves these points to global optimality.

## 6.1 SQP for Nonconvex Programs

Current research on numerical methods for the solution of Mathematical Programs with Vanishing Constraints (MPVCs) mainly focuses on regularization schemes for interior point methods, cf. [3, 105, 106, 107], as convergence analyses available for Mathematical Programs with Equilibrium Constraints (MPECs) can often be readily transferred to the case of MPVC. SQP methods for a broader class of nonlinear problems with combinatorial structure, such as MPECs and bi–level optimization problems, have been considered in [193] and have been found to require a means of finding and verifying a stationary or locally optimal solution of the nonconvex subproblems. For the SQP framework described there, [109] conjectures that MPVC strong stationarity conditions under MPVC–LICQ could be used to design an active set based method for the subproblems.

In this section, we pursue this idea and develop a parametric SQP methods on nonconvex feasible sets and show how it can be used for the solution of NLPs with vanishing constraints, arising from the application of the convexification and relaxation approach of chapter 2 to MIOCPs with constraints directly depending on binary or integer controls.

### 6.1.1 SQP on Nonconvex Feasible Sets

Extensions of SQP methods to nonconvex problems with combinatorial structure are e.g. described in [193]. We consider the NLP

$$\begin{aligned} \min_{\boldsymbol{x}\in\mathbb{R}^n} \quad & f(\boldsymbol{x}) \\ \text{s.t.} \quad & \boldsymbol{g}(\boldsymbol{x}) \in \mathcal{F}, \end{aligned} \tag{6.1}$$

where the feasible set $\mathcal{F}$ is assumed to show a combinatorial structure. Note that $\mathcal{F} \subset \mathbb{R}^{n^g}$ here is a subset of the image space of the constraint function $\boldsymbol{g}$. Given a primal–dual point $(\boldsymbol{x}, \boldsymbol{\lambda})$, the SQP subproblem reads

$$\begin{aligned} \min_{\delta\boldsymbol{x}\in\mathcal{D}\subseteq\mathbb{R}^n} \quad & \tfrac{1}{2}\delta\boldsymbol{x}^T L_{\boldsymbol{x}\boldsymbol{x}}(\boldsymbol{x},\boldsymbol{\lambda})\delta\boldsymbol{x} + f_{\boldsymbol{x}}(\boldsymbol{x})\delta\boldsymbol{x} \\ \text{s.t.} \quad & \boldsymbol{g}(\boldsymbol{x}) + \boldsymbol{g}_{\boldsymbol{x}}(\boldsymbol{x})\delta\boldsymbol{x} \in \mathcal{F}. \end{aligned} \tag{6.2}$$

We are hence interested in the assumptions that need to be imposed on the shape of the structurally nonconvex and nonlinear feasible set $\mathcal{F}$ of problem (6.1) in order to establish local convergence of the SQP method.

**Definition 6.1 (Local Star Shape)**
A set $\mathcal{F}$ is called *locally star–shaped* in a feasible point $\overline{\boldsymbol{x}} \in \mathcal{F}$ if there exists a ball $\mathcal{B}_\varepsilon(\overline{\boldsymbol{x}})$ such that it holds that

$$\forall \boldsymbol{x} \in \mathcal{F}\cap\mathcal{B}_\varepsilon(\overline{\boldsymbol{x}}),\ \alpha\in[0,1]\subset\mathbb{R}:\ \alpha\boldsymbol{x} + (1-\alpha)\overline{\boldsymbol{x}} \in \mathcal{F}\cap\mathcal{B}_\varepsilon(\overline{\boldsymbol{x}}). \tag{6.3}$$

△

It is easily verified that convex sets as well as finite intersections and finite unions of closed convex sets are locally star–shaped. Under the restriction of local starshapedness, convergence is established by the following theorem.

**Theorem 6.1 (Local Convergence of SQP)**
Let $\boldsymbol{x}^\star$ be a stationary point of (6.1), and let the feasible set $\mathcal{F}$ be locally star-shaped in $\boldsymbol{g}(\boldsymbol{x}^\star)$. Let strict complementary hold in $\boldsymbol{x}^\star$ with Lagrange multipliers $\boldsymbol{\lambda}^\star$ and let the reduced Hessian be positive definite in $(\boldsymbol{x}^\star, \boldsymbol{\lambda}^\star)$. Then the exact Hessian SQP method shows local q-quadratic convergence to the stationary point $\boldsymbol{x}^\star$. △

**Proof** A proof can be found in [193]. □

These are indeed just the strong second order sufficient conditions of theorem 3.4 and the local convergence theorem 3.7. Strict complementarity permits the extension to inequality constraints as already mentioned. Finally, local star shape of the feasible set according to definition 6.1 guarantees that the stationary point $(\boldsymbol{x}^\star, \boldsymbol{\lambda}^\star)$ indeed is unique in a local neighborhood.

### 6.1.2 Nonconvex Subproblems

The local subproblems generated by the SQP method described in the previous section are Quadratic Programs (QPs) with combinatorial structure of the feasible set. For the combinatorial NLP (6.1) we assume that there exist finitely many subsets $\mathcal{F}_i$, $1 \leqslant i \leqslant n^{\mathrm{F}}$ that are of simpler, convex shape such that a descent-based method is able to find a locally optimal solution of the subproblems

$$\begin{aligned} \min_{\boldsymbol{x}} \quad & \tfrac{1}{2}\boldsymbol{x}^T\boldsymbol{H}\boldsymbol{x} + \boldsymbol{x}^T\boldsymbol{g} \\ \text{s.t.} \quad & \boldsymbol{A}\boldsymbol{x} + \boldsymbol{b} \in \mathcal{F}_i, \end{aligned} \tag{6.4}$$

if given a feasible initializer in $\mathcal{F}_i$.

Local optimality of a candidate point $\boldsymbol{x}^\star$ for the subproblems must be linked to local optimality of the combinatorial NLP (6.1). In particular, if (6.1) has a locally optimal point $\boldsymbol{x}^\star$, this point should satisfy $\boldsymbol{A}\boldsymbol{x}^\star + \boldsymbol{b} \in \mathcal{F}_i$ for at least one of the subproblems (6.4). Vice versa, if $\boldsymbol{x}^\star$ is a locally optimal point for one of the subproblems (6.4), and if this point is also locally optimal for all subproblems $j$ that satisfy $\boldsymbol{x}^\star \in \mathcal{F}_j$, then $\boldsymbol{x}^\star$ should be a locally optimal point of (6.1), cf. [193]. An obvious sufficient condition for both requirements to hold is that

$$\mathcal{F} = \bigcup_{i=1}^{n^{\mathrm{F}}} \mathcal{F}_i. \tag{6.5}$$

Note that in general the subsets $\mathcal{F}_i$ cover $\mathcal{F}$ will not form a partition of $\mathcal{F}$, i.e., they need not be pairwise disjoint but may overlap each other. In particular, if $\mathcal{F}$ is closed the subsets $\mathcal{F}_i$ will be closed and share boundary points with adjacent ones. This gives rise to the notion of *adjacent subproblems* as follows.

**Definition 6.2 (Adjacent Subproblems)**
A local subproblem on a subset $\mathcal{F}_i \subset \mathcal{F}$ is called *adjacent to a point* $\boldsymbol{x} \in \mathbb{R}^{n^x}$ if $\boldsymbol{x} \in \mathcal{F}_i$ holds. Two subproblems on subsets $\mathcal{F}_i$, $\mathcal{F}_j$, $i \neq j$ are called adjacent if a locally optimal point $\boldsymbol{x}^\star$ of subproblem $i$ satisfies $\boldsymbol{x}^\star \in \mathcal{F}_j$. △

Starting with an initial guess and an associated initial subproblem with convex feasible set $\mathcal{F}_i \subset \mathcal{F}$ the nonconvex SQP step is computed as follows: After finding a locally optimal point $\boldsymbol{x}^\star \in \mathcal{F}_i$ using an active set method, optimality is checked for each adjacent subproblem. If optimality does not hold for one or more adjacent subproblems, one of those is chosen and the solution is continued therein. This procedure is iterated until local optimality holds for a subproblem and all adjacent ones.

**Theorem 6.2 (Finite Termination)**
For convex subproblems with regular solutions, the described decomposition procedure terminates after a finite number of iterations with either a local minimizer or an infeasible subproblem. △

**Proof** A proof can be found in [193]. □

Several points need to be addresses in more detail in order to make this approach viable. First, it is obviously critical for the performance of the nonconvex SQP method that the convex subsets $\mathcal{F}_i$ of the feasible set $\mathcal{F}$ be chosen as large as possible in order to avoid frequent continuations of the solution process. Second, as $\mathcal{F}$ is nonconvex, such continuations certainly cannot be avoided at all, and the active set method used to solve the subproblem should thus allow for efficient warm starts. Finally, both the verification of local optimality and the decision on an adjacent subproblem to continue the solution process need to exploit multiplier information in order to make efficient progress towards a locally optimal solution of (6.1).

### 6.1.3 SQP for a Special Family of MPVCs

We are concerned with applying the presented nonconvex SQP framework to the particular case of MPVCs obtained by outer convexification of discretized MIOCPs.

### Local Convergence

By theorem 5.2 the notion of stationarity in theorem 6.1 can under MPVC–LICQ be substituted by the equivalent MPVC strong stationarity condition of definition 5.11. The notion of strict complementarity then refers to the MPVC multipliers. The Hessian approximation methods of chapter 3 can be readily used also to approximate the Hessian of the MPVC Lagrangian. In the BFGS case, the well–defined MPVC multiplier step is used in place of the potentially unbounded Lagrange multiplier step. Finally, note that the feasible sets obtained for the MPVCs are indeed locally star–shaped by lemma 5.4.

### Quadratic Programs with Vanishing Constraints

The local subproblems to be solved in each iteration of the nonconvex SQP method for MPVCs obtained by outer convexification of discretized MIOCPs are referred to as Quadratic Programs with Vanishing Constraints (QPVCs), to be defined next.

**Definition 6.3 (Quadratic Program with Vanishing Constraints)**
The following extension of problem (6.8)

$$\begin{aligned} \min_{\boldsymbol{x} \in \mathcal{D} \subseteq \mathbb{R}^n} \quad & \tfrac{1}{2}\boldsymbol{x}^T \boldsymbol{H} \boldsymbol{x} + \boldsymbol{x}^T \boldsymbol{g} \\ \text{s.t.} \quad & \boldsymbol{0} \leqslant \boldsymbol{A}\boldsymbol{x} - \boldsymbol{b}, \\ & 0 \leqslant (\boldsymbol{C}_{i\star}\boldsymbol{x} - c_i)(\boldsymbol{D}_{i\star}\boldsymbol{x} - d_i) \qquad 1 \leqslant i \leqslant l, \\ & \boldsymbol{0} \leqslant \boldsymbol{D}\boldsymbol{x} - \boldsymbol{d}, \end{aligned} \tag{6.6}$$

where $\boldsymbol{C}, \boldsymbol{D} \in \mathbb{R}^{l \times n}$ and $\boldsymbol{c}, \boldsymbol{d} \in \mathbb{R}^l$, is called a *Quadratic Program with Vanishing Constraints (QPVC).*

△

Herein, the affine linear parts

$$\begin{aligned} \boldsymbol{g}(\boldsymbol{x}) &\stackrel{\text{def}}{=} \boldsymbol{C}\boldsymbol{x} - \boldsymbol{c}, \\ \boldsymbol{h}(\boldsymbol{x}) &\stackrel{\text{def}}{=} \boldsymbol{D}\boldsymbol{x} - \boldsymbol{d} \end{aligned}$$

take the role of the controlling constraint $\boldsymbol{h}(\boldsymbol{x}) \geqslant \boldsymbol{0}$ and the vanishing constraint $\boldsymbol{g}(\boldsymbol{x}) \geqslant \boldsymbol{0}$ of the previous chapter.

### A Sufficient Condition for Local Optimality

A remarkable result from [105] holds for MPVCs with convex objective and affine linear parts of the vanishing constraints, which in particular includes QPVCs with positive semidefinite Hessian $\boldsymbol{H}$. Although the problem class of QPVCs must truly be considered nonconvex due to the shape of its feasible set, the strong stationarity conditions of definition 5.11 are indeed also sufficient conditions for local optimality.

**Theorem 6.3 (QPVC First Order Sufficient Optimality Condition)**
Let the objective function $f$ be convex and let $\boldsymbol{g}$, $\boldsymbol{h}$ be affine linear in problem (5.9). Further, let $\boldsymbol{x}^\star$ be a strongly stationary point of problem (5.9). Then $\boldsymbol{x}^\star$ is a locally optimal point of problem (5.9). △

**Proof** A proof can be found in [105]. □

As equivalence to the Karush–Kuhn–Tucker (KKT) conditions strong stationarity has already been shown to be a necessary condition, the notions of stationarity and local optimality of the subproblems' solutions therefore coincide for QPVCs.

### A Sufficient Condition for Global Optimality

Figure 6.1 depicts two QPVCs that immediately reveal the difference between local and global optimality on the nonconvex feasible set. In particular, an active vanishing constraint that is not permitted to vanish may cause local minima different from the global one as can be seen from figure 6.1b. This observation is a motivation of the following sufficient condition for global optimality due to [105].

**Theorem 6.4 (Sufficient Global Optimality Condition)**
Let the objective function $f$ be convex and let $\boldsymbol{g}$, $\boldsymbol{h}$ be affine linear in problem (5.9). Further, let $\boldsymbol{x}^\star$ be a strongly stationary point of problem (5.9). Then if $\boldsymbol{\mu}_{\mathrm{h},\mathcal{I}_{0-}} \geqslant \boldsymbol{0}$ and $\boldsymbol{\mu}_{\mathrm{g},\mathcal{I}_{+0}} = \boldsymbol{0}$, it holds that $\boldsymbol{x}^\star$ is a globally optimal point of problem (5.9). △

**Proof** A proof can be found in [105]. □

### Adjacency for QPVCs

The nonconvex SQP framework of section 6.1.1 introduced the notion of adjacency for the convex subsets of the feasible set. Local optimality of a

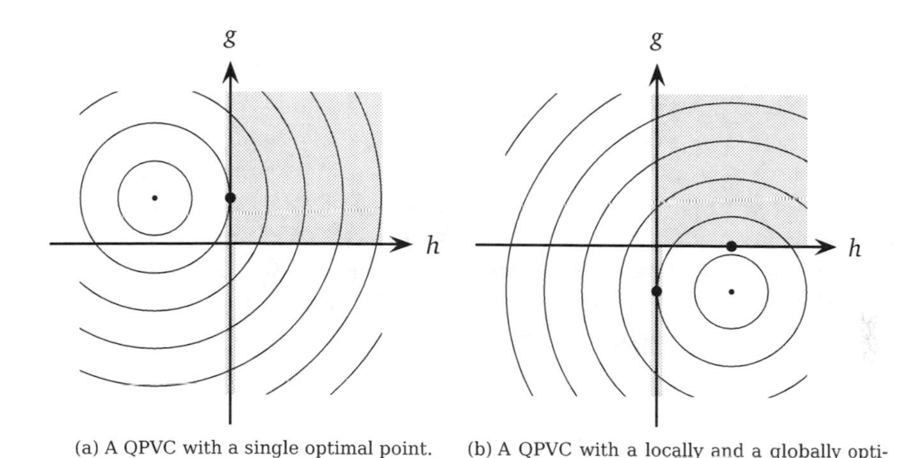


(a) A QPVC with a single optimal point. (b) A QPVC with a locally and a globally optimal point.

Figure 6.1: Locally and globally optimal points of a convex QPVC.

candidate point on the border of one convex subset needs to be verified for all adjacent convex subsets. For the case of QPVCs, borders of convex subsets are defined by active vanishing constraints $g_i(\boldsymbol{x}) = 0$, i.e., by those constraints found in the index subset $\mathcal{I}_{+0}$ of the active set, and by active controlling constraints $h_i(\boldsymbol{x}) = 0$ if the vanishing constraint is infeasible, i.e., by those constraints found in the index subset $\mathcal{I}_{0-}$. We introduce the following definition of adjacency.

**Definition 6.4 (Adjacency)**
Let $\mathcal{C}^1 \subset \mathcal{F}_{\text{QPVC}}$ and $\mathcal{C}^2 \subset \mathcal{F}_{\text{QPVC}}$ be convex subsets of the feasible set $\mathcal{F}_{\text{QPVC}}$ of problem (6.6). Assume that $\mathcal{A}^1$ and $\mathcal{A}^2$ are the active sets for all points of these subsets,

$$\forall \boldsymbol{x} \in \mathcal{C}^k : \ \mathcal{A}(\boldsymbol{x}) = \mathcal{A}^k, \qquad k \in \{1, 2\}. \tag{6.7}$$

Then the two convex subsets are called QPVC–*adjacent* if there exists a vanishing constraint $1 \leqslant i \leqslant l$ such that constraint $i$ is active in $\mathcal{A}_1$ but has vanished in $\mathcal{A}_2$, and the active sets are otherwise identical,

$$\begin{aligned} &\mathcal{I}^2_{0-} = \mathcal{I}^1_{0-} \cup \{i\}, && \mathcal{I}^2_{+0} = \mathcal{I}^1_{+0} \setminus \{i\}, \\ &\mathcal{I}^2_{00} = \mathcal{I}^1_{00}, && \mathcal{I}^2_{0+} = \mathcal{I}^1_{0+}, && \mathcal{I}^2_{++} = \mathcal{I}^1_{++}. \end{aligned}$$

△

Loosely speaking, two convex subsets are adjacent if a vanishing constraint that is active in one of them has vanished in the other.

## 6.2 Parametric Quadratic Programs

We have seen in chapter 4 that the real–time iteration scheme, by performing only one SQP iteration per control feedback, essentially solves a sequence of closely related QPs. Parametric quadratic programming offers a convenient framework for relating two subsequent QPs to each other by an affine linear homotopy. We will in the sequel of this chapter see that this relation can also be exploited to efficiently move between convex subsets of the feasible set of a QPVC. Our presentation of the most important facts about Parametric Quadratic Programs (PQPs) is guided by [25], [67], and [216].

### 6.2.1 Parametric Quadratic Programs

The constrained nonlinear optimization theory of chapter 3 obviously holds for the special case of a QP. Some results can be strengthened by convexity of the objective and affine linearity of the constraints, as detailed in this section.

**Definition 6.5 (Quadratic Program)**
The standard form of a Quadratic Program is the problem

$$\begin{aligned} \min_{x \in \mathbb{R}^n} \quad & \tfrac{1}{2} x^T H x + x^T g \\ \text{s.t.} \quad & A x \geqslant b, \end{aligned} \tag{6.8}$$

where $g \in \mathbb{R}^n$ is the linear term, $H \in \mathbb{R}^{n \times n}$ is the symmetric quadratic term or *Hessian* matrix, $A \in \mathbb{R}^{m \times n}$ is the constraint matrix, $b \in \mathbb{R}^m$ is the constraint right-hand side vector. △

**Definition 6.6 (Convexity)**
A QP is called *convex* iff the quadratic term $H \in \mathbb{R}^{n \times n}$ is positive semidefinite, and *strictly convex* iff $H$ is positive definite. It is called *nonconvex* otherwise. △

If feasible, strictly convex QPs have a unique solution. Local optimality can be determined by the KKT conditions of theorem 3.1, and global optimality follows from strict convexity. For convex QPs, this solution still exists but is not necessarily unique. For nonconvex QPs, the problem of determining whether a feasible point is a globally optimal solution of (6.8) is $\mathcal{NP}$–hard, cf. [157]. In the following, we restrict our presentation to convex QPs, and to strictly convex QPs where it is appropriate.

The problem class of PQPs is defined as follows.

**Definition 6.7 (Parametric Quadratic Program)**
The QP in standard form depending on a parameter $\boldsymbol{p} \in \mathbb{R}^{n^{\mathrm{p}}}$

$$\begin{aligned} \min_{\boldsymbol{x} \in \mathbb{R}^n} \quad & \tfrac{1}{2}\boldsymbol{x}^T \boldsymbol{H}\boldsymbol{x} + \boldsymbol{x}^T \boldsymbol{g}(\boldsymbol{p}) \\ \text{s.t.} \quad & \boldsymbol{A}\boldsymbol{x} \geqslant \boldsymbol{b}(\boldsymbol{p}), \end{aligned} \tag{6.9}$$

with gradient $\boldsymbol{g}$ and constraints right hand side $\boldsymbol{b}$ vector being affine functions of the parameter $\boldsymbol{p}$,

$$\boldsymbol{g}(\boldsymbol{p}) \stackrel{\text{def}}{=} \boldsymbol{g}_0 + \boldsymbol{G}\boldsymbol{p}, \qquad \boldsymbol{b}(\boldsymbol{p}) \stackrel{\text{def}}{=} \boldsymbol{b}_0 + \boldsymbol{B}\boldsymbol{p},$$

where $\boldsymbol{G} \in \mathbb{R}^{n \times n^{\mathrm{p}}}$, $\boldsymbol{B} \in \mathbb{R}^{m \times n^{\mathrm{p}}}$, is called a *Parametric Quadratic Program*. △

For any fixed value of $\boldsymbol{p}$ problem (6.9) becomes a QP in standard form, and the results presented in the previous section remain valid. The purpose of formulating and solving a PQP instead of an ordinary QP is to exploit prior knowledge of the QP's solution in $\boldsymbol{p}_0 \in \mathbb{R}^{n^{\mathrm{p}}}$ to infer the solution for another parameter $\boldsymbol{p}_1 \in \mathbb{R}^{n^{\mathrm{p}}}$.

**Definition 6.8 (Set of Feasible Parameters)**
The set $\mathcal{P}$ of parameters $\boldsymbol{p} \in \mathbb{R}^{n^{\mathrm{p}}}$ feasible for problem (6.9) is defined as

$$\mathcal{P} \stackrel{\text{def}}{=} \left\{ \boldsymbol{p} \in \mathbb{R}^{n^{\mathrm{p}}} \;\middle|\; \mathcal{F}(\boldsymbol{p}) \neq \emptyset \right\} \subseteq \mathbb{R}^{n^{\mathrm{p}}} \tag{6.10}$$

where $\mathcal{F}(\boldsymbol{p})$ denotes the feasible set of the QP obtained from (6.9) for the parameter $\boldsymbol{p}$. △

Affine linearity and continuity of the homotopy path in $\boldsymbol{p}$ induce convexity and closedness of the set $\mathcal{P}$ of feasible parameters.

**Theorem 6.5 (Set of Feasible Parameters)**
The set $\mathcal{P}$ of feasible parameters is convex and closed. △

**Proof** A proof can be found e.g. in [22, 67]. □

There is more structure to the set $\mathcal{P}$, which can be partitioned using the optimal solution's active set as membership criterion.

**Definition 6.9 (Critical Region)**
Let $\mathcal{P}$ be the set of feasible homotopy parameters of a strictly convex PQP. Let $\boldsymbol{x}^\star(\boldsymbol{p})$ denote the optimal solution of the QP in $\boldsymbol{p} \in \mathcal{P}$, and denote by $\mathcal{A}(\boldsymbol{x}^\star(\boldsymbol{p}))$ the associated active set. The set

$$\mathcal{C}(\mathcal{A}) \stackrel{\text{def}}{=} \{\boldsymbol{p} \in \mathcal{P} \mid \mathcal{A} = \mathcal{A}(\boldsymbol{x}^\star(\boldsymbol{p}))\} \subseteq \mathcal{P} \tag{6.11}$$

is called the *critical region* of $\mathcal{P}$ for the index set $\mathcal{A} \subseteq \{1, \ldots, m\} \subset \mathbb{N}$. △

The critical regions partition the set of feasible parameters as detailed by the following theorem.

**Theorem 6.6 (Partition of the Set of Feasible Parameters)**
Let $\mathcal{C}(\mathcal{A}_i)$ denote the critical regions of a strictly convex PQP for the active sets $\mathcal{A}_i \in \mathcal{P}(\{1,\dots,m\})$, $1 \leqslant i \leqslant 2^m$. Then the following holds:

1. The closures $\operatorname{cl}\mathcal{C}(\mathcal{A}_i)$ are closed convex polyhedra.
2. Their interiors are pairwise disjoint,

$$\forall i \neq j: \ \mathcal{C}(\mathcal{A}_i)^\circ \cap \mathcal{C}(\mathcal{A}_j)^\circ = \emptyset. \tag{6.12}$$

3. The set $\mathcal{P}$ is the union of all closures,

$$\mathcal{P} = \bigcup_{i=1}^{2^m} \operatorname{cl}\mathcal{C}(\mathcal{A}_i). \tag{6.13}$$

△

**Proof** A proof can e.g. be found in [67, 154]. □

As a result, the optimal solution $\boldsymbol{x}^\star$ and the associated objective function are piecewise affine linear functions of $\boldsymbol{p}$ as they are affine linear functions of $\boldsymbol{p}$ on every critical region $\mathcal{C}(\mathcal{A}_i)$ in $\mathcal{P}$, a result that can be found e.g. in [20, 70, 154, 223].

Restricting the considered class of PQPs to that of vector–valued homotopies is not a limitation if the destination parameter of the homotopy is known in advance. An appropriate problem transformation is provided by the following theorem.

**Theorem 6.7 (Matrix–Valued Homotopy)**
Consider a PQP with Hessian matrix $\boldsymbol{H}$ or constraints matrix $\boldsymbol{A}$ being affine functions of the homotopy parameter $\boldsymbol{p}$,

$$\begin{aligned} \min_{\boldsymbol{x}\in\mathbb{R}^n} \quad & \tfrac{1}{2}\boldsymbol{x}^T\boldsymbol{H}(\boldsymbol{p})\boldsymbol{x} + \boldsymbol{x}^T\boldsymbol{g}(\boldsymbol{p}) \\ \text{s.t.} \quad & \boldsymbol{A}(\boldsymbol{p})\boldsymbol{x} \geqslant \boldsymbol{b}(\boldsymbol{p}), \end{aligned} \tag{6.14}$$

where

$$\boldsymbol{H}(\boldsymbol{p}) \stackrel{\text{def}}{=} \boldsymbol{H}_0 + \boldsymbol{\Delta H}\boldsymbol{p}, \qquad \boldsymbol{A}(\boldsymbol{p}) \stackrel{\text{def}}{=} \boldsymbol{A}_0 + \boldsymbol{\Delta A}\boldsymbol{p},$$

with $\boldsymbol{\Delta H} \in \mathbb{R}^{n\times n\times n^{\mathrm{p}}}$, $\boldsymbol{\Delta A} \in \mathbb{R}^{m\times n\times n^{\mathrm{p}}}$. Given an arbitrary but fixed homotopy destination point $\boldsymbol{p}_1 \in \mathcal{P}$, problem (6.14) can be reformulated to a PQP in standard form with Hessian $\tilde{H} \stackrel{\text{def}}{=} H(p_1)$ and constraints matrix $\tilde{A} \stackrel{\text{def}}{=} A(p_1)$ by letting

$$\tilde{b}(p) \stackrel{\text{def}}{=} b_0 + Bp + \Delta A(p_1 - p)x, \tag{6.15a}$$

$$\tilde{g}(p) \stackrel{\text{def}}{=} g_0 + Gp - \Delta H(p_1 - p)x + (\Delta A(p_1 - p))^T \lambda, \tag{6.15b}$$

such that the optimal solutions $(x^\star(p), \lambda^\star(p))$ in $p_0$ and $p_1$ do not change. △

**Proof** The KKT conditions (3.1) of problem (6.14) in $p \in \mathbb{R}^{n^p}$ read

$$\begin{aligned}
(A_0 + \Delta A p)x - (b_0 + Bp) &\geqslant 0, \\
(H_0 + \Delta H p)x + (g_0 + Gp) - (A_0 + \Delta A p)^T \lambda &= 0, \\
\lambda &\geqslant 0, \\
\lambda^T((A_0 + \Delta A p)x - (b_0 + Bp)) &= 0.
\end{aligned}$$

Defining vectors $\tilde{g}(p)$ and $\tilde{b}(p)$ depending on the homotopy parameter $p$ as above, these conditions may equivalently be written as

$$\begin{aligned}
(A_0 + \Delta A p_1)x - \tilde{b}(p) &\geqslant 0, \\
(H_0 + \Delta H p_1)x + \tilde{g}(p) - (A_0 + \Delta A p_1)^T \lambda &= 0, \\
\lambda &\geqslant 0, \\
\lambda^T((A_0 + \Delta A p_1)x - \tilde{b}(p)) &= 0.
\end{aligned}$$

which are the KKT conditions of a PQP in standard form with gradient and constraint right hand side homotopy, but with matrices fixed in the homotopy point $p_1$. □

## 6.3 A Primal–Dual Parametric Active Set Strategy

Active set methods exploit the notion of an *active* or *working set* selecting a subset of linear independent inequality constraints that are *assumed* to be satisfied to equality. A sequence of iterates $\{x^k\}$ towards the solution of the QP is computed by solving for each step the Equality Constrained Quadratic Program (EQP) obtained from restricting the QP to the current working set $\mathcal{W}(x^k)$. This task involves the solution of a system of linear equations and appropriate techniques to accomplish this are considered in chapter 7. Based on its solution, the working set is updated by adding or removing a *blocking* constraint, until a linear independent subset of the active set belonging to the QP's optimal solution has been found.

### 6.3.1 Active Set Methods

Active set approaches usually are evaluated under the following aspects concerning their efficiency, and in this section we present a parametric primal–dual active set method due to [25, 67] for the solution of parametric QPs that is notably distinguished from more classical active set methods in several points.

#### Finding a Feasible Point

This usually is the first requirement to start an active set method. Between one third and one half of the total effort of solving a QP is typically spent for finding an initial feasible guess [88]. The task of identifying such an initial guess is usually referred to as *phase one*. Several different approaches exist, such as the solution of an auxiliary Linear Program (LP) or the relaxation of constraints in a homotopy, cf. [158].

In SQP methods and in Model Predictive Control (MPC), a series of closely related QPs is solved. For each QP, the preceding QP's solution is an obvious candidate for a feasible point, a *warm starting* technique. Changes to the constraint matrix $A$ or vector $\boldsymbol{b}$ may render this candidate infeasible, though, mandating a phase one run. Our active set algorithm is able to start in any predefined primal point.

#### Determining the Active Set Exchange

This action is necessary whenever more than one component of the dual vector $\boldsymbol{\lambda}$ indicates non–optimality. Different pivoting strategies such as choosing the constraint with the lowest index (*lexical pivoting*) or with the most negative value $\lambda_i$ are in use. Our algorithm adds and removes constraints in the order they appear on the homotopy path, thereby avoiding the need for pivoting strategies.

#### Ensuring Linear Independence of the Active Set

Whenever a constraint is added to the active set, linear dependence must be verified and its maintenance may require removing another active constraint. There may in addition be *degenerate* points in which the full active set is linearly dependent and thus will not be fully identified. Repeated addition and removal of constraints may happen with steps of zero length in between if such a point is reached, a situation that has already been investigated from the NLP point of view in section 5.2. Ultimately, the method may revisit an active set without having made significant progress in the primal–dual iterate $(\boldsymbol{x}, \boldsymbol{\lambda})$. The method is said to be *cycling*, possibly infinitely.

Depending on the class of QPs treated, the cycling phenomenon is either ignored or treated by simple heuristics. The *EXPAND* strategy [85] used by many codes [73, 84, 86] perturbs constraints in order to ensure progress in the primal–dual variables, but was shown to fail on specially crafted instances [98]. Our active set method monitors positive progress on the homotopy path and can easily detect and resolve linear dependence. In the case of *ties*, cycling may occur and resolution techniques are presented by [216].

### KKT Solution and Matrix Updates

Solving the EQP to find the primal–dual step involves a factorization and backsolve with the KKT matrix associated with the current active set. Computing this factorization is necessary after each change of the active set, and usually comes at the cost of $\mathcal{O}(n^3)$ floating-point operations where $n$ denotes the number of unknowns in the QP. Exploiting structure and sparsity is crucial for the efficiency of the active set method. For certain factorizations, it is possible to compute the factorization only once and recover it in $\mathcal{O}(n^2)$ operations by using matrix updates after each active set change. These issues are considered in chapters 7 and 8.

## 6.3.2 A Primal–Dual Parametric Active Set Strategy

We now present a parametric active set method due to [25, 67] that computes optimal solutions $(\boldsymbol{x}^\star(\tau), \boldsymbol{\lambda}^\star(\tau)) \stackrel{\text{def}}{=} (\boldsymbol{x}^\star(\boldsymbol{p}(\tau)), \boldsymbol{\lambda}^\star(\boldsymbol{p}(\tau)))$ of a PQP along a homotopy path between given parameters $\boldsymbol{p}_0$, $\boldsymbol{p}_1 \in \mathcal{P}$,

$$\boldsymbol{p}(\tau) \stackrel{\text{def}}{=} \boldsymbol{p}_0 + \tau \boldsymbol{\delta p}, \quad \boldsymbol{\delta p} \stackrel{\text{def}}{=} \boldsymbol{p}_1 - \boldsymbol{p}_0, \quad \tau \in [0, 1] \subset \mathbb{R}, \tag{6.16}$$

given the optimal solution $(\boldsymbol{x}^\star(0), \boldsymbol{\lambda}^\star(0))$ of the PQP in $\tau = 0$, i.e., for the initial parameter $\boldsymbol{p}_0 \in \mathbb{R}^{n^{\mathrm{p}}}$. Gradient and residual can be reparameterized as functions of the scalar homotopy parameter $\tau$,

$$\boldsymbol{b}(\tau) \stackrel{\text{def}}{=} \boldsymbol{b}_0 + \boldsymbol{B}(\boldsymbol{p}_0 + \tau\boldsymbol{\delta p}), \qquad \boldsymbol{g}(\tau) \stackrel{\text{def}}{=} \boldsymbol{g}_0 + \boldsymbol{G}(\boldsymbol{p}_0 + \tau\boldsymbol{\delta p}). \tag{6.17}$$

In addition, the distance to the homotopy end point is written as

$$\begin{aligned} \boldsymbol{\delta b}(\tau) &\stackrel{\text{def}}{=} \boldsymbol{b}(1) - \boldsymbol{b}(\tau) = (1-\tau)\boldsymbol{B}\boldsymbol{\delta p}, \\ \boldsymbol{\delta g}(\tau) &\stackrel{\text{def}}{=} \boldsymbol{g}(1) - \boldsymbol{g}(\tau) = (1-\tau)\boldsymbol{G}\boldsymbol{\delta p}. \end{aligned} \tag{6.18}$$

### Rationale

The basic idea of the parametric active set strategy to be presented is to move along the homotopy path from the known solution in $\tau = 0$ to the solution sought in $\tau = 1$ while maintaining both primal and dual feasibility, i.e., optimality of the iterates. This is accomplished by computing homotopies

$$\begin{aligned} &\boldsymbol{x}^\star(\tau) : [0,1] \to \mathbb{R}^n, \\ &\boldsymbol{\lambda}^\star(\tau) : [0,1] \to \mathbb{R}^m, \\ &\mathcal{W}(\tau) : [0,1] \to \{1,\dots,m\} \subset \mathbb{N}, \end{aligned} \tag{6.19}$$

that satisfy the KKT theorem 3.1 in every point $\tau \in [0,1] \subset \mathbb{R}$ and start in the known point $(\boldsymbol{x}^\star(0), \boldsymbol{\lambda}^\star(0))$ with the working set $\mathcal{W}(0) = \mathcal{W}(\boldsymbol{x}^\star(0))$. Regularity of the KKT matrix implies that $\boldsymbol{x}^\star(\cdot)$ and $\boldsymbol{\lambda}^\star(\cdot)$ are piecewise affine and that $\boldsymbol{x}^\star(\cdot)$ in addition is continuous by theorem 6.6. Hence there exist $k \geqslant 2$ primal–dual points $(\boldsymbol{x}_i, \boldsymbol{\lambda}_i) \in \mathbb{R}^{n+m}$ which for $2 \leqslant i \leqslant k-1$ are located on the common boundaries of pairs $(\mathcal{C}_{i-1}, \mathcal{C}_i)$ of critical regions, such that for $\tau \in [\tau_i, \tau_{i+1}]$, $1 \leqslant i \leqslant k-1$ it holds that

$$\begin{aligned} \boldsymbol{x}^\star(\tau) &= \boldsymbol{x}_i + (\tau - \tau_i)\boldsymbol{\delta x}_i, & \boldsymbol{x}(\tau_1) &= \boldsymbol{x}^\star(0), \\ \boldsymbol{\lambda}^\star(\tau) &= \boldsymbol{\lambda}_i + (\tau - \tau_i)\boldsymbol{\delta \lambda}_i, & \boldsymbol{\lambda}(\tau_1) &= \boldsymbol{\lambda}^\star(0), \\ \boldsymbol{x}(\tau_{i+1}) &= \boldsymbol{x}(\tau_i) + (\tau_{i+1} - \tau_i)\boldsymbol{\delta x}_i. \end{aligned} \tag{6.20}$$

Figure 6.2 depicts this situation in the space $\mathcal{P}$ of feasible homotopy parameters.

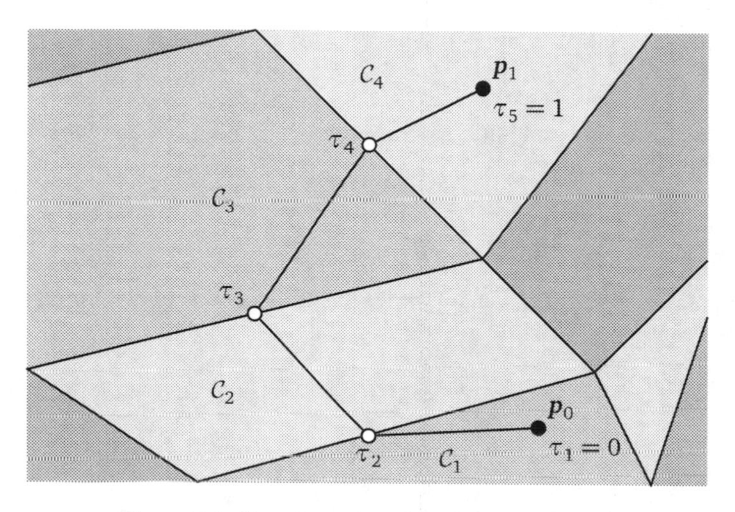


Figure 6.2: The piecewise affine homotopy path.

### A Primal–Dual Iteration

In the first iteration $k = 1$, the algorithm starts for $\tau^1 = 0$ in the given optimal solution $(\boldsymbol{x}^1, \boldsymbol{\lambda}^1) = (\boldsymbol{x}^\star(0), \boldsymbol{\lambda}^\star(0))$. The initial working set $\mathcal{W}^1$ is chosen as a linear independent subset of the active set $\mathcal{A}(\boldsymbol{x}^1)$.

In iteration $k \geqslant 1$, the primal–dual step direction $(\boldsymbol{\delta x}, \boldsymbol{\delta \lambda})$ is obtained as the solution of the linear system

$$\begin{bmatrix} \boldsymbol{H} & \boldsymbol{A}_{\mathcal{W}^k}^T \\ \boldsymbol{A}_{\mathcal{W}^k} & \end{bmatrix} \begin{bmatrix} -\boldsymbol{\delta x} \\ \boldsymbol{\delta \lambda}_{\mathcal{W}^k} \end{bmatrix} = \begin{bmatrix} \boldsymbol{\delta g}(\tau^k) \\ -\boldsymbol{\delta b}_{\mathcal{W}^k}(\tau^k) \end{bmatrix} \tag{6.21}$$

for the point $\tau^k \in [0,1] \subset \mathbb{R}$ on the homotopy path. Appropriate techniques are investigated in chapter 7. The maximum step length maintaining primal feasibility is found from

$$\alpha_{\text{prim}}^k \stackrel{\text{def}}{=} \min \left\{ \frac{b_i(\tau^k) - \boldsymbol{A}_{i\star}\boldsymbol{x}}{\boldsymbol{A}_{i\star}\boldsymbol{\delta x} - \delta b_i(\tau^k)} \;\middle|\; i \notin \mathcal{W}^k \,\wedge\, \boldsymbol{A}_{i\star}\boldsymbol{\delta x} - \delta b_i(\tau^k) < 0 \right\}. \tag{6.22}$$

Constraints $i$ satisfying this minimum are called *primal blocking constraints.* Likewise the maximum step maintaining dual feasibility is found from

$$\alpha^k_{\text{dual}} \stackrel{\text{def}}{=} \min\left\{ -\frac{\lambda_j}{\delta\lambda_j} \;\middle|\; j \in \mathcal{W}^k \ \wedge \ \delta\lambda_j < 0 \right\}, \tag{6.23}$$

and constraints $j$ satisfying this minimum are called *dual blocking constraints.* By choosing

$$\alpha^k \stackrel{\text{def}}{=} \min\left\{ 1 - \tau^k, \alpha^k_{\text{prim}}, \alpha^k_{\text{dual}} \right\} \tag{6.24}$$

we move onto the closest blocking constraint on the remainder $[\tau^k, 1]$ of the homotopy path,

$$\boldsymbol{x}^{k+1} \stackrel{\text{def}}{=} \boldsymbol{x}^k + \alpha^k \boldsymbol{\delta x}, \quad \boldsymbol{\lambda}^{k+1} \stackrel{\text{def}}{=} \boldsymbol{\lambda}^k + \alpha^k \boldsymbol{\delta\lambda}, \quad \tau^{k+1} \stackrel{\text{def}}{=} \tau^k + \alpha^k. \tag{6.25}$$

To find the new working set, we distinguish the following three cases:

1. An inactive constraint $i$ enters the working set if $\tau^{k+1} = \tau^k + \alpha^k_{\text{prim}}$: Linear independence of the new working set can be checked by solving

$$\begin{bmatrix} \boldsymbol{H} & \boldsymbol{A}^T_{\mathcal{W}^k} \\ \boldsymbol{A}_{\mathcal{W}^k} & \end{bmatrix} \begin{bmatrix} \boldsymbol{v} \\ \boldsymbol{w} \end{bmatrix} = \begin{bmatrix} \boldsymbol{a} \\ \boldsymbol{0} \end{bmatrix} \tag{6.26}$$

where $\boldsymbol{a}^T \stackrel{\text{def}}{=} \boldsymbol{A}_{i\star}$ is the constraint row $i$ of $\boldsymbol{A}$ entering the working set. If $\boldsymbol{v} \neq \boldsymbol{0}$ the new working set $\mathcal{W}^{k+1} \stackrel{\text{def}}{=} \mathcal{W}^k \cup \{i\}$ is linear independent. Linear dependence of the active set is indicated by $\boldsymbol{v} = \boldsymbol{0}$. If then $\boldsymbol{w} \leqslant \boldsymbol{0}$, the PQP is infeasible for all $\tau \geqslant \tau^{k+1}$. Otherwise, by removing the constraint

$$j = \operatorname{argmin}\left\{ \frac{\lambda^{k+1}_j}{w_j} \;\middle|\; j \in \mathcal{W}^k \ \wedge \ w_j > 0 \right\} \tag{6.27}$$

from the active set we can restore linear independence of $\mathcal{W}^{k+1} \stackrel{\text{def}}{=} \mathcal{W}^k \cup \{i\} \setminus \{j\}$. In addition, we set

$$\lambda^{k+1}_i \stackrel{\text{def}}{=} \lambda^{k+1}_j / w_j, \quad \lambda^{k+1}_j \stackrel{\text{def}}{=} 0, \quad \boldsymbol{\lambda}^{k+1}_{\mathcal{W}^k} \stackrel{\text{def}}{=} \boldsymbol{\lambda}^{k+1}_{\mathcal{W}^k} - \lambda^{k+1}_i \boldsymbol{w}_{\mathcal{W}^k}. \tag{6.28}$$

2. An active constraint $j$ leaves the working set if $\tau^{k+1} = \tau^k + \alpha^k_{\text{dual}}$: Boundedness of the new EQP can be checked by solving

$$\begin{bmatrix} H & A_{\mathcal{W}^k}^T \\ A_{\mathcal{W}^k} & \end{bmatrix} \begin{bmatrix} v \\ w \end{bmatrix} = \begin{bmatrix} 0 \\ -e_j \end{bmatrix}. \tag{6.29}$$

If $w \neq 0$, boundedness is maintained. Semidefiniteness of the Hessian is indicated by $w = 0$. If then $Av \geqslant 0$, the PQP step is unbounded for all $\tau \geqslant \tau^{k+1}$. Otherwise, by adding the constraint

$$i = \operatorname{argmin} \left\{ \frac{b_i(\tau^{k+1}) - A_{i\star} x^{(k+1)}}{A_{i\star} v} \;\middle|\; i \notin \mathcal{W}^k \;\wedge\; A_{i\star} v < 0 \right\} \tag{6.30}$$

we can restore boundedness of the EQP for the new working set $\mathcal{W}^{k+1} \stackrel{\text{def}}{=} \mathcal{W}^k \cup \{i\} \setminus \{j\}$.

3. The optimal solution $(x^\star(1), \lambda^\star(1)) = (x^{k+1}, \lambda^{k+1})$ at the end of the homotopy path is found if $\tau^{k+1} = 1$.

If $\tau^{k+1} < 1$ the algorithm continues with $k \stackrel{\text{def}}{=} k + 1$ in the new iterate.

**Remark 6.1 (Checks for Boundedness)**
In our implementation qpHPSC, the check 2. for boundedness has not been included as the feasible set $\mathcal{F}$ is known to be bounded in advance. Both the process state trajectory and the control trajectory reside in bounded domains.

### 6.3.3 Proof of the Algorithm

In this section we give proofs for the individual steps of the described parametric active set strategy. Variants can be found in [25] and [67].

**Theorem 6.8 (Linear Independence Test)**
Let the constraint matrix $A_{\mathcal{W}}$ be linear independent, and let constraint row $a^T$ enter the active set. Then the new constraint matrix is linear independent iff

$$\begin{bmatrix} H & A_{\mathcal{W}}^T \\ A_{\mathcal{W}} & \end{bmatrix} \begin{bmatrix} v \\ w \end{bmatrix} = \begin{bmatrix} a \\ 0 \end{bmatrix} \tag{6.31}$$

has a solution $v \neq 0$. △

**Proof** If $A_{\mathcal{W}}$ and $a^T$ are linear dependent, the point $(0, w)$ is a solution of (6.31). Uniqueness implies there exists no further solution with $v \neq 0$. If on the other hand $A$ and $a^T$ are linear independent, there does not exist a solution to (6.31) with $v = 0$. By theorem 3.1 a solution with $v \neq 0$ must exist. □

**Theorem 6.9 (Infeasibility Test)**
Let the constraint matrix $\boldsymbol{A}_{\mathcal{W}}$ be linear independent, and let constraint row $\boldsymbol{a}^T$ enter the active set. Let $(\boldsymbol{x}(\tau^k), \boldsymbol{\lambda}(\tau^k))$ be the optimal solution in a point $\tau^k \in [0,1] \subset \mathbb{R}$. If (6.31) has a solution $\boldsymbol{v} = \boldsymbol{0}$, $\boldsymbol{w} \leqslant \boldsymbol{0}$, the PQP is infeasible for all $\tau > \tau^k$. △

**Proof** Assume some $\boldsymbol{x} \in \mathbb{R}^n$ to satisfy $\boldsymbol{A}_{\mathcal{W}}\boldsymbol{x} \geqslant \boldsymbol{b}_{\mathcal{W}}(\tau)$ for some $\tau > \tau^k$. Then the relation

$$\boldsymbol{a}^T\boldsymbol{x} \leqslant \boldsymbol{w}^T\boldsymbol{b}_{\mathcal{W}}(\tau) \tag{6.32}$$

holds by linear dependence of $\boldsymbol{a}^T$ on $\boldsymbol{A}_{\mathcal{W}}$. On the other hand, we have infeasibility beyond $\tau^k$ of the constraint $j$ to be added,

$$\boldsymbol{b}_j(\tau) > \boldsymbol{a}^T(\boldsymbol{x} + \tau\boldsymbol{\delta x}) = \boldsymbol{w}^T\boldsymbol{A}_{\mathcal{W}}(\boldsymbol{x} + \tau\boldsymbol{\delta x}) = \boldsymbol{w}^T\boldsymbol{b}_{\mathcal{W}}(\tau) \quad \forall \tau \in (\tau^k, 1]. \tag{6.33}$$

Since $\boldsymbol{x} \in \mathbb{R}^n$ was chosen feasible but otherwise arbitrary, constraint $j$ remains infeasible for $\tau > \tau^k$ as long as all constraints in $\mathcal{W}$ remain feasible. By continuity of the piecewise affine solution and by convexity of the set $\mathcal{P}$ of feasible homotopy parameters, infeasibility of the PQP for all $\tau > \tau^k$ follows. □

**Theorem 6.10 (Restoring Linear Independence)**
Let the constraint matrix $\boldsymbol{A}_{\mathcal{W}}$ be linear independent, and let constraint row $\boldsymbol{a}^T$ enter the active set. Let $(\boldsymbol{x}(\tau^k), \boldsymbol{\lambda}(\tau^k))$ be the optimal solution in $\tau^k \in [0,1] \subset \mathbb{R}$. Let (6.31) have a solution $\boldsymbol{v} = \boldsymbol{0}$ and $\boldsymbol{w}$ with $w_j > 0$ for some $j \in \{1, \dots, m\}$. Then the matrix $\boldsymbol{A}_{\hat{\mathcal{W}}}$ with $\hat{\mathcal{W}} \stackrel{\text{def}}{=} \mathcal{W} \setminus \{j\}$ where

$$j \stackrel{\text{def}}{=} \operatorname{argmin} \left\{ \frac{\lambda_j(\tau^k)}{w_j} \;\middle|\; j \in \mathcal{W} : \; w_j > 0 \right\} \tag{6.34}$$

is linear independent from $\boldsymbol{a}^T$. △

**Proof** The stationarity condition of theorem 3.1 holds for $(\boldsymbol{x}(\tau^k), \boldsymbol{\lambda}(\tau^k))$,

$$\boldsymbol{H}\boldsymbol{x}(\tau^k) + \boldsymbol{g}(\tau^k) = \boldsymbol{A}_{\mathcal{W}}^T\boldsymbol{\lambda}(\tau^k), \tag{6.35}$$

Multiplying the linear dependence relation by some $\mu \geqslant 0$ and subtracting from (6.35) yields

$$\boldsymbol{H}\boldsymbol{x}(\tau^k) + \boldsymbol{g}(\tau^k) = \mu\boldsymbol{a} + \boldsymbol{A}_{\mathcal{W}}^T(\underbrace{\boldsymbol{\lambda}(\tau^k) - \mu\boldsymbol{w}}_{\stackrel{\text{def}}{=}\hat{\boldsymbol{\lambda}}}). \tag{6.36}$$

We find that the coefficients $(\mu, \hat{\boldsymbol{\lambda}})$ are also dual feasible solutions if $\hat{\boldsymbol{\lambda}} \geqslant \mathbf{0}$. The largest $\mu$ satisfying this requirement is found as

$$\mu \stackrel{\text{def}}{=} \min \left\{ \frac{\lambda_j(\tau^k)}{w_j} \;\middle|\; \forall j \in \mathcal{W}:\; w_j > 0 \right\}. \tag{6.37}$$

As the dual of the minimizing constraint $j$ is reduced to zero, $\lambda_j = 0$, it can be removed from the active set $\mathcal{W}$. For the active set $\hat{\mathcal{W}} \stackrel{\text{def}}{=} \mathcal{W} \setminus \{j\}$ the constraint row $\boldsymbol{a}^T$ to be added is linear independent from $\boldsymbol{A}_{\hat{\mathcal{W}}}$, as $w_j > 0$ holds.

□

**Theorem 6.11 (Continuation after Restored Linear Independence)**
Let the assumptions of theorem 6.10 be satisfied and let the primal blocking constraint in $\tau_{\text{prim}}$ be unique. Then there exists $\tau_2 \in (\tau_{\text{prim}}, 1]$ such that the constraint $j$ removed from the active set remains inactive for all $\tau \in (\tau_{\text{prim}}, \tau_2]$. △

**Proof** The step $\boldsymbol{\delta x}$ for the segment $[\tau^k, \tau_{\text{prim}}]$ of the homotopy path $[0, 1]$ satisfies

$$\begin{aligned} \boldsymbol{A}_{\mathcal{W}} \boldsymbol{x}(\tau) &= \boldsymbol{b}_{\mathcal{W}}(\tau) && \tau \in [\tau^k, 1], \\ \boldsymbol{A}_{i\star} \boldsymbol{x}(\tau) &< b_i(\tau) && \tau \in [\tau_{\text{prim}}, 1], \end{aligned}$$

such that by linear dependence of $\boldsymbol{A}_{i\star}$ on $\boldsymbol{A}_{\mathcal{W}}$

$$\boldsymbol{w}_{\mathcal{W}}^T \boldsymbol{b}_{\mathcal{W}}(\tau) < b_i(\tau) \qquad \tau \in [\tau_{\text{prim}}, 1]. \tag{6.38}$$

The next segment $[\tau_{\text{prim}}, \tau_2]$ for the working set $\tilde{\mathcal{W}} \stackrel{\text{def}}{=} \mathcal{W} \cup \{i\} \setminus \{j\}$ is chosen such that

$$\boldsymbol{A}_{\tilde{\mathcal{W}}} \boldsymbol{x}(\tau) = \boldsymbol{b}_{\tilde{\mathcal{W}}}(\tau) \qquad \tau \in [\tau_{\text{prim}}, \tau_2].$$

Again by linear dependence we have for the working set $\hat{\mathcal{W}} \stackrel{\text{def}}{=} \mathcal{W} \setminus \{j\}$

$$b_j(\tau) = \boldsymbol{w}_{\hat{\mathcal{W}}}^T \boldsymbol{b}_{\hat{\mathcal{W}}}(\tau) + w_j \boldsymbol{A}_{j\star} \boldsymbol{x}(\tau) \tau \in [\tau_{\text{prim}}, \tau_2], \tag{6.39}$$

By combining (6.38) and (6.39) and noting $w_j > 0$ by theorem 6.10 we find

$$w_j b_j(\tau) < w_j \boldsymbol{A}_{j\star} \boldsymbol{x}(\tau) \qquad \tau \in [\tau_{\text{prim}}, \tau_2],$$

which shows that the removed constraint $j$ remains feasible and strictly inactive. □

**Remark 6.2 (Ties)**
If the primal blocking constraint is not unique, i.e., more than one constraint is hit in the same point $\tau_{\text{prim}} > \tau$ on the homotopy path, the PQP is said to have *primal ties* in $\tau_{\text{prim}}$. In this situation, additional computational effort is required to identify a new active set that permits a step of nonzero length to be made before the removed constraint becomes active again.

For typical MPC problems computational evidence suggests that ties frequently can be circumvented by choice of suitable starting points and by sorting heuristics for the constraints in question. We refer to [216] for analytical methods to resolve ties in PQPs and do not further consider this situation. In our code qpHPSC we currently detect ties but do no take special action to resolve them, and have so far not experienced any numerical problems.

**Theorem 6.12 (Finite Termination)**
If the PQP has no ties, the parametric quadratic active set algorithm terminates after finitely many steps. △

**Proof** There exist at most $2^m$ critical regions associated with the active sets $\mathcal{A}_i$, $1 \leqslant i \leqslant 2^m$, and by theorem 6.6 every point $\tau \in [0,1]$ belongs to at most two critical regions as no primal ties exist. By theorem 6.11 a step of positive length in $\tau$ on the homotopy path is made after every active set change involving the addition of a primal blocking constraint to the current active set. Thus after at most $2^m$ active set changes, the end point $\tau = 1$ of the homotopy path is reached. □

In practice, exponential run time of the parametric active set method is observed on specially crafted problem instances only.

### 6.3.4 Finding a Feasible Initializer

#### A Phase One Procedure

The described algorithm assumes that the optimal solution $(\boldsymbol{x}^\star(0), \boldsymbol{\lambda}^\star(0))$ in the start $\tau = 0$ of the homotopy path be known. This allows for highly efficient hot-starting of the QP solution if the solution of a closely related QP has already been computed for some purpose. If not, the following auxiliary QP

$$\begin{aligned} \min_{x \in \mathbb{R}^n} \quad & \tfrac{1}{2} x^T H x \\ \text{s.t.} \quad & Ax \geqslant 0, \end{aligned} \tag{6.40}$$

with optimal solution $(x^\star, \lambda^\star) = (0, 0)$ can be used together with the homotopy

$$\begin{aligned} g(\tau) &\stackrel{\text{def}}{=} \tau g(p_1), \\ b(\tau) &\stackrel{\text{def}}{=} \tau b(p_1), \end{aligned}$$

from the auxiliary QP (6.40) to the one whose solution is sought. This procedure effectively describes a phase one type approach for our PQP algorithm that achieves feasibility simultaneously with optimality in the end point $\tau = 1$ of the homotopy.

#### Starting in an Arbitrary Guess

Theorem 6.7 actually hints at a more general initialization and hot starting approach that can be realized in the framework of our PQP method. Given an primal–dual point $(\hat{x}, \hat{\lambda}) \in \mathbb{R}^{n+m}$ that satisfies only dual feasibility and complementary slackness, e.g.

$$\hat{\lambda} \geqslant 0, \quad \lambda^T (A\hat{x} - b) = 0, \tag{6.41}$$

but is otherwise chosen arbitrarily, the following modification of the gradient and residual

$$\begin{aligned} \hat{b}(\tau) &= b(\tau) - A(x - \hat{x}), \\ \hat{g}(\tau) &= g(\tau) + H(x - \hat{x}) - A^T(\lambda - \hat{\lambda}) \end{aligned} \tag{6.42}$$

makes $(\hat{x}, \hat{\lambda})$ the optimal solution of the modified QP in $\tau$. This is easily verified by checking theorem 3.1. The working set $\mathcal{W}(\hat{x})$ needs to be chosen appropriately. The modification (6.42) can be applied before starting the PQP algorithm to the purpose of using $(\hat{x}, \hat{\lambda})$ as an initial guess of the PQP's solution in $\tau = 1$. Moreover, it can in principle be applied to the QP's right hand side in *any* point $\tau \in [0, 1)$ of the homotopy path *during* the execution of the described PQP algorithm. We will make extensive use of this property in section 6.4.

### 6.3.5 Parametric Quadratic Programming for SQP Methods

#### Linearizations

In an SQP method in iteration $k \geqslant 1$ the homotopy accomplishes the transition from the previous SQP iterate $(\boldsymbol{x}^k, \boldsymbol{\lambda}^k)$ found from the known optimal solution of the PQP in $\tau = 0$ to the next SQP iterate $(\boldsymbol{x}^{k+1}, \boldsymbol{\lambda}^{k+1})$ found from the optimal solution in $\tau = 1$ to be computed.

The Hessian $\boldsymbol{H}$ and constraints matrix $\boldsymbol{A}$ will in general differ in $\tau = 0$ and $\tau = 1$ as they have been updated or recomputed by a linearization in the $k$-th iterate. By virtue of theorem 6.7 the PQP can be reformulated to use the new matrix data only.

#### Bounds

In SQP methods, the quadratic subproblem usually is set up to deliver the primal step $\boldsymbol{x}^\star_{\mathrm{QP}} = \boldsymbol{\delta x}^k_{\mathrm{SQP}}$ to be added to the SQP algorithm's current iterate $\boldsymbol{x}^k_{\mathrm{SQP}}$. This is convenient for globalization methods that potentially shrink the SQP step length but maintain the direction. Roundoff errors that have accumulated during the QP's solution may lead to tiny nonzero residuals $b_i - \delta x^k_i$ for active simple bounds $i$. If a parametric active set method is used to infer the quadratic subproblem's solution from the known one of the previous SQP iteration's quadratic subproblem, one observes that these errors disturb the next right hand side homotopy step computed as $\boldsymbol{\delta b} = \boldsymbol{b}(\tau_1) - (\boldsymbol{x}^{k+1} - \boldsymbol{b}(\tau_0))$, leading to "chattering" of the bounds. The determination of the primal step length $\tau_{\mathrm{prim}}$ (6.22) on the homotopy path is then prone to flipping signs in the denominator, making the active set method numerically unstable. This leads us to the following remark.

**Remark 6.3 (Setup of the PQP Subproblem)**
If active set PQP methods are used to solve the QP subproblems in an SQP method, these subproblems should be set up to deliver the new iterate $\boldsymbol{x}^\star_{\mathrm{QP}} = \boldsymbol{x}^{k+1}_{\mathrm{SQP}}$ instead.

In this case, the right hand side homotopy step for the simple bounds need not be computed, as $\boldsymbol{\delta b}(\tau) = \boldsymbol{b}(\tau_1) - \boldsymbol{b}(\tau_0) = \boldsymbol{0}$.

## 6.4 Parametric Quadratic Programming for Nonconvex Problems

We develop an extension of the presented parametric active set strategy to the case of QPVCs. We restrict our discussion to the following special class of QPVC.

**Definition 6.10 (QPVC with Constraints Vanishing on a Bound)**
The restricted problem class of a QP with constraints vanishing on a bound is given by

$$\begin{aligned} \min_{x \in \mathbb{R}^n} \quad & \tfrac{1}{2} \boldsymbol{x}^T \boldsymbol{H} \boldsymbol{x} + \boldsymbol{x}^T \boldsymbol{g} \\ \text{s.t.} \quad & 0 \geqslant x_{\xi_i}(b_i - \boldsymbol{A}_{i\star} \boldsymbol{x}), \quad 1 \leqslant i \leqslant l, \\ & \boldsymbol{x}_{\xi} \geqslant \boldsymbol{0}, \end{aligned} \tag{6.43}$$

where $\boldsymbol{\xi} \in \{1, \ldots, n\}^l \subset \mathbb{N}$ is a fixed index vector selecting the controlling component $x_{\xi_i}$ of the unknown $\boldsymbol{x} \in \mathbb{R}^n$ for each of the vanishing constraints $1 \leqslant i \leqslant l$. △

In problem (6.43), any of the $l$ constraints $\boldsymbol{A}\boldsymbol{x} - \boldsymbol{b} \geqslant \boldsymbol{0}$ vanishes if the associated component of the unknown $\boldsymbol{x}$ is active at its lower bound $\boldsymbol{x} = \boldsymbol{0}$. Standard linear affine constraints are omitted from problem (6.43) for clarity of exposition. They do not affect the ensuing presentation of the nonconvex PQP strategy as long as the feasible set of (6.43) remains connected if it is not empty. Note that the consideration of (6.43) is not a true restriction of the applicability of our methods, as the following slack reformulation can be used to transform (6.6) to the restricted form.

**Definition 6.11 (Slack Reformulation of QPVC)**
The slack reformulation of a QPVC to the restricted form (6.43) is given by

$$\begin{aligned} \min_{x,s} \quad & \tfrac{1}{2} \boldsymbol{x}^T \boldsymbol{H} \boldsymbol{x} + \boldsymbol{x}^T \boldsymbol{g} \\ \text{s.t.} \quad & \boldsymbol{C}\boldsymbol{x} - \boldsymbol{s} \geqslant \boldsymbol{d}, \\ & -\boldsymbol{C}\boldsymbol{x} + \boldsymbol{s} \geqslant -\boldsymbol{d}, \\ & \boldsymbol{0} \geqslant s_i(b_i - \boldsymbol{A}_{i\star} \boldsymbol{x}), \quad 1 \leqslant i \leqslant l, \\ & \boldsymbol{s} \geqslant \boldsymbol{0}, \end{aligned} \tag{6.44}$$

where $\boldsymbol{s} \in \mathbb{R}^l$ is a vector of slack variables. △

Note that (6.44) still has a strictly convex objective if $\boldsymbol{H}$ is positive definite, as the slack is never a free unknown. In the case of QPVCs arising from outer convexification and relaxation of constraints direct depending on an integer control, $\boldsymbol{s}$ in (6.44) and $\boldsymbol{x}_\xi$ in (6.43) take the role of the relaxed convex multipliers $\boldsymbol{\alpha}$ of chapter 2.

### 6.4.1 Nonconvex Primal Strategy

We are concerned with devising a primal active set exchange strategy for the vanishing constraints of QPVC (6.43).

#### Primal Blockings

The feasible set $\mathcal{F}$ of (6.43) is nonconvex as detailed in section 5.4, and can be partitioned in at most $2^l$ convex subsets $\mathcal{C}_i$, each one associated with unique subset $\mathcal{V}_i \in \mathcal{P}(\{1,\dots,l\})$ of vanishing constraints that are feasible and $\mathcal{V}_i^{\mathrm{C}}$ of vanishing constraints that have vanished,

$$\mathcal{C}_i \stackrel{\text{def}}{=} \{\boldsymbol{x} \in \mathbb{R}^n \mid (i \in \mathcal{V}_i \ \wedge \ \boldsymbol{A}_{i\star}\boldsymbol{x} \geqslant b_i) \ \vee \ (i \in \mathcal{V}_i^{\mathrm{C}} \ \wedge \ \boldsymbol{x}_{\xi_i} = 0)\}. \tag{6.45}$$

An exhaustive search of these convex subsets by solving up to $2^l$ smaller standard QPs is prohibitive and would indeed lead our efforts in applying outer convexification and relaxation to the MIOCP ad absurdum, as has already been noticed in section 5.4.

We instead modify the active set exchange rule (6.22) for primal blocking constraints and adapt it to the nonconvex shape of the feasible set as follows,

$$\alpha^k_{\text{prim}} \stackrel{\text{def}}{=} \min \left\{ \frac{b_i(\tau^k) - \boldsymbol{A}_{i\star}\boldsymbol{x}}{\boldsymbol{A}_{i\star}\boldsymbol{\delta x} - \delta b_i(\tau^k)} \;\middle|\; i \notin (\mathcal{W}^k \cup \mathcal{I}_0) \ \wedge \ (\boldsymbol{A}_{i\star}\boldsymbol{\delta x} - \delta b_i(\tau^k) < 0) \right\}. \tag{6.46}$$

Using the primal test (6.46), a vanishing constraint $i$ becomes active only if it is primally blocking and the associated control variable $x_{\xi_i}$ is nonzero, i.e., the constraint $i \in \mathcal{I}_+$ is not permitted to vanish beyond its bound. If the controlling variable is zero, i.e., $i \in \mathcal{I}_0$, the vanishing constraint is free to vanish and reappear as indicated by the unknowns $\boldsymbol{x}$. In addition, as soon as its associated controlling variable hits its simple lower bound any active vanishing constraint also vanishes and is removed from the active set $\mathcal{W}^k$. This ensures $\mathcal{I}_{00} = \emptyset$ after all primal blockings, i.e., no bound or constraint ever enters the critical subset $\mathcal{I}_{00}$ of the active set. Hence these two kinds of move between adjacent convex subsets is naturally accomplished by the primal exchange strategy.

### Primal Continuation after Infeasibility

The described strategy allows to move between the convex subsets of the feasible set as long as the piecewise affine homotopy path does not cross the boundary of the nonconvex feasible set $\mathcal{F}$ into infeasibility before the end $\tau = 1$ of the homotopy has been reached. In the latter case, infeasibility of the PQP by theorem 6.9 is detected by the parametric active set method of section 6.3. For the case of a QPVC this means that for $\tau = 1$ no strongly stationary point is located in the current convex critical region. Two cases can be distinguished here.

1. The blocking part of the critical region's boundary is defined by an active regular constraint. In this case the QPVC actually is infeasible if no vanishing constraint is found in $\mathcal{I}_{+0}$.

2. The blocking constraint is a vanishing constraint becoming active and entering the index set $\mathcal{I}_{+0}$. A strongly stationary point could possibly be found if the constraint were allowed to vanish.

In both cases, let $\mathcal{A}$ and $\mathcal{X}$ denote the active sets for the constraints and simple bounds associated with this critical region, and let $j \in \mathcal{I}_{+0} \subseteq \mathcal{A}$ denote an active vanishing constraint. The solution may be continued in the adjacent critical region associated with the active set $\mathcal{A} \setminus \{j\}$ for the constraints and $\mathcal{X} \cup \{i\}$, $i = \xi_j$ for the simple bounds on the unknowns. The controlling variable $x_j$ is set to zero. In this critical region, the active vanishing constraint blocking the progress of the homotopy has vanished. Maintenance of linear independence may require removal of an additional constraint from $\mathcal{A}$ or $\mathcal{X}$.

This procedure effectively describes a primal continuation of the homotopy in a critical region that does not share a common boundary with the one the homotopy path terminated in. This involves a modification of the primal point $\boldsymbol{x}$, the active set, and consequentially also the dual point. Numerical techniques to realize this continuation in the framework of the parametric active set method of section 6.3 are developed in the sequel of this section.

### 6.4.2 Nonconvex Dual Strategy

We are concerned with devising a dual active set exchange strategy for the vanishing constraints of QPVC (6.43). In accordance with the notation of chapter 5 we denote the MPVC multipliers of the vanishing constraints $A_{i\star}x \geqslant b$ by $\mu_\mathrm{g}$ and those of the controlling variables' simple bounds $x_\xi \geqslant 0$ by $\mu_\mathrm{h}$.

#### Dual Blockings

By theorem 6.3, for a QPVC strong stationarity is equivalent to local optimality. In order to detect dual blocking constraints whose removal allows the homotopy path to make progress towards optimality, the dual blocking constraints rule (6.23) is modified according to the optimal multiplier signs known from the notion of strong MPVC stationarity introduced by definition 5.11, which is repeated here for convenience.

**Definition 6.12 (Strong Stationarity)**
A feasible point $\overline{x} \in \mathbb{R}^n$ of an MPVC is called *strongly stationary* if it holds that

$$\begin{aligned} &\Lambda_x(\overline{x}, \mu) = 0, && (6.47)\\ &\mu_{\mathrm{h},\mathcal{I}_{0+}} \geqslant 0, \quad \mu_{\mathrm{h},\mathcal{I}_{00}} \geqslant 0, \quad \mu_{\mathrm{h},\mathcal{I}_{+}} = 0, \\ &\mu_{\mathrm{g},\mathcal{I}_{+0}} \geqslant 0, \quad \mu_{\mathrm{g},\mathcal{I}_{++}} = 0, \quad \mu_{\mathrm{g},\mathcal{I}_{0}} = 0. \end{aligned}$$

△

As $\mathcal{I}_{00} = \emptyset$ is guaranteed by the primal strategy, the only index set relevant for the dual exchange of vanishing constraints is the set $\mathcal{I}_{0-}$ of vanishing constraints that are infeasible and were permitted to vanish as the associated controlling variable is zero. The following QPVC dual active set exchange rule for simple bounds prevents the affected controlling variables from leaving their lower bounds.

$$\tau^{\mathrm{dual}} \stackrel{\mathrm{def}}{=} \min\left\{ -\frac{\lambda_j}{\delta\lambda_j} \;\middle|\; j \in \mathcal{W}^k \setminus \mathcal{I}_{0-} \;\wedge\; \delta\lambda_j < 0 \right\}. \tag{6.48}$$

#### Dual Continuation

An additional modification to the dual exchange strategy is required in two places, namely whenever moving from one of the two convex subsets of a scalar vanishing constraint's nonconvex feasible set to the other one. As we will see in the next section, this modification has a close connection to the question of global optimality of the MPVC strongly stationary point identified by this QPVC active set strategy.

1. MPVC strong stationarity requires $\mu_{g,i} = 0$ for an active vanishing constraint $i \in \mathcal{I}_{+0}$ entering $\mathcal{I}_{0-}$ via $\mathcal{I}_{00}$, i.e., if the controlling variable was free and is about to enter its lower bound, causing the vanishing constraint $i$ to actually vanish beyond its bound. Then, if we had $\mu_{g,i} > 0$ in $\mathcal{I}_{+0}$, the homotopy path shows a dual discontinuity which can be treated as shown in the sequel of this section.
2. Likewise $\mu_{h,i} \geqslant 0$ is required for a controlling variable active at its lower bound in $\mathcal{I}_{0-}$ entering $\mathcal{I}_{0+}$ via $\mathcal{I}_{00}$, i.e., if the associated vanishing constraint reappears. We find that a dual discontinuity of the homotopy path occurs if we had $\mu_{h,i} < 0$ in $\mathcal{I}_{0-}$.

The resulting active set strategy is depicted in figure 6.3 as a state machine. Nodes show the index sets, subsets of the active set, and arcs show the primal and dual conditions for transition of a simple bound or a vanishing constraint from one index set to the other.

### 6.4.3 A Heuristic for Global Optimality

A sufficient condition for global optimality of a QPVCs due to [105] can be established. As a consequence, the QPVC subproblems generated by our nonconvex SQP method can actually be solved to global optimality, as is the case for standard convex QP subproblems with positive semidefinite Hessian matrices. The sequence of SQP iterates thus actually inherits the local convergence properties of the corresponding standard Lagrangian SQP method.

**Theorem 6.13 (Global Optimality Conditions)**
Let the objective function $f$ be convex and let $\boldsymbol{g}$, $\boldsymbol{h}$ be affine linear in problem (5.9). Further, let $\boldsymbol{x}^\star$ be a strongly stationary point of problem (5.9). Then if $\boldsymbol{\mu}_{h,\mathcal{I}_{0-}} \geqslant \boldsymbol{0}$ and $\boldsymbol{\mu}_{g,\mathcal{I}_{+0}} = \boldsymbol{0}$, it holds that $\boldsymbol{x}^\star$ is a globally optimal point of problem (5.9). △

**Proof** A proof can be found in [105]. □

Theorem 6.13 indicates that the active set strategy for MPVC strong stationarity presented in the previous section can potentially be improved further. Note that the global optimality conditions of the above theorem correspond to the two conditions for a dual discontinuity of the homotopy path.

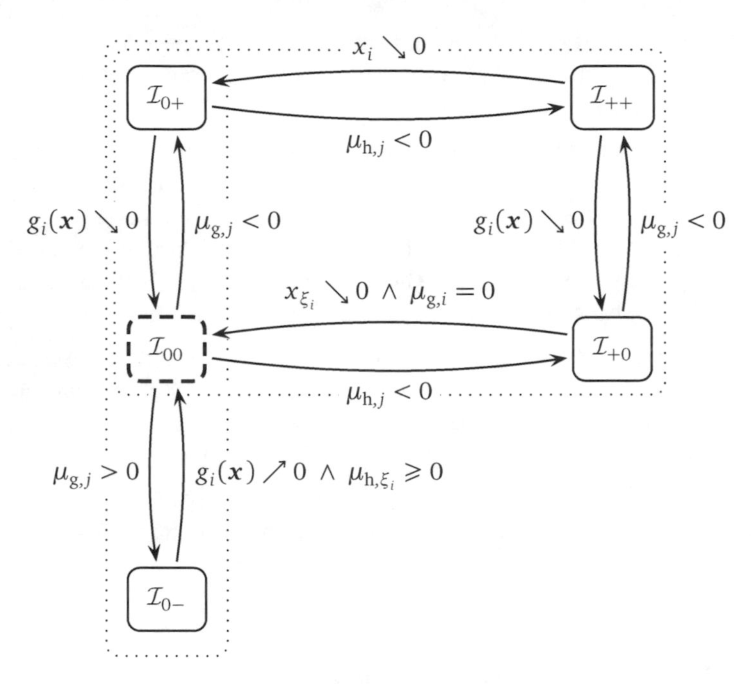


Figure 6.3: Active set exchange rules for MPVC strong stationarity. Constraints never remain in the state $\mathcal{I}_{00}$ which is shown for clarity only. Dotted lines indicate convex feasible subsets.

In detail, a dual discontinuity arises if and only if the affected active vanishing constraint or simple bound has an MPVC multiplier in violation of the sufficient conditions for global optimality.

This observation leads to the following modification of the primal active set strategy. Whenever a vanishing constraint $\boldsymbol{A}_{i\star}\boldsymbol{x} \geqslant b_i$ becomes active and is about to enter the index set $\mathcal{I}_{+0}$ because its controlling variable $x_{\xi_i}$ is nonzero, the associated MPVC multiplier $\mu_{g,i}$ would violate the sufficient global optimality conditions. We then choose to immediately continue the solution in an adjacent convex critical region of the feasible set in which the primal blocking constraint has vanished, if such a region exists. This can be realized in the same way as primal continuation after infeasibility has been realized in the previous section. The resulting active set strategy is depicted in figure 6.4 again as a state machine. Due to our global optimality heuristic, the arc from $\mathcal{I}_{+0}$ to $\mathcal{I}_{0-}$ has no counterpart in the reverse direction.

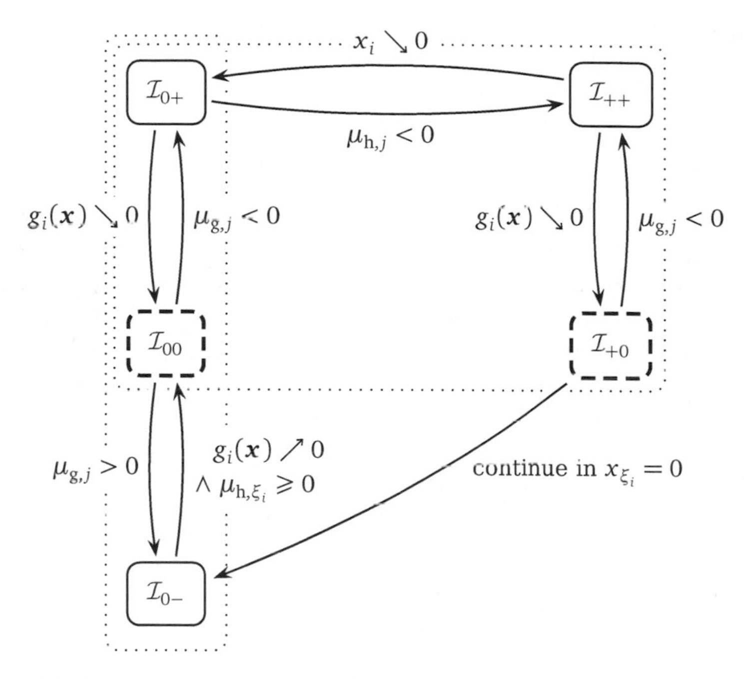


Figure 6.4: Active set exchange rules for MPVC strong stationarity with heuristic for global optimality. Constraints never remain in the states $\mathcal{I}_{00}$ and $\mathcal{I}_{+0}$ which are shown for clarity only. Due to our global optimality heuristic, the arc from $\mathcal{I}_{+0}$ to $\mathcal{I}_{0-}$ has no counterpart arc in the reverse direction.

### Resolution of Degeneracy

Forcing $x_{\xi_i} = 0$ and adding this simple lower bound to the active set may cause linear dependence of the active constraints. A first attempt at resolving this is to use the linear dependence check and resolution procedure of section 6.3.

In the case of SOS1 constraints, however, it is possible that the only active constraints potentially resolving the degeneracy are the simple lower bounds on variables $x_{\xi_j}$, $j \neq i$, controlling vanishing constraints, and it is further possible that all vanishing constraints $j \neq i$ actually have vanished. In this situation, one of these simple bounds has to be removed from the active set, and the associated vanishing constraint has to reappear, i.e., to be made feasible by continuing in an adjacent critical region after a primal discontinuity. The criterion of theorem 6.10 does not provide information on which simple bound to remove from the active set, though, as there is no signedness requirement on MPVC strong stationarity of the multipliers $\mu_{\mathrm{h}}$ in $\mathcal{I}_{0-}$. In this situation, problem dependent resolution heuristics are required to step in and indicate a vanishing constraint resolving the degeneracy.

1. Based on the sufficient optimality criterion a simple bound with multiplier $\mu_{\mathrm{h},i} < 0$ can be chosen.
2. Another possibility is to choose the vanishing constraint $j \neq i$ that would show the smallest violation of feasibility if $x_{\xi_j}$ were nonzero, and thus leads to the smallest primal discontinuity,

$$j \stackrel{\text{def}}{=} \operatorname{argmin}\{b_j - \boldsymbol{A}_{j\star}\boldsymbol{x} \mid j \in \mathcal{I}_{0-}\}. \tag{6.49}$$

3. The move to an adjacent convex feasible region could be rejected and the blocking vanishing constraint enters the set $\mathcal{I}_{+0}$ in violation of the sufficient optimality conditions.

### 6.4.4 Continuations for a Parametric Active Set Method

The primal and dual strategies of the previous section require the parametric active set method to cope with both primal and dual discontinuities of the homotopy path. As mentioned before, these can be regarded as restarts of the method in a different convex critical section of the nonconvex feasible set that is adjacent to the previous one in the sense of definition 6.4.

We now present two numerical techniques that make these continuations possible in a highly efficient way. The underlying idea is that any arbitrary primal–dual point can serve as an initializer for the parametric active set method in $\tau = 0$ as long as it satisfies all KKT conditions. The techniques to be presented guarantee this by modifying the right hand side homotopy from the initial gradient and right hand side $\boldsymbol{g}(0)$, $\boldsymbol{b}(0)$ to the final one $\boldsymbol{g}(1)$, $\boldsymbol{b}(1)$ to make sure that all KKT conditions in the new starting point are satisfied.

#### Primal Continuation

The primal continuation, used a) if infeasibility of the QPVC in the current convex subset is diagnosed, and b) in the heuristic for global optimality if a vanishing constraint is primally blocking, requires a primal variable $x_{\xi_i}$ to jump to zero. The feasibility gap after letting $x_{\xi_i} = 0$ can be computed as

$$\hat{\boldsymbol{b}}_j \overset{\text{def}}{=} \begin{cases} \max\{\boldsymbol{b}_j - \boldsymbol{A}_{j\star}\boldsymbol{x}, 0\} & \text{if } x_{\xi_j} > 0 \\ 0 & \text{otherwise.} \end{cases} \qquad 1 \leqslant j \leqslant l, \tag{6.50}$$

and the new initial right hand side is

$$\boldsymbol{b}(0) \overset{\text{def}}{=} \boldsymbol{b}(\tau^{\text{primal}}) - \hat{\boldsymbol{b}}. \tag{6.51}$$

Setting the MPVC multiplier $\mu_{\text{h},\xi_i}$ of the simple bound entering the active set as $x_{\xi_i} = 0$ to some positive value bounded away from zero has proven useful in order to avoid primal ties in the ensuing active set iterations. The introduced stationarity defect is easily covered by modifying the gradient $\boldsymbol{g}(0)$ analogously to the following dual continuation.

### Dual Continuation

The dual continuation requires either of the MPVC multipliers $\mu_{\text{g},i}$ or $\mu_{\text{h},\xi_i}$ to jump to zero. The stationarity gap after letting e.g. $\mu_{\text{g},i} = 0$ can be computed as

$$\hat{\boldsymbol{g}} \overset{\text{def}}{=} \boldsymbol{H}\boldsymbol{x} + \boldsymbol{g}(\tau^{\text{dual}}) - \boldsymbol{A}^T\boldsymbol{\mu}_{\text{g}} - \boldsymbol{\mu}_{\text{h}} \tag{6.52}$$

and the new initial gradient is

$$\boldsymbol{g}(0) \overset{\text{def}}{=} \boldsymbol{g}(\tau^{\text{dual}}) - \hat{\boldsymbol{g}} \tag{6.53}$$

satisfying

$$\boldsymbol{0} \overset{\text{def}}{=} \boldsymbol{H}\boldsymbol{x} + \boldsymbol{g}(0) - \boldsymbol{A}^T\boldsymbol{\mu}_{\text{g}} - \boldsymbol{\mu}_{\text{h}}. \tag{6.54}$$

Dual feasibility is always maintained even after a dual discontinuity.

The computational effort of continuing in an adjacent critical section is thus limited to the computation of at most three matrix–vector products and in the case of the primal continuation a recomputation or an update of the KKT system's factorization.

### 6.4.5 Active Set Exchange Rules Summary

In this section, we collect the exchange rules for the parametric active set strategy solving a QPVC.

| blocking | | index sets | | action |
|---|---|---|---|---|
| kind | type | from | to | |
| primal | bound | $\mathcal{I}_{++}$ | $\mathcal{I}_{0+}$ | conventional |
| | | $\mathcal{I}_{+0}$ | $\mathcal{I}_{00}$ | move to $\mathcal{I}_{0-}$, dual discontinuity if $\mu_{\mathrm{g},i} \neq 0$ happens if global heuristic not used |
| primal | vanishing constraint | $\mathcal{I}_{++}$ | $\mathcal{I}_{+0}$ | if global heuristic used: continue in $\mathcal{I}_{0-}$ otherwise: conventional |
| | | $\mathcal{I}_{0+}$ | $\mathcal{I}_{00}$ | move to $\mathcal{I}_{0-}$ |
| | | $\mathcal{I}_{0-}$ | $\mathcal{I}_{00}$ | move to $\mathcal{I}_{0+}$, dual discontinuity if $\mu_{\mathrm{h},i} < 0$ |
| dual | bound | $\mathcal{I}_{0+}$ | $\mathcal{I}_{++}$ | conventional |
| | | $\mathcal{I}_{00}$ | $\mathcal{I}_{+0}$ | never happens; no constraint ever is in $\mathcal{I}_{00}$ |
| | | $\mathcal{I}_{0-}$ | — | never happens due to dual step decision |
| dual | vanishing constraint | $\mathcal{I}_{+0}$ | $\mathcal{I}_{++}$ | conventional; happens if global heuristic is not used |
| | | $\mathcal{I}_{00}$ | $\mathcal{I}_{0+}$ | never happens; no constraint ever is in $\mathcal{I}_{00}$ |

Table 6.1: Possible active set exchanges in the parametric active set strategy for QPVCs.

## 6.5 Summary

In this chapter we developed a parametric nonconvex SQP method for the efficient numerical solution of discretized MIOCPs with vanishing constraints arising from outer convexification of constraints directly depending on a binary or integer control parameter. This method relies on solving the generated nonconvex subproblems to local optimality on a convex subset of their feasible sets, and continuing resp. verifying the solution in all adjacent convex subsets. For MIOCPs the local subproblems generated by our SQP method are QPs with convex objective and affine linear vanishing constraints (QPVCs). Though these are truly nonconvex problems, KKT points of QPVCs are locally optimal solutions that can be found by an appropriate active set method. We have introduced a primal–dual parametric active set method that efficiently computes solutions to a sequence of convex QPs. This method can be used in a natural way to find locally optimal solutions of a sequence of QPVCs restricted to a convex subset of the nonconvex feasible set. To this end, the shape of the feasible set and the MPVC strong stationarity conditions are accounted for in the definition of the closest primal and dual blocking constraints. Using the adjacency relation for QPVCs we developed a continuation method that allows to efficiently move between adjacent convex subsets. This enables the parametric active set strategy to find a locally optimal solution that can be repetitively verified or improved in all adjacent convex feasible subsets. We further described a heuristic for finding a globally optimal point, based on a sufficient condition for global optimality of convex MPVC. This heuristic relies on problem instance specific information in one particular case arising if moving from one convex feasible subset to another adjacent one causes linear dependence of the active set.

# 7 Linear Algebra for Block Structured QPs

In this chapter we survey linear algebra techniques for efficiently solving Quadratic Programs (QPs) and Quadratic Programs with Vanishing Constraints (QPVCs) with block structure due to direct multiple shooting. After a review of existing techniques, we present a novel algorithmic approach tailored to QPs with many control parameters due to the application of outer convexification. Our approach consists of a *block structured factorization* of the Karush–Kuhn–Tucker (KKT) systems that completes in $\mathcal{O}(mn^3)$ operations without generating any fill–in. It is derived from a combined null–space range–space method due to [202]. All operations on the KKT factorization required for the parametric active set algorithm for QPVC of chapter 6 are presented in detail, and efficient implementations are discussed. We investigate the run time complexity of this algorithm, the number of floating–point operations required for the individual steps of the presented factorization and backsolve, and give details on the memory requirements.

In chapter 8, the presented KKT factorization is further improved to a runtime complexity of $\mathcal{O}(mn^2)$ for all but the first iteration of the active set method.

## 7.1 Block Structure

In this section we have a closer look at the block structure of the QPs stemming from the direct multiple shooting discretization of the Optimal Control Problem (OCP), given in the following definition.

**Definition 7.1 (Direct Multiple Shooting Quadratic Program)**
The quadratic program obtained from a local quadratic model of the Lagrangian of the direct multiple shooting Nonlinear Program (NLP) (1.24) is

$$\begin{aligned} \min_{v} \quad & \sum_{i=0}^{m} \left( \tfrac{1}{2} v_i^T H_i v_i + v_i^T g_i \right) && (7.1) \\ \text{s.t.} \quad & r_i \leqslant R_i v_i, && 0 \leqslant i \leqslant m, \\ & h_i = G_i v_i + P_{i+1} v_{i+1}, && 0 \leqslant i \leqslant m-1. \end{aligned}$$

Herein $\boldsymbol{v}_i \in \mathbb{R}^{n_i^{\mathrm{v}}}$ denote the $m$ vectors of unknowns. The Hessian blocks are denotes by $\boldsymbol{H}_i \in \mathbb{R}^{n_i^{\mathrm{v}} \times n_i^{\mathrm{v}}}$ and gradients by $\boldsymbol{g}_i \in \mathbb{R}^{n_i^{\mathrm{v}}}$. Linearized point constraints are denoted by matrices $\boldsymbol{R}_i \in \mathbb{R}^{n_i^{\mathrm{r}} \times n_i^{\mathrm{v}}}$ and right hand side vectors $\boldsymbol{r}_i \in \mathbb{R}^{n_i^{\mathrm{r}}}$. Linearized matching conditions are written using sensitivity matrices $\boldsymbol{G}_i \in \mathbb{R}^{n^{\mathrm{h}} \times n_i^{\mathrm{v}}}$ and coupling matrices $\boldsymbol{P}_i \in \mathbb{R}^{n^{\mathrm{h}} \times n_{i+1}^{\mathrm{v}}}$ with right hand side vectors $\boldsymbol{h}_i \in \mathbb{R}^{n_i^{\mathrm{h}}}$. △

In a direct multiple shooting discretization we will usually have identical dimensions $n_i^{\mathrm{v}} = n^{\mathrm{x}} + n^{\mathrm{q}}$ for all nodes $0 \leqslant i \leqslant m$ and assume one matching condition per Ordinary Differential Equation (ODE) state, i.e., $n_i^{\mathrm{h}} = n^{\mathrm{x}}$.

Two different structure exploiting approaches at solving QP (7.1) can be distinguished. *Condensing* methods apply variants of Gaussian elimination to the block diagonal Hessian $\boldsymbol{H}$ and the block bi–diagonal constraints matrix with entries from $\boldsymbol{R}$, $\boldsymbol{G}$, and $\boldsymbol{P}$ to obtain a dense but considerably smaller QP. This is then solved with a structure–free active set method. Condensing methods for direct multiple shooting are presented in section 7.2.1.

One may also conceive active set methods on the large QP, wherein the block structure is exploited by the KKT system factorization in *every iteration* of the active set loop. The challenge here lies with the varying size and structure of the KKT matrix over the course of active set changes. For a given active set $\mathcal{A}(\boldsymbol{v})$, the QP (7.1) is reduced to a purely equality constrained QP as follows.

**Definition 7.2 (Direct Multiple Shooting Equality Constrained QP)**
The Equality Constrained Quadratic Program (EQP) obtained from the direct multiple shooting QP (7.1) for a given active set $\mathcal{A}(\boldsymbol{v})$ (cf. definition 3.3) is

$$\begin{aligned} \min_{\boldsymbol{v}} \quad & \sum_{i=0}^{m} \left( \tfrac{1}{2} \boldsymbol{v}_i^T \boldsymbol{H}_i \boldsymbol{v}_i + \boldsymbol{v}_i^T \boldsymbol{g}_i \right) && && (7.2) \\ \text{s.t.} \quad & \boldsymbol{r}_i^{\mathcal{A}} = \boldsymbol{R}_i^{\mathcal{A}} \boldsymbol{v}_i, && 0 \leqslant i \leqslant m, \\ & \boldsymbol{h}_i = \boldsymbol{G}_i \boldsymbol{v}_i + \boldsymbol{P}_{i+1} \boldsymbol{v}_{i+1}, && 0 \leqslant i \leqslant m-1. \end{aligned}$$

The notation $\boldsymbol{R}_i^{\mathcal{A}}$ and $\boldsymbol{r}_i^{\mathcal{A}}$ denotes the restriction of $\boldsymbol{R}_i$ and $\boldsymbol{r}_i$ on the rows resp. elements of the constraints in the active set $\mathcal{A}(\boldsymbol{v})$. △

The KKT system associated with this EQP according to theorem 3.1 is given in definition 7.3 on page 177. The performance of active set methods crucially depends on the ability to efficiently exploit simple bounds on the unknown $\boldsymbol{v}$. In particular when treating Mixed–Integer Optimal Control Problems (MIOCPs) by outer convexification, many relaxed binary control parameters in resulting NLPs and QP will be active at their lower or upper bound. Active simple bounds correspond to components $v_i$ that can immediately be eliminated from the KKT system (7.3). This reduces the computational effort of the ensuing factorization.

**Definition 7.3 (KKT System of the Direct Multiple Shooting EQP)**
For a given active set $\mathcal{A}(v)$ the KKT system of problem (7.2) reads

$$\begin{bmatrix} H_0 & R_0^{\mathcal{A}T} & G_0^T & & & & & & \\ R_0^{\mathcal{A}} & & & & & & & & \\ G_0 & & & P_1 & & & & & \\ & & P_1^T & H_1 & R_1^{\mathcal{A}T} & G_1^T & & & \\ & & & R_1^{\mathcal{A}} & & \ddots & & & \\ & & & G_1 & \ddots & & P_m & & \\ & & & & & P_m^T & H_m & R_m^{\mathcal{A}T} \\ & & & & & & R_m^{\mathcal{A}} & \end{bmatrix} \begin{bmatrix} -v_0 \\ \mu_0^{\mathcal{A}} \\ \lambda_0 \\ -v_1 \\ \mu_1^{\mathcal{A}} \\ \vdots \\ -v_m \\ \mu_m^{\mathcal{A}} \end{bmatrix} = \begin{bmatrix} g_0 \\ -r_0^{\mathcal{A}} \\ -h_0 \\ g_1 \\ -r_1^{\mathcal{A}} \\ \vdots \\ g_m \\ -r_m^{\mathcal{A}} \end{bmatrix} \tag{7.3}$$

with Lagrange multipliers $\mu_i^{\mathcal{A}} \in \mathbb{R}^{n_i^{r\mathcal{A}}}$ of the linearized decoupled point constraints and $\lambda_i \in \mathbb{R}^{n^x}$ of the linearized matching conditions. △

To gain additional insight into the structure of simple bounds, we define the multiple shooting QP with simple bounds on $v$ and derive the more detailed KKT system.

**Definition 7.4 (Direct Multiple Shooting QP with Simple Bounds)**
The quadratic program obtained from a local quadratic model of the Lagrangian of the direct multiple shooting NLP (1.24) is

$$\begin{aligned} \min_{v} \quad & \sum_{i=0}^{m} \left( \tfrac{1}{2} v_i^T H_i v_i + v_i^T g_i \right) && \qquad (7.4) \\ \text{s.t.} \quad & l_i \leqslant v_i \leqslant u_i, && 0 \leqslant i \leqslant m, \\ & r_i \leqslant R_i v_i, && 0 \leqslant i \leqslant m, \\ & h_i = G_i v_i + P_{i+1} v_{i+1}, && 0 \leqslant i \leqslant m-1. \end{aligned}$$

△

In the following, the active set $\mathcal{A}(v)$ refers to the active linear inequality constraints only. We introduce the active set $\mathcal{X}(v)$ of active simple bounds on the unknown $v$ by straightforward extension of definition 3.3, and denote by $\mathcal{F}(v)$ its complement set in the index set of all simple bounds.

**Definition 7.5 (Active Simple Bound, Active Set of Simple Bounds)**
Let $\overline{x} \in \mathbb{R}^n$ be a feasible point of problem (7.4). A simple bound on $\overline{v}_{ij}$, $0 \leqslant i \leqslant m$, $1 \leqslant j \leqslant n_i^{\text{v}}$ is called *active* if $\overline{v}_{ij} = l_{ij}$ or $\overline{v}_{ij} = u_{ij}$ holds. It is called *inactive* otherwise.

The set of indices of all active simple bounds

$$\mathcal{X}(\overline{\boldsymbol{v}}) \stackrel{\text{def}}{=} \{(i,j) \mid \overline{v}_{ij} = l_{ij} \ \vee \ \overline{v}_{ij} = u_{ij}\} \subset \mathbb{N}_0 \times \mathbb{N} \tag{7.5}$$

is called the *active set of the simple bounds* associated with $\overline{\boldsymbol{v}}$. We denote by $\mathcal{F}(\overline{\boldsymbol{v}})$ the complement of $\mathcal{X}(\overline{\boldsymbol{v}})$ in the set $\{(i,j) \mid 0 \leqslant i \leqslant m,\ 1 \leqslant j \leqslant n_i^{\mathrm{v}}\}$. △

**Definition 7.6 (Direct Multiple Shooting EQP with Simple Bounds)**
The EQP obtained from the direct multiple shooting QP (7.4) for a given active set $\mathcal{A}(\boldsymbol{v})$ is

$$\begin{aligned}
\min_{\boldsymbol{v}} \quad & \sum_{i=0}^{m} \left( \tfrac{1}{2} \boldsymbol{v}_i^T \boldsymbol{H}_i \boldsymbol{v}_i + \boldsymbol{v}_i^T \boldsymbol{g}_i \right) \\
\text{s.t.} \quad & \boldsymbol{b}_i = \boldsymbol{I}_i^{\mathcal{X}} \boldsymbol{v}_i, && 0 \leqslant i \leqslant m, \\
& \boldsymbol{r}_i = \boldsymbol{R}_i^{\mathcal{A}} \boldsymbol{v}_i, && 0 \leqslant i \leqslant m, \\
& \boldsymbol{h}_i = \boldsymbol{G}_i \boldsymbol{v}_i + \boldsymbol{P}_{i+1} \boldsymbol{v}_{i+1}, && 0 \leqslant i \leqslant m-1.
\end{aligned} \tag{7.6}$$

The matrices $\boldsymbol{I}_i^{\mathcal{X}}$ contain the subsets of rows of the identity matrix $\boldsymbol{I}$ belonging to active simple bounds. The vectors $\boldsymbol{b}_i$ contain an element–wise combination of the entries of the vectors $\boldsymbol{l}_i$ and $\boldsymbol{u}_i$ as indicated by the current set $\mathcal{X}$ of fixed unknowns, i.e., active simple bounds. △

**Definition 7.7 (KKT System of Direct Multiple Shooting EQP with Bounds)**
For a given active set $\mathcal{A}(\boldsymbol{v})$ the KKT system of problem (7.6) reads

$$\left[\begin{array}{ccccccccccccccc}
\boldsymbol{H}_0^{\mathcal{FF}} & \boldsymbol{H}_0^{\mathcal{FX}} & & \boldsymbol{R}_0^{\mathcal{AF}T} & \boldsymbol{G}_0^{\star\mathcal{F}T} & & & & & & & & & & \\
\boldsymbol{H}_0^{\mathcal{XF}} & \boldsymbol{H}_0^{\mathcal{XX}} & \boldsymbol{I}_0^{\mathcal{XX}T} & \boldsymbol{R}_0^{\mathcal{AX}T} & \boldsymbol{G}_0^{\star\mathcal{X}T} & & & & & & & & & & \\
 & \boldsymbol{I}_0^{\mathcal{XX}} & & & & & & & & & & & & & \\
\boldsymbol{R}_0^{\mathcal{AF}} & \boldsymbol{R}_0^{\mathcal{AX}} & & & & & & & & & & & & & \\
\boldsymbol{G}_0^{\star\mathcal{F}} & \boldsymbol{G}_0^{\star\mathcal{X}} & & & & \boldsymbol{P}_1^{\star\mathcal{F}} & \boldsymbol{P}_1^{\star\mathcal{X}} & & & & & & & & \\
 & & & & \boldsymbol{P}_1^{\star\mathcal{F}T} & \boldsymbol{H}_1^{\mathcal{FF}} & \boldsymbol{H}_1^{\mathcal{FX}} & & \boldsymbol{R}_1^{\mathcal{AF}T} & \boldsymbol{G}_1^{\star\mathcal{F}T} & & & & & \\
 & & & & \boldsymbol{P}_1^{\star\mathcal{X}T} & \boldsymbol{H}_1^{\mathcal{XF}} & \boldsymbol{H}_1^{\mathcal{XX}} & \boldsymbol{I}_1^{\mathcal{XX}T} & \boldsymbol{R}_1^{\mathcal{AX}T} & \boldsymbol{G}_1^{\star\mathcal{X}T} & & & & & \\
 & & & & & & \boldsymbol{I}_1^{\mathcal{XX}} & & & & & & & & \\
 & & & & & \boldsymbol{R}_1^{\mathcal{AF}} & \boldsymbol{R}_1^{\mathcal{AX}} & & & & \ddots & & & & \\
 & & & & & \boldsymbol{G}_1^{\star\mathcal{F}} & \boldsymbol{G}_1^{\star\mathcal{X}} & & & \ddots & & \boldsymbol{P}_m^{\star\mathcal{F}} & \boldsymbol{P}_m^{\star\mathcal{X}} & & \\
 & & & & & & & & & & \boldsymbol{P}_m^{\star\mathcal{F}T} & \boldsymbol{H}_m^{\mathcal{FF}} & \boldsymbol{H}_m^{\mathcal{FX}} & & \boldsymbol{R}_m^{\mathcal{AF}T} \\
 & & & & & & & & & & \boldsymbol{P}_m^{\star\mathcal{X}T} & \boldsymbol{H}_m^{\mathcal{XF}} & \boldsymbol{H}_m^{\mathcal{XX}} & \boldsymbol{I}_m^{\mathcal{XX}T} & \boldsymbol{R}_m^{\mathcal{AX}T} \\
 & & & & & & & & & & & & \boldsymbol{I}_m^{\mathcal{XX}} & & \\
 & & & & & & & & & & & \boldsymbol{R}_m^{\mathcal{AF}} & \boldsymbol{R}_m^{\mathcal{AX}} & &
\end{array}\right]
\begin{bmatrix} -\boldsymbol{v}_0^{\mathcal{F}} \\ -\boldsymbol{v}_0^{\mathcal{X}} \\ \boldsymbol{\nu}_0^{\mathcal{X}} \\ \boldsymbol{\mu}_0^{\mathcal{A}} \\ \boldsymbol{\lambda}_0 \\ -\boldsymbol{v}_1^{\mathcal{F}} \\ -\boldsymbol{v}_1^{\mathcal{X}} \\ \boldsymbol{\nu}_1^{\mathcal{X}} \\ \boldsymbol{\mu}_1^{\mathcal{A}} \\ \vdots \\ -\boldsymbol{v}_m^{\mathcal{F}} \\ -\boldsymbol{v}_m^{\mathcal{X}} \\ \boldsymbol{\nu}_m^{\mathcal{X}} \\ \boldsymbol{\mu}_m^{\mathcal{A}} \end{bmatrix}
=
\begin{bmatrix} \boldsymbol{g}_0^{\mathcal{F}} \\ \boldsymbol{g}_0^{\mathcal{X}} \\ -\boldsymbol{b}_0^{\mathcal{X}} \\ -\boldsymbol{r}_0^{\mathcal{A}} \\ -\boldsymbol{h}_0 \\ \boldsymbol{g}_1^{\mathcal{F}} \\ \boldsymbol{g}_1^{\mathcal{X}} \\ -\boldsymbol{b}_1^{\mathcal{X}} \\ -\boldsymbol{r}_1^{\mathcal{A}} \\ \vdots \\ \boldsymbol{g}_m^{\mathcal{F}} \\ \boldsymbol{g}_m^{\mathcal{X}} \\ -\boldsymbol{b}_m^{\mathcal{X}} \\ -\boldsymbol{r}_m^{\mathcal{A}} \end{bmatrix}$$

Here, $\boldsymbol{\nu}_i^{\mathcal{X}} \in \mathbb{R}^{n^{\mathcal{X}}}$ are the Lagrange multipliers of the active simple bounds. The set superscripts $\mathcal{A}$, $\mathcal{F}$, and $\mathcal{X}$ denote the subsets of rows (first) and columns (second) of the matrices, while an asterisk ($\star$) indicates that all rows are chosen. △

From system (7.7) we can already see that $v_i^{\mathcal{X}}$ can be easily determined. Possible factorizations of the remainder of system (7.7) should exploit its symmetry and block banded structure, and generate as little fill–in as possible outside the band of block matrices.

## 7.2 Survey of Existing Methods

In this section we survey existing structure exploiting methods for either the preprocessing of QP (7.1) or for the factorization of the EQP's KKT system (7.3). We present *condensing* methods and Riccati *recursion* as block structured techniques, and generic sparse approaches such as $LBL^T$ and $LU$ *factorizations*. All presented methods will be evaluated with regard to their runtime complexity in terms of the number of control parameters $n^{\text{q}}$ and the length $m$ of the discretization grid. Both may be large in mixed–integer Model Predictive Control (MPC) problems. Emphasis is put on the applicability of the structure exploitation in conjunction with active set methods as discussed in chapters 4 and 6. Where available, we mention representative implementations of the presented techniques.

### 7.2.1 Condensing

Condensing algorithms for direct multiple shooting have been first described by [34, 36, 162, 167] and later extended in e.g. [134, 191]. A detailed description can be found in [132]. We start by reordering the unknown

$$v = (s_0, q_0, \ldots, s_{m-1}, q_{m-1}, s_m)$$

of QP (7.1) to separate the additionally introduced node values $v_2$ from the single shooting values $v_1$ as follows,

$$v_1 \stackrel{\text{def}}{=} (s_0, q_0, \ldots, q_{m-1}), \qquad v_2 \stackrel{\text{def}}{=} (s_1, \ldots, s_m). \tag{7.7}$$

Assuming classical multiple shooting coupling matrices $\boldsymbol{P}_i = [\,\boldsymbol{P}_i^{\mathrm{s}}\ \boldsymbol{P}_i^{\mathrm{q}}\,] = [\,-\boldsymbol{I}\ \boldsymbol{0}\,]$ this yields for QP (7.1) the reordered constraints matrix

$$\left[\begin{array}{cccc|cccc} \boldsymbol{G}_0^{\mathrm{s}} & \boldsymbol{G}_0^{\mathrm{q}} & & & -\boldsymbol{I} & & & \\ & \boldsymbol{G}_1^{\mathrm{q}} & & & \boldsymbol{G}_1^{\mathrm{s}} & -\boldsymbol{I} & & \\ & & \ddots & & & \ddots & \ddots & \\ & & & \boldsymbol{G}_{m-1}^{\mathrm{q}} & & & \boldsymbol{G}_{m-1}^{\mathrm{s}} & -\boldsymbol{I} \\ \hline \boldsymbol{R}_0^{\mathrm{s}} & \boldsymbol{R}_0^{\mathrm{q}} & & & & & & \\ & \boldsymbol{R}_1^{\mathrm{q}} & & & \boldsymbol{R}_1^{\mathrm{s}} & & & \\ & & \ddots & & & \ddots & & \\ & & & \boldsymbol{R}_{m-1}^{\mathrm{q}} & & & \boldsymbol{R}_{m-1}^{\mathrm{s}} & \\ & & & & & & & \boldsymbol{R}_m^{\mathrm{s}} \end{array}\right]. \tag{7.8}$$

Here, superscripts $^{\mathrm{s}}$ and $^{\mathrm{q}}$ denote the subset of columns of the matrices belonging to the states and controls respectively. We may now use the negative identity matrix blocks of the matching condition equalities as pivots to formally eliminate the additionally introduced multiple shooting state values $\boldsymbol{v}_2$ from system (7.8), analogous to Gaussian elimination. From this elimination procedure the dense constraint matrix

$$\left[\begin{array}{cc} \overline{\boldsymbol{G}} & -\boldsymbol{I} \\ \overline{\boldsymbol{R}} & \end{array}\right] \stackrel{\text{def}}{=} \left[\begin{array}{ccccc|cccc} \boldsymbol{G}_0^{\mathrm{s}} & \boldsymbol{G}_0^{\mathrm{q}} & & & & -\boldsymbol{I} & & & \\ \boldsymbol{G}_1^{\mathrm{s}}\boldsymbol{G}_0^{\mathrm{s}} & \boldsymbol{G}_1^{\mathrm{s}}\boldsymbol{G}_0^{\mathrm{q}} & \boldsymbol{G}_1^{\mathrm{q}} & & & & -\boldsymbol{I} & & \\ \vdots & \vdots & \vdots & \ddots & & & & \ddots & \\ \boldsymbol{\Gamma}_0^{m-1} & \boldsymbol{\Gamma}_1^{m-1}\boldsymbol{G}_0^{\mathrm{q}} & \boldsymbol{\Gamma}_2^{m-1}\boldsymbol{G}_1^{\mathrm{q}} & \cdots & \boldsymbol{G}_{m-1}^{\mathrm{q}} & & & & -\boldsymbol{I} \\ \hline \boldsymbol{R}_0^{\mathrm{s}} & \boldsymbol{R}_0^{\mathrm{q}} & & & & & & & \\ \boldsymbol{R}_1^{\mathrm{s}}\boldsymbol{G}_0^{\mathrm{s}} & \boldsymbol{R}_1^{\mathrm{s}}\boldsymbol{G}_0^{\mathrm{q}} & \boldsymbol{R}_1^{\mathrm{q}} & & & & & & \\ \vdots & \vdots & \vdots & \ddots & & & & & \\ \boldsymbol{R}_m^{\mathrm{s}}\boldsymbol{\Gamma}_0^{m-1} & \boldsymbol{R}_m^{\mathrm{s}}\boldsymbol{\Gamma}_1^{m-1}\boldsymbol{G}_0^{\mathrm{q}} & \boldsymbol{R}_m^{\mathrm{s}}\boldsymbol{\Gamma}_2^{m-1}\boldsymbol{G}_1^{\mathrm{q}} & \cdots & \boldsymbol{R}_m^{\mathrm{s}}\boldsymbol{G}_{m-1}^{\mathrm{q}} & & & & \end{array}\right] \tag{7.9}$$

is obtained, with sensitivity matrix products $\boldsymbol{\Gamma}_i^j$ defined as

$$\boldsymbol{\Gamma}_i^j \stackrel{\text{def}}{=} \begin{cases} \prod_{l=i}^{j} \boldsymbol{G}_l^{\mathrm{s}} \stackrel{\text{def}}{=} \boldsymbol{G}_j^{\mathrm{s}} \cdot \ldots \cdot \boldsymbol{G}_i^{\mathrm{s}} & \text{if } 0 \leqslant i \leqslant j \leqslant m-1, \\ \boldsymbol{I} & \text{if } i > j. \end{cases} \tag{7.10}$$

From (7.9) we deduce that, after this elimination step, the transformed QP in terms of the two parts $\boldsymbol{v}_1$ and $\boldsymbol{v}_2$ of the unknown reads

$$\begin{aligned} \min_{\boldsymbol{v}_1,\boldsymbol{v}_2} \quad & \frac{1}{2}\begin{bmatrix}\boldsymbol{v}_1\\ \boldsymbol{v}_2\end{bmatrix}^T \begin{bmatrix}\overline{\boldsymbol{H}}_{11} & \overline{\boldsymbol{H}}_{12}\\ \overline{\boldsymbol{H}}_{12}^T & \overline{\boldsymbol{H}}_{22}\end{bmatrix} \begin{bmatrix}\boldsymbol{v}_1\\ \boldsymbol{v}_2\end{bmatrix} + \begin{bmatrix}\boldsymbol{v}_1\\ \boldsymbol{v}_2\end{bmatrix}^T \begin{bmatrix}\overline{\boldsymbol{g}}_1\\ \overline{\boldsymbol{g}}_2\end{bmatrix} \\ \text{s.t.} \quad & 0 = \overline{\boldsymbol{G}}\boldsymbol{v}_1 - \boldsymbol{I}\boldsymbol{v}_2 - \overline{\boldsymbol{h}}, \\ & 0 \leqq \overline{\boldsymbol{R}}\boldsymbol{v}_1 - \overline{\boldsymbol{r}}, \end{aligned} \tag{7.11}$$

wherein the Hessian blocks and the gradient are obtained by reordering $\boldsymbol{H}$ and $\boldsymbol{g}$. The right hand side vectors $\overline{\boldsymbol{h}}$ and $\overline{\boldsymbol{r}}$ are computed by applying the Gaussian elimination steps to $\boldsymbol{h}$ and $\boldsymbol{r}$. System (7.11) lends itself to the elimination of the unknown $\boldsymbol{v}_2$. By this step we arrive at the condensed QP

$$\begin{aligned} \min_{\boldsymbol{v}_1} \quad & \tfrac{1}{2}\boldsymbol{v}_1{}^T\overline{\overline{\boldsymbol{H}}}\boldsymbol{v}_1 + \boldsymbol{v}_1^T\overline{\overline{\boldsymbol{g}}} \\ \text{s.t.} \quad & 0 \leqslant \overline{\boldsymbol{R}}\boldsymbol{v}_1 - \overline{\boldsymbol{r}} \end{aligned} \tag{7.12}$$

with the following dense Hessian matrix and gradient obtained from substitution of $\boldsymbol{v}_2$ in the objective of (7.11)

$$\overline{\overline{\boldsymbol{H}}} = \overline{\boldsymbol{H}}_{11} + \overline{\boldsymbol{H}}_{12}\overline{\boldsymbol{G}} + \overline{\boldsymbol{G}}^T\overline{\boldsymbol{H}}_{12}^T + \overline{\boldsymbol{G}}^T\overline{\boldsymbol{H}}_{22}\overline{\boldsymbol{G}}, \tag{7.13a}$$

$$\overline{\overline{\boldsymbol{g}}} = \overline{\boldsymbol{g}}_1 + \overline{\boldsymbol{G}}^T\overline{\boldsymbol{g}}_2 - 2(\overline{\boldsymbol{H}}_{12}^T\overline{\boldsymbol{h}} + \overline{\boldsymbol{G}}^T\overline{\boldsymbol{H}}_{22}\overline{\boldsymbol{h}}). \tag{7.13b}$$

The matrix multiplications required for the computation of these values are easily laid out to exploit the block structure of $\overline{\boldsymbol{G}}$ and $\overline{\boldsymbol{H}}$, cf. [132]. The dense QP (7.12) can be solved using one of the widely available numerical codes for dense Quadratic Programming, such as `QPOPT` [86] or `qpOASES` [68]. In addition, from the above elimination steps one easily recovers $\boldsymbol{v}_2$ from the solution $\boldsymbol{v}_1$ of the condensed QP (7.12),

$$\boldsymbol{v}_2 = \overline{\boldsymbol{G}}\boldsymbol{v}_1 - \overline{\boldsymbol{h}}. \tag{7.14}$$

Implementations of condensing techniques can e.g. be found in the multiple shooting code for optimal control `MUSCOD-II` [132, 134].

**Remark 7.1 (Runtime Complexity of Condensing)**
The computation of (7.9) and of $\overline{\overline{\boldsymbol{H}}}$ evidently takes $\mathcal{O}(m^3n^3)$ operations, where $n = n^{\mathrm{x}} + n^{\mathrm{q}}$. Hence condensing techniques are efficient for problems with short horizons, i.e., small values of $m$. The resulting condensed QP (7.12) has $n^{\mathrm{x}} + mn^{\mathrm{q}}$ un-

knowns, which makes condensing techniques attractive for QPs with considerably more states than control parameters, i.e., $n^{\mathrm{q}} \ll n^{\mathrm{x}}$.

**Remark 7.2 (Variant for the Case of Many Controls)**
If condensing methods are used at all for MIOCPs with $n^{\mathrm{q}} > n^{\mathrm{x}}$ after outer convexification, the computation of the chains of sensitivity matrices should at least be rearranged to postpone the prolongations onto the control space as long as possible, e.g.

$$\boldsymbol{\Gamma}_{j+1}^{k} \boldsymbol{G}_{j}^{\mathrm{q}} = \left( \boldsymbol{G}_{k}^{\mathrm{s}} \cdot \ldots \cdot \boldsymbol{G}_{j+1}^{\mathrm{s}} \right) \boldsymbol{G}_{j}^{\mathrm{q}} \quad \text{instead of} \quad \boldsymbol{\Gamma}_{j+1}^{k} \boldsymbol{G}_{j}^{\mathrm{q}} = \boldsymbol{G}_{k}^{\mathrm{s}} \left( \boldsymbol{G}_{k+1}^{\mathrm{s}} \left( \ldots \boldsymbol{G}_{j+1}^{\mathrm{s}} \left( \boldsymbol{G}_{j}^{\mathrm{q}} \right) \ldots \right) \right)$$

This usually is *not* the case for existing implementations of condensing methods, which use the second formulation preferable in the case $n^{\mathrm{q}} \ll n^{\mathrm{x}}$.

For comparative studies, the above variant of condensing is realized in the multiple shooting real–time online optimization method `MuShROOM`, see appendix A and the numerical results in chapter 9.

### Potentially Active Bounds Strategy

We have so far ignored simple bounds on the additionally introduced direct multiple shooting states $\boldsymbol{s}_1, \ldots, \boldsymbol{s}_m$. Under the condensing elimination procedure, these become linear constraints depending on all predecessor node states and controls. The computation of their Jacobians comes at the cost of $\mathcal{O}(m^2 n^3)$ and their inclusion in the condensed QP would lead to an unnecessarily large constraints dimension. In [133] it has therefore been proposed to use a *potentially active bounds* strategy. In each Sequential Quadratic Programming (SQP) iteration, the condensed quadratic subproblem is solved without computing or including the condensed bounds on $\boldsymbol{s}_1, \ldots, \boldsymbol{s}_m$. If the optimal solution after the blowup step is found to violate one of the simple bounds – this test is fast –, only the condensed Jacobians for the violating simple state bounds are computed and the QP is resolved with the old solution serving as a hot starting guess. The included bounds are remembered for the subsequent SQP iterations, allowing the QP subproblems to gradually grow as required to maintain feasibility of the multiple shooting node states.

## 7.2.2 Riccati Recursion

The solution of QP (7.1) can be attacked in the framework of dynamic programming, which leads to a Riccati recursion scheme. Our presentation can be found in [51] and we refer to [202, 176] for algorithmic variants and further details.

Starting with the cost function

$$\Phi_m(\boldsymbol{s}_m) \stackrel{\text{def}}{=} \tfrac{1}{2}\boldsymbol{s}_m^T \boldsymbol{H}_m^{\text{ss}} \boldsymbol{s}_m + \boldsymbol{s}_m^T \boldsymbol{g}_m \tag{7.15}$$

for the last shooting node, the cost–to–go function for any node $0 \leqslant i \leqslant m-1$ is constructed from that of the successor node $i+1$ as the solution of the small QP in the following unknown $\boldsymbol{v}_i = (\boldsymbol{s}_i, \boldsymbol{q}_i)$

$$\begin{aligned} \min_{\boldsymbol{q}_i, \boldsymbol{s}_{i+1}} \quad & \tfrac{1}{2}\boldsymbol{v}_i^T \boldsymbol{H}_i \boldsymbol{v}_i + \boldsymbol{v}_i^T \boldsymbol{g}_i + \Phi_{i+1}(\boldsymbol{s}_{i+1}) \\ \text{s.t.} \quad & \boldsymbol{G}_i \boldsymbol{v}_i + \boldsymbol{P}_{i+1} \boldsymbol{v}_{i+1} = \boldsymbol{h}_i. \end{aligned} \tag{7.16}$$

which yields for any given state $\boldsymbol{s}_i$ the optimal control choice $\boldsymbol{q}_i$ minimizing the cost–to–go to $\boldsymbol{s}_{i+1}$ and recursively to the final node $m$. As the objective function of the $m-1$ QPs remains quadratic, this procedure is efficient.

In a backward sweep running in node $i = m-1$ and running to the initial node $i = 0$, the state vectors $\boldsymbol{s}_{i+1}$ are eliminated using the matching condition. This allows to derive an analytic expression for the solution $\boldsymbol{q}_i(\boldsymbol{s}_i)$ of the small equality constrained QP (7.16), and thus for the cost to go functions $\Phi_i(\boldsymbol{s}_i)$. The actual value $\boldsymbol{s}_0$ is the measured or estimated process state, and needs to be known only after the backward sweep has been completed. From $\boldsymbol{s}_0$, the sequence of optimal control parameters $\boldsymbol{q}$ and the new state trajectory values $\boldsymbol{s}$ can be determined. We refer to [51] for details on the involved linear algebra operations.

**Remark 7.3 (Applicability of Riccati recursion)**
As Riccati recursion makes no provisions for treating inequality constraints, it can either be applied to solve purely equality constrained QPs, or it can be used as a KKT solver inside the active set loop by applying it to the equality constrained subproblem (EQP) for a given active set. Matrix update techniques are not applicable, however, and the runtime complexity remains at $\mathcal{O}(mn^3)$.

Riccati recursion techniques have primarily found application in interior point methods, cf. [14, 102, 176, 177]. The dual active set code `QPSchur` [12] applies related Schur complement techniques that allow for matrix updates. In section 7.3 we present in detail a factorization related to Riccati recursion. Matrix updates for this factorization are derived in chapter 8.

### 7.2.3 Symmetric Indefinite Factorization

Direct methods for the solution of system (7.3) have gained increased interest because arbitrary sparsity patterns of the KKT matrix to be factorized can be exploited more easily. An appropriate choice for system (7.3) are symmetric indefinite factorizations $\boldsymbol{K} = \boldsymbol{L}\boldsymbol{B}\boldsymbol{L}^T$, where $\boldsymbol{L}$ is a lower triangular factor and $\boldsymbol{B}$ is a block diagonal matrix with $2 \times 2$ pivot blocks on the diagonal that capture the indefiniteness of the KKT system [44, 158].

Symmetric indefinite factorizations have become the workhorses for most interior point methods, e.g. [82, 212, 214, 215]. The drawback that usually makes these factorizations ineffective for use in active set methods is the lack of fast matrix updates procedures, cf. [129]. This means that the factors $\boldsymbol{L}$ and $\boldsymbol{B}$ of the KKT matrix $\boldsymbol{K}$ cannot be recovered in a computationally inexpensive way after an active set change, but have to be recomputed from scratch.

The dense symmetric indefinite factorization code DSYTRF is available as part of the publicly available *Linear Algebra Package (LAPACK)* [9]. Highly efficient sparse codes are the multifrontal symmetric indefinite solvers MA27 [57] and MA57 [56] which are part of the *Harwell Subroutine Library (HSL)*.

### 7.2.4 LU Factorization

LU factorization codes do not take advantage of the symmetry of the KKT system, but provide facilities for matrix updates after active set changes. Appropriate techniques can be found in [62, 220].

The banded dense LU factorization code DGBTRF is available as part of the publicly available *Linear Algebra Package (LAPACK)* [9] and can exploit a considerable part of the sparsity present in the multiple shooting structured KKT system (7.3) for small sizes $n$ and larger numbers $m$ of the KKT blocks, i.e., when the block banded matrix is nearly component–wise banded. If unsymmetric pivoting is used, the bandwidth of the $L$ factor cannot be bounded and the factorization fills in, cf. [89].

Multifrontal LU factorizations exploting arbitrary sparsity patterns are provided e.g. by the software package UMFPACK [50]. Matrix updates to sparse factorizations tend to fill in over the course of a few active set changes, then requiring recomputation of the LU decomposition from scratch. This approach is explored in the code QPBLU for large–scale Quadratic Programming by [108] and coworkers. Here, updates of the LU factorization are computed using LUMOD, cf. [62] and also [220] for the underlying elementary transformations.

## 7.3 A Factorization for Structured KKT Systems with Many Controls

In this section we present a factorization of the KKT matrix of the direct multiple shooting QP (7.6) that is particularly suitable for the case of many control parameters. It works on the block structure of the KKT matrix, and is free of fill–in effects. The approach to be presented is based on previous work on a family of combined Hessian projection and Schur complement reductions of the multiple shooting QP in [201, 202, 203]. Extension of these works to tree–structured QPs can be found in [204]. The application of the presented factorization to MIOCPs has been first shown in [119, 120].

We present the individual steps of this new factorization for one selected shooting node indexed by $i$. The described steps have to be performed for all nodes $0 \leqslant i \leqslant m$. To avoid having to specially treat the first ($i = 0$) and last ($i = m$) shooting nodes, we define the following empty extra matrices and vectors with appropriate dimensions,

$$\boldsymbol{P}_0 \overset{\text{def}}{=} \boldsymbol{0}, \qquad \boldsymbol{\lambda}_{-1} \overset{\text{def}}{=} \boldsymbol{0}, \quad \tilde{\boldsymbol{\lambda}}_{-1} \overset{\text{def}}{=} \boldsymbol{0},$$
$$\boldsymbol{G}_m \overset{\text{def}}{=} \boldsymbol{0}, \quad \boldsymbol{P}_{m+1} \overset{\text{def}}{=} \boldsymbol{0}, \quad \boldsymbol{h}_m \overset{\text{def}}{=} \boldsymbol{0}, \quad \boldsymbol{\lambda}_m \overset{\text{def}}{=} \boldsymbol{0}.$$

### 7.3.1 Fixed Variables Step

We start by solving in system (7.7) for the step of the subset $\boldsymbol{v}_i^{\mathcal{X}}$ of the primal unknowns $\boldsymbol{v}_i$ fixed to their simple upper or lower bounds. This yields

$$\boldsymbol{I}_i^{\mathcal{X}}(-\boldsymbol{v}_i^{\mathcal{X}}) = -\boldsymbol{b}_i^{\mathcal{X}} \tag{7.17}$$

which trivially reduces to $\boldsymbol{v}_i^{\mathcal{X}} = \boldsymbol{b}_i^{\mathcal{X}}$. Hence, any fixed component of $\boldsymbol{v}_i$ moves exactly onto the bound it is fixed to. From the stationarity conditions for $\boldsymbol{g}_i^{\mathcal{X}}$ in system (7.7) we may recover the bounds multipliers $\boldsymbol{\nu}_i^{\mathcal{X}}$ once $\boldsymbol{v}_i$, $\boldsymbol{\lambda}_{i-1}$, $\boldsymbol{\lambda}_i$, and $\boldsymbol{\mu}_i^{\mathcal{A}}$ are known

$$\boldsymbol{\nu}_i^{\mathcal{X}} = \boldsymbol{g}_i^{\mathcal{X}} - \boldsymbol{R}_i^{\mathcal{A}\mathcal{X}^T}\boldsymbol{\mu}_i^{\mathcal{A}} - \boldsymbol{G}_i^{\star\mathcal{X}^T}\boldsymbol{\lambda}_i - \boldsymbol{P}_i^{\star\mathcal{X}^T}\boldsymbol{\lambda}_{i-1} + \boldsymbol{H}_i^{\mathcal{X}\mathcal{F}}\boldsymbol{v}_i^{\mathcal{F}} + \boldsymbol{H}_i^{\mathcal{X}\mathcal{X}}\boldsymbol{v}_i^{\mathcal{X}}. \tag{7.18}$$

### 7.3.2 Hessian Projection Step

Assuming regularity of the current active set $\mathcal{A}(\boldsymbol{v})$, the number of active decoupled point constraints does not exceed the number of free unknowns. This allows us to perform a QR decomposition of the transpose of the decoupled point constraints matrix $\boldsymbol{R}_i^{\mathcal{AF}}$. We write this decomposition as a TQ decomposition that is obtained from a QR decomposition by transposition reversal of the order of columns of the QR factors,

$$\boldsymbol{R}_i^{\mathcal{AF}}\boldsymbol{Q}_i = \begin{bmatrix} \boldsymbol{0} & \boldsymbol{T}_i \end{bmatrix}, \quad \boldsymbol{Q}_i = \begin{bmatrix} \boldsymbol{Z}_i & \boldsymbol{Y}_i \end{bmatrix} \tag{7.19}$$

The orthogonal matrix $\boldsymbol{Q}_i \in \mathbb{R}^{n_i^{\mathrm{v}} \times n_i^{\mathrm{v}}}$ is separated into column orthogonal bases $\boldsymbol{Y}_i \in \mathbb{R}^{n_i^{\mathrm{v}} \times n^{\mathcal{Y}}}$ and $\boldsymbol{Z}_i \in \mathbb{R}^{n_i^{\mathrm{v}} \times n^{\mathcal{Z}}}$ of the range space $\mathcal{Y}$ and the null space $\mathcal{Z}$ of $\boldsymbol{R}_i^{\mathcal{AF}}$ respectively. The matrix $\boldsymbol{T}_i \in \mathbb{R}^{n^{\mathcal{Y}} \times n^{\mathcal{Y}}}$ is a southeast triangular factor. Accordingly, we partition the step of the free subset $\boldsymbol{v}_i^{\mathcal{F}}$ of the primal unknowns $\boldsymbol{v}_i$ into components $\boldsymbol{v}_i^{\mathcal{Y}}$ and $\boldsymbol{v}_i^{\mathcal{Z}}$,

$$\boldsymbol{v}_i^{\mathcal{F}} = \boldsymbol{Y}_i\boldsymbol{v}_i^{\mathcal{Y}} + \boldsymbol{Z}_i\boldsymbol{v}_i^{\mathcal{Z}}. \tag{7.20}$$

Using this relation to solve the primal feasibility condition for decoupled point constraints in system (7.7) for the range space part of $\boldsymbol{v}_i^{\mathcal{F}}$ yields

$$\boldsymbol{T}_i\boldsymbol{v}_i^{\mathcal{Y}} = \boldsymbol{r}_i^{\mathcal{A}} - \boldsymbol{R}_i^{\mathcal{AX}}\boldsymbol{v}_i^{\mathcal{X}} \tag{7.21}$$

which requires a backsolve with the southeast triangular factor $\boldsymbol{T}_i$. We continue by projecting the remaining equations of system (7.7) onto the null space of the free variables part $\boldsymbol{R}_i^{\mathcal{AF}}$ of the decoupled point constraints by multiplication with $\boldsymbol{Y}_i^T$ from the left. Substituting $\boldsymbol{Y}_i\boldsymbol{v}_i^{\mathcal{Y}}$ for $\boldsymbol{v}_i^{\mathcal{F}}$ in the stationarity conditions for $\boldsymbol{g}_i^{\mathcal{F}}$ of system (7.7) we find

$$\begin{aligned} \boldsymbol{Y}_i{}^T\boldsymbol{P}_i^{\star\mathcal{F}T}\boldsymbol{\lambda}_{i-1} - \boldsymbol{Y}_i{}^T\boldsymbol{H}_i^{\mathcal{FF}}\boldsymbol{Y}_i\boldsymbol{v}_i^{\mathcal{Y}} - \boldsymbol{Y}_i{}^T\boldsymbol{H}_i^{\mathcal{FF}}\boldsymbol{Z}_i\boldsymbol{v}_i^{\mathcal{Z}} + \boldsymbol{Y}_i{}^T\boldsymbol{R}_i^{\mathcal{AF}T}\boldsymbol{\mu}_i^{\mathcal{A}} + \boldsymbol{Y}_i{}^T\boldsymbol{G}_i^{\star\mathcal{F}T}\boldsymbol{\lambda}_i \\ = \boldsymbol{Y}_i{}^T\boldsymbol{g}_i^{\mathcal{F}} + \boldsymbol{Y}_i{}^T\boldsymbol{H}_i^{\mathcal{FX}}\boldsymbol{v}_i^{\mathcal{X}} \end{aligned} \tag{7.22}$$

from which we may recover the point constraint multipliers $\boldsymbol{\mu}_i^{\mathcal{A}}$ using a backsolve with $\boldsymbol{T}_i^T$ once $\boldsymbol{v}_i^{\mathcal{Z}}$, $\boldsymbol{\lambda}_{i-1}$, and $\boldsymbol{\lambda}_i$ are known

$$\boldsymbol{T}_i^T\boldsymbol{\mu}_i^{\mathcal{A}} = \boldsymbol{Y}_i^T\left(\boldsymbol{g}_i^{\mathcal{F}} + \boldsymbol{H}_i^{\mathcal{FF}}\boldsymbol{v}_i^{\mathcal{F}} + \boldsymbol{H}_i^{\mathcal{FX}}\boldsymbol{v}_i^{\mathcal{X}} - \boldsymbol{G}_i^{\star\mathcal{F}T}\boldsymbol{\lambda}_i - \boldsymbol{P}_i^{\star\mathcal{F}T}\boldsymbol{\lambda}_{i-1}\right). \tag{7.23}$$

Finally, substituting $\boldsymbol{Y}_i\boldsymbol{v}_i^{\mathcal{Y}} + \boldsymbol{Z}_i\boldsymbol{v}_i^{\mathcal{Z}}$ for $\boldsymbol{v}_i^{\mathcal{F}}$ in the matching conditions and $\boldsymbol{Z}_i\boldsymbol{v}_i^{\mathcal{Z}}$ for $\boldsymbol{v}_i^{\mathcal{F}}$ in the stationarity conditions of system (7.7) we find

$$\begin{aligned} \boldsymbol{Z}_i^T \boldsymbol{P}_i^{\star\mathcal{F}^T} \boldsymbol{\lambda}_{i-1} - \boldsymbol{Z}_i^T \boldsymbol{H}_i^{\mathcal{F}\mathcal{F}} \boldsymbol{Z}_i \boldsymbol{v}_i^{\mathcal{Z}} + \boldsymbol{Z}_i^T \boldsymbol{R}_i^{\mathcal{A}\mathcal{F}^T} \boldsymbol{\mu}_i^{\mathcal{A}} + \boldsymbol{Z}_i^T \boldsymbol{G}_i^{\star\mathcal{F}^T} \boldsymbol{\lambda}_i \qquad (7.24a) \\ = \boldsymbol{Z}_i^T \boldsymbol{g}_i^{\mathcal{F}} + \boldsymbol{Z}_i^T \boldsymbol{H}_i^{\mathcal{F}\mathcal{F}} \boldsymbol{Y}_i \boldsymbol{v}_i^{\mathcal{Y}} + \boldsymbol{Z}_i^T \boldsymbol{H}_i^{\mathcal{F}\mathcal{X}} \boldsymbol{v}_i^{\mathcal{X}}, \\ \boldsymbol{G}_i^{\star\mathcal{F}} \boldsymbol{Z}_i \boldsymbol{v}_i^{\mathcal{Z}} + \boldsymbol{P}_{i+1}^{\star\mathcal{F}} \boldsymbol{Z}_{i+1} \boldsymbol{v}_{i+1}^{\mathcal{Z}} \qquad (7.24b) \\ = \boldsymbol{h}_i - \boldsymbol{G}_i^{\star\mathcal{F}} \boldsymbol{Y}_i \boldsymbol{v}_i^{\mathcal{Y}} - \boldsymbol{G}_i^{\star\mathcal{X}} \boldsymbol{v}_i^{\mathcal{X}} - \boldsymbol{P}_{i+1}^{\star\mathcal{F}} \boldsymbol{Y}_{i+1} \boldsymbol{v}_{i+1}^{\mathcal{Y}} - \boldsymbol{P}_{i+1}^{\star\mathcal{X}} \boldsymbol{v}_{i+1}^{\mathcal{X}}. \end{aligned}$$

For further treatment of these equations in the Schur complement step to follow, let us define null space projections as follows:

$$\begin{aligned} \tilde{\boldsymbol{H}}_i &\overset{\text{def}}{=} \boldsymbol{Z}_i^T \boldsymbol{H}_i^{\mathcal{F}\mathcal{F}} \boldsymbol{Z}_i, \qquad (7.25a) \\ \tilde{\boldsymbol{G}}_i &\overset{\text{def}}{=} \boldsymbol{G}_i^{\star\mathcal{F}} \boldsymbol{Z}_i, \\ \tilde{\boldsymbol{P}}_i &\overset{\text{def}}{=} \boldsymbol{P}_i^{\star\mathcal{F}} \boldsymbol{Z}_i, \\ \tilde{\boldsymbol{g}}_i &\overset{\text{def}}{=} \boldsymbol{Z}_i^T \left( \boldsymbol{g}_i^{\mathcal{F}} + \boldsymbol{H}_i^{\mathcal{F}\mathcal{F}} \boldsymbol{Y}_i \boldsymbol{v}_i^{\mathcal{Y}} + \boldsymbol{H}_i^{\mathcal{F}\mathcal{X}} \boldsymbol{v}_i^{\mathcal{X}} \right), \qquad (7.25b) \\ \tilde{\boldsymbol{h}}_i &\overset{\text{def}}{=} \boldsymbol{h}_i - \boldsymbol{G}_i^{\star\mathcal{F}} \boldsymbol{Y}_i \boldsymbol{v}_i^{\mathcal{Y}} - \boldsymbol{P}_{i+1}^{\star\mathcal{F}} \boldsymbol{Y}_{i+1} \boldsymbol{v}_{i+1}^{\mathcal{Y}} - \boldsymbol{G}_i^{\star\mathcal{X}} \boldsymbol{v}_i^{\mathcal{X}} - \boldsymbol{P}_{i+1}^{\star\mathcal{X}} \boldsymbol{v}_{i+1}^{\mathcal{X}}. \qquad (7.25c) \end{aligned}$$

With this notation, the system of equations (7.24a) reduces to

$$\begin{aligned} \tilde{\boldsymbol{P}}_i^T \boldsymbol{\lambda}_{i-1} - \tilde{\boldsymbol{H}}_i \boldsymbol{v}_i^{\mathcal{Z}} + \tilde{\boldsymbol{G}}_i^T \boldsymbol{\lambda}_i &= \tilde{\boldsymbol{g}}_i, \qquad (7.26a) \\ \tilde{\boldsymbol{G}}_i \boldsymbol{v}_i^{\mathcal{Z}} + \tilde{\boldsymbol{P}}_{i+1} \boldsymbol{v}_{i+1}^{\mathcal{Z}} &= \tilde{\boldsymbol{h}}_i. \qquad (7.26b) \end{aligned}$$

The remaining system to be solved for $\boldsymbol{v}^{\mathcal{Z}}$ and $\boldsymbol{\lambda}$ can be put in matrix form,

$$\begin{bmatrix} \tilde{\boldsymbol{H}}_0 & \tilde{\boldsymbol{G}}_0^T & & & & \\ \tilde{\boldsymbol{G}}_0 & & \tilde{\boldsymbol{P}}_1 & & & \\ & \tilde{\boldsymbol{P}}_1^T & \tilde{\boldsymbol{H}}_1 & \tilde{\boldsymbol{G}}_1^T & & \\ & & \tilde{\boldsymbol{G}}_1 & \ddots & & \\ & & & \ddots & & \tilde{\boldsymbol{P}}_m \\ & & & & \tilde{\boldsymbol{P}}_m^T & \tilde{\boldsymbol{H}}_m \end{bmatrix} \begin{bmatrix} -\boldsymbol{v}_0^{\mathcal{Z}} \\ \boldsymbol{\lambda}_0 \\ -\boldsymbol{v}_1^{\mathcal{Z}} \\ \boldsymbol{\lambda}_1 \\ \vdots \\ -\boldsymbol{v}_m^{\mathcal{Z}} \end{bmatrix} = \begin{bmatrix} \tilde{\boldsymbol{g}}_0 \\ -\tilde{\boldsymbol{h}}_0 \\ \tilde{\boldsymbol{g}}_1 \\ -\tilde{\boldsymbol{h}}_1 \\ \vdots \\ \tilde{\boldsymbol{g}}_m \end{bmatrix}. \qquad (7.27)$$

### 7.3.3 Schur Complement Step

Solving (7.26a) for the null space part $\boldsymbol{v}_i^{\mathcal{Z}}$ of the free subset $\boldsymbol{v}_i^{\mathcal{F}}$ of the primal unknowns, we find

$$\tilde{\boldsymbol{H}}_i \boldsymbol{v}_i^{\mathcal{Z}} = \left( \tilde{\boldsymbol{P}}_i^T \boldsymbol{\lambda}_{i-1} + \tilde{\boldsymbol{G}}_i^T \boldsymbol{\lambda}_i - \tilde{\boldsymbol{g}}_i \right), \qquad (7.28)$$

which yields $\boldsymbol{v}_i^{\mathcal{Z}}$, assuming knowledge of the matching condition duals $\boldsymbol{\lambda}$. Inserting this relation into (7.26b) and collecting for $\boldsymbol{\lambda}_i$ yields

$$\tilde{\boldsymbol{G}}_i\tilde{\boldsymbol{H}}_i^{-1}\tilde{\boldsymbol{P}}_i^T\boldsymbol{\lambda}_{i-1} + (\tilde{\boldsymbol{G}}_i\tilde{\boldsymbol{H}}_i^{-1}\tilde{\boldsymbol{G}}_i^T + \tilde{\boldsymbol{P}}_{i+1}\tilde{\boldsymbol{H}}_{i+1}^{-1}\tilde{\boldsymbol{P}}_{i+1}^T)\boldsymbol{\lambda}_i + \tilde{\boldsymbol{P}}_{i+1}\tilde{\boldsymbol{H}}_{i+1}^{-1}\tilde{\boldsymbol{G}}_{i+1}^T\boldsymbol{\lambda}_{i+1}$$
$$= \tilde{\boldsymbol{G}}_i\tilde{\boldsymbol{H}}_i^{-1}\tilde{\boldsymbol{g}}_i + \tilde{\boldsymbol{P}}_{i+1}\tilde{\boldsymbol{H}}_{i+1}^{-1}\tilde{\boldsymbol{g}}_{i+1} + \tilde{\boldsymbol{h}}_i. \tag{7.29}$$

Assuming positive definiteness of the Hessian blocks $\boldsymbol{H}_i$ on the null space $\mathcal{Z}$ of the active simple bounds and point constraints, we may employ a Cholesky decomposition $\boldsymbol{U}_i^T\boldsymbol{U}_i = \tilde{\boldsymbol{H}}_i$ of the projected Hessian. We define the following symbols to simplify the notation of (7.29):

$$\hat{\boldsymbol{G}}_i \stackrel{\text{def}}{=} \tilde{\boldsymbol{G}}_i\boldsymbol{U}_i^{-1}, \qquad \hat{\boldsymbol{P}}_i \stackrel{\text{def}}{=} \tilde{\boldsymbol{P}}_i\boldsymbol{U}_i^{-1}, \quad \hat{\boldsymbol{g}}_i \stackrel{\text{def}}{=} \boldsymbol{U}_i^{-T}\tilde{\boldsymbol{g}}_i, \tag{7.30}$$
$$\boldsymbol{A}_i \stackrel{\text{def}}{=} \hat{\boldsymbol{G}}_i\hat{\boldsymbol{G}}_i^T + \hat{\boldsymbol{P}}_{i+1}\hat{\boldsymbol{P}}_{i+1}^T, \quad \boldsymbol{B}_i \stackrel{\text{def}}{=} \hat{\boldsymbol{G}}_i\hat{\boldsymbol{P}}_i^T, \quad \boldsymbol{a}_i \stackrel{\text{def}}{=} \hat{\boldsymbol{G}}_i\hat{\boldsymbol{g}}_i + \hat{\boldsymbol{P}}_{i+1}\hat{\boldsymbol{g}}_{i+1} + \tilde{\boldsymbol{h}}_i.$$

With this, relation (7.28) for $\boldsymbol{v}_i^{\mathcal{Z}}$ can be written as

$$\boldsymbol{U}_i\boldsymbol{v}_i^{\mathcal{Z}} = \left(\hat{\boldsymbol{P}}_i^T\boldsymbol{\lambda}_{i-1} + \hat{\boldsymbol{G}}_i^T\boldsymbol{\lambda}_i - \hat{\boldsymbol{g}}_i\right) \tag{7.31}$$

and (7.29) simplifies to

$$\boldsymbol{B}_i\boldsymbol{\lambda}_{i-1} + \boldsymbol{A}_i\boldsymbol{\lambda}_i + \boldsymbol{B}_{i+1}^T\boldsymbol{\lambda}_{i+1} = \boldsymbol{a}_i \tag{7.32}$$

which can be put in matrix form as

$$\begin{bmatrix} \boldsymbol{A}_0 & {\boldsymbol{B}_1}^T & & \\ \boldsymbol{B}_1 & \boldsymbol{A}_1 & \ddots & \\ & \ddots & & {\boldsymbol{B}_{m-1}}^T \\ & & \boldsymbol{B}_{m-1} & \boldsymbol{A}_{m-1} \end{bmatrix} \boldsymbol{\lambda} = \boldsymbol{a}. \tag{7.33}$$

### 7.3.4 Block Tridiagonal Factorization and Backsolve

The linear system (7.33) is symmetric by construction and positive definite under reasonable assumptions, see theorem 7.1 on page 196. It has block tridiagonal shape and can be factorized by a tailored Cholesky decomposition [9] as follows. We may assume w.l.o.g. the following shape

$$\begin{bmatrix} \boldsymbol{V}_0 & \boldsymbol{D}_1 \\ & \boldsymbol{V}_{1,\dots,m-1} \end{bmatrix}$$

of the Cholesky factor of matrix (7.33). Here we have separated the upper triangular diagonal block $V_0 \in \mathbb{R}^{n_0^h \times n_0^h}$ and its off–diagonal block row $D_1 \in \mathbb{R}^{n_0^h \times n_1^h}$ from the Cholesky factor $V_{1,\dots,m-1}$ of the remainder of the system. From this we obtain for system (7.33) the representation

$$\begin{bmatrix} V_0^T & 0 \\ D_1^T & V_{1,\dots,m-1}^T \end{bmatrix} \begin{bmatrix} V_0 & D_1 \\ 0 & V_{1,\dots,m-1} \end{bmatrix} = \begin{bmatrix} V_0^T V_0 & V_0^T D_1 \\ D_1^T V_0 & D_1^T D_1 + V_{1,\dots,m-1}^T V_{1,\dots,m-1} \end{bmatrix}. \tag{7.34}$$

This yields the identities

$$V_0^T V_0 = A_0, \qquad V_0^T D_1 = [\, B_1^T \; 0 \dots 0 \,], \tag{7.35}$$

and the recursion

$$V_{1,\dots,m-1}^T V_{1,\dots,m-1} = A_{1,\dots,m-1} - D_1^T D_1. \tag{7.36}$$

Observe now that the block column $D_1$ consists of zero blocks except for the single side–diagonal block entry $V_0^{-T} B_1^T$, and observe further that the block product $D_1^T D_1$ affects the block $A_1$ of the remainder $A_{1,\dots,m-1}$ of system (7.33) only. Hence the block structure is carried over to the Cholesky factor, which allows us to apply the presented identities to the subsequent blocks 1 to $m-1$ as well. This is summarized in the following block tridiagonal Cholesky decomposition procedure to be carried out in ascending order for all nodes $0 \leqslant i \leqslant m-1$:

$$V_i^T V_i = A_i, \tag{7.37a}$$

$$V_i^T D_{i+1} = \begin{bmatrix} B_{i+1}^T & 0 & \dots & 0 \end{bmatrix} \qquad \text{if } i \leqslant m-2, \tag{7.37b}$$

$$A_{i+1} = A_{i+1} - D_{i+1} D_{i+1}^T \qquad \text{if } i \leqslant m-2. \tag{7.37c}$$

Using this decomposition the solution of system (7.33) to find the vector $\boldsymbol{\lambda}$ of matching condition duals is now straightforward according to the following recurrence relation:

$$V_i^T \tilde{\lambda}_i = a_i - D_{i-1}^T \tilde{\lambda}_{i-1} \qquad 0 \leqslant i \leqslant m-1, \tag{7.38a}$$

$$V_i \lambda_i = \tilde{\lambda}_i - D_i \lambda_{i+1} \qquad m-1 \geqslant i \geqslant 0. \tag{7.38b}$$

### 7.3.5 Efficient Implementation

The arrangement and number of matrix and vector operations required to compute the operations of the Hessian Projection Schur Complement (HPSC) factorization and backsolve can be further improved, as detailed in this section.

The performance of a literal implementation of the presented relations suffers from two problems. First, hidden common subexpressions are not exploited. Second and more significant, linear algebra operations on subsets of the rows and columns of the involved matrices as selected by the current active set result in widely inferior performance as highly efficient *Level 2* and *Level 3 BLAS* routines, e.g. available from the *ATLAS* library [219], cannot be employed. To ameliorate this situation, we introduce intermediates

$$\boldsymbol{j}_{i,1} \stackrel{\text{def}}{=} \boldsymbol{H}_i \left( \boldsymbol{Y}_i \boldsymbol{v}_i^{\mathcal{Y}} + \boldsymbol{v}_i^{\mathcal{X}} \right) + \boldsymbol{g}_i, \tag{7.39a}$$

$$\boldsymbol{j}_{i,2} \stackrel{\text{def}}{=} \boldsymbol{j}_{i,1} - \boldsymbol{P}_i^T \boldsymbol{\lambda}_{i-1} - \boldsymbol{G}_i^T \boldsymbol{\lambda}_i, \tag{7.39b}$$

$$\boldsymbol{J}_i \stackrel{\text{def}}{=} \boldsymbol{Z}_i \boldsymbol{U}_i^{-1}, \tag{7.39c}$$

which can be computed efficiently. Precomputing (7.39c) additionally saves two backsolves with $\boldsymbol{U}_i$ in the computation of the Schur complements $\hat{\boldsymbol{G}}_i$ and $\hat{\boldsymbol{P}}_i$. Using these intermediate values, the right hand side of the block tridiagonal system reads

$$\hat{\boldsymbol{g}}_i \stackrel{\text{def}}{=} \boldsymbol{J}_i^T \boldsymbol{j}_{i,1}^{\mathcal{F}}, \tag{7.40}$$

$$\boldsymbol{a}_i \stackrel{\text{def}}{=} \hat{\boldsymbol{G}}_i \hat{\boldsymbol{g}}_i + \hat{\boldsymbol{P}}_{i+1} \hat{\boldsymbol{g}}_{i+1} + \boldsymbol{G}_i \boldsymbol{v}_i + \boldsymbol{P}_{i+1} \boldsymbol{v}_{i+1} - \boldsymbol{h}_i \tag{7.41}$$

and the backsolve steps (7.18), (7.23), and (7.31) reduce to

$$\boldsymbol{Z}_i \boldsymbol{v}_i^{\mathcal{Z}} = \boldsymbol{J}_i \boldsymbol{J}_i^T \boldsymbol{j}_{i,2}^{\mathcal{F}}, \tag{7.42a}$$

$$\boldsymbol{j}_{i,3} \stackrel{\text{def}}{=} \boldsymbol{j}_{i,2} + \begin{bmatrix} \boldsymbol{H}_i^{\mathcal{F}\mathcal{F}} \\ \boldsymbol{H}_i^{\mathcal{X}\mathcal{F}} \end{bmatrix} \boldsymbol{Z}_i \boldsymbol{v}_i^{\mathcal{Z}}, \tag{7.42b}$$

$$\boldsymbol{T}_i^T \boldsymbol{\mu}_i^{\mathcal{A}} = \boldsymbol{Y}_i^T \boldsymbol{j}_{i,3}^{\mathcal{F}}, \tag{7.42c}$$

$$\boldsymbol{v}_i = \boldsymbol{j}_{i,3}^{\mathcal{X}} - \boldsymbol{R}_i^{\mathcal{A}\mathcal{X}^T} \boldsymbol{\mu}_i. \tag{7.42d}$$

with an additional intermediate term $\boldsymbol{j}_3$. In (7.42b), indexed access to a subset of the columns of $\boldsymbol{H}_i$ is required, but the accessed columns are continuous, allowing at least the use of *Level 1 BLAS*. Indexed linear algebra operations on the subsets of the intermediates $\boldsymbol{j}_1$, $\boldsymbol{j}_2$, and $\boldsymbol{j}_3$ can be avoided by copying the affected vector elements to continuous storage in only $\mathcal{O}(n)$ additional memory access operations.

The HPSC factorization is given in algorithm 7.2. The described improved backsolve with the factorization is summarized in algorithm 7.1. Both algorithms are implemented in our structure exploiting QP code qpHPSC, see appendix A.

### 7.3.6 Testing for Degeneracy and Boundedness

The tests for linear independence of the new active set and for positive definiteness of the new Hessian, given in theorem 6.8, require a backsolve with a special right hand side that has only a single nonzero block. In this section we describe a tailored backsolve procedure for the degeneracy test.

#### Testing for Degeneracy

Testing for linear independence of the new active set requires a backsolve with the following special right hand side:

1. If the constraint entering the active set is point constraint row $k$ of node $j$, we have for $0 \leqslant i \leqslant m$

$$\boldsymbol{g} = \boldsymbol{0}, \quad \boldsymbol{h} = \boldsymbol{0}, \quad \boldsymbol{b} = \boldsymbol{0}, \quad \boldsymbol{r}_i = \begin{cases} \boldsymbol{R}^{\mathcal{A}}_{j,k\star} & \text{if } i = j, \\ \boldsymbol{0} & \text{otherwise.} \end{cases} \tag{7.43}$$

2. If the constraint entering the active set is a simple bound on variable $x_k$ of node $j$, we have $0 \leqslant i \leqslant m$

$$\boldsymbol{g} = \boldsymbol{0}, \quad \boldsymbol{h} = \boldsymbol{0}, \quad \boldsymbol{r} = \boldsymbol{0}, \quad \boldsymbol{b}_i = \begin{cases} \boldsymbol{e}_k & \text{if } i = j, \\ \boldsymbol{0} & \text{otherwise.} \end{cases} \tag{7.44}$$

As constraints only affect a single node in the block structured QP (7.4), the block vector has a single nonzero block only. This property may be exploited to speed up the linear independence test as follows.

Let a point constraint in node $j$ become active, and let $\boldsymbol{r}_j \stackrel{\text{def}}{=} \boldsymbol{R}^{\mathcal{A}}_{j,k\star}$ denote the new constraint row $k$ of $\boldsymbol{R}^{\mathcal{A}}_j$. We find $\boldsymbol{v}^{\mathcal{X}} = \boldsymbol{0}$ for all blocks, $\boldsymbol{v}^{\mathcal{Y}}_i = \boldsymbol{0}$ for all

$i \neq j$, and $\boldsymbol{v}_j^{\mathcal{Y}} = \boldsymbol{T}_j^{-1}\boldsymbol{r}_j$. The intermediate terms introduced in section 7.3.5 simplify to

$$\boldsymbol{j}_{i,1} \stackrel{\text{def}}{=} \begin{cases} \boldsymbol{H}_j\boldsymbol{Y}_j\boldsymbol{v}_j^{\mathcal{Y}} + \boldsymbol{g}_j & \text{if } i = j, \\ \boldsymbol{0} & \text{otherwise,} \end{cases} \tag{7.45a}$$

$$\boldsymbol{j}_{i,2} \stackrel{\text{def}}{=} \boldsymbol{j}_{i,1} - \boldsymbol{P}_i^T\boldsymbol{\lambda}_{i-1} + \boldsymbol{G}_i^T\boldsymbol{\lambda}_i. \tag{7.45b}$$

For the block tridiagonal system's right hand side we find

$$\boldsymbol{a}_i = \begin{cases} \boldsymbol{0} & \text{if } i < j-1 \text{ or } i > j, \\ \hat{\boldsymbol{P}}_j\hat{\boldsymbol{g}}_j + \boldsymbol{P}_j^{\mathcal{F}}\boldsymbol{Y}_j\boldsymbol{v}_j^{\mathcal{Y}} & \text{if } i = j-1, \\ \hat{\boldsymbol{G}}_j\hat{\boldsymbol{g}}_j + \boldsymbol{G}_j^{\mathcal{F}}\boldsymbol{Y}_j\boldsymbol{v}_j^{\mathcal{Y}} & \text{if } i = j. \end{cases} \tag{7.46}$$

The forward sweep then yields $\tilde{\boldsymbol{\lambda}}_i = \boldsymbol{0}$ for all $i < j-1$, while we cannot infer $\boldsymbol{\lambda}_i = \boldsymbol{0}$ from any of the backward sweep steps. At this point no further significant savings can be made.

If a simple bound on the unknown $x_k$ of node $j$ becomes active, the new constraint row to be tested is $\boldsymbol{e}_k$, $0 \leqslant k \leqslant n^{\mathrm{x}}$. We find $\boldsymbol{v}_i^{\mathcal{X}} = \boldsymbol{0}$ for all $i \neq j$, and $\boldsymbol{v}_j^{\mathcal{X}} = \boldsymbol{e}_k$. Further, $\boldsymbol{v}_i^{\mathcal{Y}} = \boldsymbol{0}$ for all $i \neq j$, and $\boldsymbol{v}_j^{\mathcal{Y}} = -\boldsymbol{T}_j^{-1}\boldsymbol{r}_k$. Thus, we effectively test for linear independence after adding the previously fixed *column* $k$ of the point constraints $\boldsymbol{R}_i^{\mathcal{A}}$ of this node. The intermediate term $\boldsymbol{j}_{i,1}$ introduced in section 7.3.5 simplifies to

$$\boldsymbol{j}_{i,1} \stackrel{\text{def}}{=} \begin{cases} \boldsymbol{H}_i\left(\boldsymbol{Y}_i\boldsymbol{v}_i^{\mathcal{Y}} + \boldsymbol{e}_k\right) + \boldsymbol{g}_i & \text{if } i = j, \\ \boldsymbol{0} & \text{otherwise.} \end{cases} \tag{7.47}$$

With this, the argument continues as above.

### Testing for Boundedness

The HPSC factorization does not allow for significant savings to be made when testing for boundedness of the Parametric Quadratic Program (PQP) after a constraint has left the active set.

In our implementation of the block structured parametric active set code qpHPSC we assume the null space Hessian matrix blocks $\boldsymbol{Z}_i^T\boldsymbol{H}_i^{\mathcal{F}\mathcal{F}}\boldsymbol{Z}_i$ to be positive definite for all occurring active sets. Tests for positive definiteness, i.e., for boundedness of the PQP are thus omitted. This is additionally justified by noting that in an MPC problem formulation, all system states and control parameters typically reside in bounded domains.

**Algorithm 7.1:** The backsolve with the HPSC factorization.

**input** : HPSC factorization $\boldsymbol{T}, \boldsymbol{Y}, \boldsymbol{Z}, \boldsymbol{U}, \boldsymbol{V}, \boldsymbol{D}$,
KKT system blocks $\boldsymbol{H}$, $\boldsymbol{G}$, $\boldsymbol{P}$, $\boldsymbol{R}^{\mathcal{AX}}$,
KKT right hand side $\boldsymbol{g}, \boldsymbol{b}, \boldsymbol{r}, \boldsymbol{h}$.
**output** : KKT solution $\boldsymbol{v}, \boldsymbol{\lambda}, \boldsymbol{\mu}, \boldsymbol{\nu}$.
**for** $i = 0 : m$ **do**
  $\boldsymbol{v}_i^{\mathcal{X}} = \boldsymbol{b}_i^{\mathcal{X}}$;
  $\boldsymbol{v}_i^{\mathcal{F}} = \boldsymbol{Y}_i \boldsymbol{T}_i \backslash \left(\boldsymbol{r}_i - \boldsymbol{R}_i^{\mathcal{AX}} \boldsymbol{b}_i^{\mathcal{X}}\right)$;
  $\boldsymbol{j}_1 = \boldsymbol{H}_i \boldsymbol{v}_i + \boldsymbol{g}_i$;
  $\hat{\boldsymbol{g}}_i = \boldsymbol{V}_i^T \boldsymbol{j}_i^{\mathcal{F}}$;
  **if** $i > 0$ **then** $\boldsymbol{\lambda}_{i-1} = \hat{\boldsymbol{G}}_{i-1} \hat{\boldsymbol{g}}_{i-1} + \hat{\boldsymbol{P}}_i \hat{\boldsymbol{g}}_i + \boldsymbol{G}_{i-1} \boldsymbol{v}_{i-1} + \boldsymbol{P}_i \boldsymbol{v}_i - \boldsymbol{h}_{i-1}$;
**end**
$\boldsymbol{\lambda}_0 = \boldsymbol{V}_0^T \backslash \boldsymbol{\lambda}_0$;
**for** $i = 1 : m-1$ **do**
  $\boldsymbol{\lambda}_i = \boldsymbol{V}_i^T \backslash \left(\boldsymbol{\lambda}_i - \boldsymbol{D}_i^T \boldsymbol{\lambda}_{i-1}\right)$;
**end**
$\boldsymbol{\lambda}_{m-1} = \boldsymbol{V}_{m-1} \backslash \boldsymbol{\lambda}_{m-1}$;
**for** $i = m-2 : 0$ **do**
  $\boldsymbol{\lambda}_i = \boldsymbol{V}_i \backslash \left(\boldsymbol{\lambda}_i - \boldsymbol{D}_{i+1} \boldsymbol{\lambda}_{i+1}\right)$;
**end**
**for** $i = 0 : m$ **do**
  **if** $i < m$ **then** $\boldsymbol{j}_1 \mathrel{-}= \boldsymbol{G}_i^T \boldsymbol{\lambda}_i$;
  **if** $i > 0$ **then** $\boldsymbol{j}_1 \mathrel{-}= \boldsymbol{P}_i^T \boldsymbol{\lambda}_{i-1}$;
  $\tilde{\boldsymbol{v}}_i^{\mathcal{Z}} = \boldsymbol{J}_i \boldsymbol{J}_i^T \boldsymbol{j}_1^{\mathcal{F}}$;
  $\boldsymbol{v}_i \mathrel{+}= \boldsymbol{v}_i^{\mathcal{Z}}$;
  $\boldsymbol{j}_1 \mathrel{+}= [\, \boldsymbol{H}_i^{\mathcal{FF}} \ \boldsymbol{H}_i^{\mathcal{XF}} \,]^T \tilde{\boldsymbol{v}}_i^{\mathcal{Z}}$;
  $\boldsymbol{\mu}_i^{\mathcal{A}} = \boldsymbol{T}_i^T \backslash \boldsymbol{Y}_i^T \boldsymbol{j}_1^{\mathcal{F}}$;
  $\boldsymbol{v}_i = \boldsymbol{j}_1^{\mathcal{X}} - \boldsymbol{R}_i^{\mathcal{AX}^T} \boldsymbol{\mu}_i$;
**end**

**Algorithm 7.2:** The HPSC factorization for the multiple shooting EQP's KKT system.

**input** : KKT system blocks $H^{\mathcal{FF}}, G^{\star\mathcal{F}}, P^{\star\mathcal{F}}, R^{\mathcal{AF}}$.
**output** : HPSC factorization $T, Y, Z, U, V, D$.
**for** $i = 0 : m$ **do**
  $[T_i, Y_i, Z_i] = \mathrm{tq}(R_i^{\mathcal{AF}^T})$;
  $U_i = \mathrm{chol}(Z_i^T H_i^{\mathcal{FF}} Z_i)$;
  $J_i = Z_i / U_i$;
  **if** $i < m$ **then** $\hat{G}_i = G_i^{\star\mathcal{F}} J_i$;
  **if** $i > 0$ **then** $\hat{P}_i = P_i^{\star\mathcal{F}} J_i$;
**end**
$A_0 = 0$;
**for** $i = 0 : m-1$ **do**
  $A_i \mathrel{+}= \hat{G}_i \hat{G}_i^T + \hat{P}_{i+1} \hat{P}_{i+1}^T$;
  $V_i = \mathrm{chol}(A_i)$;
  **if** $i < m-1$ **then**
    $D_{i+1} = V_i^T \backslash \hat{P}_{i+1} \hat{G}_i^T$;
    $A_{i+1} = -D_{i+1}^T D_{i+1}$;
  **end**
**end**

### 7.3.7 A Simplification based on Conjugacy

If the Hessian blocks $H_i^{\mathcal{FF}}$ are positive definite for all occuring active sets, and the QP thus is strictly convex, a simplification of the presented factorization and backsolve procedure can be derived based on conjugacy, cf. [158].

Instead of computing a TQ decomposition of the constraints matrices $R_i$ as in section 7.3.2, the idea here is to compute a column orthogonal base $Q_i$ of $R_i^{\mathcal{AF}}$ satisfying

$$Q_i^T H_i^{\mathcal{FF}} Q_i = I, \qquad R_i^{\mathcal{AF}} Q_i = \begin{bmatrix} 0 & T_i \end{bmatrix}. \tag{7.48}$$

This means that the columns of $Q_i$ are additionally required to be conjugate with respect to the Hessian block $H_i^{\mathcal{FF}}$ of the free variables. The matrix $Q_i$ can be constructed easily from a TQ decomposition of $R_i$

$$R_i^{\mathcal{AF}} \tilde{Q}_i = \begin{bmatrix} 0 & \tilde{T}_i \end{bmatrix} \tag{7.49}$$

and a Cholesky decomposition of the symmetric positive definite product $\tilde{\boldsymbol{Q}}_i^T \boldsymbol{H}_i^{\mathcal{F}\mathcal{F}} \tilde{\boldsymbol{Q}}_i = \boldsymbol{L}_i \boldsymbol{L}_i^T$. Then (7.48) is satisfied with

$$\boldsymbol{Q}_i \stackrel{\text{def}}{=} \tilde{\boldsymbol{Q}}_i \boldsymbol{L}^{-T}, \qquad \boldsymbol{T}_i \stackrel{\text{def}}{=} \tilde{\boldsymbol{T}}_i \boldsymbol{L}^{-T}. \tag{7.50}$$

Partitioning $\boldsymbol{Q}_i$ into null–space and range–space column orthogonal bases $\boldsymbol{Z}_i$ and $\boldsymbol{Y}_i$ as in section 7.3.2 we get the following identities to be exploited in both the factorization and the backsolve steps:

$$\boldsymbol{Z}_i^T \boldsymbol{H}_i^{\mathcal{F}\mathcal{F}} \boldsymbol{Z}_i = \boldsymbol{I}, \qquad \boldsymbol{Z}_i^T \boldsymbol{H}_i^{\mathcal{F}\mathcal{F}} \boldsymbol{Y}_i = \boldsymbol{0}, \qquad \boldsymbol{Y}_i^T \boldsymbol{H}_i^{\mathcal{F}\mathcal{F}} \boldsymbol{Y}_i = \boldsymbol{I}. \tag{7.51}$$

The HPSC factorization simplifies as follows. In the Hessian projection step we find

$$\tilde{\boldsymbol{H}}_i = \boldsymbol{I}, \qquad \boldsymbol{j}_{i,1}^{\mathcal{F}} \stackrel{\text{def}}{=} \boldsymbol{H}_i^{\mathcal{F}\mathcal{X}} \boldsymbol{v}_i^{\mathcal{X}} + \boldsymbol{g}_i, \qquad \tilde{\boldsymbol{g}}_i \stackrel{\text{def}}{=} \boldsymbol{Z}_i^T \boldsymbol{j}_{i,1}^{\mathcal{F}}. \tag{7.52}$$

As the projected Hessian is the identity, the Schur complement step including the Cholesky decomposition of the projected Hessian vanishes completely,

$$\hat{\boldsymbol{G}}_i = \tilde{\boldsymbol{G}}_i, \qquad \hat{\boldsymbol{P}}_i = \tilde{\boldsymbol{P}}_i, \qquad \hat{\boldsymbol{g}}_i = \tilde{\boldsymbol{g}}_i, \tag{7.53}$$

and the block tridiagonal system factorization can proceed immediately. An efficient implementation of the backsolve with the computed simplified HPSC factorization differs from section 7.3.5 in

$$\boldsymbol{Z}_i \boldsymbol{v}_i^{\mathcal{Z}} = \boldsymbol{Z}_i \boldsymbol{Z}_i^T \boldsymbol{j}_{i,2}^{\mathcal{F}}, \tag{7.54a}$$

$$\boldsymbol{T}_i^T \boldsymbol{\mu}_i = \boldsymbol{v}_i^{\mathcal{Y}} + \boldsymbol{Y}_i^T \boldsymbol{j}_{i,2}^{\mathcal{F}}, \tag{7.54b}$$

$$\boldsymbol{v}_i = \boldsymbol{t}_{i,2}^{\mathcal{X}} - \boldsymbol{R}_i^{\mathcal{X}^T} \boldsymbol{\mu}_i + \boldsymbol{H}_i^{\mathcal{X}\mathcal{F}} \boldsymbol{Z}_i \boldsymbol{v}_i^{\mathcal{Z}}. \tag{7.54c}$$

## 7.4 Properties and Extensions

In this section we investigate the applicability and numerical stability of the presented HPSC factorization. Pivoting of the applied factorizations as well as iterative refinement of the obtained solution are mentioned. A dynamic programming interpretation of the HPSC factorization as given in [202] is stated.

### 7.4.1 Applicability

The following theorem shows that, given a KKT system with direct multiple shooting block structure, the HPSC factorization is as widely applicable as the popular null space method for solving the KKT system of a dense QP.

**Theorem 7.1 (Applicability of the HPSC Factorization)**
The HPSC factorization is applicable to a KKT system with

1. direct multiple shooting block structure,
2. linear independent active constraints (LICQ, definition 3.4),
3. positive definite Hessian on the null space of the active set. △

**Proof** Assumption (2.) implies regularity of the $\boldsymbol{R}_i^{\mathcal{AF}}$ and thus existence of the TQ decompositions. Assumption (3.) guarantees the existence of the Cholesky decompositions of the projected Hessian blocks $\tilde{\boldsymbol{H}}_i = \boldsymbol{Z}_i^T \boldsymbol{H}_i^{\mathcal{FF}} \boldsymbol{Z}_i$. It remains to be shown that the block tridiagonal system (7.33) which we denote by $\boldsymbol{K}$ is positive definite. To this end, observe that we have for (7.33) the representation

$$
\begin{aligned}
\boldsymbol{K} &= \begin{bmatrix} \boldsymbol{A}_0 & \boldsymbol{B}_1^T & & \\ \boldsymbol{B}_1 & \boldsymbol{A}_1 & \boldsymbol{B}_2^T & \\ & \boldsymbol{B}_2 & \boldsymbol{A}_2 & \ddots \\ & & \ddots & \ddots \end{bmatrix} \qquad (7.55) \\
&= \begin{bmatrix} \boldsymbol{G}_0 & \boldsymbol{P}_1 & & \\ & \boldsymbol{G}_1 & \boldsymbol{P}_2 & \\ & & \boldsymbol{G}_2 & \ddots \\ & & & \ddots \end{bmatrix} \begin{bmatrix} \tilde{\boldsymbol{H}}_0 & & & \\ & \tilde{\boldsymbol{H}}_1 & & \\ & & \tilde{\boldsymbol{H}}_2 & \\ & & & \ddots \end{bmatrix} \begin{bmatrix} \boldsymbol{G}_0^T & & & \\ \boldsymbol{P}_1^T & \boldsymbol{G}_1^T & & \\ & \boldsymbol{P}_2^T & \boldsymbol{G}_2^T & \\ & & \ddots & \ddots \end{bmatrix} \\
&\stackrel{\text{def}}{=} \boldsymbol{M}\tilde{\boldsymbol{H}}\boldsymbol{M}^T.
\end{aligned}
$$

By assumption (3.) we have positive definiteness of all diagonal blocks $\tilde{\boldsymbol{H}}_i$ of $\tilde{\boldsymbol{H}}$ and hence of $\tilde{\boldsymbol{H}}$ itself, i.e., it holds that

$$\forall \boldsymbol{w} \neq \boldsymbol{0} : \ \boldsymbol{w}^T \tilde{\boldsymbol{H}} \boldsymbol{w} > 0. \qquad (7.56)$$

By assumption (2.) the matrix $\boldsymbol{M}$ of equality matching conditions has full row rank and for all $\boldsymbol{v} \neq \boldsymbol{0}$ it holds that $\boldsymbol{w} = \boldsymbol{M}^T \boldsymbol{v} \neq \boldsymbol{0}$. Hence

$$\forall v \neq \mathbf{0}: \ (\boldsymbol{M}^T \boldsymbol{v})^T \tilde{\boldsymbol{H}} (\boldsymbol{M}^T \boldsymbol{v}) = \boldsymbol{v}^T (\boldsymbol{M} \tilde{\boldsymbol{H}} \boldsymbol{M}^T) \boldsymbol{v} = \boldsymbol{v}^T \boldsymbol{K} \boldsymbol{v} > 0 \tag{7.57}$$

which is the condition for positive definiteness of the system $\boldsymbol{K}$. This completes the proof. □

### 7.4.2 Uniqueness

We investigate the uniqueness of the HPSC factorization.

**Theorem 7.2 (Uniqueness of the HPSC Factorization)**
The HPSC factorization is unique up to the choice of the signs of the reverse diagonal entries of the southeast triangular factors $\boldsymbol{T}_i$, and up to the choice of the orthonormal null space column basis vectors $\boldsymbol{Z}_i$. △

**Proof** The employed Cholesky factorizations are unique. Thus the uniqueness properties of the initial block QR factorizations carry over to the HPSC factorization. The thin TQ factorizations $\boldsymbol{R}_i^{\mathcal{AF}^T} \boldsymbol{Y}_i = \boldsymbol{T}_i$ are unique up to the signs of the reverse diagonal elements of the $\boldsymbol{T}_i$ and the choice of $\boldsymbol{Z}_i$ is free subject to orthonormality of $\boldsymbol{Q}_i = [\, Z_i \ Y_i \,]$. For proofs of the uniqueness properties of Cholesky and QR factorizations we refer to e.g. [89]. □

### 7.4.3 Stability

In this section, we address the stability of the HPSC factorization. We are interested in the propagation of roundoff errors in the gradient $\boldsymbol{g}$ and right hand side $(\boldsymbol{b}, \boldsymbol{r}, \boldsymbol{h})$ through the backsolve with a HPSC factorization to the primal–dual step $(\boldsymbol{v}, \boldsymbol{\lambda}, \boldsymbol{\mu}, \boldsymbol{\nu})$.

Like condensing methods and Riccati iterations, the HPSC factorization fixes parts of the pivoting sequence, which may possibly lead to stability problems for ill–conditioned KKT systems. The Mathematical Program with Vanishing Constraints (MPVC) Lagrangian formalism introduced in chapter 6 has been introduced specifically to eliminate the major source of ill–conditioning in the targeted class of problems, QPs resulting from MIOCPs treated by outer convexification. Furthermore, all employed factorizations of the matrix blocks are stable under the assumptions of theorem 7.1. The use of a Schur complement step in section 7.3.3 still mandates caution for problems with ill–conditioned matching condition Jacobians. This may potentially be the case for processes with highly nonlinear dynamics on different time scales. For the numerical results presented in chapter 9, no problems were observed after use of the MPVC Lagrangian formalism. In the following we briefly mention iterative refinement and opportunities for pivoting of the involved factorizations to improve the backsolve's accuracy, should the need arise.

### Pivoting

Pivoting algorithms can be incorporated into the HPSC factorization in several places to help with the issue of ill–conditioning. Possible extensions include a pivoted QR decomposition of the point constraints,

$$\boldsymbol{\Pi}_i^{\mathrm{R}} \boldsymbol{R}_i^{\mathcal{AF}T} \boldsymbol{Q}_i = \begin{bmatrix} \boldsymbol{0} & \boldsymbol{T}_i \end{bmatrix}. \tag{7.58}$$

and a symmetrically pivoted block cholesky decomposition of the null space Hessian

$$\boldsymbol{\Pi}_i^{\mathrm{H}T} \tilde{\boldsymbol{H}}_i \boldsymbol{\Pi}_i^{\mathrm{H}} = \boldsymbol{U}_i^T \boldsymbol{U}_i. \tag{7.59}$$

We refer to [89] and the references found therein for details. The most promising option probably is symmetric block pivoting of the block tridiagonal Cholesky decomposition of system (7.33). This last option requires cheap condition estimates of the diagonal blocks $\boldsymbol{A}_i$ and can be shown to produce at most one additional off-diagonal block in system (7.33).

### Iterative Refinement

A different possibility to diminish the error in the KKT system's solution found using the backsolve algorithm 7.1 is to apply iterative refinement, cf. [89]. This allows to increase the number of significant digits of the primal–dual step from $n$ to $N \cdot n$ at the expense of $N-1$ additional backsolves with the residuals. The procedure is given in algorithm 7.3. Iterative refinement has been included in our implementation qpHPSC, see appendix A.

## 7.4.4 A Dynamic Programming Interpretation

In [202] a dynamic programming interpretation of system (7.1) is given as shown in figure 7.1. The KKT factorization determines the unknowns $(\boldsymbol{x}_i, \boldsymbol{u}_i)$ on the range spaces of the point constraints defined by $\boldsymbol{R}_i$, $\boldsymbol{e}_i$. The null space

**Algorithm 7.3:** Iterative refinement of a backsolve with the HPSC factorization.

**input** : HPSC factorization $\mathcal{H} = (\boldsymbol{T}, \boldsymbol{Y}, \boldsymbol{Z}, \boldsymbol{U}, \boldsymbol{V}, \boldsymbol{D})$,
KKT system blocks $\mathcal{K} = (\boldsymbol{H}, \boldsymbol{G}, \boldsymbol{P}, \boldsymbol{R})$,
KKT right hand side $\boldsymbol{k} = (\boldsymbol{g}, \boldsymbol{b}, \boldsymbol{r}, \boldsymbol{h})$,
$N$

**output** : KKT solution $\boldsymbol{v}, \boldsymbol{\lambda}, \boldsymbol{\mu}, \boldsymbol{\nu}$.

$[\boldsymbol{v}, \boldsymbol{\lambda}, \boldsymbol{\mu}, \boldsymbol{\nu}] = \boldsymbol{0}$;
$\delta \boldsymbol{k} = \boldsymbol{k}$;
**for** $i = 1 : N$ **do**
  $[\boldsymbol{v}, \boldsymbol{\lambda}, \boldsymbol{\mu}, \boldsymbol{\nu}] += \textbf{hpsc\_backsolve}(\mathcal{H}, \mathcal{K}, \delta \boldsymbol{k})$;
  $\delta \boldsymbol{k} = \textbf{kkt\_multiply}(\mathcal{H}, [\boldsymbol{v}, \boldsymbol{\lambda}, \boldsymbol{\mu}, \boldsymbol{\nu}]) - \delta \boldsymbol{k}$;
**end**

part remains free and are defined as the result of optimization problems on the manifolds

$$\mathcal{N}_i(\boldsymbol{x}) \stackrel{\text{def}}{=} \left\{ \boldsymbol{u} \in \mathbb{R}^{n^{\text{u}}} \mid \boldsymbol{R}_i^{\text{x}} \boldsymbol{x} + \boldsymbol{R}_i^{\text{u}} \boldsymbol{u} = \boldsymbol{e}_i \right\}.$$

We further define the manifolds of feasible states under the mapping defined by $\boldsymbol{G}_i$, $\boldsymbol{P}_{i+1}$, $\boldsymbol{h}_i$,

$$\mathcal{S}_i(\boldsymbol{x}_{i+1}, \boldsymbol{u}_{i+1}) \stackrel{\text{def}}{=} \left\{ \boldsymbol{x} \in \mathbb{R}^{n^{\text{x}}} \mid \exists \boldsymbol{u} \in \mathcal{N}(\boldsymbol{x}) : \ \boldsymbol{G}_i^{\text{x}} \boldsymbol{x} + \boldsymbol{G}_i^{\text{u}} \boldsymbol{u} + \boldsymbol{P}_{i+1}^{\text{x}} \boldsymbol{x}_{i+1} + \boldsymbol{P}_{i+1}^{\text{u}} \boldsymbol{u}_{i+1} = \boldsymbol{h}_i \right\},$$

and the manifold of feasible controls for a given feasible initial state $\boldsymbol{x}_i$ alike,

$$\mathcal{U}_i(\boldsymbol{x}_i, \boldsymbol{x}_{i+1}, \boldsymbol{u}_{i+1}) \stackrel{\text{def}}{=} \left\{ \boldsymbol{u} \in \mathcal{N}(\boldsymbol{x}_i) \mid \boldsymbol{G}_i^{\text{x}} \boldsymbol{x}_i + \boldsymbol{G}_i^{\text{u}} \boldsymbol{u} + \boldsymbol{P}_{i+1}^{\text{x}} \boldsymbol{x}_{i+1} + \boldsymbol{P}_{i+1}^{\text{u}} \boldsymbol{u}_{i+1} = \boldsymbol{h}_i \right\}.$$

For a given state $\boldsymbol{x}_{m-1}$, the optimal control $\boldsymbol{u}_{m-1}$, steering the process to the terminal state $\boldsymbol{x}_m \in \mathcal{S}_m$ is now chosen according to Bellman's principle as minimizer of the objective $\varphi$

$$\boldsymbol{u}_{m-1}(\boldsymbol{x}_{m-1}) = \underset{\boldsymbol{u}}{\operatorname{argmin}} \left\{ \varphi_{m-1}(\boldsymbol{u}, \boldsymbol{x}_{m-1}, \boldsymbol{x}_m) \mid \boldsymbol{u} \in \mathcal{U}_{m-1}(\boldsymbol{x}_{m-1}, \boldsymbol{x}_m) \right\}.$$

As can be seen, this control can be determined *locally*, i.e., without consideration of the unknowns $0 \leqslant i \leqslant m-2$, once we have found $\boldsymbol{x}_{m-1}$. In the same spirit, all further values can be found during a backward sweep starting with $i = m-2$ as solutions of local optimization problems depending on

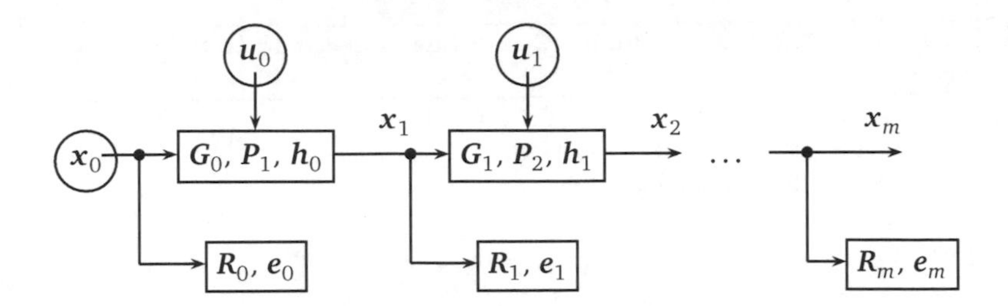


Figure 7.1: Dynamic programming interpretation of the HPSC factorization.

the preceeding state. For $0 \leqslant i \leqslant m-2$ we have

$$u_i(x_i) = \underset{u}{\operatorname{argmin}} \{\varphi_i(u, x_i, x_{i+1}) \mid u \in \mathcal{U}_i(x_i, x_{i+1}, u_{i+1})\}.$$

The initial state $x_0$ is finally found by minimizing over $\mathcal{S}_0$, and determines all other unknowns. In the case of Nonlinear Model Predictive Control (NMPC), $\mathcal{S}_0$ only contains one element, the estimated or measured system state embedded by the initial value embedding constraint.

## 7.5 Computational Complexity

In this section we investigate the computational effort in computing the HPSC factorization and performing a backsolve to find the primal–dual step.

### 7.5.1 Floating–Point Operations

We investigate the runtime complexity of the HPSC factorization in terms of the number of shooting nodes $m$, the number of control parameters $n^{\mathrm{q}}$ which may be high after application of outer convexification to a MIOCP, and the number of states $n^{\mathrm{x}}$ which is assumed to be reasonably small.

We denote for a single shooting node by $n$ the sum $n^{\mathrm{x}}+n^{\mathrm{q}}$, by $n^{\mathrm{F}}$ and $n^{\mathrm{X}}$ the numbers of free and fixed unknowns, by $n^{\mathrm{R}}$ and $n^{\mathrm{R}}_{\mathcal{A}}$ the total number and the number of active point constraints, and finally by $n^{\mathrm{Z}}$ and $n^{\mathrm{Y}}$ the dimensions of the null space and the range space of the active point constraints. For the purpose of complexity estimates, it obviously holds that $n^{\mathrm{R}}_{\mathcal{A}} = n^{\mathrm{Y}}$, $n^{\mathrm{Z}} + n^{\mathrm{Y}} = n^{\mathrm{F}}$ and $n^{\mathrm{F}} + n^{\mathrm{X}} = n$.

In the following, a Floating–Point Operation (FLOP) is understood to comprise a floating–point multiplication and addition at once. The exact number of FLOPs spent in a numerical code is implementation dependent and may vary by a constant factor, but not in complexity, from the numbers given below. In all FLOP counts, we drop lower order terms that are independent of the problem's dimensions.

### Cost of a Factorization

The QR decomposition of the transposed active point constraints Jacobians $\boldsymbol{R}_i^{\mathcal{AF}}$ takes ${n^{\mathrm{r}}_{\mathcal{A}}}^2(n-\frac{1}{3}n^{\mathrm{r}}_{\mathcal{A}})$ FLOPs, whereafter $n^{\mathrm{Y}}=n^{\mathrm{r}}_{\mathcal{A}}$ under rank assumptions. The particular case of MIOCPs with constraints treated by outer convexification mandates discussion. We may assume $n^{\mathrm{r}}=Cn^{\mathrm{q}}$ vanishing constraints where $C$ is a small constant, e.g. $C=2$. Under MPVC–LICQ, at most $n^{\mathrm{r}}_{\mathcal{A}}=C$ of these can be active for each SOS1 set per shooting node, which results in $C^2(n-\frac{1}{3}C)\in\mathcal{O}(n)$ FLOPs. The operation counts for all further steps of algorithm 7.2 can be found in table 7.1. Overall, the runtime complexity of the factorization is $\mathcal{O}(mn^3)$ and in particular $\mathcal{O}(mn)+\mathcal{O}(mn^2n^{\mathrm{x}})$ under MPVC–LICQ for MIOCPs treated by outer convexification.

| Step | FLOPs |
|---|---|
| QR decomposition | $C^2(n-\frac{C}{3})$ |
| Projected Hessian | $n^{\mathrm{Z}}n^{\mathrm{F}}(n^{\mathrm{Z}}+n^{\mathrm{F}})$ |
| Cholesky decomposition | $\frac{1}{3}n^{\mathrm{Z}3}$ |
| Temporary $J_1$ | $n^{\mathrm{F}}n^{\mathrm{Z}2}$ |
| Schur complements | $2n^{\mathrm{x}}n^{\mathrm{F}}n^{\mathrm{Z}}$ |
| Tridiagonal blocks | $2n^{\mathrm{x}3}$ |
| Cholesky decomposition | $\frac{7}{3}n^{\mathrm{x}3}$ |

Table 7.1: FLOP counts per shooting node for the HPSC factorization, algorithm 7.2.

### Cost of a Backsolve

The operation counts for all steps of the backsolve algorithm 7.1 with the HPSC factorization can be found in table 7.2. Overall, the computational effort is bounded by $m(15n^2 + \mathcal{O}(n))$. It is obvious that the computational effort crucially depends on the relative size of the range space and null space of the active constraints as well as on the number of free and fixed unknowns. In table 7.3, upper bounds on the runtime complexity for different assumptions on the active set's dimensions $n^{\mathrm{X}}$ and $n^{\mathrm{Y}}$ are listed.

The backsolve's runtime complexity grows quadratically in the number of states and controls. The growth rate is significantly lower for the dependency on the number $n^{\mathrm{q}}$ of control parameters. This is appropriate for MIOCPs treated by outer convexification, which tend to have more control parameters than differential states.

| Step | FLOPs |
|---|---|
| Fixed variables step | — |
| Range space step | $n^{\mathrm{F}} n^{\mathrm{X}} + n^{\mathrm{Y}2} + n^{\mathrm{F}} n^{\mathrm{Y}} + n^{\mathrm{Y}}$ |
| Temporary $\boldsymbol{j}_1$ | $n^2 + n$ |
| Temporary $\hat{g}_i$ | $n^{\mathrm{Z}} n^{\mathrm{F}}$ |
| Right hand side $\boldsymbol{a}_i$ | $2n^{\mathrm{x}} n^{\mathrm{Z}} + 2n^{\mathrm{x}} n + n$ |
| Matching condition multipliers | $4n^{\mathrm{x}2} + n^{\mathrm{x}}$ |
| Temporary $\boldsymbol{j}_2$ | $2n^{\mathrm{x}} n$ |
| Null space step | $2n^{\mathrm{Z}} n^{\mathrm{F}}$ |
| Temporary $\boldsymbol{j}_3$ | $n^{\mathrm{Z}} n$ |
| Point constraint multipliers | $n^{\mathrm{Y}2} + n^{\mathrm{Y}} n^{\mathrm{F}}$ |
| Simple bounds multipliers | $n^{\mathrm{X}} n^{\mathrm{Y}} + n^{\mathrm{X}}$ |

Table 7.2: FLOP counts per shooting node for a backsolve with the HPSC factorization, algorithm 7.1.

The backsolve completes faster as more variables are fixed to at their upper or lower bounds, i.e., as $n^{\mathrm{X}}$ approaches $n$ and $n^{\mathrm{F}}$ approaches zero. For MIOCPs treated by outer convexification, most control parameters will be active at either bound, cf. also theorem 2.1. Note that the limit case $n^{\mathrm{x}} = n$, $n^{\mathrm{Y}} = 0$ violates Linear Independence Constraint Qualification (LICQ) and allows further savings as we have $\boldsymbol{\lambda} = \mathbf{0}$ for the matching conditions multipliers. The FLOP bound in table 7.3 is put in parentheses. Concerning point constraints in the case of MIOCPs treated by outer convexification, the majority of constraints are likely of the vanishing constraint type and will not enter the active set. The gains in the backsolve runtime for increasing numbers of active constraints are small, though.

| Active Set | | FLOP bound in terms of $n^x$, $n^q$ |
|---|---|---|
| $n^X = 0$, | $n^Y = 0$ | $15n^{x2} + 5n^{q2} + 16n^x n^q$ |
| $n^X = 0$, | $n^Y = \frac{n}{3}$ | $13.9n^{x2} + 4.6n^{q2} + 14.4n^x n^q$ |
| $n^X = \frac{n}{3}$, | $n^Y = 0$ | $12.6n^{x2} + 3.3n^{q2} + 11.8n^x n^q$ |
| $n^X = \frac{n}{3}$, | $n^Y = \frac{n}{3}$ | $11.7n^{x2} + 3n^{q2} + 10.7n^x n^q$ |
| $n^X = n$, | $n^Y = 0$ | $(9n^{x2} + n^{q2} + 6n^x n^q)$ |

Table 7.3: FLOP bounds for a backsolve with the HPSC factorization depending on the active set. Lower order terms are dropped.

### 7.5.2 Memory Requirements

The memory requirements of the matrices and factors computed by the HPSC factorization according to algorithm 7.2 can be found in table 7.4. All matrices are allocated with their worst case dimensions such that re-allocations during the active set iterations are not necessary. For simplicity and as the dimension of the block local factors can be expected to be small, triangular factors are held in square matrix storage where one half of the storage space is never touched. The overall memory footprint of the HPSC factorization is $m(4n^2 + 2n^x n + 2n^{x2} + \mathcal{O}(n))$ which is bounded by $m(8n^2 + \mathcal{O}(n))$.

| Matrix | $\boldsymbol{T}_i$ | $\boldsymbol{Q}_i$ | $\boldsymbol{U}_i$ | $\boldsymbol{J}_i$ | $\hat{\boldsymbol{G}}_i$ | $\hat{\boldsymbol{P}}_i$ | $\boldsymbol{V}_i$ | $\boldsymbol{D}_i$ |
|---|---|---|---|---|---|---|---|---|
| rows | $n^Y$ | $n^F$ | $n^Z$ | $n^F$ | $n^x$ | $n^x$ | $n^x$ | $n^x$ |
| columns | $n^Y$ | $n^F$ | $n^Z$ | $n^Z$ | $n^Z$ | $n^Z$ | $n^x$ | $n^x$ |
| doubles allocated | $n^2$ | $n^2$ | $n^2$ | $n^2$ | $n^x n$ | $n^x n$ | $n^{x2}$ | $n^{x2}$ |

Table 7.4: Memory requirements per shooting node for the matrices and factors computed by the HPSC factorization, algorithm 7.1.

## 7.6 Summary

In this section, we have examined in detail the block structure of the quadratic subproblems induced by the multiple shooting discretization. We have surveyed block structured algorithms for its solution, such as condensing that preprocesses the block structured QP into a smaller but dense one, and Riccati iterations which can be derived by a dynamic programming argument. Factorizations of the QP's KKT system that exploit arbitrary sparsity patterns, such as $LDL^T$ and $LU$ decompositions, have been mentioned. Examination of these menthods showed that they either are inappropriate for OCPs with many control parameters, cannot easily be incorporated in active set methods, or are likely to suffer from fill–in after a few iterations of the active set method.

To address this issue, we have presented a new block structured factorization of the QP's KKT system that can for the case of MIOCPs be computed efficiently in $\mathcal{O}(mn) + \mathcal{O}(mn^2n^{\text{x}})$ operations, and is thus ideally suited for long horizons and problems with many control parameters. We have investigated a computationally efficient implementation of this factorization in the context of the parametric active set method for QPVCs of chapter 6 that has been realized in our block structured QP code qpHPSC. We have derived a simplification based on conjugacy applicable to problems with positive definite Hessian. Memory requirements and floating point operations for all parts of the factorization algorithm and the backsolve algorithm have been presented in detail.

# 8 Matrix Updates for the Block Structured Factorization

The ability to update the Karush–Kuhn–Tucker (KKT) system's factorization after addition or deletion of a constraint or simple bound is of vital importance for the efficiency of any active–set method, as described in section 6.3. The Hessian Projection Schur Complement (HPSC) factorization of the KKT system introduced in the previous chapter combines block local TQ decompositions, Cholesky decompositions, and Schur complements. A block tridiagonal Cholesky decomposition of the remaining symmetric positive definite system completes the factorization. In [201, 202, 203] a closely related factorization was used in an interior–point method. These methods typically perform few but expensive iterations using a modification of the KKT system of the entire Quadratic Program's in each iteration, and thus by design do not require matrix updates.

In this chapter we show how established *matrix update* techniques, also referred to as *basis repair* techniques, can be transferred from dense matrices and active–set methods to the block structure of the KKT system of direct multiple shooting Quadratic Programs (QPs) and to the HPSC factorization. The aim is to make the HPSC factorization applicable for use in a fast block structured active–set method. We derive matrix updates for all four cases of active set changes, namely adding or deleting a simple bound and for adding or deleting a point constraint. These update techniques allow to infer a factorization of the KKT matrix from the preceding one after the active set has changed. Using these updates we design a block structured active set method that computes the feedback control parameters with a run time complexity of only $\mathcal{O}(mn^2)$ after an initial factorization has been computed.

## 8.1 Matrix Updates Overview

Techniques for updating the factorizations of matrices with various properties have been studied for many years, and a multitude of updates tailored to special situations have been developed. A good overview over QR and Cholesky techniques is already found in [83]. In this section we briefly introduce the fundamental ideas behind some selected techniques to promote an initial understanding of the matrix updates issue. We mention Givens plane rotations and orthogonal eliminations as an important tool to modify the pattern of nonzero entries of an arbitrary matrix in a numerically stable way. They will be used throughout this chapter in order to restore the triangular shape of certain matrix factors.

### 8.1.1 Existing Techniques

In this section we briefly present the principal ideas behind selected existing matrix updates for Cholesky and QR factorizations to familiarize the reader with the issues and techniques of updating matrix factors. We refer to [83] for proofs of correctness and details on numerical stability, alternative approaches, and a comparison of their computational effort.

#### Appending a Row and Column to a Cholesky Factorization

A Cholesky factorization $\boldsymbol{A} = \boldsymbol{R}^T\boldsymbol{R}$ of a symmetric positive definite matrix $\boldsymbol{A}$ can be updated after adding a row and column to $\boldsymbol{A}$,

$$\begin{bmatrix} \boldsymbol{A} & \boldsymbol{a} \\ \boldsymbol{a}^T & \alpha \end{bmatrix} = \begin{bmatrix} \boldsymbol{R}^T & \boldsymbol{0} \\ \boldsymbol{r}^T & \varrho \end{bmatrix} \begin{bmatrix} \boldsymbol{R} & \boldsymbol{r} \\ \boldsymbol{0}^T & \varrho \end{bmatrix} = \begin{bmatrix} \boldsymbol{R}^T\boldsymbol{R} & \boldsymbol{R}^T\boldsymbol{r} \\ \boldsymbol{r}^T\boldsymbol{R} & \boldsymbol{r}^T\boldsymbol{r} + \varrho^2 \end{bmatrix}. \tag{8.1}$$

From this relation we easily determine expressions for the new column entries $\boldsymbol{r}$ and $\varrho$ of the updated Cholesky factor,

$$\begin{aligned} \boldsymbol{r} &= \boldsymbol{R}^{-T}\boldsymbol{a}, \\ \varrho &= (\alpha - \boldsymbol{r}^T\boldsymbol{r})^{\frac{1}{2}}. \end{aligned} \tag{8.2}$$

Positive definiteness of $\boldsymbol{A}$ is maintained only if $\alpha > \boldsymbol{r}^T\boldsymbol{r}$. From (8.1) it can also be seen that removing the last row and column of $\boldsymbol{A}$ is virtually free as the Cholesky factor $\boldsymbol{R}$ simply looses the last row and column as well.

#### Rank 1 Modifications of a Cholesky Factorization

Another frequently needed modification of Cholesky factorization is a rank one modification $\boldsymbol{A} \pm \alpha \cdot \boldsymbol{a}\boldsymbol{a}^T$, $\alpha > 0$ to the entire symmetric positive definite matrix. This modification is called an *update* if the dyadic product is added, and a *downdate* if it is subtracted. For a rank one update, the identity

$$\begin{bmatrix} \alpha^{\frac{1}{2}}\boldsymbol{a} & \boldsymbol{R}^T \end{bmatrix} \begin{bmatrix} \alpha^{\frac{1}{2}}\boldsymbol{a}^T \\ \boldsymbol{R} \end{bmatrix} = \boldsymbol{R}^T\boldsymbol{R} + \alpha\boldsymbol{a}\boldsymbol{a}^T \tag{8.3}$$

provides factors of the updated matrix which are rectangular and whose pattern of nonzero entries does not show upper triangular shape. Orthogonal eliminations can be used to restore the shape and yield an updated Cholesky factor. A downdate can be realized by observing

$$\begin{bmatrix} \boldsymbol{r}^T & \varrho \\ \boldsymbol{R}^T & \boldsymbol{0} \end{bmatrix} \begin{bmatrix} \boldsymbol{r} & \boldsymbol{R} \\ \varrho & \boldsymbol{0} \end{bmatrix} = \begin{bmatrix} \alpha & \boldsymbol{a}^T \\ \boldsymbol{a} & \boldsymbol{R}^T\boldsymbol{R} \end{bmatrix} \tag{8.4}$$

wherein the entries $\boldsymbol{r}$ and $\varrho$ of the first column of the extended factor are chosen as in (8.2). The extended factor is specifically constructed to allow the use of orthogonal eliminations for the transformation of the factors on the left hand side to the new shape

$$\begin{bmatrix} \boldsymbol{0}^T & \alpha^{\frac{1}{2}} \\ \boldsymbol{R}^{\star T} & \alpha^{\frac{1}{2}}\boldsymbol{a} \end{bmatrix} \begin{bmatrix} \boldsymbol{0} & \boldsymbol{R}^\star \\ \alpha^{\frac{1}{2}} & \alpha^{\frac{1}{2}}\boldsymbol{a}^T \end{bmatrix} = \begin{bmatrix} \alpha & \alpha\boldsymbol{a}^T \\ \alpha\boldsymbol{a} & \boldsymbol{R}^{\star T}\boldsymbol{R}^\star + \alpha\boldsymbol{a}\boldsymbol{a}^T \end{bmatrix} \tag{8.5}$$

which yields the new Cholesky factor $\boldsymbol{R}^\star$ satisfying the desired identity $\boldsymbol{R}^{\star T}\boldsymbol{R}^\star = \boldsymbol{R}^T\boldsymbol{R} - \alpha\boldsymbol{a}\boldsymbol{a}^T$. Positive definiteness is again maintained only if $\alpha > \boldsymbol{r}^T\boldsymbol{r}$.

### Appending a Row to a QR Factorization

For the QR factorization we start by discussing the addition of a row to the matrix $\boldsymbol{A} \in \mathbb{R}^{m\times n}$ with $m \geqslant n$ and full column rank. We can easily extend the QR factorization of $\boldsymbol{A}$ to include the new row $\boldsymbol{a}^T$,

$$\begin{bmatrix} \boldsymbol{A} \\ \boldsymbol{a}^T \end{bmatrix} = \begin{bmatrix} \boldsymbol{Q}_1 & \boldsymbol{Q}_2 & \boldsymbol{0} \\ \boldsymbol{0}^T & \boldsymbol{0}^T & 1 \end{bmatrix} \begin{bmatrix} \boldsymbol{R} \\ \boldsymbol{0} \\ \boldsymbol{a}^T \end{bmatrix}. \tag{8.6}$$

Orthogonality of the new matrix $\boldsymbol{Q}^\star$ is obviously maintained, and we can again apply orthogonal eliminations to obtain the upper triangular factor $\boldsymbol{R}^\star$ and clear the row $\boldsymbol{a}^T$ in the null space block below. Deleting an arbitrary row is possible for example by reversal of this process. The row of $\boldsymbol{Q}$ in question is transformed to the unit vector $\boldsymbol{e}_n$ by applying orthogonal eliminations, and the triangular shape of $\boldsymbol{R}$ is restored by the same means after removal of the row.

### Appending a Column to a QR Factorization

The addition of a column $\boldsymbol{a}$ to the matrix $\boldsymbol{A} \in \mathbb{R}^{m \times n}$ with $m \geqslant n$ and full column rank of the extended matrix $\begin{bmatrix} \boldsymbol{A} & \boldsymbol{a} \end{bmatrix}$ is possible by observing

$$\begin{bmatrix} \boldsymbol{A} & \boldsymbol{a} \end{bmatrix} = \begin{bmatrix} \boldsymbol{Q}_1^\star & \boldsymbol{Q}_2^\star \end{bmatrix} \begin{bmatrix} \boldsymbol{R} & \boldsymbol{r} \\ \boldsymbol{0}^T & \varrho \\ \boldsymbol{0} & \boldsymbol{0} \end{bmatrix} \tag{8.7}$$

where $\boldsymbol{Q}_1^\star$ has gained a row from $\boldsymbol{Q}_2$. This allows to compute the new elements $\boldsymbol{r}$ and $\varrho$ of the triangular factor by exploiting orthogonality of $\boldsymbol{Q}^\star$ to find

$$\boldsymbol{Q}_1^{\star T} \boldsymbol{a} = \begin{bmatrix} \boldsymbol{r} \\ \varrho \end{bmatrix}. \tag{8.8}$$

Deleting an arbitrary column from $\boldsymbol{A}$ destroys the triangular shape of the factor $\boldsymbol{R}$ after its corresponding column has been deleted. The shape can again be restored by applying orthogonal eliminations to $\boldsymbol{Q}$ and $\boldsymbol{R}$.

## 8.1.2 Orthogonal Eliminations

As seen in the previous section, orthogonal eliminations realized by Givens plane rotations are an important tool used to modify the pattern of nonzero elements in matrix factors. They play a central role in the construction of updated matrix factors from existing ones.

**Definition 8.1 (Givens Matrix)**
For a given angle $\varphi \in [0, 2\pi)$ and indices $i, j$ with $1 \leqslant i < j \leqslant n$, the Givens *matrix* $\boldsymbol{O}_i^j(\varphi) \in \mathbb{R}^{n \times n}$ is defined as

$$\boldsymbol{O}_i^j(\varphi) \overset{\text{def}}{=} \begin{bmatrix} \boldsymbol{I}_{i-1} & & & & \\ & \cos\varphi & & \sin\varphi & \\ & & \boldsymbol{I}_{j-i-1} & & \\ & -\sin\varphi & & \cos\varphi & \\ & & & & \boldsymbol{I}_{n-j} \end{bmatrix}. \tag{8.9}$$

△

Multiplication of $\boldsymbol{O}_i^j$ with a vector from the right ($\boldsymbol{O}_i^j \boldsymbol{v}$) represents a clockwise rotation of the $(i,j)$ plane by the angle $\varphi$, while multiplication from the left ($\boldsymbol{v}^T \boldsymbol{O}_i^j$) represents a counterclockwise rotation in a standard right–handed coordinate system. Givens matrices can be constructed to zero out a single element of a vector by modifying another single element only.

**Definition 8.2 (Orthogonal Eliminations $\boldsymbol{O}_i^j$)**
The *orthogonal elimination* matrix $\boldsymbol{O}_i^j(\boldsymbol{v}) \in \mathbb{R}^{n\times n}$, $1 \leqslant i,j \leqslant n$, $i \neq j$ for a vector $\boldsymbol{v} \in \mathbb{R}^n$ is defined element–wise as

$$\left(\boldsymbol{O}_i^j(\boldsymbol{v})\right)_{kl} \overset{\text{def}}{=} \begin{cases} v_i/\varrho & \text{if } (k,l) = (i,i) \vee (k,l) = (j,j), \\ v_j/\varrho & \text{if } (k,l) = (i,j), \\ -v_j/\varrho & \text{if } (k,l) = (j,i), \\ 1 & \text{if } k = l \wedge k \neq i \wedge k \neq j, \\ 0 & \text{otherwise,} \end{cases} \tag{8.10}$$

with $\varrho \overset{\text{def}}{=} \sqrt{v_i^2 + v_j^2}$.

△

From the above definition it can be seen that orthogonal elimination matrices can actually be constructed without the need to evaluate trigonometric functions in order to compute the angle $\varphi$. The following lemma formally states the properties of orthogonal elimination matrices.

**Lemma 8.1 (Identities for the Orthogonal Eliminations)**
The orthogonal elimination matrix $\boldsymbol{O}_i^j(\boldsymbol{v})$ of definition 8.2 satisfies the following identities:

1. Orthogonality: $\boldsymbol{O}_i^j(\boldsymbol{v})\boldsymbol{O}_i^j(\boldsymbol{v})^T = \boldsymbol{I}$.
2. Modification of element $i$: $\left(\boldsymbol{O}_i^j(\boldsymbol{v})\boldsymbol{v}\right)_i = \varrho$.
3. Elimination of element $j$: $\left(\boldsymbol{O}_i^j(\boldsymbol{v})\boldsymbol{v}\right)_j = 0$.
4. All elements except $i$ and $j$ remain unmodified: $\left(\boldsymbol{O}_i^j(\boldsymbol{v})\boldsymbol{v}\right)_k = v_k$, $k \neq i$, $k \neq j$. △

**Proof** Easily verified by direct calculation. □

In particular, the multiplication from the left $\boldsymbol{O}_i^j\boldsymbol{A}$ applies the Givens rotation to all columns of $\boldsymbol{A}$, while the multiplication from the right $\boldsymbol{A}\boldsymbol{O}_i^{jT}$ applies it to all rows of $\boldsymbol{A}$. Algorithm 8.1 is an exemplary way of computing and storing an orthogonal elimination that eliminates $v_j$ by modifying $v_i$. It requires four multiplications and a square root. In [89, 99] details on the fast and numerically stable computation of orthogonal elimination matrices can be found that improve algorithm 8.1 and are used in our implementation qpHPSC.

**Algorithm 8.1:** Computing and storing an orthogonal elimination matrix.

**input** : $v \in \mathbb{R}^m$, $i$, $j$
**output** : $c$, $s$
$\varrho = \sqrt{v_i^2 + v_j^2}$;
$c = v_i/\varrho$;
$s = v_j/\varrho$;

Orthogonal eliminations can be applied to all rows or columns of a matrix $\boldsymbol{A} \in \mathbb{R}^{m\times n}$ in only $4mn$ multiplications by exploiting their special structure. Algorithm 8.2 exemplarily shows how an orthogonal elimination can be applied to all columns of a matrix. A row–wise version of this algorithm is easily derived. Alternatively, algorithm 8.2 can be applied to $\boldsymbol{A}^T$ and then also yields the transpose of the desired result. In [99], a more elaborate storage and multiplication scheme is discussed that requires only $3mn$ multiplications.

**Algorithm 8.2:** Orthogonal elimination in all columns of a matrix.

**input** : $\boldsymbol{A} \in \mathbb{R}^{m\times n}$, $c$, $s$, $i$, $j$
**output** : $\boldsymbol{A} = \boldsymbol{O}_i^j(\boldsymbol{v})\boldsymbol{A}$
**for** $k = 1 : n$ **do**
  $a = A_{ik}$;
  $b = A_{jk}$;
  $A_{ik} = ac + bs$;
  $A_{jk} = bc - as$;
**end**

### 8.1.3 Applications

For dense active set range space and null space methods it is known that a sequence of a QR decomposition and a Cholesky decomposition can be updated properly after active set changes. A description of the necessary steps for a dense null space active set method can be found e.g. in [67, 158]. Update techniques for the LU factorization can be found in [62] and are used in an active set method in [108], but are not relevant for the HPSC factorization. Schur complement updates are used in a dual active set method by [12] that assumes a block diagonal Hessian.

## 8.2 Updating the Block Local Reductions

In this section, we derive matrix updates for the block local reductions of the HPSC factorization for all four cases of active set changes. The first steps of the matrix updates concerning the TQ decomposition and the Cholesky decomposition of the null space Hessian are known from the dense null space method, cf. [158] and the description of an actual implementation in [67]. The extensions to the Schur complement step and the block tridiagonal Cholesky decomposition are new contributions and have first been published in [121]. Matrix updates for the block tridiagonal Cholesky decomposition of the reduced symmetric positive definite system (7.33) are treated in the next section.

### 8.2.1 Preliminaries

#### Notation

In the following, we denote by the list of matrices

$$\mathcal{K}(\mathcal{A}) \stackrel{\text{def}}{=} (\boldsymbol{H}, \boldsymbol{R}, \boldsymbol{G}, \boldsymbol{P})$$

the block structured KKT system 7.7 on page 178 for a given active set $\mathcal{A}$. We further denote by the list of matrices

$$\mathcal{H}(\mathcal{A}) \stackrel{\text{def}}{=} (\boldsymbol{T}, \boldsymbol{Z}, \boldsymbol{Y}, \boldsymbol{U}, \hat{\boldsymbol{G}}, \hat{\boldsymbol{P}}, \boldsymbol{V}, \boldsymbol{D})$$

an associated HPSC factorization of the block structured KKT system $\mathcal{K}(\mathcal{A})$. We further distinguish by an asterisk ($^\star$) a factorization or matrix after the update from its counterpart before the update.

### Permutations

We are concerned with modifications of the HPSC factorization of chapter 7 after a permutation of the vector of unknowns. Such permutations will allow us to make assumptions about the index position of the unknown affected by a matrix update.

**Theorem 8.1 (HPSC Factorization after Permutation of the Unknowns)**
Let $\mathcal{H}$ be a HPSC factorization of the block structured KKT system $\mathcal{K}(\mathcal{A}) = (\boldsymbol{H}, \boldsymbol{R}, \boldsymbol{G}, \boldsymbol{P})$. Further, let $\boldsymbol{\Pi}_i \in \mathbb{R}^{n_i^{\mathcal{F}} \times n_i^{\mathcal{F}}}$ be permutation matrices such that $\boldsymbol{x}_i^{\mathcal{F}\star} \stackrel{\text{def}}{=} \boldsymbol{\Pi}_i \boldsymbol{x}_i^{\mathcal{F}}$ are the permuted vectors of free unknowns. Then a HPSC factorization $\mathcal{H}^\star$ of the permuted KKT system $\mathcal{K}^\star(\mathcal{A})$ is given by

$$\boldsymbol{Y}_i^\star \stackrel{\text{def}}{=} \boldsymbol{\Pi}_i \boldsymbol{Y}_i, \qquad \boldsymbol{Z}_i^\star \stackrel{\text{def}}{=} \boldsymbol{\Pi}_i \boldsymbol{Z}_i, \tag{8.11}$$

while the matrices $\boldsymbol{T}^\star, \boldsymbol{U}^\star, \hat{\boldsymbol{G}}^\star, \hat{\boldsymbol{P}}^\star, \boldsymbol{V}^\star$ and $\boldsymbol{D}^\star$ of the new factorization $\mathcal{H}^\star$ are identical to those of the old one $\mathcal{H}$. △

**Proof** We first consider the block matrix entries of the permuted KKT system $\mathcal{K}^\star$. For invariance of the KKT system under the permutations $\boldsymbol{\Pi}_i$ of the free unknowns $\boldsymbol{x}_i^{\mathcal{F}}$, it holds that

$$\begin{aligned}
\boldsymbol{R}_i^{\mathcal{AF}\star} &= \boldsymbol{R}_i^{\mathcal{AF}} \boldsymbol{\Pi}_i^T, &&\Longrightarrow \boldsymbol{R}_i^{\mathcal{AF}\star} \boldsymbol{x}_i^{\mathcal{F}\star} = \boldsymbol{R}_i^{\mathcal{AF}} \boldsymbol{\Pi}_i^T \boldsymbol{\Pi}_i \boldsymbol{x}_i^{\mathcal{F}} = \boldsymbol{R}_i^{\mathcal{AF}} \boldsymbol{x}_i^{\mathcal{F}}, \\
\boldsymbol{G}_i^{\mathcal{F}\star} &= \boldsymbol{G}_i^{\mathcal{F}} \boldsymbol{\Pi}_i^T, &&\Longrightarrow \boldsymbol{G}_i^{\mathcal{F}\star} \boldsymbol{x}_i^{\mathcal{F}\star} = \boldsymbol{G}_i^{\mathcal{F}} \boldsymbol{\Pi}_i^T \boldsymbol{\Pi}_i \boldsymbol{x}_i^{\mathcal{F}} = \boldsymbol{G}_i^{\mathcal{F}} \boldsymbol{x}_i^{\mathcal{F}}, \\
\boldsymbol{P}_i^{\mathcal{F}\star} &= \boldsymbol{P}_i^{\mathcal{F}} \boldsymbol{\Pi}_i^T, &&\Longrightarrow \boldsymbol{P}_i^{\mathcal{F}\star} \boldsymbol{x}_i^{\mathcal{F}\star} = \boldsymbol{P}_i^{\mathcal{F}} \boldsymbol{\Pi}_i^T \boldsymbol{\Pi}_i \boldsymbol{x}_i^{\mathcal{F}} = \boldsymbol{P}_i^{\mathcal{F}} \boldsymbol{x}_i^{\mathcal{F}}, \\
\boldsymbol{H}_i^{\mathcal{FF}\star} &= \boldsymbol{\Pi}_i \boldsymbol{H}_i^{\mathcal{FF}} \boldsymbol{\Pi}_i^T, &&\Longrightarrow \boldsymbol{x}_i^{\mathcal{F}\star T} \boldsymbol{H}_i^{\mathcal{FF}\star} \boldsymbol{x}_i^{\mathcal{F}\star} = \boldsymbol{x}_i^{\mathcal{F}T} \boldsymbol{\Pi}_i^T \boldsymbol{\Pi} \boldsymbol{H}_i^{\mathcal{FF}} \boldsymbol{\Pi}_i^T \boldsymbol{\Pi}_i \boldsymbol{x}_i^{\mathcal{F}} \\
& && \qquad = \boldsymbol{x}_i^{\mathcal{F}T} \boldsymbol{H}_i^{\mathcal{FF}} \boldsymbol{x}_i^{\mathcal{F}}.
\end{aligned}$$

For the first step of the HPSC factorization, the block local TQ factorizations of the free unknowns part of the active point constraints' Jacobians $\boldsymbol{R}_i^{\mathcal{AF}}$, observe

$$\begin{aligned}
\begin{bmatrix} \boldsymbol{0} & \boldsymbol{T}_i \end{bmatrix} &= \boldsymbol{R}_i^{\mathcal{AF}} \begin{bmatrix} \boldsymbol{Z}_i & \boldsymbol{Y}_i \end{bmatrix} = \boldsymbol{R}_i^{\mathcal{AF}} \underbrace{\boldsymbol{\Pi}_i^T \boldsymbol{\Pi}_i}_{=\boldsymbol{I}} \begin{bmatrix} \boldsymbol{Z}_i & \boldsymbol{Y}_i \end{bmatrix} \\
&= \boldsymbol{R}_i^{\mathcal{AF}\star} \begin{bmatrix} \boldsymbol{\Pi}_i \boldsymbol{Z}_i & \boldsymbol{\Pi}_i \boldsymbol{Y}_i \end{bmatrix} = \boldsymbol{R}_i^{\mathcal{AF}\star} \begin{bmatrix} \boldsymbol{Z}_i^\star & \boldsymbol{Y}_i^\star \end{bmatrix} = \begin{bmatrix} \boldsymbol{0} & \boldsymbol{T}_i^\star \end{bmatrix},
\end{aligned} \tag{8.12}$$

which proves the relations (8.11) for $Y_i^\star$, $Z_i^\star$ and invariance of the southest triangular factor $T_i^\star = T_i$. For the Cholesky factors $U_i$ of the projected Hessians $\tilde{H}_i$, we find

$$U_i^{\star T} U_i^\star = Z_i^{\star T} H_i^{\mathcal{F}\mathcal{F}\star} Z_i^\star = Z_i^T \underbrace{\Pi_i^T (\Pi_i}_{=I} H_i^{\mathcal{F}\mathcal{F}} \underbrace{\Pi_i^T) \Pi_i}_{=I} Z_i = Z_i^T H_i^{\mathcal{F}\mathcal{F}} Z_i = U_i^T U_i, \tag{8.13}$$

hence the Cholesky factors $U_i^\star$ of the permuted KKT system's HPSC factorization are identical to the factors $U_i$ of the original one. The Schur complements $\hat{G}_i$ and $\hat{P}_i$ are unaffected as well,

$$\begin{aligned} \hat{G}_i^\star &\stackrel{\text{def}}{=} G_i^{\mathcal{F}\star} Z_i^\star U_i^{-1\star} = (G_i^{\mathcal{F}} \underbrace{\Pi_i^T) \Pi_i}_{=I} Z_i U_i^{-1} = G_i^{\mathcal{F}} Z_i U_i^{-1} = \hat{G}_i, \\ \hat{P}_i^\star &\stackrel{\text{def}}{=} P_i^{\mathcal{F}\star} Z_i^\star U_i^{-1\star} = (P_i^{\mathcal{F}} \underbrace{\Pi_i^T) \Pi_i}_{=I} Z_i U_i^{-1} = P_i^{\mathcal{F}} Z_i U_i^{-1} = \hat{P}_i. \end{aligned} \tag{8.14}$$

This evidently carries over to the blocks $A_i^\star$ and $B_i^\star$ of the positive definite block tridiagonal system (7.33) on page 188,

$$\begin{aligned} A_i^\star &\stackrel{\text{def}}{=} \hat{G}_i^\star \hat{G}_i^{\star T} + \hat{P}_i^\star \hat{P}_i^{\star T} = \hat{G}_i \hat{G}_i^T + \hat{P}_i \hat{P}_i^T = A_i, \\ B_i^\star &\stackrel{\text{def}}{=} \hat{G}_i^\star \hat{P}_i^{\star T} = \hat{G}_i \hat{P}_i^T = B_i, \end{aligned} \tag{8.15}$$

and hence also to the Cholesky factor blocks $V_i$, $D_i$ of this system. This completes the proof. □

Note finally that permutations of the fixed part $v_i^{\mathcal{X}}$ of the unknowns do not affect the factorization. The Lagrange multipliers $\nu_i^{\mathcal{X}}$ of the active simple bounds must be permuted accordingly.

### Projectors

We will frequently need to remove the last row or column of a matrix, reflecting the fact that the number of free unknowns, or the size of the range space or null space of a TQ decomposition has decreased by one. To this end, we introduce the projector $\mathfrak{I}$ that serves to cut a column or a row off a matrix $A$, and give a proof of two useful properties of this projector.

**Definition 8.3 (Projector $\mathfrak{I}$)**
Let $\boldsymbol{A} \in \mathbb{R}^{m \times n}$ be an arbitrary matrix. The *column cutting projection* $\boldsymbol{A}\mathfrak{I}$ and the *row cutting projection* $\mathfrak{I}^T\boldsymbol{A}$ are defined as

$$\mathfrak{I} \stackrel{\text{def}}{=} \begin{bmatrix} \boldsymbol{I}_{n-1} \\ \boldsymbol{0}^T \end{bmatrix} \in \mathbb{R}^{n \times n-1}, \qquad \mathfrak{I}^T \stackrel{\text{def}}{=} \begin{bmatrix} \boldsymbol{I}_{m-1} & \boldsymbol{0} \end{bmatrix} \in \mathbb{R}^{m-1 \times m}. \tag{8.16}$$

△

**Lemma 8.2 (Identities for the Projectors)**
The projector $\mathfrak{I}$ of definition 8.3 satisfies the following identities.

1. For all $\boldsymbol{A} = \begin{bmatrix} \tilde{\boldsymbol{A}} & \boldsymbol{a} \end{bmatrix} \in \mathbb{R}^{m \times n}$ it holds that $\boldsymbol{A}\mathfrak{I} = \tilde{\boldsymbol{A}} \in \mathbb{R}^{m \times n-1}$, i.e., the matrix $\boldsymbol{A}$ looses the last column.
2. For all $\boldsymbol{A}^T = \begin{bmatrix} \tilde{\boldsymbol{A}}^T \\ \boldsymbol{a}^T \end{bmatrix} \in \mathbb{R}^{n \times m}$ it holds that $\mathfrak{I}^T\boldsymbol{A}^T = \tilde{\boldsymbol{A}} \in \mathbb{R}^{m-1 \times n}$, i.e., the matrix $\boldsymbol{A}$ looses the last row.
3. For regular triangular $\boldsymbol{A} \in \mathbb{R}^{n \times n}$ it holds that if $\mathfrak{I}^T\boldsymbol{A}\mathfrak{I}$ is regular then $(\mathfrak{I}^T\boldsymbol{A}\mathfrak{I})^{-1} = \mathfrak{I}^T\boldsymbol{A}^{-1}\mathfrak{I}$.
4. For $\boldsymbol{O} \in \mathcal{O}(n, \mathbb{R})$ it holds $\boldsymbol{O}\mathfrak{I}\mathfrak{I}^T\boldsymbol{O}^T = \boldsymbol{I} - \boldsymbol{o}\boldsymbol{o}^T$ where $\boldsymbol{o}$ is the last column of $\boldsymbol{O}$.△

**Proof** Identities 1. and 2. are easily verified by direct calculation. To prove 3. we write

$$\boldsymbol{A} = \begin{bmatrix} \tilde{\boldsymbol{A}} & \boldsymbol{a}_1 \\ \boldsymbol{a}_2^T & a_3 \end{bmatrix}, \qquad \boldsymbol{A}^{-1} \stackrel{\text{def}}{=} \boldsymbol{B} = \begin{bmatrix} \tilde{\boldsymbol{B}} & \boldsymbol{b}_1 \\ \boldsymbol{b}_2^T & b_3 \end{bmatrix}$$

such that $\mathfrak{I}^T\boldsymbol{A}\mathfrak{I} = \tilde{\boldsymbol{A}}$ and $\mathfrak{I}^T\boldsymbol{A}^{-1}\mathfrak{I} = \tilde{\boldsymbol{B}}$. For the inverse we have the defining relation $\boldsymbol{A}\boldsymbol{B} = \boldsymbol{I}$, implying $\tilde{\boldsymbol{A}}\tilde{\boldsymbol{B}} + \boldsymbol{a}_1\boldsymbol{b}_2^T = \boldsymbol{I}$ and $\tilde{\boldsymbol{A}}^{-1} = \tilde{\boldsymbol{B}}$ holds iff $\boldsymbol{a}_1 = \boldsymbol{0}$ or $\boldsymbol{b}_2 = \boldsymbol{0}$. This includes, but is not limited to, the case of lower or upper triangular matrices $\boldsymbol{A}$, as claimed. Finally to prove 4. we let $\boldsymbol{O} = [\, \tilde{\boldsymbol{o}}\ \boldsymbol{o} \,]$ such that $\boldsymbol{O}\mathfrak{I} = \tilde{\boldsymbol{O}}$ and find

$$\boldsymbol{I} = \boldsymbol{O}\boldsymbol{O}^T = \tilde{\boldsymbol{O}}\tilde{\boldsymbol{O}}^T + \boldsymbol{o}\boldsymbol{o}^T = \boldsymbol{O}\mathfrak{I}\mathfrak{I}^T\boldsymbol{O}^T + \boldsymbol{o}\boldsymbol{o}^T.$$

□

### Algorithms

In the following, all matrix updates are also summarized in the form of algorithms, on which we have several remarks to be made concerning their presentation and efficiency.

We make use of a function **givens**$(v,i,j)$ that computes a Givens rotation eliminating $v_j$ by modifying $v_i$, and a function **apply**$(L,s,c,i,j)$ that applies a Givens rotation defined by $s$, $c$ to the elements $i$ and $j$ of all rows of a given list $L$ of matrices. Any extra checks required for the first node $i=1$ and the last node $i=m$ have been omitted for clarity of exposition. The numbers $n^{\mathcal{F}}$, $n^{\mathcal{Y}}$, and $n^{\mathcal{Z}}$ are understood to refer to the dimensions prior to the active set change. In the same spirit, all KKT blocks refer to the KKT matrix associated with the active set prior to the update.

Truly efficient implementations require some further modifications that have been excluded in order to improve readability. For example, one would frequently make use of a temporary rolling column to avoid having to enlarge certain matrices prior to applying a sequence of Givens rotations. Also, the eliminations of those elements that define the Givens rotations would be applied already during computation of the Givens rotation in order to guarantee exact zeros in the eliminated components. These improvements are realized in our implementation qpHPSC. The runtime complexity bound of $\mathcal{O}(mn^2)$ is still satisfied by all simplified algorithms.

### 8.2.2 Adding a Simple Bound

If a simple bound $v_{ij}^{\mathcal{F}} = b_{ij}$ becomes active, we may assume w.l.o.g. that the unknown to be fixed is the last component $j = n_i^{\mathcal{F}}$ of $\boldsymbol{v}_i^{\mathcal{F}}$. This can be ensured by applying a suitable permutation to the unknown $\boldsymbol{v}_i^{\mathcal{F}}$ and the matrix $\boldsymbol{Q}_i$, cf. theorem 8.1. In the following theorem we show how to restore the block local reductions of a HPSC factorization for the new KKT matrix with active simple bound on $v_{ij}^{\mathcal{F}}$.

**Theorem 8.2 (Adding a Simple Bound to the Block Local Reductions)**
Let $\mathcal{H}(\mathcal{A})$ be a HPSC factorization of the KKT system $\mathcal{K}(\mathcal{A})$. Let the simple bound $v_{ij} = b_{ij}$ on the last free component $j = n_i^{\mathcal{F}}$ of the unknown of node $0 \leqslant i \leqslant m$ be inactive in $\mathcal{A}$, and denote by $\mathcal{A}^\star$ the otherwise identical active set with activated simple bound. Assume further that $\mathcal{A}^\star$ satisfies Linear Independence Constraint Qualification (LICQ). Then there exists a HPSC factorization $\mathcal{H}^\star(\mathcal{A}^\star)$ of the KKT matrix $\mathcal{K}^\star(\mathcal{A}^\star)$ that satisfies

$$\boldsymbol{Q}_i^\star = \mathfrak{I}^T \boldsymbol{Q}_i \mathfrak{O}_{\mathrm{ZT}} \mathfrak{I}, \qquad \boldsymbol{T}_i^\star = \boldsymbol{T}_i \mathfrak{O}_{\mathrm{T}} \mathfrak{I}, \tag{8.17a}$$

$$\boldsymbol{U}_i^\star = \mathfrak{I}^T \mathfrak{O}_{\mathrm{U}} \boldsymbol{U}_i \mathfrak{O}_{\mathrm{Z}} \mathfrak{I}, \tag{8.17b}$$

$$\begin{bmatrix} \hat{\boldsymbol{G}}_i^\star & \hat{\boldsymbol{g}} \end{bmatrix} = \hat{\boldsymbol{G}}_i \mathfrak{O}_{\mathrm{U}}^T, \qquad \begin{bmatrix} \hat{\boldsymbol{P}}_i^\star & \hat{\boldsymbol{p}} \end{bmatrix} = \hat{\boldsymbol{P}}_i \mathfrak{O}_{\mathrm{U}}^T, \tag{8.17c}$$

$$\boldsymbol{A}_{i-1}^\star = \boldsymbol{A}_{i-1} - \hat{\boldsymbol{p}} \hat{\boldsymbol{p}}^T, \qquad \boldsymbol{A}_i^\star = \boldsymbol{A}_i - \hat{\boldsymbol{g}} \hat{\boldsymbol{g}}^T, \boldsymbol{B}_i^\star = \boldsymbol{B}_i - \hat{\boldsymbol{g}} \hat{\boldsymbol{p}}^T \tag{8.17d}$$

where $\mathfrak{O}_{\mathrm{ZT}}$ with subsequences $\mathfrak{O}_{\mathrm{Z}}$ and $\mathfrak{O}_{\mathrm{T}}$, and $\mathfrak{O}_{\mathrm{U}}$ are appropriately chosen sequences of Givens rotations. △

**Proof** We first consider relation (8.17a) for the TQ factorization matrices $\boldsymbol{Z}_i^\star$, $\boldsymbol{Y}_i^\star$, and $\boldsymbol{T}_i^\star$. We add the simple bound's constraint row vector $\boldsymbol{e} = (0,\dots,0,1)$ to the *extended constraints matrix* of node $i$, comprising the simple bounds and the decoupled point constraints

$$\begin{bmatrix} \boldsymbol{e}^T & \boldsymbol{0}^T \\ \boldsymbol{0} & \boldsymbol{I}_i^{\mathcal{X}} \\ \boldsymbol{R}_i^{\mathcal{AF}} & \boldsymbol{R}_i^{\mathcal{AX}} \end{bmatrix} \begin{bmatrix} \boldsymbol{Z}_i & \boldsymbol{Y}_i & \\ & & \boldsymbol{I}_i^{\mathcal{X}} \end{bmatrix} = \begin{bmatrix} \boldsymbol{t}_{\mathrm{Z}}^T & \boldsymbol{t}_{\mathrm{Y}}^T & \\ & & \boldsymbol{I}_i^{\mathcal{X}} \\ & \boldsymbol{T}_i & \boldsymbol{R}_i^{\mathcal{AX}} \end{bmatrix}. \tag{8.18}$$

Herein, $\boldsymbol{t}^T \overset{\text{def}}{=} \boldsymbol{e}^T\boldsymbol{Q}_i$, i.e., $\boldsymbol{t}_{\mathrm{Z}}^T \overset{\text{def}}{=} \boldsymbol{e}^T\boldsymbol{Z}_i$ and $\boldsymbol{t}_{\mathrm{Y}}^T \overset{\text{def}}{=} \boldsymbol{e}^T\boldsymbol{Y}_i$. The right hand side of (8.18) has lost the southeast triangular shape present in $\boldsymbol{T}_i$, hence (8.18) does not yet provide a proper TQ factorization of $\boldsymbol{R}_i^{\mathcal{AF}\star}$. In order to restore this shape we eliminate the elements of $\boldsymbol{t}$ using a sequence $\mathfrak{O}_{\mathrm{Z}}\mathfrak{O}_{\mathrm{T}}$ of $n_i^{\mathcal{F}}-1$ Givens rotations

$$\begin{aligned} \mathfrak{O}_{\mathrm{Z}} &\overset{\text{def}}{=} \boldsymbol{O}_2^{1\,T}\boldsymbol{O}_3^{2\,T}\cdot\ldots\cdot\boldsymbol{O}_{n^{\mathcal{Z}}}^{n^{\mathcal{Z}}-1\,T}, \\ \mathfrak{O}_{\mathrm{T}} &\overset{\text{def}}{=} \boldsymbol{O}_{n^{\mathcal{Z}}+1}^{n^{\mathcal{Z}}\;T}\boldsymbol{O}_3^{2\,T}\cdot\ldots\cdot\boldsymbol{O}_{n^{\mathcal{F}}}^{n^{\mathcal{F}}-1\,T}, \\ \mathfrak{O}_{\mathrm{ZT}} &\overset{\text{def}}{=} \begin{bmatrix} \mathfrak{O}_{\mathrm{Z}} & \\ & \boldsymbol{I} \end{bmatrix}\mathfrak{O}_{\mathrm{T}}, \end{aligned} \tag{8.19}$$

that serve to transform $\boldsymbol{t}^T = (\boldsymbol{t}_{\mathrm{Z}}^T, \boldsymbol{t}_{\mathrm{Y}}^T)$ into the unit row vector $\boldsymbol{e}^T$,

$$\begin{bmatrix} \boxed{\boldsymbol{t}_{\mathcal{Z}}^T \;\; \boldsymbol{t}_{\mathcal{Y}}^T} & \\ & \boldsymbol{I}_i^{\mathcal{X}} \\ \boldsymbol{0} \;\; \boldsymbol{T}_i & \boldsymbol{R}_i^{\mathcal{AX}} \end{bmatrix} \begin{bmatrix} \mathfrak{O}_{\mathrm{ZT}} & \\ & \boldsymbol{I} \end{bmatrix} = \begin{bmatrix} \boldsymbol{0}^T & \boldsymbol{0}^T & 1 & \\ & & & \boldsymbol{I}_i^{\mathcal{X}} \\ & \boldsymbol{T}_i^\star & \boldsymbol{r} & \boldsymbol{R}_i^{\mathcal{AX}} \end{bmatrix} \overset{\text{def}}{=} \begin{bmatrix} & & \boldsymbol{I}_i^{\mathcal{X}\star} \\ \boldsymbol{0} & \boldsymbol{T}_i^\star & \boldsymbol{R}_i^{\mathcal{AX}\star} \end{bmatrix}.$$

The last $n_i^{\mathcal{F}} - n^{\mathcal{Z}}$ rotations of the sequence $\boldsymbol{O}_{\mathrm{T}}$ introduce a reverse subdiagonal into $\boldsymbol{T}_i$. By shifting $\boldsymbol{T}_i$ to the left we obtain the new TQ factor $\boldsymbol{T}_i^\star$ and the null space dimension shrinks by one. The remaining column $\boldsymbol{r}$ belongs the now fixed component $v_{ij}$ of the unknown and enters $\boldsymbol{R}_i^{\mathcal{AX}}$ to form the new fixed part of the active point constraints matrix $\boldsymbol{R}_i^{\mathcal{AX}\star}$. Having applied the Givens rotations to the right hand side of (8.18) we do so in the same way for the left hand side to recover equality and find the new null space and range space bases $\boldsymbol{Z}_i^\star$ and $\boldsymbol{Y}_i^\star$,

$$\begin{bmatrix} \boxed{Z_i \;\; Y_i} & \\ & I_i^{\mathcal{X}} \end{bmatrix} \begin{bmatrix} \mathfrak{O}_{\mathrm{ZT}} & \\ & I \end{bmatrix} = \begin{bmatrix} Z_i^\star & \boxed{y \;\; Y} & 0 & \\ & 0 \;\; 0 & 1 & \\ & & & I_i^{\mathcal{X}} \end{bmatrix} \stackrel{\text{def}}{=} \begin{bmatrix} Z_i^\star & Y_i^\star & \\ & & I_i^{\mathcal{X}\star} \end{bmatrix}. \tag{8.20}$$

This yields the identity

$$\begin{bmatrix} Z_i^\star & Y_i^\star \end{bmatrix} = Q_i^\star = \mathfrak{I}^T Q_i \mathfrak{O}_{\mathrm{ZT}} \mathfrak{I}, \tag{8.21}$$

which proves the relations (8.17a) for the TQ factorization matrices $Z_i^\star$, $Y_i^\star$, and $T_i^\star$.

We next consider the reduced Hessian's Cholesky factor $U_i^\star$ (8.17b). We have separated the first $n^{\mathcal{Z}} - 1$ rotations of the sequence $\mathfrak{O}_{\mathrm{ZT}}$ as they affect the new null space basis matrix $Z_i^\star$ only, such that (8.21) can be for the new null space basis $Z_i^\star$ as

$$Z_i^\star = \mathfrak{I}^T Z_i \mathfrak{O}_{\mathrm{Z}} \mathfrak{I}. \tag{8.22}$$

The projected Hessian's new Cholesky factor $U_i^\star$ is found from

$$\begin{aligned} U_i^{\star T} U_i^\star = Z_i^{\star T} H_i^{\mathcal{F}\mathcal{F}\star} Z_i^\star &= \mathfrak{I}^T \mathfrak{O}_{\mathrm{Z}}^T Z_i^T \mathfrak{I} (\mathfrak{I}^T H_i^{\mathcal{F}\mathcal{F}} \mathfrak{I}) \mathfrak{I}^T Z_i \mathfrak{O}_{\mathrm{Z}} \mathfrak{I} \\ &= \mathfrak{I}^T \mathfrak{O}_{\mathrm{Z}}^T Z_i^T \begin{bmatrix} I & 0 \\ 0^T & 0 \end{bmatrix} H_i^{\mathcal{F}\mathcal{F}} \begin{bmatrix} I & 0 \\ 0^T & 0 \end{bmatrix} Z_i \mathfrak{O}_{\mathrm{Z}} \mathfrak{I}. \end{aligned} \tag{8.23}$$

Apparently the terms $\mathfrak{I}\mathfrak{I}^T$ obstruct the reuse of the existing Cholesky factorization $Z_i^T H_i^{\mathcal{F}\mathcal{F}} Z_i = U_i^T U_i$. Observe now that the last row of the matrix $Z_i \mathfrak{O}_{\mathrm{Z}} \mathfrak{I}$ is zero as can be seen from (8.20) and is cut off in the definition of $Z_i^\star$ in (8.22). We may therefore replace the terms $\mathfrak{I}\mathfrak{I}^T$ in (8.23) by $I$ without impacting equality and find

$$\begin{aligned} U_i^{\star T} U_i^\star = Z_i^{\star T} H_i^{\mathcal{F}\mathcal{F}\star} Z_i^\star &= \mathfrak{I}^T \mathfrak{O}_{\mathrm{Z}}^T Z_i^T H_i^{\mathcal{F}\mathcal{F}} Z_i \mathfrak{O}_{\mathrm{Z}} \mathfrak{I} \\ &= \mathfrak{I}^T \mathfrak{O}_{\mathrm{Z}}^T U_i^T U_i \mathfrak{O}_{\mathrm{Z}} \mathfrak{I} = (U_i \mathfrak{O}_{\mathrm{Z}} \mathfrak{I})^T U_i \mathfrak{O}_{\mathrm{Z}} \mathfrak{I}. \end{aligned} \tag{8.24}$$

Hence the new factor would be $U_i \mathfrak{O}_{\mathrm{Z}} \mathfrak{I}$ which is no longer a Cholesky factor as it is non–square and has an additional subdiagonal of nonzero elements introduced by $\mathfrak{O}_{\mathrm{Z}}$. The upper triangular shape is restored by a second sequence of $n^{\mathcal{Z}} - 1$ Givens rotations denoted by $\mathfrak{O}_{\mathrm{U}}$,

$$\mathfrak{O}_{\mathrm{U}} \stackrel{\text{def}}{=} O_{n^{\mathcal{Z}}}^{n^{\mathcal{Z}}-1} \cdot \ldots \cdot O_3^2 O_2^1, \tag{8.25}$$

and constructed to eliminate each subdiagonal element in $\boldsymbol{U}_i\mathfrak{O}_{\mathrm{Z}}\mathfrak{I}$ using the diagonal element located directly above. The last row becomes zero and is cut off, which yields the Cholesky factor $\boldsymbol{U}_i^{\star}$ of $\boldsymbol{Z}_i^{\star T}\boldsymbol{H}_i^{\mathcal{F}\mathcal{F}\star}\boldsymbol{Z}_i^{\star}$,

$$\boldsymbol{U}_i^{\star} = \mathfrak{I}^T\mathfrak{O}_{\mathrm{U}}\boldsymbol{U}_i\mathfrak{O}_{\mathrm{Z}}\mathfrak{I}. \tag{8.26}$$

This proves relation (8.17b) for the Cholesky factor $\boldsymbol{U}_i^{\star}$ of the reduced Hessian $\boldsymbol{H}_i^{\mathcal{F}\mathcal{F}\star}$.

The updates (8.17c) to the projected sensitivity matrices $\hat{\boldsymbol{G}}_i$ and $\hat{\boldsymbol{P}}_i$ can be computed constructively from $\boldsymbol{Z}_i^{\star}$ and $\boldsymbol{U}_i^{\star}$ as

$$\begin{aligned}\hat{\boldsymbol{G}}_i^{\star} = \boldsymbol{G}_i^{\mathcal{F}\star}\boldsymbol{Z}_i^{\star}\boldsymbol{U}_i^{\star-1} &= \boldsymbol{G}_i^{\mathcal{F}}\mathfrak{I}\left(\mathfrak{I}^T\boldsymbol{Z}_i\mathfrak{O}_{\mathrm{Z}}\mathfrak{I}\right)\left(\mathfrak{I}^T\mathfrak{O}_{\mathrm{U}}\boldsymbol{U}_i\mathfrak{O}_{\mathrm{Z}}\mathfrak{I}\right)^{-1} \\ &= \boldsymbol{G}_i^{\mathcal{F}}(\mathfrak{I}\mathfrak{I}^T)\boldsymbol{Z}_i\mathfrak{O}_{\mathrm{Z}}\mathfrak{I}\left(\mathfrak{I}^T\mathfrak{O}_{\mathrm{U}}\boldsymbol{U}_i\mathfrak{O}_{\mathrm{Z}}\mathfrak{I}\right)^{-1}\end{aligned} \tag{8.27}$$

which by replacing again the first occurrence of $\mathfrak{I}\mathfrak{I}^T$ by $\boldsymbol{I}$ becomes

$$= \boldsymbol{G}_i^{\mathcal{F}}\boldsymbol{Z}_i\mathfrak{O}_{\mathrm{Z}}\mathfrak{I}\left(\mathfrak{I}^T\mathfrak{O}_{\mathrm{U}}\boldsymbol{U}_i\mathfrak{O}_{\mathrm{Z}}\mathfrak{I}\right)^{-1} \tag{8.28}$$

and according to 3. in lemma 8.2 becomes,

$$\begin{aligned}&= \boldsymbol{G}_i^{\mathcal{F}}\boldsymbol{Z}_i\mathfrak{O}_{\mathrm{Z}}\mathfrak{I}\mathfrak{I}^T\left(\mathfrak{O}_{\mathrm{U}}\boldsymbol{U}_i\mathfrak{O}_{\mathrm{Z}}\right)^{-1}\mathfrak{I} \\ &= \boldsymbol{G}_i^{\mathcal{F}}\boldsymbol{Z}_i\left(\mathfrak{O}_{\mathrm{Z}}\mathfrak{I}\mathfrak{I}^T\mathfrak{O}_{\mathrm{Z}}^T\right)\boldsymbol{U}_i^{-1}\mathfrak{O}_{\mathrm{U}}^T\mathfrak{I}\end{aligned} \tag{8.29}$$

and with 4. in lemma 8.2, letting $\boldsymbol{z} \overset{\text{def}}{=} (\mathfrak{O}_{\mathrm{Z}})_{[:,n^z]}$, this becomes

$$\begin{aligned}&= \boldsymbol{G}_i^{\mathcal{F}}\boldsymbol{Z}_i\left(\boldsymbol{I} - \boldsymbol{z}\boldsymbol{z}^T\right)\boldsymbol{U}_i^{-1}\mathfrak{O}_{\mathrm{U}}^T\mathfrak{I} \\ &= \left(\boldsymbol{G}_i^{\mathcal{F}}\boldsymbol{Z}_i\boldsymbol{U}_i^{-1}\right)\mathfrak{O}_{\mathrm{U}}^T\mathfrak{I} - \boldsymbol{G}_i^{\mathcal{F}}\boldsymbol{Z}_i\boldsymbol{z}\left(\boldsymbol{z}^T\boldsymbol{U}_i^{-1}\mathfrak{O}_{\mathrm{U}}^T\mathfrak{I}\right) \\ &= \hat{\boldsymbol{G}}_i\mathfrak{O}_{\mathrm{U}}^T\mathfrak{I} - \boldsymbol{G}_i^{\mathcal{F}}\boldsymbol{Z}_i\boldsymbol{z}\left(\boldsymbol{z}^T\boldsymbol{U}_i^{-1}\mathfrak{O}_{\mathrm{U}}^T\mathfrak{I}\right).\end{aligned} \tag{8.30}$$

Consider now that $\boldsymbol{z}^T\boldsymbol{U}_i^{-1}\mathfrak{O}_{\mathrm{U}}^T$ is the last row of the inverse of the updated upper triangular factor $\mathfrak{O}_{\mathrm{U}}\boldsymbol{U}_i\mathfrak{O}_{\mathrm{Z}}$. This row is zero by construction of $\mathfrak{O}_{\mathrm{U}}$, except for the last element which is cut off, hence

$$\hat{\boldsymbol{G}}_i^{\star} = \hat{\boldsymbol{G}}_i\mathfrak{O}_{\mathrm{U}}^T\mathfrak{I}. \tag{8.31}$$

The corresponding relation $\hat{\boldsymbol{P}}_i^{\star} = \hat{\boldsymbol{P}}_i\mathfrak{O}_{\mathrm{U}}^T\mathfrak{I}$ is shown in exactly the same way.

This finally proves the claimed relations (8.17c). For the tridiagonal system blocks $A_{i-1}^\star$, $A_i^\star$ and $B_i^\star$ in (8.17d) affected by the updates to $\hat{G}_i$ and $\hat{P}_i$ we find constructively that

$$A_i^\star = \hat{G}_i^\star \hat{G}_i^{\star T} + \hat{P}_{i+1}\hat{P}_{i+1}^T = \hat{G}_i \mathfrak{D}_R^T \mathfrak{I} \left(\hat{G}_i \mathfrak{D}_U^T \mathfrak{I}\right)^T + \hat{P}_{i+1}\hat{P}_{i+1}^T \tag{8.32}$$
$$= \hat{G}_i \left(\mathfrak{D}_U^T \mathfrak{I}\mathfrak{I}^T \mathfrak{D}_U\right) \hat{G}_i^T + \hat{P}_{i+1}\hat{P}_{i+1}^T$$

which again by 4. in lemma 8.2 and $u^T \stackrel{\text{def}}{=} (\mathfrak{D}_U)_{[n^z,:]}$ is

$$= \hat{G}_i \left(I - uu^T\right) \hat{G}_i^T + \hat{P}_{i+1}\hat{P}_{i+1}^T \tag{8.33}$$
$$= \hat{G}_i \hat{G}_i^T - (\hat{G}_i u)(\hat{G}_i u)^T + \hat{P}_{i+1}\hat{P}_{i+1}^T$$
$$= A_i - (\hat{G}_i u)(\hat{G}_i u)^T$$
$$= A_i - \hat{g}\hat{g}^T, \qquad \hat{g} \stackrel{\text{def}}{=} \hat{G}_i u.$$

By the same argument we find

$$A_{i-1}^\star = A_{i-1} - \hat{p}\hat{p}^T, \quad \hat{p} \stackrel{\text{def}}{=} \hat{P}_i u, \qquad B_i^\star = B_i - \hat{g}\hat{p}^T. \tag{8.34}$$

This proves the relations (8.17d) for the tridiagonal system blocks $A_{i-1}^\star$, $A_i^\star$, and $B_i^\star$.

□

The resulting factorization update procedure is summarized in in algorithm 8.3 on page 227. The modifications to $A$ and $B$ of system (7.7) take the form of an subtraction of a dyadic product from a $2 \times 2$ subblock as can be seen by writing

$$\begin{bmatrix} \ddots & \ddots & & & \\ \ddots & A_{i-2}^\star & B_{i-1}^{\star\,T} & & \\ & B_{i-1}^\star & A_{i-1}^\star & B_i^{\star T} & \\ & & B_i^\star & A_i & B_{i+1}^{\star\,T} \\ & & & B_{i+1} & A_{i+1} & \ddots \\ & & & & \ddots & \ddots \end{bmatrix} = \begin{bmatrix} \ddots & \ddots & & & \\ \ddots & A_{i-2} & B_{i-1}^{\;T} & & \\ & B_{i-1} & A_{i-1} & B_i^T & \\ & & B_i & A_i & B_{i+1}^{\;T} \\ & & & B_{i+1} & A_{i+1} & \ddots \\ & & & & \ddots & \ddots \end{bmatrix} + \begin{bmatrix} \vdots \\ 0 \\ \hat{p} \\ \hat{g} \\ 0 \\ \vdots \end{bmatrix} \cdot \begin{bmatrix} \vdots \\ 0 \\ \hat{p} \\ \hat{g} \\ 0 \\ \vdots \end{bmatrix}^T.$$

A suitable downdate to the block tridiagonal Cholesky factorization is derived in section 8.3.2.

### 8.2.3 Adding a Point Constraint

If a point constraint on the unknown $\boldsymbol{v}_i$ of node $i$ becomes active, it can be appended to the list of previously active point constraints, as their relative order is of no significance. In the following theorem we show how to restore the block local reductions of a HPSC factorization for the new KKT matrix with active point constraint on $\boldsymbol{v}_i$.

**Theorem 8.3 (Adding a Point Constraint to the Block Local Reductions)**
Let $\mathcal{H}(\mathcal{A})$ be a HPSC factorization of the KKT system $\mathcal{K}(\mathcal{A})$. Let the point constraint $(\boldsymbol{R}_i)_{j\star}\boldsymbol{v}_i \geqslant r_{ij}$ on the unknown $\boldsymbol{v}_i$ of node $0 \leqslant i \leqslant m$ be inactive in the active set $\mathcal{A}$, and denote by $\mathcal{A}^\star$ the otherwise identical active set with activeated point constraint. Assume further that $\mathcal{A}^\star$ satisfies LICQ. Then there exists a HPSC factorization $\mathcal{H}^\star(\mathcal{A}^\star)$ of the KKT matrix $\mathcal{K}^\star(\mathcal{A}^\star)$ that satisfies

$$\boldsymbol{Z}_i^\star \begin{bmatrix} \boldsymbol{y} \end{bmatrix} = \boldsymbol{Z}_i \mathfrak{O}_{\mathrm{Z}}, \qquad \boldsymbol{Y}_i^\star = \begin{bmatrix} \boldsymbol{y} & \boldsymbol{Y}_i \end{bmatrix} \qquad \boldsymbol{T}_i^\star = \begin{bmatrix} \boldsymbol{0} & \boldsymbol{T}_i \\ \tau & (\boldsymbol{R}_i^{\star\mathcal{F}})_{j\star}\boldsymbol{Y}_i \end{bmatrix} \tag{8.35a}$$

$$\boldsymbol{U}_i^\star = \mathfrak{I}^T \mathfrak{O}_{\mathrm{U}} \boldsymbol{U}_i \mathfrak{O}_{\mathrm{Z}} \mathfrak{I} \tag{8.35b}$$

$$\begin{bmatrix} \hat{\boldsymbol{G}}_i^\star & \hat{\boldsymbol{g}} \end{bmatrix} = \hat{\boldsymbol{G}}_i \mathfrak{O}_{\mathrm{U}}^T \qquad \begin{bmatrix} \hat{\boldsymbol{P}}_i^\star & \hat{\boldsymbol{p}} \end{bmatrix} = \hat{\boldsymbol{P}}_i \mathfrak{O}_{\mathrm{U}}^T \tag{8.35c}$$

$$\boldsymbol{A}_{i-1}^\star = \boldsymbol{A}_{i-1} - \hat{\boldsymbol{p}}\hat{\boldsymbol{p}}^T, \qquad \boldsymbol{A}_i^\star = \boldsymbol{A}_i - \hat{\boldsymbol{g}}\hat{\boldsymbol{g}}^T, \quad \boldsymbol{B}_i^\star = \boldsymbol{B}_i - \hat{\boldsymbol{g}}\hat{\boldsymbol{p}}^T, \tag{8.35d}$$

where $\mathfrak{O}_{\mathrm{Z}}$ and $\mathfrak{O}_{\mathrm{U}}$ are appropriately chosen sequences of Givens rotations, and $\tau = \left\| \boldsymbol{R}_{i,j\star}^{\star\mathcal{F}} \boldsymbol{Z}_i \mathfrak{O}_{\mathrm{Z}} \right\|$. △

**Proof** We first consider the relations (8.35a) for the matrices $\boldsymbol{Z}_i^\star$, $\boldsymbol{Y}_i^\star$, and $\boldsymbol{T}_i^\star$ of the TQ factorization of the active point constraints. If an inactive point constraint $1 \leqslant j \leqslant n_i^{\mathrm{r}}$ on node $i$ becomes active, the row vector $\boldsymbol{r}^T \stackrel{\mathrm{def}}{=} (\boldsymbol{R}_i^{\star\mathcal{F}})_{j\star}$ is appended to the bottom of the matrix of active point constraints $\boldsymbol{R}_i^{\mathcal{AF}}$ and its TQ factorization

$$\begin{bmatrix} \boldsymbol{R}_i^{\mathcal{AF}} \\ \boldsymbol{r}^T \end{bmatrix} \begin{bmatrix} \boldsymbol{Z}_i & \boldsymbol{Y}_i \end{bmatrix} = \begin{bmatrix} \boldsymbol{0} & \boldsymbol{T}_i \\ \boldsymbol{t}_{\mathrm{Z}}^T & \boldsymbol{t}_{\mathrm{Y}}^T \end{bmatrix}, \tag{8.36}$$

wherein $\boldsymbol{t}^T = \boldsymbol{r}^T \boldsymbol{Q}_i$, i.e., $\boldsymbol{t}_{\mathrm{Z}}^T = \boldsymbol{r}^T \boldsymbol{Z}_i$ and $\boldsymbol{t}_{\mathrm{Y}}^T = \boldsymbol{r}^T \boldsymbol{Y}_i$. To restore the southeast triangular shape of the right hand side of (8.36), a series of $n^{\mathcal{Z}} - 1$ Givens rotations

$$\mathfrak{O}_{\mathrm{Z}} \stackrel{\mathrm{def}}{=} \boldsymbol{O}_2^{1\,T} \boldsymbol{O}_3^{2\,T} \cdot \ldots \cdot \boldsymbol{O}_{n^{\mathcal{Z}}}^{n^{\mathcal{Z}}-1\,T} \tag{8.37}$$

is applied, eliminating all entries of $\boldsymbol{t}_{\mathrm{Z}}$ outside the triangular shape,

$$\begin{bmatrix} \mathbf{0} & \boldsymbol{T}_i \\ \boldsymbol{t}_{\mathrm{Z}}^T & \boldsymbol{t}_{\mathrm{Y}}^T \end{bmatrix} \begin{bmatrix} \mathfrak{O}_{\mathrm{Z}} & \\ & \boldsymbol{I} \end{bmatrix} = \left[ \begin{array}{c|cc} \mathbf{0} & \mathbf{0} & \boldsymbol{T}_i \\ \mathbf{0}^T & \tau & \boldsymbol{t}_{\mathrm{Y}}^T \end{array} \right] \stackrel{\text{def}}{=} \begin{bmatrix} \mathbf{0} & \boldsymbol{T}_i^\star \end{bmatrix}. \tag{8.38}$$

The factor $\boldsymbol{T}_i^\star$ gains a row and column as the range space dimension increases and the null space dimension decreases, To maintain equality in (8.36) we apply the sequence $\mathfrak{O}_{\mathrm{Z}}$ also to the left hand side,

$$\begin{aligned} &\boldsymbol{R}_i^{\mathcal{AF}\star} \begin{bmatrix} \boldsymbol{Z}_i & \boldsymbol{Y}_i \end{bmatrix} \begin{bmatrix} \mathfrak{O}_{\mathrm{Z}} & \\ & \boldsymbol{I} \end{bmatrix} \\ &= \boldsymbol{R}_i^{\mathcal{AF}\star} \left[ \boldsymbol{Z}_i^\star \; \boxed{\boldsymbol{y} \;\; \boldsymbol{Y}_i} \right] \stackrel{\text{def}}{=} \boldsymbol{R}_i^{\mathcal{AF}\star} \begin{bmatrix} \boldsymbol{Z}_i^\star & \boldsymbol{Y}_i^\star \end{bmatrix} \end{aligned} \tag{8.39}$$

from which find the bases $\boldsymbol{Z}_i^\star$ and $\boldsymbol{Y}_i^\star$,

$$\begin{bmatrix} \boldsymbol{Z}_i^\star & \boldsymbol{y} \end{bmatrix} = \boldsymbol{Z}_i \mathfrak{O}_{\mathrm{Z}}, \qquad \boldsymbol{Y}_i^\star = \begin{bmatrix} \boldsymbol{y} & \boldsymbol{Y}_i \end{bmatrix}. \tag{8.40}$$

This proves the relations (8.35a) for the TQ factorization matrices $\boldsymbol{Z}_i^\star$, $\boldsymbol{Y}_i^\star$, and $\boldsymbol{T}_i^\star$. The projected Hessian factor $\boldsymbol{U}_i^\star$ would be

$$\boldsymbol{U}_i^{\star T} \boldsymbol{U}_i^\star = \boldsymbol{Z}_i^{\star T} \boldsymbol{H}_i^{\mathcal{FF}\star} \boldsymbol{Z}_i^\star = \mathfrak{I}^T \mathfrak{O}_{\mathrm{Z}}^T \boldsymbol{Z}_i^T \boldsymbol{H}_i^{\mathcal{FF}} \boldsymbol{Z}_i \mathfrak{O}_{\mathrm{Z}} \mathfrak{I} = \mathfrak{I}^T \mathfrak{O}_{\mathrm{Z}}^T \boldsymbol{U}_i^T \boldsymbol{U}_i \mathfrak{O}_{\mathrm{Z}} \mathfrak{I}, \tag{8.41}$$

and similar to section 8.2.2 the upper triangular shape of $\boldsymbol{U}_i \mathfrak{O}_{\mathrm{Z}} \mathfrak{I}$ is lost in (8.41) and needs to be recovered by the sequence

$$\mathfrak{O}_{\mathrm{U}} \stackrel{\text{def}}{=} \boldsymbol{O}_{n^{\mathcal{Z}}-1}^{n^{\mathcal{Z}}} \cdot \ldots \cdot \boldsymbol{O}_2^3 \boldsymbol{O}_1^2, \tag{8.42}$$

again resulting in

$$\boldsymbol{U}_i^\star \stackrel{\text{def}}{=} \mathfrak{I}^T \mathfrak{O}_{\mathrm{U}} \boldsymbol{U}_i \mathfrak{O}_{\mathrm{Z}} \mathfrak{I}. \tag{8.43}$$

This shows relation (8.35b) for the reduced Hessian's Cholesky factor $\boldsymbol{U}_i^\star$. With this result the proof of the remaining relations (8.35c) for $\hat{\boldsymbol{G}}_i^\star$ and $\hat{\boldsymbol{P}}_i^\star$ and (8.35d) for $\hat{\boldsymbol{A}}_{i-1}^\star$, $\hat{\boldsymbol{A}}_i^\star$, and $\hat{\boldsymbol{B}}_i^\star$ proceeds exactly as in section 8.2.2. □

The resulting factorization update procedure is summarized in algorithm 8.4 on page 228.

### 8.2.4 Deleting a Simple Bound

If a simple bound on $v_{ij}^{\mathcal{X}}$ becomes inactive on node $i$, we may assume w.l.o.g. that the unknown to be freed from its bound is the first component $j = 1$ of the fixed unknowns $\boldsymbol{v}_i^{\mathcal{X}}$, and that it becomes the new last component $n^{\mathcal{F}} + 1$ of the free unknowns $\boldsymbol{v}_i^{\mathcal{F}}$. This can again be ensured by applying a suitable permutation to the unknown $\boldsymbol{v}_i^{\mathcal{X}}$ and the matrix $\boldsymbol{Q}_i$, cf. theorem 8.1. In the following theorem we show how to restore the block local reductions of a HPSC factorization for the new KKT matrix with inactive simple bound on $v_{ij}^{\mathcal{X}}$.

**Theorem 8.4 (Deleting a Simple Bound from the Block Local Reductions)** Let $\mathcal{H}(\mathcal{A})$ be a HPSC factorization of the KKT system $\mathcal{K}(\mathcal{A})$. Let the simple bound on the first fixed component $v_{i1}^{\mathcal{X}}$ of the unknown $\boldsymbol{v}_i$ of node $0 \leqslant i \leqslant m$ be active in $\mathcal{A}$, and denote by $\mathcal{A}^\star$ the otherwise identical active set with inactivated simple bound. Then there exists a HPSC factorization $\mathcal{H}^\star(\mathcal{A}^\star)$ of the KKT matrix $\mathcal{K}^\star(\mathcal{A}^\star)$ that satisfies

$$\boldsymbol{Z}_i^\star = \begin{bmatrix} \boldsymbol{Z}_i & \boldsymbol{z} \\ \boldsymbol{0}^T & \zeta \end{bmatrix}, \qquad \begin{bmatrix} \boldsymbol{z} & \boldsymbol{Y}_i^\star \\ \zeta & \boldsymbol{0}^T \end{bmatrix} = \begin{bmatrix} \boldsymbol{Y}_i & \\ & 1 \end{bmatrix} \mathfrak{O}_{\mathrm{T}}, \tag{8.44a}$$

$$\begin{bmatrix} \boldsymbol{0} & \boldsymbol{T}_i^\star \end{bmatrix} = \begin{bmatrix} \boldsymbol{T}_i & (\boldsymbol{R}_i^{\mathcal{X}})_{\star 1} \end{bmatrix} \mathfrak{O}_{\mathrm{T}}, \tag{8.44b}$$

$$\boldsymbol{U}_i^\star = \begin{bmatrix} \boldsymbol{U}_i & \boldsymbol{u} \\ \boldsymbol{0}^T & \varrho \end{bmatrix}, \qquad \boldsymbol{u} = \boldsymbol{U}_i^{-T} \boldsymbol{Z}_i^{\,T} (\boldsymbol{H}_i^{\mathcal{F}\mathcal{F}} \boldsymbol{z} + \boldsymbol{h}\zeta), \tag{8.44c}$$

$$\varrho = \sqrt{\boldsymbol{z}^T (\boldsymbol{H}_i^{\mathcal{F}\mathcal{F}} \boldsymbol{z} + \boldsymbol{h}\zeta) + \zeta(\boldsymbol{h}^T \boldsymbol{z} + \eta\zeta) - \boldsymbol{u}^T \boldsymbol{u}}$$

$$\hat{\boldsymbol{G}}_i^\star = \begin{bmatrix} \hat{\boldsymbol{G}}_i & \hat{\boldsymbol{g}} \end{bmatrix} \qquad \text{with } \hat{\boldsymbol{g}} = (\boldsymbol{G}_i^{\mathcal{F}^\star} \boldsymbol{z} - \hat{\boldsymbol{G}} \boldsymbol{u})/\varrho, \tag{8.44d}$$

$$\hat{\boldsymbol{P}}_i^\star = \begin{bmatrix} \hat{\boldsymbol{P}}_i & \hat{\boldsymbol{p}} \end{bmatrix} \qquad \text{with } \hat{\boldsymbol{p}} = (\boldsymbol{P}_i^{\mathcal{F}^\star} \boldsymbol{z} - \hat{\boldsymbol{P}} \boldsymbol{u})/\varrho,$$

$$\boldsymbol{A}_{i-1}^\star = \boldsymbol{A}_{i-1} + \hat{\boldsymbol{p}} \hat{\boldsymbol{p}}^T, \tag{8.44e}$$

$$\boldsymbol{A}_i^\star = \boldsymbol{A}_i + \hat{\boldsymbol{g}} \hat{\boldsymbol{g}}^T,$$

$$\boldsymbol{B}_i^\star = \boldsymbol{B}_i + \hat{\boldsymbol{g}} \hat{\boldsymbol{p}}^T,$$

where $\mathfrak{O}_{\mathrm{T}}$ is an appropriately chosen sequence of Givens rotations. △

**Proof** We again consider the TQ factorization of $\boldsymbol{R}_i^{\mathcal{A}\mathcal{F}^\star}$ first. In the extended constraints matrix

$$\begin{bmatrix} & \boldsymbol{I}_i^{\mathcal{X}} \\ \boldsymbol{R}_i^{\mathcal{F}} & \boldsymbol{R}_i^{\mathcal{X}} \end{bmatrix} \begin{bmatrix} \boldsymbol{Z}_i & \boldsymbol{Y}_i & \\ & & \boldsymbol{I}_i^{\mathcal{X}} \end{bmatrix} = \begin{bmatrix} & & \boldsymbol{I}_i^{\mathcal{X}} \\ \boldsymbol{0} & \boldsymbol{T}_i & \boldsymbol{R}_i^{\mathcal{X}} \end{bmatrix}, \tag{8.45}$$

the row belonging to the simple bound is removed on the left hand side. Consequentially, the first column of $\boldsymbol{R}_i^{\mathcal{X}}$ belonging to the unknown $v_{i1}^{\mathcal{X}}$ to be freed from its bound becomes the last one of $\boldsymbol{T}_i$ on the right hand side,

$$\left[\begin{array}{cc|c} & & \boldsymbol{I}_i^{\mathcal{X}\star} \\ \boldsymbol{R}_i^{\mathcal{F}} & \boldsymbol{r}_1^{\mathcal{X}} & \boldsymbol{R}_i^{\mathcal{X}\star} \end{array}\right] \begin{bmatrix} \boldsymbol{Z}_i & \boldsymbol{Y}_i & & \\ & & 1 & \\ & & & \boldsymbol{I}_i^{\mathcal{X}\star} \end{bmatrix} = \left[\begin{array}{c|cc|c} & & & \boldsymbol{I}_i^{\mathcal{X}\star} \\ \boldsymbol{0} & \boldsymbol{T}_i & \boldsymbol{r}_1^{\mathcal{X}} & \boldsymbol{R}_i^{\mathcal{X}\star} \end{array}\right]. \tag{8.46}$$

The southeast triangular shape of the factor $\begin{bmatrix} \boldsymbol{T}_i & \boldsymbol{r}_1^{\mathcal{X}} \end{bmatrix}$ is restored by a sequence of $n^{\mathcal{Y}}$ Givens rotations

$$\mathfrak{O}_{\mathrm{T}} \overset{\text{def}}{=} \boldsymbol{O}_{n^{\mathcal{F}}}^{n^{\mathcal{F}}+1^T} \cdot \ldots \cdot \boldsymbol{O}_{n^{\mathcal{Z}}+1}^{n^{\mathcal{Z}}+2^T} \tag{8.47}$$

by eliminating each element on the reverse subdiagonal using the element to the right thereby transforming the first column to zero,

$$\begin{bmatrix} \boldsymbol{T}_i & \boldsymbol{r}_1^{\mathcal{X}} \end{bmatrix} \mathfrak{O}_{\mathrm{T}} \overset{\text{def}}{=} \begin{bmatrix} \boldsymbol{0} & \boldsymbol{T}_i^{\star} \end{bmatrix}. \tag{8.48}$$

This proves relation (8.44b) for the new southeast triangular TQ factor $\boldsymbol{T}_i^{\star}$.

To maintain equivalence in (8.46) we apply the sequence $\mathfrak{O}_{\mathrm{T}}$ to the left hand side as well. By construction this sequence leaves $\boldsymbol{Z}_i$ unaffected, but affects the first column of $\boldsymbol{Y}_i$ which becomes the new last one of $\boldsymbol{Z}_i^{\star}$ as the null space dimension grows by one,

$$\left[\begin{array}{c|c} \boldsymbol{Z}_i & \boldsymbol{Y}_i \\ & \quad 1 \end{array}\right] \begin{bmatrix} \boldsymbol{I} & \\ & \mathfrak{O}_{\mathrm{T}} \end{bmatrix} = \left[\begin{array}{cc|c} \boldsymbol{Z} & \boldsymbol{z} & \boldsymbol{Y}_i^{\star} \\ \boldsymbol{0}^T & \zeta & \boldsymbol{y}^T \end{array}\right] \overset{\text{def}}{=} \begin{bmatrix} \boldsymbol{Z}_i^{\star} & \boldsymbol{Y}_i^{\star} \end{bmatrix} \tag{8.49}$$

This proves relations (8.44a) and (8.44b) for the null space and range space bases $\boldsymbol{Z}_i^{\star}$ and $\boldsymbol{Y}_i^{\star}$.

In order to prove relation (8.44c) for $\boldsymbol{U}_i^{\star}$, we denote the elements of the new last row and column of the Hessian $\boldsymbol{H}_i^{\mathcal{F}\mathcal{F}\star}$ by a known row/column vector $\boldsymbol{h}$ and a known scalar $\eta$ for the new diagonal element. For the Hessian's Cholesky factor $\boldsymbol{U}_i^{\star}$ yet to be determined, these are denoted by an unknown column vector $\boldsymbol{u}$ and an unknown scalar $\varrho$ for the new diagonal element,

$$\boldsymbol{H}_i^{\mathcal{F}\mathcal{F}\star} \overset{\text{def}}{=} \begin{bmatrix} \boldsymbol{H}_i^{\mathcal{F}\mathcal{F}} & \boldsymbol{h} \\ \boldsymbol{h}^T & \eta \end{bmatrix}, \qquad \boldsymbol{U}_i^{\star} \overset{\text{def}}{=} \begin{bmatrix} \boldsymbol{U}_i & \boldsymbol{u} \\ \boldsymbol{0}^T & \varrho \end{bmatrix}. \tag{8.50}$$

For the new projected Hessian factor $\boldsymbol{U}_i^{\star}$ we find from expanding the projected Hessian $\boldsymbol{Z}_i^{\star T} \boldsymbol{H}_i^{\mathcal{F}\mathcal{F}\star} \boldsymbol{Z}_i^{\star} = \boldsymbol{U}_i^{\star T} \boldsymbol{U}_i^{\star}$ to

$$\begin{bmatrix} Z_i^T & 0 \\ z^T & \zeta \end{bmatrix} \begin{bmatrix} H_i^{\mathcal{FF}} & h \\ h^T & \eta \end{bmatrix} \begin{bmatrix} Z_i & z \\ 0^T & \zeta \end{bmatrix} = \begin{bmatrix} U_i^T U_i & U_i^T u \\ u^T U_i & u^T u + \varrho^2 \end{bmatrix} \tag{8.51}$$

that we can compute the factor's new column entries $u$ and $\varrho$ from the entries $h$ and $\eta$ of $H_i^{\mathcal{FF}\star}$ as follows,

$$U_i^T u = Z_i^T (H_i^{\mathcal{FF}} z + h\zeta), \tag{8.52a}$$

$$\varrho = \sqrt{z^T (H_i^{\mathcal{FF}} z + h\zeta) + \zeta(h^T z + \eta\zeta) - u^T u}. \tag{8.52b}$$

This proves relation (8.44c) for the projected Hessian Cholesky factor $U_i^\star$.

We next consider the Schur complement relations (8.44d). Since the initial $n^{\mathcal{Z}}$ columns of $Z_i^\star$ and $U_i^\star$ are identical to those of $Z_i$ and $U_i$, we find for the Schur complements $\hat{G}_i$ and $\hat{P}_i$ that $\hat{G}_i^\star = \begin{bmatrix} \hat{G}_i & \hat{g} \end{bmatrix}$ where $\hat{g}$ is an additional column that can be found as follows:

$$\hat{G}_i^\star U_i^\star = G_i^{\mathcal{F}\star} Z_i^\star \tag{8.53}$$

$$\iff \begin{bmatrix} \hat{G}_i & \hat{g} \end{bmatrix} \begin{bmatrix} U_i & u \\ 0^T & \varrho \end{bmatrix} = \begin{bmatrix} G_i^{\mathcal{F}} & g_1^{\mathcal{X}} \end{bmatrix} \begin{bmatrix} Z_i & z \\ 0^T & \zeta \end{bmatrix}$$

$$\iff \begin{bmatrix} \hat{G}_i U_i & \hat{G}_i u + \hat{g}\varrho \end{bmatrix} = \begin{bmatrix} G_i^{\mathcal{F}} Z_i & G_i^{\mathcal{F}} z + g_1^{\mathcal{X}} \zeta \end{bmatrix}$$

with $g_1^{\mathcal{X}}$ denoting the first column of the fixed variables part $G_i^{\mathcal{X}}$ of $G_i$. Solving for $\hat{g}$ yields

$$\hat{g}^\star = (G_i^{\mathcal{F}} z + g_1^{\mathcal{X}} \zeta - \hat{G} u)/\varrho. \tag{8.54}$$

This proves relation (8.44d) for $\hat{G}_i^\star$ and the proof for $\hat{P}_i^\star$ can be carried out in the same way.

Finally, to show relation (8.44e) for the tridiagonal system blocks $A_{i-1}^\star$, $A_i^\star$, and $B_i^\star$ we compute

$$A_i^\star = \hat{G}_i^\star \hat{G}_i^{\star T} + \hat{P}_{i+1} \hat{P}_{i+1}^T = \hat{G}_i \hat{G}_i^T + \hat{g}\hat{g}^T + \hat{P}_{i+1} \hat{P}_{i+1}^T = A_i + \hat{g}\hat{g}^T. \tag{8.55}$$

The block $A_i$ is thus affected by a rank one update, and identical relations hold for $A_{i-1}$ and $B_i$. This completes the proof. □

The resulting factorization update procedure is summarized in algorithm 8.5 on page 228. Observe now that the rank one modification to the $2 \times 2$ subblock of the block tridiagonal system has positive sign as the active set shrinks. A suitable update to the block tridiagonal Cholesky factorization is derived in section 8.3.

### 8.2.5 Deleting a Point Constraint

If a decoupled point constraint on node $i$ becomes inactive, a row $j$ is removed from $R_i^{\mathcal{AF}}$. In the following theorem we show how to restore the block local reductions of a HPSC factorization for the new KKT matrix with inactive point constraint.

**Theorem 8.5 (Deleting a Point Constraint from the Block Local Reductions)**
Let $\mathcal{H}(\mathcal{A})$ be a HPSC factorization of the KKT system $\mathcal{K}(\mathcal{A})$. Let the point constraint $(R_i)_{[j,:]}v_i \geqslant r_{ij}$ on the unknown $v_i$ of node $0 \leqslant i \leqslant m$ be active in $\mathcal{A}$, and denote by $\mathcal{A}^\star$ the otherwise identical active set with inactive point constraint. Then there exists a HPSC factorization $\mathcal{H}^\star(\mathcal{A}^\star)$ of the KKT matrix $\mathcal{K}^\star(\mathcal{A}^\star)$ satisfying

$$Z_i^\star = \begin{bmatrix} Z_i & z \end{bmatrix}, \qquad \begin{bmatrix} z & Y_i^\star \end{bmatrix} = Y_i\mathfrak{O}_\mathrm{T}, \begin{bmatrix} 0 & T_i^\star \end{bmatrix} = T_i\mathfrak{O}_\mathrm{T}, \tag{8.56a}$$

$$U_i^\star = \begin{bmatrix} U_i & u \\ 0^T & \varrho \end{bmatrix} \quad \text{with} \begin{bmatrix} u \\ \varrho \end{bmatrix} = \begin{bmatrix} U_i^{-T} Z_i^T H_i^{\mathcal{FF}} z \\ \sqrt{z^T H_i^{\mathcal{FF}} z - u^T u} \end{bmatrix}, \tag{8.56b}$$

$$\hat{G}_i^\star = \begin{bmatrix} \hat{G}_i & \hat{g}^\star \end{bmatrix} \qquad \text{with } \hat{g}^\star = (G_i^{\mathcal{F}} z - \hat{G}u)/\varrho, \tag{8.56c}$$

$$\hat{P}_i^\star = \begin{bmatrix} \hat{P}_i & \hat{p}^\star \end{bmatrix} \qquad \text{with } \hat{p}^\star = (P_i^{\mathcal{F}} z - \hat{P}u)/\varrho,$$

$$A_{i-1}^\star = A_{i-1} + \hat{p}^\star \hat{p}^{\star T}, \qquad A_i^\star = A_i + \hat{g}^\star \hat{g}^{\star T}, B_i^\star = B_i + \hat{g}^\star \hat{p}^{\star T}, \tag{8.56d}$$

where $\mathfrak{O}_\mathrm{T}$ is an appropriately chosen sequence of Givens rotations. △

**Proof** We again start with the proof of relation (8.56a) for the TQ factorization matrices $Z_i^\star$, $Y_i^\star$, and the southeast triangular factor $T_i^\star$. The row $j$ of $R_i^{\mathcal{AF}}$ belonging to the point constraint to be inactivated is removed from both $R_i^{\mathcal{AF}}$ and the triangular factor $T_i$ in the TQ factorization. This yields

$$R_i^{\mathcal{AF}\star} \begin{bmatrix} Z_i & Y_i \end{bmatrix} = \begin{bmatrix} 0 & \tilde{T}_i \end{bmatrix} \tag{8.57}$$

where the triangular shape of $\tilde{T}_i$ has been destroyed by the removal of row $j$. We restore it using the series of $n^{\mathcal{Y}} - j$ Givens rotations

$$\mathfrak{O}_\mathrm{T} \stackrel{\text{def}}{=} O_{n^{\mathcal{Y}}-j}^{n^{\mathcal{Y}}-j+1\,T} \cdot \ldots \cdot O_1^{2\,T}. \tag{8.58}$$

Applied to the right hand side of (8.57) this transformation results in

$$\begin{bmatrix} 0 & \tilde{T}_i \end{bmatrix} \begin{bmatrix} I & \\ & \mathfrak{O}_\mathrm{T} \end{bmatrix} \stackrel{\text{def}}{=} \begin{bmatrix} 0 & 0 & T_i^\star \end{bmatrix}. \tag{8.59}$$

We find that $T_i$ shrinks by one row and column reflecting the increased dimension of the null space. Applying $\mathfrak{O}_\mathrm{T}$ to the right hand side of (8.57) to maintain equality, the null space basis $Z_i$ remains unaffected we obtain

$$\begin{bmatrix} Z_i & Y_i \end{bmatrix} \begin{bmatrix} I_{n^Z} & \\ & \mathfrak{O}_\mathrm{T} \end{bmatrix} = \begin{bmatrix} \boxed{Z_i \ \ z} & Y_i^\star \end{bmatrix} \stackrel{\text{def}}{=} \begin{bmatrix} Z_i^\star & Y_i^\star \end{bmatrix}. \tag{8.60}$$

This proves relation (8.56a) for the TQ factorization matrices $Z_i^\star$, $Y_i^\star$, and $T_i^\star$.

Relation (8.56b) for the Cholesky factor $U_i^\star$ of the projected Hessian $Z_i^{\star T} H_i^{\mathcal{F}\mathcal{F}\star} Z_i^\star$ is derived as follows. Denoting the new elements of the Hessian factor again with a vector $u$ and a scalar $\varrho$,

$$U_i^\star = \begin{bmatrix} U_i & u \\ 0^T & \varrho \end{bmatrix}, \tag{8.61}$$

we determine the new column entries of the Hessian factor $U_i^\star$ similar to 8.2.4 from

$$\begin{bmatrix} Z_i & z \end{bmatrix}^T H_i^{\mathcal{F}\mathcal{F}\star} \begin{bmatrix} Z_i & z \end{bmatrix} = \begin{bmatrix} U_i^T U_i & U_i^T u \\ u^T U_i & u^T u + \varrho^2 \end{bmatrix}. \tag{8.62}$$

Observing $H_i^{\mathcal{F}\mathcal{F}\star} = H_i^{\mathcal{F}\mathcal{F}}$ as the set of active simple bounds did not change, this yields

$$u = U_i^{-T} Z_i^T H_i^{\mathcal{F}\mathcal{F}} z, \qquad \varrho = \sqrt{z^T H_i^{\mathcal{F}\mathcal{F}} z - u^T u}. \tag{8.63}$$

This proves relation (8.56b) for the Cholesky factor $U_i^\star$ of the reduced Hessian.

Finally, relation (8.56c) for the Schur complements $\hat{G}_i^\star$ and $\hat{P}_i^\star$ can again be found constructively in the form $\hat{G}_i^\star = \begin{bmatrix} \hat{G}_i & \hat{g} \end{bmatrix}$ as the initial $n^Z$ columns of $Z_i$ and $U_i$ remained unchanged. Here again $\hat{g}$ is an additional column that is determined as follows. From

$$\begin{bmatrix} \hat{G}_i & \hat{g} \end{bmatrix} u_i^\star = G_i^\mathcal{F} Z_i^\star \iff \begin{bmatrix} \hat{G}_i U_i & \hat{G}_i u + \hat{g}\varrho \end{bmatrix} = \begin{bmatrix} G_i^\mathcal{F} Z_i & G_i^\mathcal{F} z \end{bmatrix} \tag{8.64}$$

we determine that new column by solving for $\hat{g}$,

$$\hat{g} = (G_i^\mathcal{F} z - \hat{G}_i u)/\varrho. \tag{8.65}$$

With this result for $\hat{G}_i^\star$ and the analogous one for $\hat{P}_i^\star$ relation (8.56c) is shown and the proof of relation (8.56d) for the tridiagonal system blocks $A_{i-1}^\star$, $A_i^\star$, and $B_i^\star$ proceeds as in section 8.2.4. □

The resulting update procedure is summarized in algorithm 8.6 on page 229.

---

**Algorithm 8.3:** Matrix updates adding a simple bound to the active set.

---

**input** : $\mathcal{K}$, $\mathcal{H}$, i
**output** : $\mathcal{H}^\star$
$t = Q_{i[n^\mathcal{F},:]}$;
**for** $k = 1 : n^\mathcal{Z} - 1$ **do**
  $[s, c] =$ **givens**$(t, k+1, k)$;
  $[t, Q_i, U_i] =$ **apply**$([t, Q_i, U_i], s, c, k+1, k)$;
**end**
$T_i = \begin{bmatrix} \textbf{zeros}(n^\mathcal{Y}, n^\mathcal{Z}) & T_i \end{bmatrix}$;
**for** $k = n^\mathcal{Z} : n^\mathcal{F} - 1$ **do**
  $[s, c] =$ **givens**$(t, k+1, k)$;
  $[t, Q_i, T_i] =$ **apply**$([t, Q_i, T_i], s, c, k+1, k)$;
**end**
$T_i = T_{i[:, n^\mathcal{Z} : n^\mathcal{F} - 1]}$;
**for** $k = 1 : n^\mathcal{Z} - 1$ **do**
  $[s, c] =$ **givens**$(U_{[:,k]}, k, k+1)$;
  $[U_i^T, \hat{G}_i, \hat{P}_i] =$ **apply**$([U_i^T, \hat{G}_i, \hat{P}_i], s, c, k, k+1)$;
**end**
$U_i = U_{i[1:n^\mathcal{Z}-1, 1:n^\mathcal{Z}-1]}$;
$\hat{g} = \hat{G}_{i[:, n^\mathcal{Z}]}$; $\hat{p} = \hat{P}_{i[:, n^\mathcal{Z}]}$;
$\hat{G}_i = \hat{G}_{i[:, 1:n^\mathcal{Z}-1]}$; $\hat{P}_i = \hat{P}_{i[:, 1:n^\mathcal{Z}-1]}$;
$[V, D] =$ **block_tri_choldown**$(V, D, \hat{g}, \hat{p}, i)$;

---

**Algorithm 8.4:** Matrix updates adding a point constraint.

**input** : $\mathcal{K}$, $\mathcal{H}$, i, j
**output** : $\mathcal{H}^\star$
$t = R_{i[j,\mathcal{F}]} Q_i$;
**for** $k = 1 : n^{\mathcal{Z}} - 1$ **do**
  $[s, c] =$ **givens**$(t, k+1, k)$;
  $[t, Q_i, U_i] =$ **apply**$([t, Q_i, U_i], s, c, k+1, k)$;
**end**
$T_i = \left[\textbf{zeros}(n^{\mathcal{Y}}, 1) \;\; T_i; \;\; t_{[n^{\mathcal{Z}}+1:n^{\mathcal{F}}]}\right]$;
**for** $k = 1 : n^{\mathcal{Z}} - 1$ **do**
  $[s, c] =$ **givens**$(U_{[:,k]}, k, k+1)$;
  $[U_i^T, \hat{G}_i, \hat{P}_i] =$ **apply**$([U_i^T, \hat{G}_i, \hat{P}_i], s, c, k, k+1)$;
**end**
$U_i = U_{i[1:n^{\mathcal{Z}}-1, 1:n^{\mathcal{Z}}-1]}$;
$\hat{g} = \hat{G}_{i[:,n^{\mathcal{Z}}]}$; $\hat{p} = \hat{P}_{i[:,n^{\mathcal{Z}}]}$;
$\hat{G}_i = \hat{G}_{i[:,1:n^{\mathcal{Z}}-1]}$; $\hat{P}_i = \hat{P}_{i[:,1:n^{\mathcal{Z}}-1]}$;
$[V, D] =$ **block_tri_choldown**$(V, D, \hat{g}, \hat{p}, i)$;

**Algorithm 8.5:** Matrix updates deleting a simple bound.

**input** : $\mathcal{K}$, $\mathcal{H}$, i, j
**output** : $\mathcal{H}^\star$
$T_i = \left[T_i \;\; R_i^{\mathcal{AX}}{}_{[:,1]}\right]$;
$Q_i = \left[Q_i \;\; \textbf{zeros}(n^{\mathcal{F}}, 1); \;\; \textbf{zeros}(1, n^{\mathcal{F}}) \;\; 1\right]$;
**for** $k = 1 : n^{\mathcal{Y}}$ **do**
  $[s, c] =$ **givens**$(T_{i[:,k]}, n^{\mathcal{Y}} + 2 - k, n^{\mathcal{Y}} + 1 - k)$;
  $[T_i, Y_i] =$ **apply**$([T_i, Y_i], s, c, n^{\mathcal{Y}} + 2 - k, n^{\mathcal{Y}} + 1 - k)$;
**end**
$T_i = T_{i[:,2:n^{\mathcal{Y}}+1]}$;
$[z; \zeta] = Q_{i[:,n^{\mathcal{Z}}+1]}$;
$[h; \eta] = [H_i^{\mathcal{FX}}{}_{[:,1]}; \; H_i^{\mathcal{XX}}{}_{[1,1]}]$;
$u = U_i^T \backslash (Z_i^T H_i^{\mathcal{FF}} z + h\zeta)$;
$\varrho =$ **sqrt**$(z^T H_i^{\mathcal{FF}} z + \zeta(h^T z + \eta\zeta) - u^T u)$;
$U_i = \left[U_i \;\; u; \;\; \textbf{zeros}(1, n^{\mathcal{Z}}) \;\; \varrho\right]$;
$\hat{g} = (G_i^{\mathcal{F}} z + G_i^{\mathcal{X}}{}_{[:,1]}\zeta - \hat{G}u)/\varrho$; $\hat{p} = (P_i^{\mathcal{F}} z + P_i^{\mathcal{X}}{}_{[:,1]}\zeta - \hat{P}u)/\varrho$;
$\hat{G}_i = \left[\hat{G}_i \;\; \hat{g}\right]$; $\hat{P}_i = \left[\hat{P}_i \;\; \hat{p}\right]$;
$[V, D] =$ **block_tri_cholup**$(V, D, \hat{g}, \hat{p}, i)$;

**Algorithm 8.6:** Matrix updates deleting a point constraint.

**input** : $\mathcal{K}$, $\mathcal{H}$, i, j
**output** : $\mathcal{H}^\star$
$T_i = \begin{bmatrix} T_{i\,[1:j-1,:]} & T_{i\,[j+1:n^{\mathcal{Y}},:]} \end{bmatrix}$;
**for** $k = j : n^{\mathcal{Y}} - 1$ **do**
  $[s,c] = \textbf{givens}(T_{i\,[k,:]}, n^{\mathcal{Y}} - k + 1, n^{\mathcal{Y}} - k)$;
  $[T_i, Y_i] = \textbf{apply}([T_i, Y_i], s, c, n^{\mathcal{Y}} - k + 1, n^{\mathcal{Y}} \;\; k)$;
**end**
$T_i = T_{i\,[:,2:n^{\mathcal{Y}}]}$;
$[z;\zeta] = Q_{i\,[:,n^{\mathcal{Z}}+1]}$;
$u = U_i^T \backslash (Z_i^T H_i^{\mathcal{F}\mathcal{F}} z)$;
$\varrho = \textbf{sqrt}(z^T H_i^{\mathcal{F}\mathcal{F}} z - u^T u)$;
$U_i = \begin{bmatrix} U_i & u; & \textbf{zeros}(1, n^{\mathcal{Z}}) & \varrho \end{bmatrix}$;
$\hat{g} = (G_i^{\mathcal{F}} z - \hat{G} u)/\varrho$; $\hat{p} = (P_i^{\mathcal{F}} z - \hat{P} u)/\varrho$;
$\hat{G}_i = \begin{bmatrix} \hat{G}_i & \hat{g} \end{bmatrix}$; $\hat{P}_i = \begin{bmatrix} \hat{P}_i & \hat{p} \end{bmatrix}$;
$[V, D] = \textbf{block\_tri\_cholup}(V, D, \hat{g}, \hat{p}, i)$;

## 8.3 Modifying the Block Tridiagonal Factorization

We conclude the presentation of the block structured update procedures by deriving the update and a downdate procedure for the tridiagonal block Cholesky factorization of system (7.33). In detail, our update will treat a rank one modification of some or all blocks $A_i$ together with the appropriate rank one modification of the affected subdiagonal blocks $B_i$. As we have seen, this situation arises as common final part of all four cases of active set changes, where two blocks $A_{i-1}$ and $A_i$ are updated or downdated together with the interleaving subdiagonal block $B_i$.

To this end, the shape restoration approach initially presented in section 8.1.1 for a dense Cholesky factorization cannot be applied to the block tridiagonal system, as the zero pattern cannot be represented by a single dyadic product. Hence, efficient exploitation of the tridiagonal structure is necessary to individually apply the rank one update to every diagonal and side diagonal block.

### 8.3.1 A Rank 1 Update

In order to derive $O(mn^2)$ algorithms for both procedures instead, we carry out a single step of the block tridiagonal Cholesky factorization of section 7.3.4, incorporating the addition or subtraction of the dyadic product.

**Theorem 8.6 (Update to a Block Tridiagonal Cholesky Factorization)**
Let $(\boldsymbol{A},\boldsymbol{B})$ with $\boldsymbol{A}_i \in \mathbb{R}^{n\times n}$, $0 \leqslant i \leqslant m-1$ and $\boldsymbol{B}_i \in \mathbb{R}^{n\times n}$, $1 \leqslant i \leqslant m-1$ be the diagonal and subdiagonal blocks of a positive definite block tridiagonal system. Let $(\boldsymbol{A}^\star,\boldsymbol{B}^\star)$ be a positive rank one modification of $(\boldsymbol{A},\boldsymbol{B})$ defined by vectors $\boldsymbol{y}_i \in \mathbb{R}^n$,

$$\begin{aligned} \boldsymbol{A}_i^\star &= \boldsymbol{A}_i + \boldsymbol{y}_i\boldsymbol{y}_i^T, && 0 \leqslant i \leqslant m-1, \\ \boldsymbol{B}_i^\star &= \boldsymbol{B}_i + \boldsymbol{y}_{i-1}\boldsymbol{y}_i^T, && 1 \leqslant i \leqslant m-1. \end{aligned} \tag{8.66}$$

Further, let $(\boldsymbol{V},\boldsymbol{D})$ be the upper triangular and subdiagonal blocks of the Cholesky factorization of $(\boldsymbol{A},\boldsymbol{B})$. Then it holds that the Cholesky factorization $(\boldsymbol{V}^\star,\boldsymbol{D}^\star)$ of $(\boldsymbol{A}^\star,\boldsymbol{B}^\star)$ is obtained from $(\boldsymbol{V},\boldsymbol{D})$ as

$$\begin{aligned} \boldsymbol{V}_i^\star &= \boldsymbol{O}_i^T\boldsymbol{V}_i + \boldsymbol{o}_i\boldsymbol{z}_i^T, && 0 \leqslant i \leqslant m-1, \\ \boldsymbol{D}_i^\star &= \boldsymbol{O}_{i-1}^T(\boldsymbol{D}_i + \boldsymbol{V}_{i-1}^{-T}\boldsymbol{z}_{i-1}\boldsymbol{y}_i^T), && 1 \leqslant i \leqslant m-1, \end{aligned} \tag{8.67}$$

with vectors $\boldsymbol{z}_i$ defined by the recursion formula

$$\boldsymbol{z}_0 = \boldsymbol{y}_0, \qquad \boldsymbol{z}_i = \delta_{i-1}(\boldsymbol{D}_i^T\boldsymbol{V}_{i-1}^{-T}\boldsymbol{z}_{i-1} - \boldsymbol{y}_i), \qquad 1 \leqslant i \leqslant m-2, \tag{8.68}$$

and the sequences $\mathfrak{O}_{\mathrm{V_i}}$ of Givens rotations eliminating the diagonal of $\boldsymbol{V}_i^T$ denoted by the matrix

$$\mathfrak{O}_{\mathrm{V_i}} \stackrel{\text{def}}{=} \begin{bmatrix} \boldsymbol{o}_i^T & \delta_i \\ \boldsymbol{O}_i & -\delta_i\boldsymbol{V}_i^{-T}\boldsymbol{y}_i \end{bmatrix}. \tag{8.69}$$

△

**Proof** For updating the diagonal blocks we employ a variant of method C3 described in [83]. Forming a sequence $\mathfrak{O}_{\mathrm{V}}$ of Givens rotations

$$\mathfrak{O}_{\mathrm{V}} \stackrel{\text{def}}{=} \boldsymbol{O}_1^2\boldsymbol{O}_2^3 \cdot \ldots \cdot \boldsymbol{O}_{n-1}^n \tag{8.70}$$

to eliminate the diagonal elements of $\boldsymbol{V}_0^T$, we restore the lower triangular shape of the following system:

$$\begin{bmatrix} \boldsymbol{z}_0 & \boldsymbol{V}_0^T \\ 1 & \boldsymbol{0}^T \end{bmatrix} \underbrace{\begin{bmatrix} \boldsymbol{o}_{21}^T & \varrho_{22} \\ \boldsymbol{O}_{11} & \boldsymbol{o}_{12} \end{bmatrix}}_{=\mathcal{O}_{\mathrm{V}}} = \begin{bmatrix} \boldsymbol{V}_0^{\star T} & \boldsymbol{0} \\ \boldsymbol{r}^T & \delta_{\mathrm{n}} \end{bmatrix}. \tag{8.71}$$

Correctness can be verified by multiplying each side of (8.71) by its transpose and comparing entries, which yields the identity $V_0^T V_0 + z_0 z_0^T = V_0^{\star T} V_0^\star$. For the new Cholesky factor we find $V_0^{\star T} = V_0^T O_{11} + z_0 o_{21}^T$ as claimed. Besides, the following identities hold:

$$o_{12} = -\delta_n V_0^{-T} z_0 \qquad o_{21} = r \qquad \varrho_{22} = \delta_n \tag{8.72}$$

From the block tridiagonal Cholesky algorithm we have the identity $D_1^\star = V_0^{\star -T} B_1^{\star T}$ which gives $D_1^\star = O_{11}^{-1} V_0^{-T}(B_1^T + z_0 y_1^T)$. The difficulty here lies with finding $O_{11}^{-1}$, as $O_{11}$ is not orthogonal. To address this issue, we expand the above identity to

$$\begin{bmatrix} D_1^\star \\ d \end{bmatrix} = \begin{bmatrix} o_{21} & O_{11}^T \\ \varrho_{22} & o_{12}^T \end{bmatrix} \begin{bmatrix} 0 & 1 \\ V_0^{-T} & -V_0^{-T} z_0 \end{bmatrix} \begin{bmatrix} B_1^T + z_0 y_1^T \\ 0 \end{bmatrix} = \begin{bmatrix} O_{11}^T D_1 + O_{11}^T V_0^{-T} z_0 y_1^T \\ o_{12}^T D_1 + o_{12}^T V_0^{-T} z_0 y_1^T \end{bmatrix}.$$

Herein, we have exploited orthogonality of $\mathfrak{O}_V$ and formed the inverse explicitly. We find that the columns of $D_1$ are affected by the Givens sequence like the rows of $V_1$,

$$D_1^\star = O_{11}^T (D_1 + V_0^{-T} z_0 y_1^T) \tag{8.73}$$

as claimed. We now compute $D_1^{\star T} D_1^\star$, the new downdate to the following diagonal block $A_1$. By orthogonality of $\mathfrak{O}_V$ we have with $v \overset{\text{def}}{=} V_0^{-T} z_0$

$$O_{11} O_{11}^T = I - o_{12} o_{12}^T = I - \delta_n^2 v v^T \tag{8.74}$$

and can express $D_1^{\star T} D_1^\star$ as

$$\begin{aligned} D_1^{\star T} D_1 &= (D_1^T + y_1 v^T) O_{11} O_{11}^T (D_1 + v y_1^T) \\ &= (D_1^T + y_1 v^T)(D_1 + v y_1^T) - \delta_n^2 (D_1^T v + y_1 v^T v)(v^T D_1 + v^T v y_1^T). \end{aligned} \tag{8.75}$$

By expanding, collecting identical terms, and using the identity $1 - \delta_n^2 v^T v = \delta_n^2$ which is easily verified using (8.72) and orthogonality of $\mathfrak{O}_V$, we find

$$D_1^{\star T} D_1 = D_1^T D_1 + \delta_n^2 (D_1^T v - y_1)(D_1^T v - y_1)^T + y_1 y_1^T. \tag{8.76}$$

Using this relation, the update of the next diagonal block $A_1$ reads

$$\begin{aligned} V_1^{\star T} V_1^\star &= A_1^\star - D_1^{\star T} D_1^\star + y_1 y_1^T = A_1 - D_1^T D_1 - z_1 z_1^T \\ &= V_1^T V_1 - z_1 z_1^T \end{aligned} \tag{8.77}$$

with a vector $\boldsymbol{z}_1 = \delta_n(\boldsymbol{D}_1^T \boldsymbol{V}_0^{-T} \boldsymbol{z}_0 - \boldsymbol{y}_1)$ as claimed. Closing the loop by continuing with the update of $\boldsymbol{V}_1$, we eventually obtain the claimed relations for all further blocks by repeating the argument for the nodes $1, \ldots, m-1$. This completes the proof. □

All required computations can be carried out in at most $O(n^2)$ operations per block, where $n$ is the size of the square matrices $\boldsymbol{A}_i$ or $\boldsymbol{B}_i$. The update sequence obviously can start at any block $i \geqslant 0$, but must always run until the last block $m-1$ has been updated. The number of nodes affected is $m-i$, thus the total runtime complexity is bounded by $O(mn^2)$. A summary of this update procedure for the block tridiagonal Cholesky factorization is given in algorithm 8.7.

**Algorithm 8.7:** Updating a block tridiagonal Cholesky decomposition.

```
input  : V, D, g, p, i
output : V*, D*
for j = i : m do
    v = V_{j-1}^T \ p;
    V_{j-1} = [p^T; V_{j-1}];
    p = D_j^T v - g;
    D_j = [zeros(1, n^x); D_j + v g^T];
    d = [1; zeros(n^x, 1)];
    for k = 1 : n^x do
        [s, c] = givens(V_{j-1 [k,:]}, k, k+1);
        [V_{j-1}^T, D_j^T, d^T] = apply([V_{j-1}^T, D_j^T, d^T], s, c, k, k+1);
    end
    V_{j-1} = V_{j-1 [1:n^x,:]};
    D_j = D_{j [1:n^x,:]};
    p = d_{[n^x]} p;
    g = 0;
end
```

### 8.3.2 A Rank 1 Downdate

In this section, we finally derive a rank one downdate procedure for the block tridiagonal Cholesky factorization of system (7.33), to be used when adding a simple bound or decoupled point constraint to the active set. The derivation is technically equivalent to that for the rank one update of section 8.3 and differs only in several minor steps. As the obtained downdate formulas are slightly different, we give the full derivation for completeness here.

**Theorem 8.7 (Downdate to a Block Tridiagonal Cholesky Factorization)**
Let $(\boldsymbol{A},\boldsymbol{B})$ with $\boldsymbol{A}_i \in \mathbb{R}^{n\times n}$, $0 \leqslant i \leqslant m-1$ and $\boldsymbol{B}_i \in \mathbb{R}^{n\times n}$, $1 \leqslant i \leqslant m-1$ be the diagonal and subdiagonal blocks of a positive definite block tridiagonal system. Let $(\boldsymbol{A}^\star,\boldsymbol{B}^\star)$ be a negative rank one modification of $(\boldsymbol{A},\boldsymbol{B})$ defined by vectors $\boldsymbol{y}_i \in \mathbb{R}^n$,

$$\begin{aligned} \boldsymbol{A}_i^\star &= \boldsymbol{A}_i - \boldsymbol{y}_i\boldsymbol{y}_i^T, & 0 \leqslant i \leqslant m-1, \\ \boldsymbol{B}_i^\star &= \boldsymbol{B}_i - \boldsymbol{y}_{i-1}\boldsymbol{y}_i^T, & 1 \leqslant i \leqslant m-1. \end{aligned} \tag{8.78}$$

Further, let $(\boldsymbol{V},\boldsymbol{D})$ be the upper triangular and subdiagonal blocks of the Cholesky factorization of $(\boldsymbol{A},\boldsymbol{B})$. Then it holds that the Cholesky factorization $(\boldsymbol{V}^\star,\boldsymbol{D}^\star)$ of $(\overline{\boldsymbol{A}},\overline{\boldsymbol{B}})$ is obtained from $(\boldsymbol{V},\boldsymbol{D})$ as

$$\begin{aligned} \boldsymbol{V}_i^\star &= \boldsymbol{O}_i\boldsymbol{V}_i, & 0 \leqslant i \leqslant m-1, \\ \boldsymbol{D}_i^\star &= (\boldsymbol{O}_{i-1} - \boldsymbol{o}_{i-1}\tfrac{1}{\delta_\mathrm{n}}(\boldsymbol{V}_{i-1}^{-T}\boldsymbol{z}_{i-1})^T)(\boldsymbol{D}_i - \boldsymbol{V}_{i-1}^{-T}\boldsymbol{z}_{i-1}\boldsymbol{y}_i), & 1 \leqslant i \leqslant m-1, \end{aligned} \tag{8.79}$$

with vectors $\boldsymbol{z}_i$ defined by the recursion formula

$$\boldsymbol{z}_0 = \boldsymbol{y}_0, \quad \boldsymbol{z}_i = \tfrac{1}{\delta_\mathrm{n}}(\boldsymbol{V}_{i-1}^{-T}\boldsymbol{z}_{i-1})^T(\boldsymbol{D}_i - \boldsymbol{V}_{i-1}^{-T}\boldsymbol{z}_{i-1}\boldsymbol{y}_i) - \delta_\mathrm{n}\boldsymbol{y}_i, \quad 1 \leqslant i \leqslant m-2, \tag{8.80}$$

and the sequences $\mathfrak{O}_{\mathrm{V}_i}$ of Givens rotations eliminating the diagonal of $\boldsymbol{V}_i^T$ denoted by the matrix

$$\mathfrak{O}_{\mathrm{V}_i} \stackrel{\mathrm{def}}{=} \begin{bmatrix} \boldsymbol{O}_i & \boldsymbol{o}_i \\ \boldsymbol{z}_i^T\boldsymbol{V}_i^{-1} & \delta_i \end{bmatrix}. \tag{8.81}$$

△

**Proof** For downdating the diagonal blocks we again employ method C3 of [83], who for a downdate now form

$$\begin{bmatrix} \boldsymbol{v} & \boldsymbol{V}_0 \\ \delta_\mathrm{n} & \boldsymbol{0}^T \end{bmatrix} \tag{8.82}$$

with $\boldsymbol{v} \stackrel{\mathrm{def}}{=} \boldsymbol{V}_0^{-T}\boldsymbol{z}_0$, $\delta_\mathrm{n} \stackrel{\mathrm{def}}{=} \sqrt{1-\boldsymbol{v}^T\boldsymbol{v}}$, and project $\begin{bmatrix}\boldsymbol{v}^T & \delta_\mathrm{n}\end{bmatrix}$ onto the unit row vector $\boldsymbol{e}_n^T$ by applying a sequence of Givens plane rotations

$$\mathfrak{O}_\mathrm{V} \stackrel{\mathrm{def}}{=} \boldsymbol{O}_{n+1}^1 \cdot \ldots \cdot \boldsymbol{O}_{n+1}^{n-1} \cdot \boldsymbol{O}_{n+1}^n \tag{8.83}$$

to (8.82), yielding

$$\underbrace{\begin{bmatrix} \boldsymbol{O}_{11} & \boldsymbol{o}_{12} \\ \boldsymbol{o}_{21}^T & \varrho_{22} \end{bmatrix}}_{=\mathfrak{O}_\mathrm{V}} \begin{bmatrix} \boldsymbol{v} & \boldsymbol{V}_0 \\ \delta_\mathrm{n} & \boldsymbol{0}^T \end{bmatrix} = \begin{bmatrix} \boldsymbol{0} & \boldsymbol{V}_0^\star \\ 1 & \boldsymbol{z}_0^T \end{bmatrix}. \tag{8.84}$$

Correctness can be verified by multiplying each side of (8.84) by its transpose and comparing entries, which yields the identity and $\boldsymbol{V}_0^{\star T}\boldsymbol{V}_0^\star = \boldsymbol{V}_0^T\boldsymbol{V}_0 - \boldsymbol{z}_0\boldsymbol{z}_0^T$. From (8.84) we get for $\boldsymbol{V}_0^\star$ the expression $\boldsymbol{V}_0^\star = \boldsymbol{O}_{11}\boldsymbol{V}_0$ and the identity $\boldsymbol{o}_{21} = \boldsymbol{v}$ to be used in the next section. As in section 8.3 we expand the identity $\boldsymbol{D}_1^\star = \boldsymbol{V}_0^{\star -T}\boldsymbol{B}_1^{\star T}$ to exploit orthogonality of $\mathfrak{O}_\mathrm{V}$ and by forming the inverse of (8.82) explicitly we find

$$\begin{bmatrix} \boldsymbol{D}_1^\star \\ \boldsymbol{d} \end{bmatrix} = \mathfrak{O}_\mathrm{V} \begin{bmatrix} \boldsymbol{0} & \boldsymbol{V}_0^{-T} \\ \frac{1}{\delta_\mathrm{n}} & -\frac{1}{\delta_\mathrm{n}}\boldsymbol{v}^T\boldsymbol{V}_0^{-T} \end{bmatrix} \begin{bmatrix} \boldsymbol{0} \\ \boldsymbol{B}_1^T - \boldsymbol{z}_0\boldsymbol{y}_1^T \end{bmatrix} = \mathfrak{O}_\mathrm{V} \begin{bmatrix} \boldsymbol{D}_1 - \boldsymbol{v}\boldsymbol{y}_1^T \\ -\frac{1}{\delta_\mathrm{n}}\boldsymbol{v}^T(\boldsymbol{D}_1 - \boldsymbol{v}\boldsymbol{y}_1^T) \end{bmatrix} \tag{8.85}$$

As claimed, $\boldsymbol{D}_1$ is updated and its rows are affected by the Givens sequence like the rows of $\boldsymbol{V}_0$. A nontrivial dyadic term is added, though, due to non–orthogonality of $\boldsymbol{O}_{11}$. We now compute $\boldsymbol{D}_1^{\star T}\boldsymbol{D}_1^\star$ for the third step. We note $1/(1-\boldsymbol{v}^T\boldsymbol{v}) = 1/\delta_n^2 = 1/\varrho_{22}^2$.

$$\begin{aligned} \boldsymbol{D}_1^{\star T}\boldsymbol{D}_1^\star &= \boldsymbol{B}_1^\star\boldsymbol{V}_0^{-1}\boldsymbol{O}_{11}^{-1}\boldsymbol{O}_{11}^{-T}\boldsymbol{V}_0^{-T}\boldsymbol{B}_1^{\star T} \\ &= \boldsymbol{B}_1^\star\boldsymbol{V}_0^{-1}(\boldsymbol{I} - \boldsymbol{o}_{21}\boldsymbol{o}_{21}^T)^{-1}\boldsymbol{V}_0^{-T}\boldsymbol{B}_1^{\star T} \\ &= \boldsymbol{B}_1^\star\boldsymbol{V}_0^{-1}\left(\boldsymbol{I} + \delta_\mathrm{n}^{-2}\boldsymbol{v}\boldsymbol{v}^T\right)\boldsymbol{V}_0^{-T}\boldsymbol{B}_1^{\star T} \\ &= (\boldsymbol{D}_1^T - \boldsymbol{y}_1\boldsymbol{v}^T)(\boldsymbol{D}_1 - \boldsymbol{v}\boldsymbol{y}_1^T) + \delta_\mathrm{n}^{-2}(\boldsymbol{D}_1^T\boldsymbol{v} - \boldsymbol{y}_1\boldsymbol{v}^T\boldsymbol{v})(\boldsymbol{v}^T\boldsymbol{D}_1 - \boldsymbol{v}^T\boldsymbol{v}\boldsymbol{y}_1^T) \end{aligned} \tag{8.86}$$

Here we have exploited orthogonality of $\mathfrak{O}_\mathrm{V}$ and applied the Sherman–Morrison formula to find $(\boldsymbol{O}_{11}^T\boldsymbol{O}_{11})^{-1}$. Expanding, collecting identical terms and using the identity $1 + \delta_\mathrm{n}^{-2}\boldsymbol{v}^T\boldsymbol{v} = \delta_\mathrm{n}^{-2}$ we find

$$\boldsymbol{D}_1^{\star T}\boldsymbol{D}_1^\star = \boldsymbol{D}_1^T\boldsymbol{D}_1 + \delta_\mathrm{n}^{-2}(\boldsymbol{D}_1^T\boldsymbol{v} - \boldsymbol{y}_1)(\boldsymbol{D}_1^T\boldsymbol{v} - \boldsymbol{y}_1)^T - \boldsymbol{y}_1\boldsymbol{y}_1^T \tag{8.87}$$

Using this relation, the downdate of the next diagonal block $\boldsymbol{A}_1$ reads

$$\begin{aligned} \boldsymbol{V}_1^{\star T}\boldsymbol{V}_1^\star &= \boldsymbol{A}_1^\star - \boldsymbol{D}_1^{\star T}\boldsymbol{D}_1^\star - \boldsymbol{y}_1\boldsymbol{y}_1^T \\ &= \boldsymbol{A}_1 - \boldsymbol{D}_1^T\boldsymbol{D}_1 - \boldsymbol{z}_1\boldsymbol{z}_1^T = \boldsymbol{V}_1^T\boldsymbol{V}_1 - \boldsymbol{z}_1\boldsymbol{z}_1^T \end{aligned} \tag{8.88}$$

with a vector $\boldsymbol{z}_1 = \frac{1}{\delta_\mathrm{n}}(\boldsymbol{D}_1^T\boldsymbol{v} - \boldsymbol{y}_1)$. This vector can be written more conveniently as

$$z_1^T = \frac{1}{\delta_n} v^T (D_1 - v y_1^T) - \delta_n y_1 \tag{8.89}$$

which allows to reuse intermediate terms from (8.85). Closing the loop by continuing with the downdate of $V_1$, we eventually obtain the claimed relations for all further blocks by repeating the argument for the nodes $1, \ldots, m-1$. This completes the proof. □

A summary of this downdate procedure for the block tridiagonal Cholesky factorization is given in algorithm 8.8 on page 235.

**Algorithm 8.8:** Downdating a block tridiagonal Cholesky decomposition.

```
input  : V, D, g, p, i
output : V*, D*
for j = i : m do
    v = V_{j-1}^T \ p;
    if v^T v > 1 then error("positive definiteness lost");
    δ_n = sqrt(1 − v^T v);
    p = v^T D_j / δ_n − g δ_n;
    g = 0;
    V_{j-1} = [V_{j-1}; zeros(1, n^x)];
    D_j = [D_j; −v^T D_j / δ_n];
    v = [v; δ_n];
    for k = n^x : −1 : 1 do
        [s, c] = givens(v, n^x + 1, k);
        [v^T, V_{j-1}^T, D_j^T] = apply([v^T, V_{j-1}^T, D_j^T], s, c, n^x + 1, k);
    end
    V_{j-1} = V_{j-1 [1:n^x, :]};
    D_j = D_{j [1:n^x, :]};
end
```

## 8.4 Summary

In this section we derived fast matrix updates for the HPSC factorization presented in chapter 7. We covered all four cases of active set changes that can arise when solving the QP with direct multiple shooting block structure by an active set method, namely addition or removal of a simple bound and addition or removal of a decoupled point constraint. We gave formal proofs of the matrix update procedures based on established techniques due to [83, 158], and provided efficient algorithmic realizations. All presented matrix updates have a runtime complexity that is bounded by $\mathcal{O}(mn^2)$ and are implemented in our numerical code qpHPSC, see appendix A. The algorithmic techniques presented in this chapter are key to giving fast control feedback in Nonlinear Model Predictive Control (NMPC) problems with many control parameters, such as mixed–integer problems treated by outer convexification.

# 9 Numerical Results

In this chapter we present mixed–integer optimal control problems and mixed–integer model predictive control problems of the form of definition 2.1. They serve as example applications used to demonstrate the various theoretical results and the applicability and performance of the new algorithms presented in this thesis.

In section 9.1 a switched–mode dynamic system is considered to study both the sum–up rounding approximation theorem and the convexified switch costs formulation of chapter 2 at its example. The problem's relaxed optimal solution after partial outer convexification features bang–bang arcs, a path–constrained arc, and a singular arc and sum–up rounding solutions chatter on the later ones. Solutions obtained by penalization and constraining of switches of the integer control are presented.

In section 9.2 we study the contractivity estimate for the mixed–integer real–time iteration scheme of chapter 4 at the example of a nonlinear system with instable steady state. We derive bounds, Lipschitz constants, and estimates of the contractivity constants required to evaluate the sampling time estimate. The behavior of the mixed–integer Nonlinear Model Predictive Control (NMPC) scheme is examined for various sampling times and is found to be in good accordance with our estimate.

In section 9.3 we investigate an autonomous robot path–following and communication problem that is frequently studied in the literature and give an Mathematical Program with Vanishing Constraints (MPVC) reformulation of this problem. We show that the nonconvex active set method developed in this thesis is indeed able solve this problem to optimality. We study the consequences of constraint violation at the example of an interior point method that fails to solve a significant number of problem instances if no appropriate reformulation of the MPVC is used. Our implementation is competitive both in terms of computation time and quality of the obtained locally optimal solutions.

In section 9.4, a Mixed–Integer Optimal Control Problem (MIOCP) modelling a time–optimal test driving scenario is presented and the newly developed structure exploiting linear algebra techniques are compared against various alternative possibilities of solving the block structured Karush–Kuhn–Tucker (KKT) systems. A detailed investigation of the run times for a large number of problem instances reveals that our algorithmic techniques have significant performance advantages for all but the smallest problem instances.

Finally, we consider a nonlinear model–predictive cruise controller including predictive gear shifts, a challenging real–world industrial problem, in section 9.5. A vehicle model is derived from first principles and parameterized by real–world data. The combinatorial nature of the engine speed constraint depending on the selected gear is identified and a vanishing constraint formulation is adopted after examination of the reformulations proposed in chapter 5. The algorithmic techniques for long prediction horizons and many control parameters developed in chapter 7 and 8 are applied to this problem in order to achieve sampling times and control feedback delays small enough to verify that our techniques are indeed capable of solving this problem on–board the truck under demanding real–time constraints even on a supposed hardware platform with limited computational power.

## 9.1 Mixed–Integer Optimal Control with Switch Costs

In this section we investigate a nonlinear switched dynamic system introduced in [60]. We present a MIOCP formulation due to [186] and solve it using the outer convexification and relaxation approach. The problem's relaxed optimal solution features bang–bang arcs, a path constrained arc, and a singular arc. Convergence of the integer feasible sum–up rounding solution's objective to that of the relaxed Optimal Control Problem (OCP) with increasingly fine discretizations of the control trajectory is investigated. The sum–up rounding solutions show chattering behavior on the path–constrained and the singular arc. We apply our convexified switch cost formulation to this problem in order to penalize frequent switching and to obtain solutions that switch only a predefined number of times.

### 9.1.1 Problem Formulation

We consider a nonlinear MIOCP formulation of the investigated problem on the time horizon $\mathcal{T} \overset{\text{def}}{=} [t_0, t_f] \overset{\text{def}}{=} [0, 1] \subset \mathbb{R}$ as presented in [186] is given in (9.1).

$$
\begin{aligned}
\min_{\boldsymbol{x}(\cdot),\boldsymbol{w}(\cdot)} \quad & x_3(t_f) && (9.1)\\
\text{s.t.} \quad & \dot{x}_1(t) = -\frac{x_1(t)}{\sin(1)}\sin(w_1(t)) \\
& \qquad + \big(x_1(t)+x_2(t)\big)\, w_2^2(t) + \big(x_1(t)-x_2(t)\big)\, w_3^3(t), && t\in\mathcal{T},\\
& \dot{x}_2(t) = \big(x_1(t)+2x_2(t)\big)\, w_1(t) \\
& \qquad + \big(x_1(t)-2x_2(t)\big)\, w_2(t) \\
& \qquad + \big(x_1(t)x_2(t)-x_3(t)\big)\big(w_2^2(t)-w_2^3(t)\big) \\
& \qquad + \big(x_1(t)+x_2(t)\big)\, w_3(t), && t\in\mathcal{T},\\
& \dot{x}_3(t) = x_1^2(t)+x_2^2(t), && t\in\mathcal{T},\\
& x_1(t) \geqslant 0.4, && t\in\mathcal{T},\\
& \boldsymbol{w}(t) \in \{(1,0,0),(0,1,0),(0,0,1)\} \stackrel{\text{def}}{=} \Omega, && t\in\mathcal{T},\\
& \boldsymbol{x}(t_0) = \left(\tfrac{1}{2},\tfrac{1}{2},0\right).
\end{aligned}
$$

This MIOCP can be relaxed by letting $\boldsymbol{w}(t) \in [0,1]^3$ and requiring $\sum_{i=1}^{3} w_i(t) = 1$ for all $t \in \mathcal{T}$. This relaxation amounts to inner convexification of the problem and yields a nonlinear continuous OCP.

The nonlinear Ordinary Differential Equation (ODE) system (9.1) has been specially construction to include a number of nonlinearities that become apparent in an inner convexification formulation that relaxes the binary control vector $\boldsymbol{w}(t)$, but vanish if the ODE dynamics are evaluated in binary feasible choices $\boldsymbol{\omega}^i \in \Omega$ only. In the outer convexification reformulation, we introduce one binary control trajectory $\omega_i(t) \in \{0,1\}$ for all $t \in \mathcal{T}$ per choice $\boldsymbol{\omega}^i$ found in the set $\Omega$, and relax these trajectories to $\alpha_i(t) \in [0,1]$ for all $t \in \mathcal{T}$. The resulting problem is given in (9.2).

$$
\begin{aligned}
\min_{\boldsymbol{x}(\cdot),\boldsymbol{\alpha}(\cdot)} \quad & x_3(t_f) && (9.2)\\
\text{s.t.} \quad & \dot{x}_1(t) = -x_1(t)\alpha_1(t) \\
& \qquad + (x_1(t)+x_2(t))\alpha_2(t) \\
& \qquad + (x_1(t)-x_2(t))\alpha_3(t), && t\in\mathcal{T},\\
& \dot{x}_2(t) = (x_1(t)+2x_2(t))\alpha_1(t) \\
& \qquad + (x_1(t)-2x_2(t))\alpha_2(t) \\
& \qquad + (x_1(t)+x_2(t))\alpha_3(t), && t\in\mathcal{T},
\end{aligned}
$$

$$
\begin{aligned}
\dot{x}_3(t) &= x_1^2(t) + x_2^2(t), && t \in \mathcal{T}, \\
x_1(t) &\geqslant 0.4, && t \in \mathcal{T}, \\
\boldsymbol{\alpha}(t) &\in [0,1]^3, && t \in \mathcal{T}, \\
\sum_{i=1}^{3} \alpha_i(t) &= 1, && t \in \mathcal{T}, \\
\boldsymbol{x}(t_0) &= \left(\tfrac{1}{2}, \tfrac{1}{2}, 0\right).
\end{aligned}
$$

Problem (9.2) is by construction identical to the switched system introduced in [60] with an additional path constraint on $x_1(t)$ introduced to obtain a path–constrained arc.

### 9.1.2 Optimal Solutions

In table 9.1 the objective function values $x_3(t_f)$ and the remaining infeasibility of the optimal solution for the relaxed problem after outer convexification with respect to the integer control $\boldsymbol{w}(t)$ (9.2), and for the integer feasible solution obtained from the partially convexified relaxed one by application of sum–up rounding can be found for increasingly fine control discretizations $m$. All problems have been solved to an acceptable KKT tolerance of $10^{-10}$. As can be seen clearly from table 9.1, the sum–up rounding solution's objective converges to that of the partially convexified relaxed problem while the infeasibility of the control–independent path constraint converges to zero.

| | Convexified relaxed | | Sum–up rounding | | |
|---|---|---|---|---|---|
| $m$ | Objective | Infeasibility | Objective | Infeasibility | Switches |
| 20 | 0.9976458 | $1.19 \cdot 10^{-13}$ | 1.050542 | $5.29 \cdot 10^{-2}$ | 9 |
| 40 | 0.9956212 | $3.93 \cdot 10^{-13}$ | 0.9954084 | $2.13 \cdot 10^{-4}$ | 12 |
| 80 | 0.9955688 | $1.50 \cdot 10^{-14}$ | 0.9957063 | $1.37 \cdot 10^{-4}$ | 23 |
| 160 | 0.9955637 | $1.66 \cdot 10^{-14}$ | 0.9956104 | $4.66 \cdot 10^{-5}$ | 47 |
| 320 | 0.9955615 | $1.16 \cdot 10^{-12}$ | 0.9958528 | $2.91 \cdot 10^{-4}$ | 93 |

Table 9.1: Objective function values and infeasibilities of the outer convexified relaxed problem (9.2), and the integer feasible solutions obtained from the latter by sum–up rounding.

This would not be the case if we had applied sum–up rounding to the solution of the original nonlinear problem (9.1). Due to sum–up rounding of

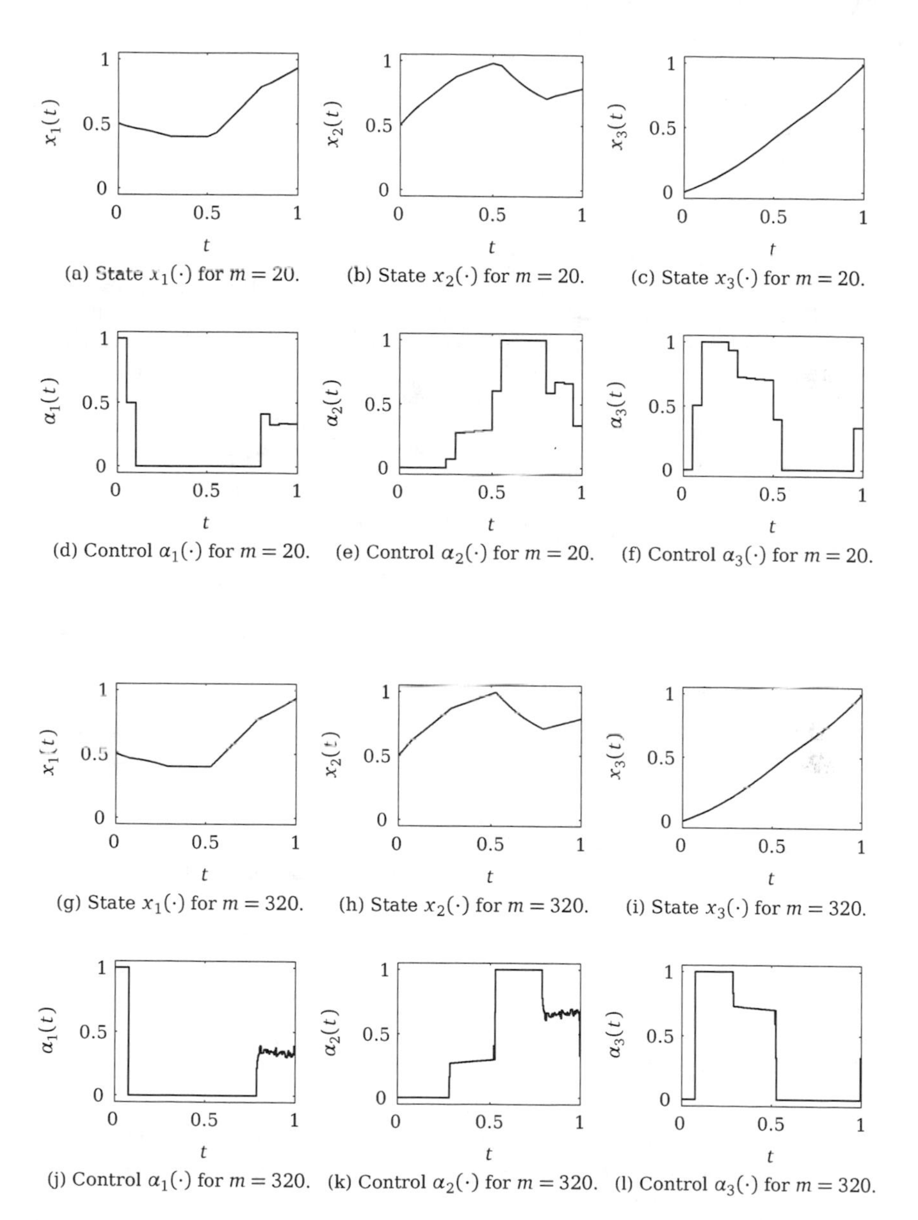


(a) State $x_1(\cdot)$ for $m = 20$. (b) State $x_2(\cdot)$ for $m = 20$. (c) State $x_3(\cdot)$ for $m = 20$.

(d) Control $\alpha_1(\cdot)$ for $m = 20$. (e) Control $\alpha_2(\cdot)$ for $m = 20$. (f) Control $\alpha_3(\cdot)$ for $m = 20$.

(g) State $x_1(\cdot)$ for $m = 320$. (h) State $x_2(\cdot)$ for $m = 320$. (i) State $x_3(\cdot)$ for $m = 320$.

(j) Control $\alpha_1(\cdot)$ for $m = 320$. (k) Control $\alpha_2(\cdot)$ for $m = 320$. (l) Control $\alpha_3(\cdot)$ for $m = 320$.

Figure 9.1: Relaxed optimal solutions of the partially convexified problem (9.2) for discretizations $m = 20$ and $m = 320$ of the control trajectory.

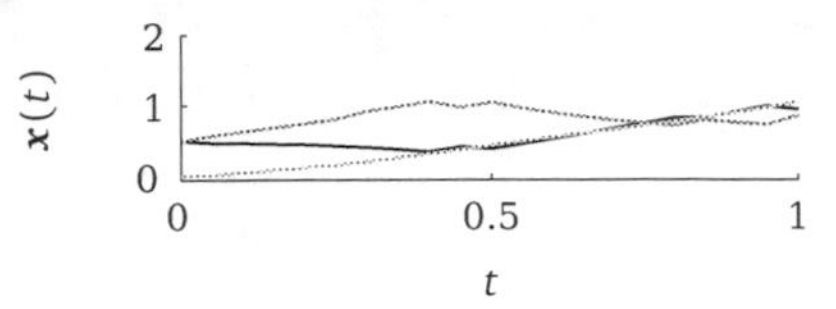


(a) SUR-0.5 state trajectory $\boldsymbol{x}(\cdot)$ for $m = 20$.

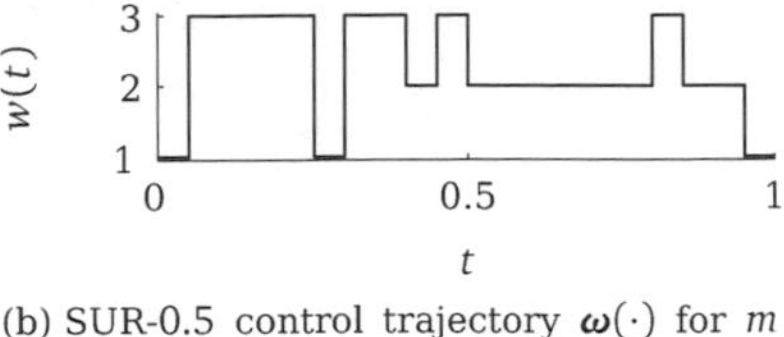


(b) SUR-0.5 control trajectory $\boldsymbol{\omega}(\cdot)$ for $m = 20$. Control has 9 switches.

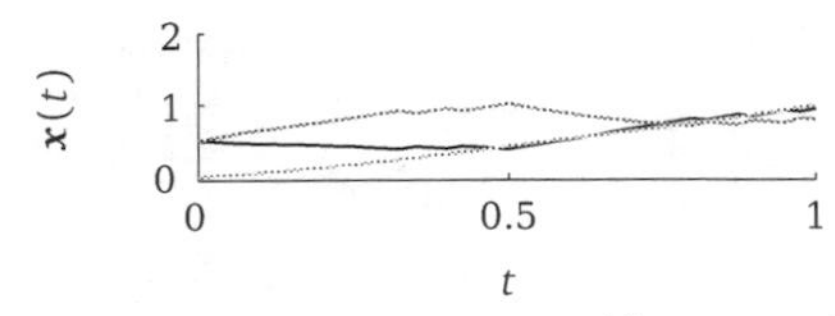

(c) SUR-0.5 state trajectory $\boldsymbol{x}(\cdot)$ for $m = 40$.

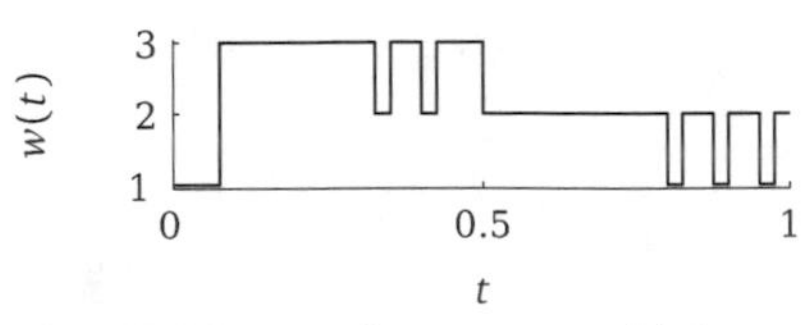

(d) SUR-0.5 control trajectory $\boldsymbol{\omega}(\cdot)$ for $m = 40$. Control has 12 switches.

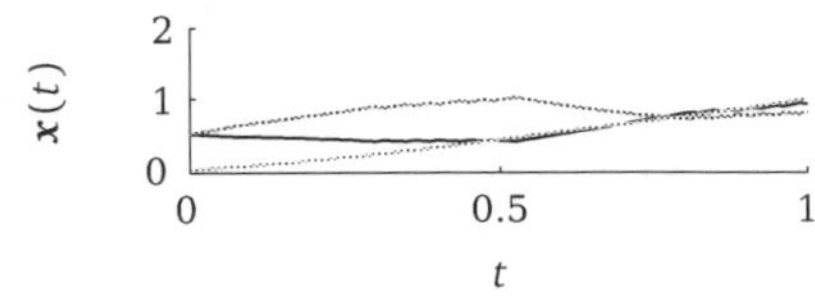

(e) SUR-0.5 state trajectory $\boldsymbol{x}(\cdot)$ for $m = 80$.

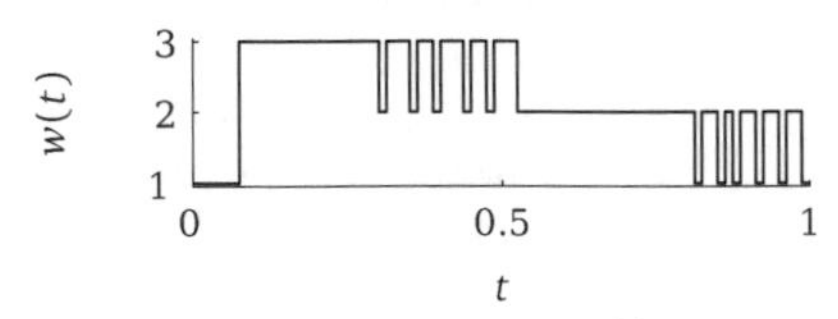

(f) SUR-0.5 control trajectory $\boldsymbol{\omega}(\cdot)$ for $m = 80$. Control has 23 switches.

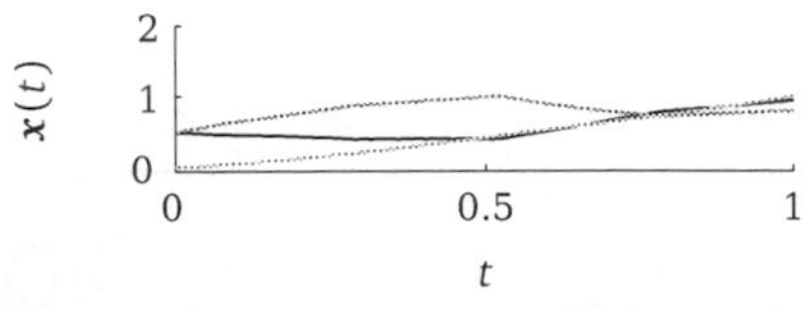

(g) SUR-0.5 state trajectory $\boldsymbol{x}(\cdot)$ for $m = 160$.

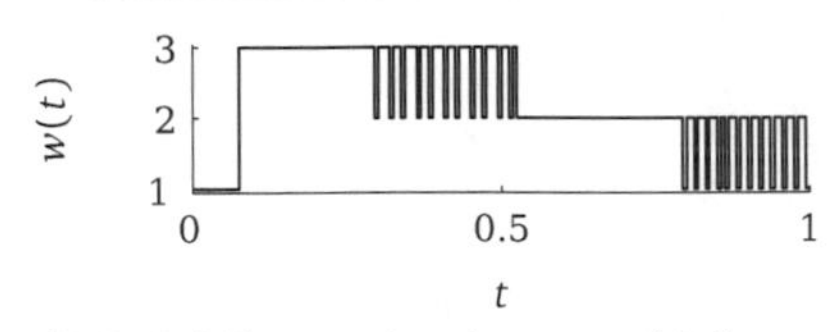

(h) SUR-0.5 control trajectory $\boldsymbol{\omega}(\cdot)$ for $m = 160$. Control has 47 switches.

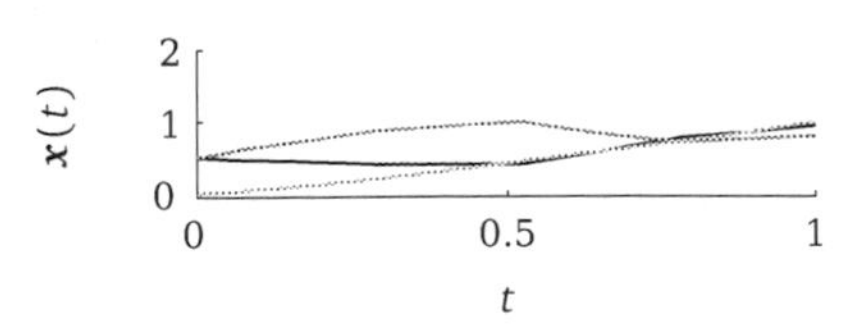

(i) SUR-0.5 state trajectory $\boldsymbol{x}(\cdot)$ for $m = 320$.

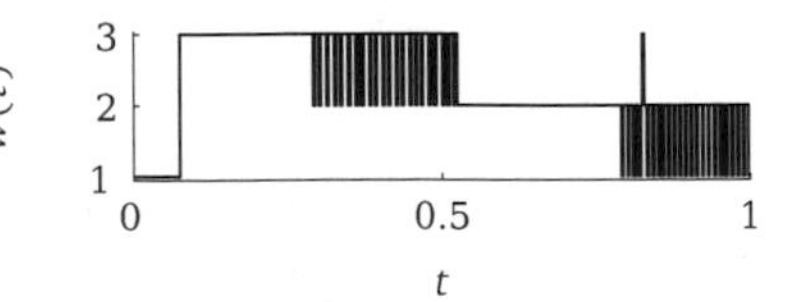

(j) SUR-0.5 control trajectory $\boldsymbol{\omega}(\cdot)$ for $m = 320$. Control has 93 switches.

Figure 9.2: Sum–up rounding solutions of problem (9.2) for increasingly fine control discretizations $m$ = 20, 40, 80, 160, 320. Sum–up rounding state trajectories are colored as $x_1(t)$ (—), $x_2(t)$ (—), and $x_3(t)$ ( ). Chattering between $w(t) = 2$ and $w(t) = 3$ occurs on the path constrained arc and chattering between $w(t) = 1$ and $w(t) = 2$ on the singular arc. The integer control trajectory $w(t) = \sum_{i=1}^{3} i\alpha_i(t)$ is shown in the right column.

the control, the state trajectory $x_1(t)$ violates the path constraint $x_1(t) \geqslant 0.4$ on the path–constrained arc.

The solutions to the partially convexified and relaxed problem (9.2) are also depicted in figure 9.1 for two choices $m = 20$ and $m = 320$ of the control trajectory. Figure 9.2 shows the sum–up rounding solutions corresponding to table 9.1. On the path–constrained arc and the singular arc, sum–up rounding leads to chattering of the integer control trajectory.

### 9.1.3 Switch Cost Formulation

Being interested in solutions to problem (9.1) that avoid frequent switchs of the integer control trajectory $\boldsymbol{w}(t)$, we address the chattering behavior of the solutions to problem (9.2) using our convex switch cost formulation of section 2.5. The appropriate formulation after a direct multiple shooting discretization for problem (9.2) on a horizon $\mathcal{T} \stackrel{\text{def}}{=} [0,1]$ divided into $m$ intervals $[t_i, t_{i+1}] \subset \mathcal{T}$, $0 \leqslant i \leqslant m-1$, of length $1/m$ is given in equation (9.3),

$$\min_{\boldsymbol{x},\boldsymbol{\alpha},\boldsymbol{\sigma},\boldsymbol{a}} \; x_{m,3} + \pi \sum_{i=0}^{m-1} \sum_{j=1}^{3} \sigma_{i,j} \tag{9.3a}$$

$$\begin{aligned}
\text{s.t.} \quad & \boldsymbol{0} = \boldsymbol{x}_i(t_{i+1};\, \boldsymbol{x}_i, \boldsymbol{\alpha}_i) - \boldsymbol{x}_{i+1}, && 0 \leqslant i \leqslant m-1, \quad (9.3\text{b}) \\
& x_{i,1} \geqslant 0.4, && 0 \leqslant i \leqslant m, \\
& \boldsymbol{\alpha}_i \in [0,1]^3, && 0 \leqslant i \leqslant m-1, \\
& \sum_{j=1}^{3} \alpha_{ij} = 1, && 0 \leqslant i \leqslant m-1, \\
& \boldsymbol{x}_0 = \left(\tfrac{1}{2}, \tfrac{1}{2}, 0\right), && \\
& \sigma_{i,j} = a_{i,j}(\alpha_{i,j} + \alpha_{i+1,j}) && 0 \leqslant i \leqslant m-1, \\
& \qquad + (1 - a_{i,j})(2 - \alpha_{i,j} - \alpha_{i+1,j}), && 1 \leqslant j \leqslant 3, \quad (9.3\text{c}) \\
& \sigma_{\max} \geqslant \sum_{i=0}^{m-1} \sum_{j=1}^{3} \sigma_{i,j}, && \quad (9.3\text{d}) \\
& \boldsymbol{a}_i \in [0,1]^3, && 0 \leqslant i \leqslant m-1.
\end{aligned}$$

Therein, the ODE representation of this switched nonlinear system is hidden in the matching condition notation (9.3b). Changes in one of the three relaxed convex multipliers $\alpha_j$ at time $t_i \in \mathcal{T}$ is indicated by $\sigma_{i,j} > 0$ according to our convex switch cost formulation (9.3c). The accumulated sum of switch costs is penalized by a penalty factor $\pi > 0$ in the objective (9.3a) and constrained by an admissible maximum $\sigma_{\max} > 0$ in (9.3d). Obviously the penalty factor $\pi$ may be chosen as zero or the admissible maximum $\sigma_{\max}$ may be chosen sufficiently large if it is desired to remove either of the two possibilities from the formulation.

### 9.1.4 Switch Cost Penalizing and Constraining Solutions

In its convexified switch cost formulation, problem (9.3) is solved using a discretization of $m = 100$ control intervals and for various penalizations $\pi = 2^p$, $-10 \leqslant p \leqslant 3$, and switch cost constraints $8 \leqslant \sigma_{\max} \leqslant 12$. Objective function values, the number of switches, and the number of fractional convex relaxed control multipliers $\alpha_{i,j}$ are listed in table 9.3 for the penalization and in table 9.2 for the constraining of switches. For comparison, the sum–up rounding solution obtained for (9.2) and $m = 100$ yields an objective function value of 0.956409 and switches 29 times.

The penalization approach (9.3) to reduce the total number of switches yields integer feasible results, consequentially with an integer number of switches, for $\pi \geqslant 2^{-7}$. For smaller penalizations, fractional relaxed results are obtained and integer feasible ones are computed by sum–up rounding. The number of switches coincides with the rounded–up number of switches computed for the relaxed solution. As has already been observed in [183] for the case of relaxations of MIOCPs, the penalization approach does not allow immediate conclusions on the relation between the penalization $\pi$ and the resulting optimal solution to be established. In particular, heavy penalizations $p \geqslant 1$ attracted solutions switching more frequently than those obtained for $2^{-4} \leqslant p \leqslant 2^{-1}$.

Numerical results obtained using the constraining approach enforcing a strict upper bound $\sigma_{\max}$ on the total number of switches are listed in table 9.2 for $9 \leqslant \sigma_{\max} \leqslant 12$. For tighter constraints, our Sequential Quadratic Programming (SQP) method failed to converge due to infeasible Quadratic Program (QP) subproblems. Ill–conditioning of the L–BFGS approximation of the Hessian of the Lagrangian is one reason for this behavior. Approximating the curvature of the convex switch cost formulation (9.3c) using a secant update accumulates ill–conditioned information similar to the situation investigated in chapter 5 for the Jacobian of complementary inequalities. We may expect significantly improved convergence behavior of our convex switch cost formulation constraining the total number of switches if an exact Hessian SQP method is used.

An overview over all computed switch cost penalizing or constraining solutions to problem (9.3) is shown in table 9.3. For example using a penalization of $p = 2^{-8}$ for the total switch cost, the number of switches is reduced to 5 or 17% of the original 29 switches, at the cost of increasing the objective function value $x_2(t_f)$ to 1.124183 or 118% of the original value 0.956409.

### 9.1.5 Summary

In this section we presented a MIOCP formulation for a nonlinear switched dynamic system which in its outer convexification reformulation coincides with a problem due to [60]. Its relaxed solution featured bang–bang arcs, a path–constrained arc, and a singular arc. Sum–up rounding on increasingly fine control discretizations has been investigated and shown to approximate the relaxed solution's objective as expected from theorem 2.4. The obtained integer feasible control trajectories showed chattering behavior on the path–constrained and the singular arc. The convex switch cost formulation proposed in chapter 2 was used to compute relaxed solutions to the (partially) convexified reformulation that penalize or constrain switches of the control trajectory. With both approaches we obtained integer feasible control trajectories with only a limited number of switches and analyzed the increase in the objective function's value. Future work includes the computation of tighter switch cost constraining solutions using an exact Hessian SQP method avoiding ill–conditioned secant updates.

| Constraint | Convexified relaxed | | | Sum–up rounding | | |
|---|---|---|---|---|---|---|
| $\sigma_{\max} =$ | Objective | # Sw. | # Frac. | Objective | Infeasibility | # Sw. |
| 9 | 1.025428 | 9 | 5 | 1.043467 | 0 | 10 |
| 10 | 1.020497 | 10 | 9 | 1.017587 | $1.80 \cdot 10^{-4}$ | 11 |
| 11 | 1.015846 | 11 | 10 | 1.017741 | $7.00 \cdot 10^{-5}$ | 11 |
| 12 | 1.011306 | 12 | 13 | 1.007177 | $3.27 \cdot 10^{-4}$ | 13 |

Table 9.2: Solutions constraining the number of switches found for problem (9.3) and $m = 100$ using our convex switch cost formulation.

| Penalty $\pi$ | Convexified relaxed | | | Sum–up rounding | | |
|---|---|---|---|---|---|---|
| $= 2^p$, $p =$ | Objective | # Sw. | # Frac. | Objective | Infeasibility | # Sw. |
| 3 | 1.337647 | 3 | 0 | dto. | – | dto. |
| 2 | 1.385266 | 3 | 0 | dto. | – | dto. |
| 1 | 1.232517 | 2 | 0 | dto. | – | dto. |
| 0 | 1.456469 | 2 | 0 | dto. | – | dto. |
| −1 | 1.693815 | 1 | 0 | dto. | – | dto. |
| −2 | 1.813649 | 1 | 0 | dto. | – | dto. |
| −3 | 1.847532 | 1 | 0 | dto. | – | dto. |
| −4 | 1.813649 | 1 | 0 | dto. | – | dto. |
| −5 | 1.456299 | 2 | 0 | dto. | – | dto. |
| −6 | 1.362813 | 2 | 0 | dto. | – | dto. |
| −7 | 1.310834 | 3 | 0 | dto. | – | dto. |
| −8 | 1.129062 | 5.50 | 2 | 1.124183 | $1.36 \cdot 10^{-4}$ | 5 |
| −9 | 1.021471 | 8.51 | 6 | 1.020104 | $2.17 \cdot 10^{-4}$ | 9 |
| −10 | 1.004172 | 9.78 | 7 | 1.003216 | $3.29 \cdot 10^{-4}$ | 10 |

Table 9.3: Solutions penalizing the number of switches found for problem (9.3) and $m = 100$ using our convex switch cost formulation.

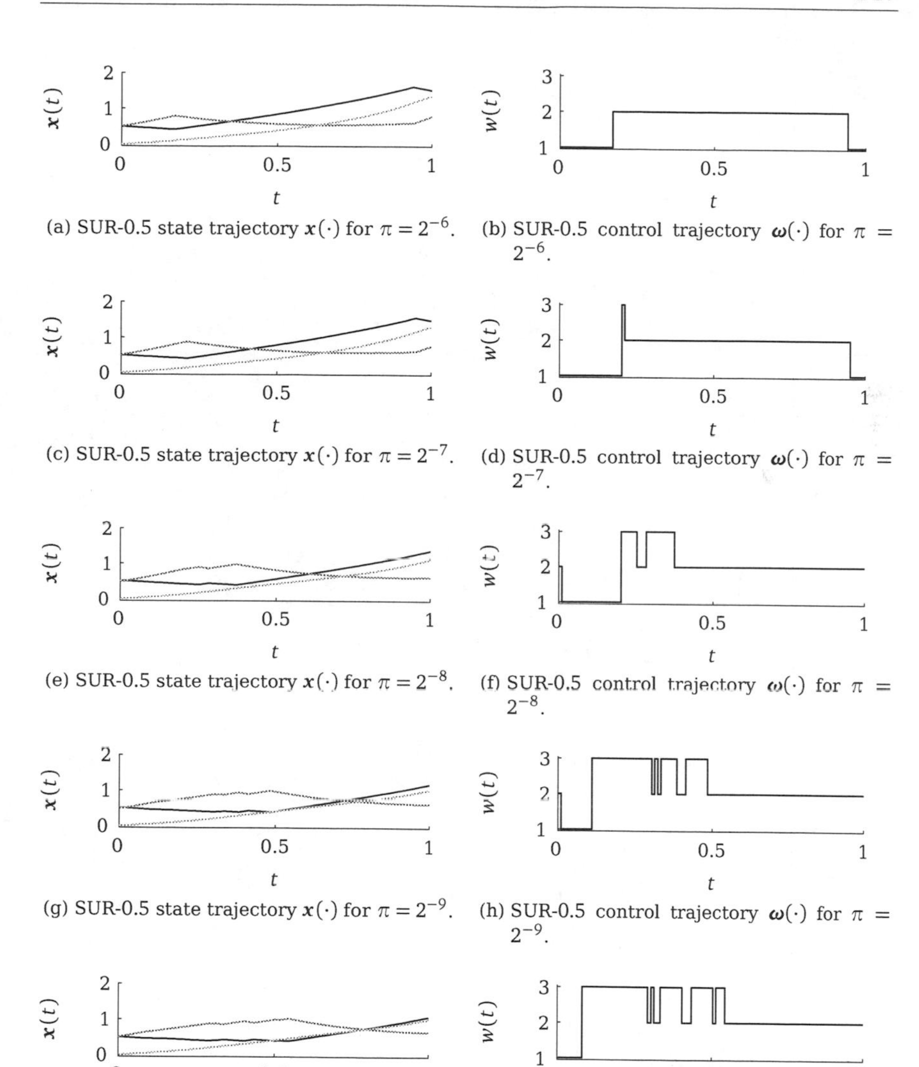


(a) SUR-0.5 state trajectory $\boldsymbol{x}(\cdot)$ for $\pi = 2^{-6}$.

(b) SUR-0.5 control trajectory $\boldsymbol{\omega}(\cdot)$ for $\pi = 2^{-6}$.

(c) SUR-0.5 state trajectory $\boldsymbol{x}(\cdot)$ for $\pi = 2^{-7}$.

(d) SUR-0.5 control trajectory $\boldsymbol{\omega}(\cdot)$ for $\pi = 2^{-7}$.

(e) SUR-0.5 state trajectory $\boldsymbol{x}(\cdot)$ for $\pi = 2^{-8}$.

(f) SUR-0.5 control trajectory $\boldsymbol{\omega}(\cdot)$ for $\pi = 2^{-8}$.

(g) SUR-0.5 state trajectory $\boldsymbol{x}(\cdot)$ for $\pi = 2^{-9}$.

(h) SUR-0.5 control trajectory $\boldsymbol{\omega}(\cdot)$ for $\pi = 2^{-9}$.

(i) SUR-0.5 state trajectory $\boldsymbol{x}(\cdot)$ for $\pi = 2^{-10}$.

(j) SUR-0.5 control trajectory $\boldsymbol{\omega}(\cdot)$ for $\pi = 2^{-10}$.

Figure 9.3: Selected sum–up rounding solutions of problem (9.3) penalizing the total switch cost for $m = 100$ and penalties of $\pi = 2^{-6}$ to $\pi = 2^{-10}$ on the switch cost of the convexified relaxed problem. Sum–up rounding state trajectories are colored as $x_1(t)$ (—), $x_2(t)$ (—), and $x_3(t)$ (—).

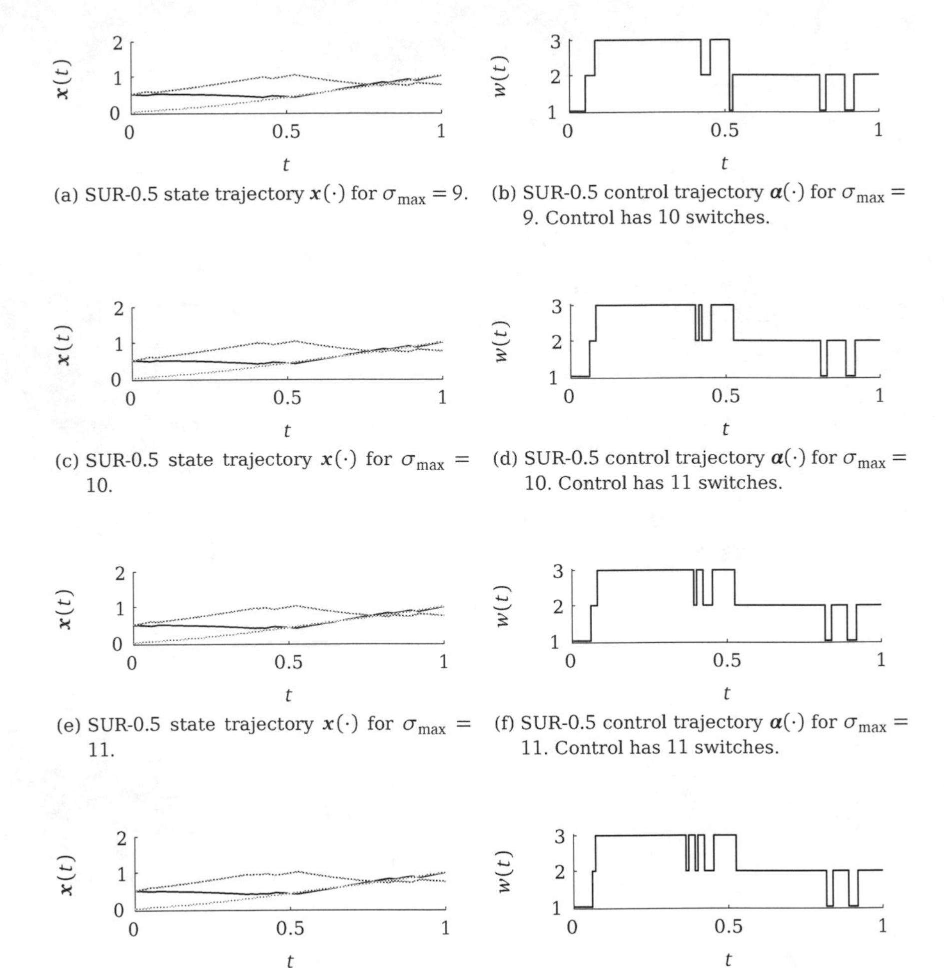


(a) SUR-0.5 state trajectory $\boldsymbol{x}(\cdot)$ for $\sigma_{\max} = 9$.

(b) SUR-0.5 control trajectory $\boldsymbol{\alpha}(\cdot)$ for $\sigma_{\max} = 9$. Control has 10 switches.

(c) SUR-0.5 state trajectory $\boldsymbol{x}(\cdot)$ for $\sigma_{\max} = 10$.

(d) SUR-0.5 control trajectory $\boldsymbol{\alpha}(\cdot)$ for $\sigma_{\max} = 10$. Control has 11 switches.

(e) SUR-0.5 state trajectory $\boldsymbol{x}(\cdot)$ for $\sigma_{\max} = 11$.

(f) SUR-0.5 control trajectory $\boldsymbol{\alpha}(\cdot)$ for $\sigma_{\max} = 11$. Control has 11 switches.

(g) SUR-0.5 state trajectory $\boldsymbol{x}(\cdot)$ for $\sigma_{\max} = 12$.

(h) SUR-0.5 control trajectory $\boldsymbol{\alpha}(\cdot)$ for $\sigma_{\max} = 12$. Control has 13 switches.

Figure 9.4: Selected sum–up rounding solutions of problem (9.3) constraining the total number of switches for $m = 100$ and constraints $\sigma_{\max} = 9$ to $\sigma_{\max} = 12$ on the number of switches of the convexified relaxed problem's solution. If the relaxed solutions are not integer feasible, sum–up rounding solutions may switch slightly more often. Sum–up rounding state trajectories are colored as $x_1(t)$ (—), $x_2(t)$ (……), and $x_3(t)$ (……).

## 9.2 Mixed–Integer NMPC Scheme Contractivity

In this section we study the application of the contractivity estimates for the mixed–integer real–time iteration scheme of chapter 4 at the example of a small but nonlinear and instable problem due to [51]. We demonstrate again the outer convexification reformulation and the sum–up rounding approximation. We derive certain bounds, Lipschitz constants, and contractivity constants for this example. The mixed–integer real–time iteration contractivity theorem and the resulting sampling time bound are evaluated and numerical computations are performed for its verification.

### 9.2.1 Problem Formulation

We consider this problem in a mixed–integer variant (9.4) in which the control $w(t)$ is allowed to assume one of the discrete choices $\{-1, 0, 1\}$ only,

$$\begin{aligned} \min_{x(\cdot), w(\cdot)} \quad & \tfrac{1}{2} \int_0^3 x^2(t) + w^2(t) \, \mathrm{d}t \\ \text{s.t.} \quad & \dot{x}(t) = (1 + x(t))\, x(t) + w(t), \quad && t \in [0, 3], \\ & x(0) = x_{\text{meas}}, \\ & x(3) = 0, \\ & x(t) \in [-1, 1], && t \in [0, 3], \\ & w(t) \in \{-1, 0, 1\}, && t \in [0, 3]. \end{aligned} \tag{9.4}$$

Given an estimated or observed initial state $x_{\text{meas}}$ of the process $x(t)$ on $0\text{s} \leqslant t \leqslant 3\text{s}$ the purpose of the formulation (9.4) is to steer $x(\cdot)$ back into its steady–state $x(\cdot) = 0$ by applying an additive control $w(\cdot)$. The optimal control trajectory achieving this for an initial value of $x(0) = 0.05$ is depicted in figure 9.5 on page 250.

The steady–state is instable, i.e., any slight deviation $\varepsilon > 0$ will trigger a run–away of the process. This is depicted in figure 9.6 for an initial value of $x(0) = 0.05$ and no control applied. The process can be steered back into its steady state as long as

$$\dot{x}(t) = (1 + x(t))x(t) + w(t) \begin{cases} \leqslant 0 & \text{if } x(t) > 0, \\ \geqslant 0 & \text{if } x(t) < 0, \end{cases} \tag{9.5}$$

can be satisfied within the bounds of $w(t)$. The applicable control $w(\cdot)$ is bounded by $-1$ and $1$, hence the steady state can be reached only for

$$\begin{aligned} & |(1+x(t))x(t)| < 1 \\ \iff \quad & x(t) \in \left[-\tfrac{1}{2} - \tfrac{\sqrt{5}}{2}, -\tfrac{1}{2} + \tfrac{\sqrt{5}}{2}\right] \approx [-1.618, 0.618]. \end{aligned}$$

In the real–time iteration scheme, the first preparation and feedback phase pass without an appropriate control being available for feedback to

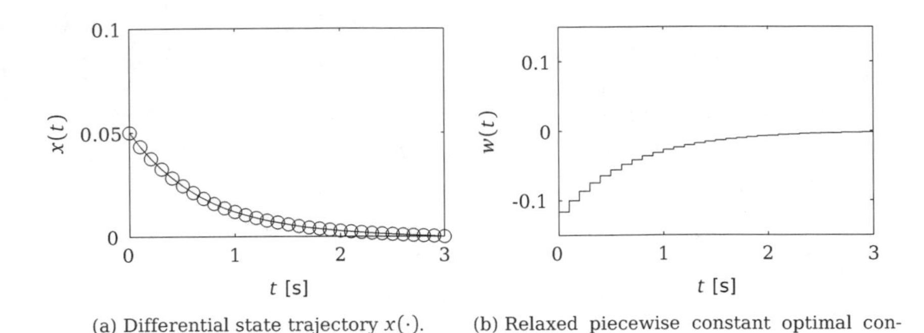


(a) Differential state trajectory $x(\cdot)$.

(b) Relaxed piecewise constant optimal control trajectory $w(\cdot)$.

Figure 9.5: Relaxed optimal solution for problem (9.4), steering the process back from an initial value $x(0) = 0.05$ into its unstable steady state.

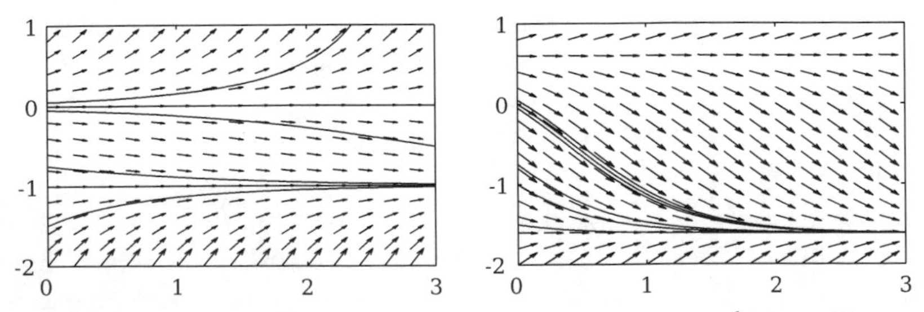

(a) Instable steady state $x(t) = 0$ and the stable one $x(t) = -1$; no control applied

(b) States below $x(t) = \frac{1}{2}(-1 + \sqrt{5})$ can be steered back to the steady state using $u(t) \geqslant -1$.

Figure 9.6: Vector fields showing stable and instable steady state of problem (9.4).

the process. In this case, the runaway behavior further reduces the set of admissible initial values that can be steered back to the instable steady state.

We apply the outer convexification reformulation to the objective function and the dynamics of problem (9.4) that depend on the integer control $w(\cdot)$. In problem (9.6) three convex control multipliers named $w_{-1}(\cdot)$, $w_0(\cdot)$ and $w_1(\cdot)$ are introduced for the admissible integer choices of $w(\cdot)$ as follows,

$$\begin{aligned}
\min_{x(\cdot),\boldsymbol{w}(\cdot)} \quad & \frac{1}{2}\int_0^3 x^2(t)+w_{-1}(t)+w_1(t)\,\mathrm{d}t && (9.6)\\
\text{s.t.} \quad & \dot{x}(t)=(1+x(t))\,x(t)-w_{-1}(t)+w_1(t), && t\in[0,3],\\
& x(0)=x_{\text{meas}}, \\
& x(3)=0, \\
& x(t)\in[-1,1], && t\in[0,3],\\
& \boldsymbol{w}(t)\in\{0,1\}^3, && t\in[0,3],\\
& \sum_{i=-1}^{1} w_i(t)=1, && t\in[0,3].
\end{aligned}$$

The convex multiplier $w_0(t)$ lends itself to straightforward elimination, making the SOS–1 constraint an inequality constraint.

As the steady state is inherently instable, we cannot expect an integer control in $\{-1,0,1\}$ applied on a fixed discretization grid to steer the process back to its steady state. We instead desire the process to be kept in a close neighborhood of the steady state and expect that frequent integer control feedback needs to be applied to counteract the runaway behavior. This is shown in table 9.4 and figure 9.7 again for an initial value of $x(0)=0.05$ and choices of increasingly fine equidistant control discretizations. The integer feasible SUR–0.5 solution succeeds in approximating the solution of the relaxed convexified problem (9.6), while the relaxed nonlinear one shows a much lower objective, as expected from theorem 2.4. Note that for $m=320$ the sum–up rounding objective is actually better than that of the convexified relaxed problem, but at the price of violating the end–point constraint. For all relaxed convexified solutions, one of the $m$ relaxed controls is fractional as the optimal time for switching from $-1$ to 0 does not coincide with the control discretization grid point times. This could be further improved by a switching time optimization, see e.g. [122, 183], in which the shooting interval durations are subject to optimization while the controls are fixed to the SUR-0.5 solution values.

| $m$ | Relaxed Non-linear Objective | Relaxed Convexified Objective | RC and SUR–0.5 | |
|---|---|---|---|---|
| | | | Objective | Infeasibility |
| 20 | $3.1952 \cdot 10^{-3}$ | $2.7054 \cdot 10^{-2}$ | $2.3484 \cdot 10^{+1}$ | $1.7060 \cdot 10^{+1}$ |
| 40 | $3.1397 \cdot 10^{-3}$ | $2.6014 \cdot 10^{-2}$ | $6.7006 \cdot 10^{-2}$ | $3.1365 \cdot 10^{-1}$ |
| 80 | $3.1140 \cdot 10^{-3}$ | $2.5774 \cdot 10^{-2}$ | $4.3250 \cdot 10^{-2}$ | $3.5548 \cdot 10^{-1}$ |
| 160 | $3.1018 \cdot 10^{-3}$ | $2.5708 \cdot 10^{-2}$ | $3.0076 \cdot 10^{-2}$ | $8.5880 \cdot 10^{-2}$ |
| 320 | $3.0958 \cdot 10^{-3}$ | $2.5696 \cdot 10^{-2}$ | $2.5471 \cdot 10^{-2}$ | $9.2865 \cdot 10^{-2}$ |
| 640 | $3.0928 \cdot 10^{-3}$ | $2.5691 \cdot 10^{-2}$ | $2.5810 \cdot 10^{-2}$ | $4.2590 \cdot 10^{-3}$ |
| 1280 | $3.0913 \cdot 10^{-3}$ | $2.5691 \cdot 10^{-2}$ | $2.5809 \cdot 10^{-2}$ | $4.2590 \cdot 10^{-3}$ |

Table 9.4: Objective functions and 1–norms of infeasibilities of the solutions to problem (9.6) shown in figure 9.7. With increasingly finer granularity of the control discretization, the SUR-0.5 solution succeeds in approaching feasibility of the end point constraint and approximating the objective of the convexified relaxed problem. The relaxed convexified solution showed one fractional control for all investigated choices of $m$.

### 9.2.2 Constants

Problem (9.4) is simple enough to permit carrying out an explicit derivation of certain Lipschitz constants and bounds that will be required in the following to demonstrate the contractivity estimate developed for our mixed–integer real–time iteration scheme.

- The bound on the ODE system's right hand side after convexification $\tilde{f}(t,x(t)) = x^2(t) + x(t)$ on the domain $(t,x) \in \mathcal{D}(x_0) \stackrel{\text{def}}{=} [0,3] \times [-1,x_0]$ depending on the parameter $x_0 > -1$ is

$$\sup_{\mathcal{D}} |x^2 + x| = |x_0^2 + x_0| \stackrel{\text{def}}{=} b_{\tilde{f}}(x_0), \qquad b_{\tilde{f}}(-\tfrac{1}{2} + \tfrac{1}{2}\sqrt{5}) = 1. \tag{9.7}$$

- The local Lipschitz constant on the ODE system's right hand side after convexification $\tilde{f}(t,x(t))$ on $\mathcal{D}(x_0)$ is

$$\sup_{\mathcal{D}} |2x + 1| = |2x_0 + 1| \stackrel{\text{def}}{=} A_{\tilde{f}}(x_0), \qquad A_{\tilde{f}}(-\tfrac{1}{2} + \tfrac{1}{2}\sqrt{5}) = \sqrt{5}. \tag{9.8}$$

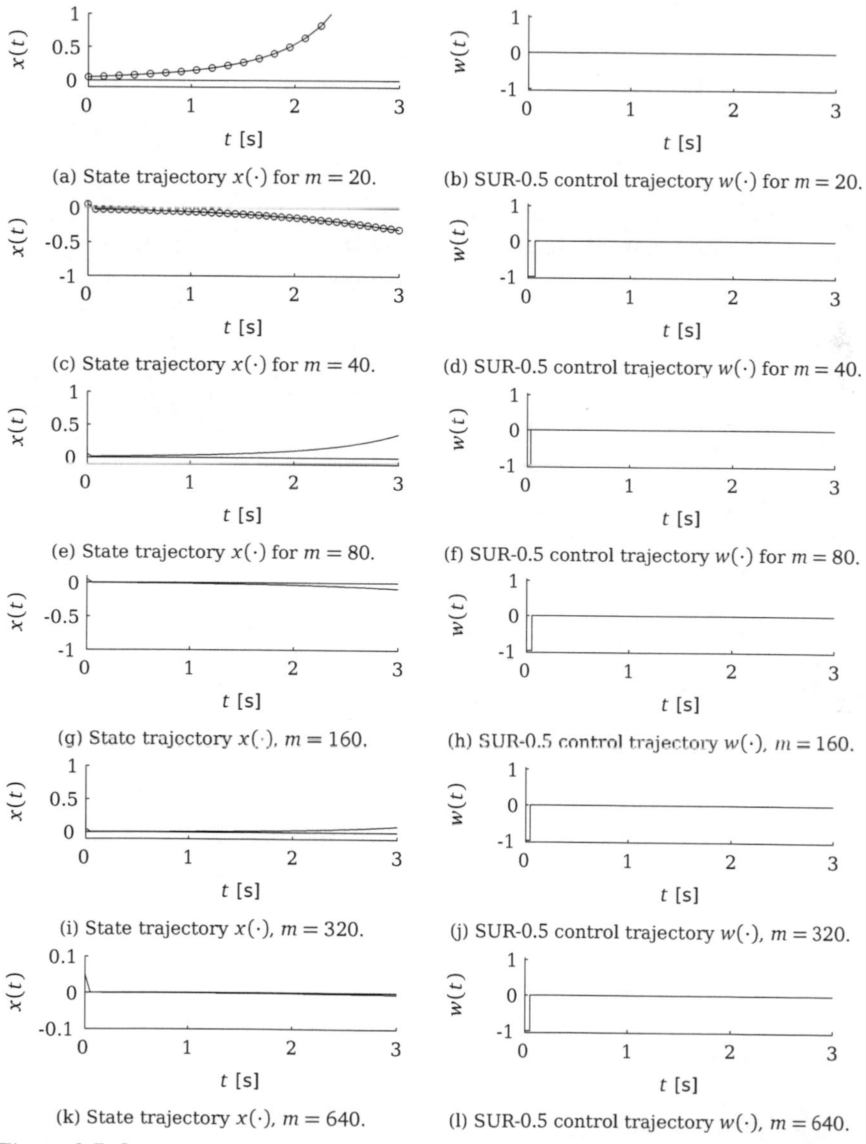


(a) State trajectory $x(\cdot)$ for $m = 20$.

(b) SUR-0.5 control trajectory $w(\cdot)$ for $m = 20$.

(c) State trajectory $x(\cdot)$ for $m = 40$.

(d) SUR-0.5 control trajectory $w(\cdot)$ for $m = 40$.

(e) State trajectory $x(\cdot)$ for $m = 80$.

(f) SUR-0.5 control trajectory $w(\cdot)$ for $m = 80$.

(g) State trajectory $x(\cdot)$, $m = 160$.

(h) SUR-0.5 control trajectory $w(\cdot)$, $m = 160$.

(i) State trajectory $x(\cdot)$, $m = 320$.

(j) SUR-0.5 control trajectory $w(\cdot)$, $m = 320$.

(k) State trajectory $x(\cdot)$, $m = 640$.

(l) SUR-0.5 control trajectory $w(\cdot)$, $m = 640$.

Figure 9.7: Sum–up rounding integer solution for problem (9.6) attempting to keep the process in a neighborhood of its instable steady state, starting from an initial value of $x(0) = 0.05$ Solutions are shown for different granularities $m = 20, 40, 80, 160, 320, 640$ of the control discretization.

- The two contraction constants $\kappa$ and $\omega$ of theorem 4.3 and the bound $\beta$ on the norm of $\tilde{\boldsymbol{M}}^{k+1}(\boldsymbol{y}^{k+1})$ are not easily derived explicitly. We instead compute underestimates after every SQP iteration of the classical real–time iteration scheme on the problem (9.6) using definitions 3.12 and 3.13. The maximal underestimators obtained over 100 real–time iterations for various initial values and control discretizations are shown in figure 9.8 from which we determine $\kappa \leqslant 0.34$ and $\omega \leqslant 0.56$. The norm bound $\beta$ depends on $m$ and we find $\beta = 26.67$ for $m = 20$.

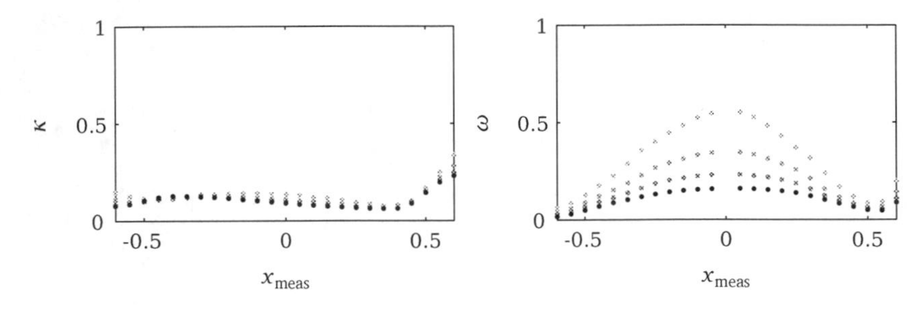


(a) Estimates of contraction constant $\kappa$. (b) Estimates of contraction constant $\omega$.

Figure 9.8: Maximal underestimators of the contraction constants $\kappa$ and $\omega$ according to definitions 3.12 and 3.13 for the convexified example (9.6), computed for various initial values $x_{\text{meas}}$ and control discretizations $m = 20$ ($\cdots$), $m = 40$ ($\cdots$), $m = 80$ ($\cdots$), and $m = 160$ ($\cdots$).

It is worth noting that figure 9.8 reveals a frequently observed advantage of direct multiple shooting. As the number $m$ of introduced multiple shooting nodes increases, the nonlinearity of the boundary value problem estimated by $\omega$ decreases, which helps to improve the speed of convergence. For a special case proof of this property see [6].

### 9.2.3 Sampling Time Estimate

In figure 9.9 primal state and control component and dual matching condition Lagrange multiplier components of the Newton step modification $\tilde{\boldsymbol{M}}^{k+1}(\boldsymbol{y}^{k+1})\boldsymbol{J}^k(\boldsymbol{y}^k)\boldsymbol{e}^k(\boldsymbol{y}^k)$ can be seen for the worst–case choice $\boldsymbol{e}_k^{\mathrm{q}} = 1$ and increasingly fine control discretizations, i.e., shorter sampling times $\delta t$. Clearly, the primal component $s_{k+1}$ and the dual one $\lambda_k$ belonging to the matching condition coupling to shooting node $k+1$ dominate the step modification.

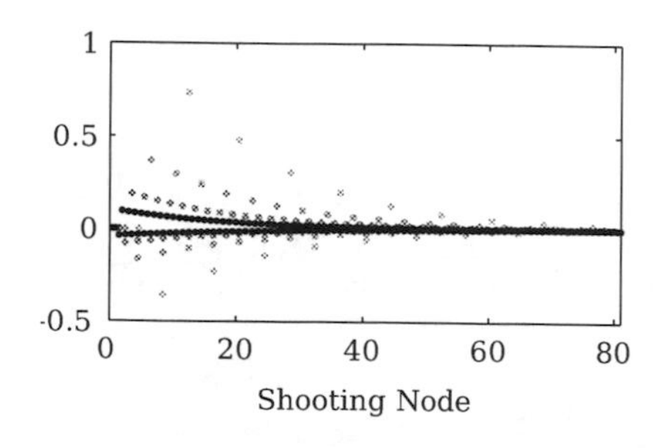


(a) Primal part; state and control components alternate.

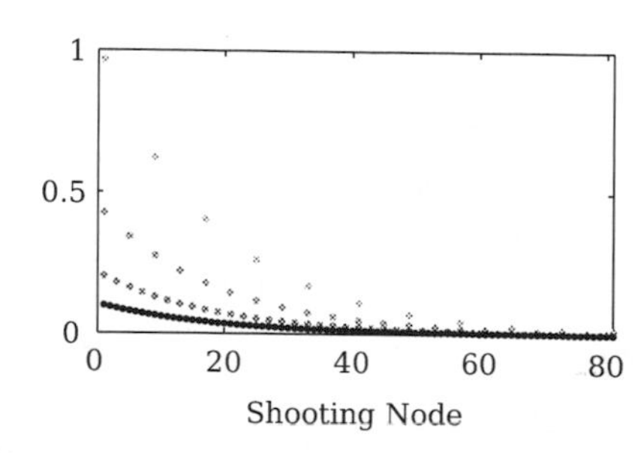


(b) Matching condition Lagrange multiplier part.

Figure 9.9: State and control part and matching condition Lagrange multiplier part of the term $\tilde{M}^{k+1}(y^{k+1})J^k(y^k)e^k(y^k)$ for the first mixed–integer real–time iteration of the example problem (9.4). The modification $e_k^q = 1$ and $m = 10$ ($\cdots$), $m = 20$ ($\cdots$), $m = 40$ ($\cdots$), $m = 80$ ($\cdots$) shooting intervals were chosen. The components of the additional vector term arising in the $\kappa$–condition due to the inexactness introduced to cover rounding of the control approach zero as the length of the first shooting interval decreases.

With the determined constants, the estimate provided by theorem 4.7 in chapter 4 yields an upper bound on $\delta t$ determined by the solution of

$$\beta b_{\tilde{f}} \delta t \exp(A_{\tilde{f}} \delta t) < \tfrac{2}{\omega}(1-\kappa)^2, \tag{9.9}$$

for $\delta t$, which is not analytically available. For $m = 20$ shooting intervals and using the worst–case constants at hand, i.e., $\kappa = 0.34$, $\omega = 0.56$, $\beta = 26.67$, and if choosing $x(t) \leqslant -\frac{1}{2} + \frac{1}{2}\sqrt{5}$ for all $t$ which leads to $b_{\tilde{f}} = 1$ and $A_{\tilde{d}} = \sqrt{5}$, this estimate evaluates to $\delta t < 0.052$s. Hence we may expect the mixed–integer real–time iteration scheme to successfully control the process for a choice of $m = 20$ and e.g. $\delta t = 0.05$. Figure 9.11 shows the trajectory and controls produced by 2000 seconds of mixed–integer real–time iterations with $\delta t = 0.05$, i.e., 40,000 iterations. The process is successfully kept in a small neighborhood of the instable steady state $x = 0$ (figure 9.10a) with a maximum deviation of about 0.03 (figure 9.10b). The computed integer control trajectory chatters between the three admissible values (figure

9.10d). A discrete Fourier transformation of the control trajectory (figure 9.10c) clearly reveals periodicity of the obtained control.

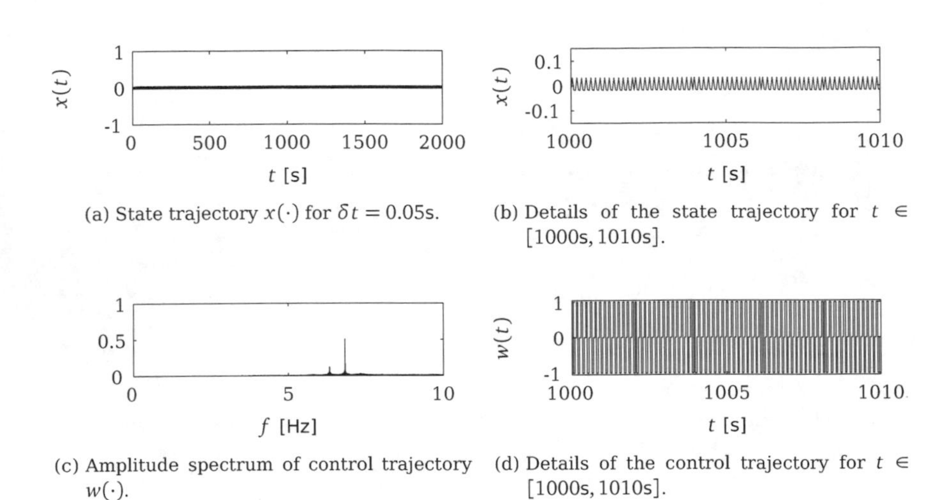


(a) State trajectory $x(\cdot)$ for $\delta t = 0.05$s.

(b) Details of the state trajectory for $t \in [1000\text{s}, 1010\text{s}]$.

(c) Amplitude spectrum of control trajectory $w(\cdot)$.

(d) Details of the control trajectory for $t \in [1000\text{s}, 1010\text{s}]$.

Figure 9.10: 2000 seconds of successfully contracting mixed–integer real–time iterations that keep the process (9.6) in a neighborhood of its instable steady state, starting from an initial value of $x(0) = 0.05$. As expected from the contractivity estimate, the mixed–integer real–time iteration scheme is able to control the process for a sampling time of $\delta t = 0.05$s. The chattering integer control exhibits two significant frequencies as shown by the amplitude spectrum obtained by discrete Fourier transformation of the computed integer control trajectory in figure 9.10c.

Note, however, that the bound $\delta t < 0.052$s is neither a necessary nor a sufficient condition and will in general not be tight, as we have underestimated the contraction constants $\kappa$ and $\omega$, overestimated the quantity $(1-\kappa)||\boldsymbol{\Delta}\tilde{\boldsymbol{y}}^k||$, and overestimated the actually applying bounds and Lipschitz constants by choice of $\mathcal{D}$. In addition, larger bounds on the sampling times $\delta t$ are obtained if more restrictive upper bounds imposed on the deviation of $x(\cdot)$ from zero can be satisfied. Figures 9.11 and 9.12 show trajectories and controls produced with gradually growing sampling times $\delta t$. It is easily seen that the mixed–integer real–time iteration scheme is able to keep the process in a close neighborhood of the instable steady state as long as the perturbation of the system state due to rounding does indeed satisfy the increasingly strict bounds imposed by the larger sampling times.

For the example (9.4, 9.6) at hand, we have chosen to track the instable steady state for which, due to the exponential runaway behavior of the process, we may expect the derived estimate to be quite close to the actual bounds on $\delta t$ that can be observed by numerical computations. Indeed we can analyze the solution obtained for the first sampling time $\delta t = 0.15$s that lead to failure of the scheme. We find from figure 9.12 that the offending perturbation of the process is $x_0 = 0.3875$, leading to the maximal underestimators $\kappa = 0.14$ and $\omega = 0.56$ for the contraction constants as seen from figure 9.8, and to $b_{\tilde{f}} = 0.54$ and $A_{\tilde{f}} = 1.78$ derived from equations (9.7) and (9.8). Indeed, for these constants our mixed–integer real–time iteration contractivity estimate yields the (certainly not tight) upper bound $\delta t < 0.140$s, indicating in agreement with our numerical observations that the chosen sampling time of 0.15s is too large to expect contractivity of the scheme if faced with a deviation of 0.3875.

### 9.2.4 Observations

Two noteworthy observations can be made about the obtained integer feedback control trajectories. First, in figures 9.10, 9.11 and 9.12 discrete Fourier transformations of the chattering control trajectories are shown in place of the actual trajectory that could not possibly be represented in print. From these figures, it can be easily seen that all stable feedback schemes generate controls that exhibit two distinct periodic chattering frequencies. In contrary, the control does not exhibit such a periodicity for all diverging schemes, as indicated by the obvious lack of a distinct peak in the amplitude spectrum.

| $x_{0,\max}$ | 0.618 | 0.5 | 0.4 | 0.3875 | 0.3 | 0.2 | 0.1 |
|---|---|---|---|---|---|---|---|
| $\kappa$ | 0.34 | 0.17 | 0.15 | 0.15 | 0.15 | 0.15 | 0.15 |
| $b_{\tilde{f}}$ | 1.00 | 0.75 | 0.56 | 0.54 | 0.39 | 0.24 | 0.11 |
| $A_{\tilde{f}}$ | 2.24 | 2.00 | 1.80 | 1.78 | 1.60 | 1.40 | 1.20 |
| $\delta t_{\max}$ in s | 0.052 | 0.101 | 0.136 | 0.140 | 0.185 | 0.275 | 0.489 |

Table 9.5: Upper bounds on $\delta t$ for example (9.6), depending on upper bounds on $x(t)$. The values $\omega = 0.56$ and $\beta = 26.67$ are independent of the bound on $x(t)$.

Second, phase diagrams of the process state and integer feedback control as shown in figures 9.13 lead us to the observation that for all stable schemes, one or more small neighborhoods of process states can be associated with each admissible integer control. The chattering feedback control trajectories could essentially be generated by a finite state machine with four states (figure 9.13a for $\delta t = 0.05\text{s}$) or three states (figures 9.13c and 9.13d for $\delta t \geqslant 0.1\text{s}$). In figure 9.13b for $\delta t = 0.075\text{s}$ an overlay of both is observed as the sampling time increases. In contrary, this phenomenon cannot be observed for any of the diverging schemes in figure 9.14. Both observations certainly merit further detailed investigations.

Figure 9.11 shows 2000 seconds of successfully contracting mixed–integer real–time iterations that keep the process (9.6) in a neighborhood of its instable steady state, starting from an initial value of $x(0) = 0.05$. The mixed–integer real–time iteration scheme is able to control the process also for the larger sampling times $\delta t = 0.075\text{s}$, $\delta t = 0.1\text{s}$, and $\delta t = 0.125\text{s}$. The maximum deviation of the process trajectory from the instable steady state $x(t) = 0$ remains bounded by $\pm 0.1$, such that the obtained estimate $\delta t < 0.052$ is overly conservative.

Figure 9.12 shows non–contracting mixed–integer real–time iterations failing to keep the process (9.6) in a neighborhood of its instable steady state, starting from an initial value of $x(0) = 0.05$. As expected from the contractivity estimate, the mixed–integer real–time iteration scheme fails for sampling times $\delta t \geqslant 0.15$. Note how the amplitude spectrum no longer exhibits any significant frequency of the erratically chattering integer control. The scheme fails to contract and runaway of the process is observed within the time bound of 2000 seconds once the respective bound seen from table 9.5 is violated. Starting with $\delta t = 0.15\text{s}$ in figure 9.12, the system's behavior becomes more and more erratic and fails to track the instable steady state.

### 9.2.5 Summary

In this section we investigated the newly developed mixed–integer real–time iteration scheme of chapter 4 at the example of a small nonlinear system due to [51] that shows finite–time blowup of the solution if left uncontrolled. This system was to be kept in a neighborhood of its instable steady state applying one of three possible integer control choices. The simplicity of the problem allowed to explicitly derive certain bounds and Lipschitz constants and we provided numerical estimates of required contractivity constants. The developed sampling time estimate was applied to the problem and the obtained sampling time of 0.05s yielded a highly periodic chattering integer control that kept the system stable over the whole of the examined period of 2000 seconds. Tighter bounds on the maximum deviation from the instable steady state led to larger admissible sampling times and corresponding solutions were presented. In accordance with the predictions made by our developed sampling time estimate, the mixed–integer real–time iteration scheme failed to control the investigated system as soon as its deviation from the instable steady state violated the boundedness assumption used to derive the sampling time estimate. Future work includes the more detailed investigation of the chattering behavior of the obtained integer feedback controls.

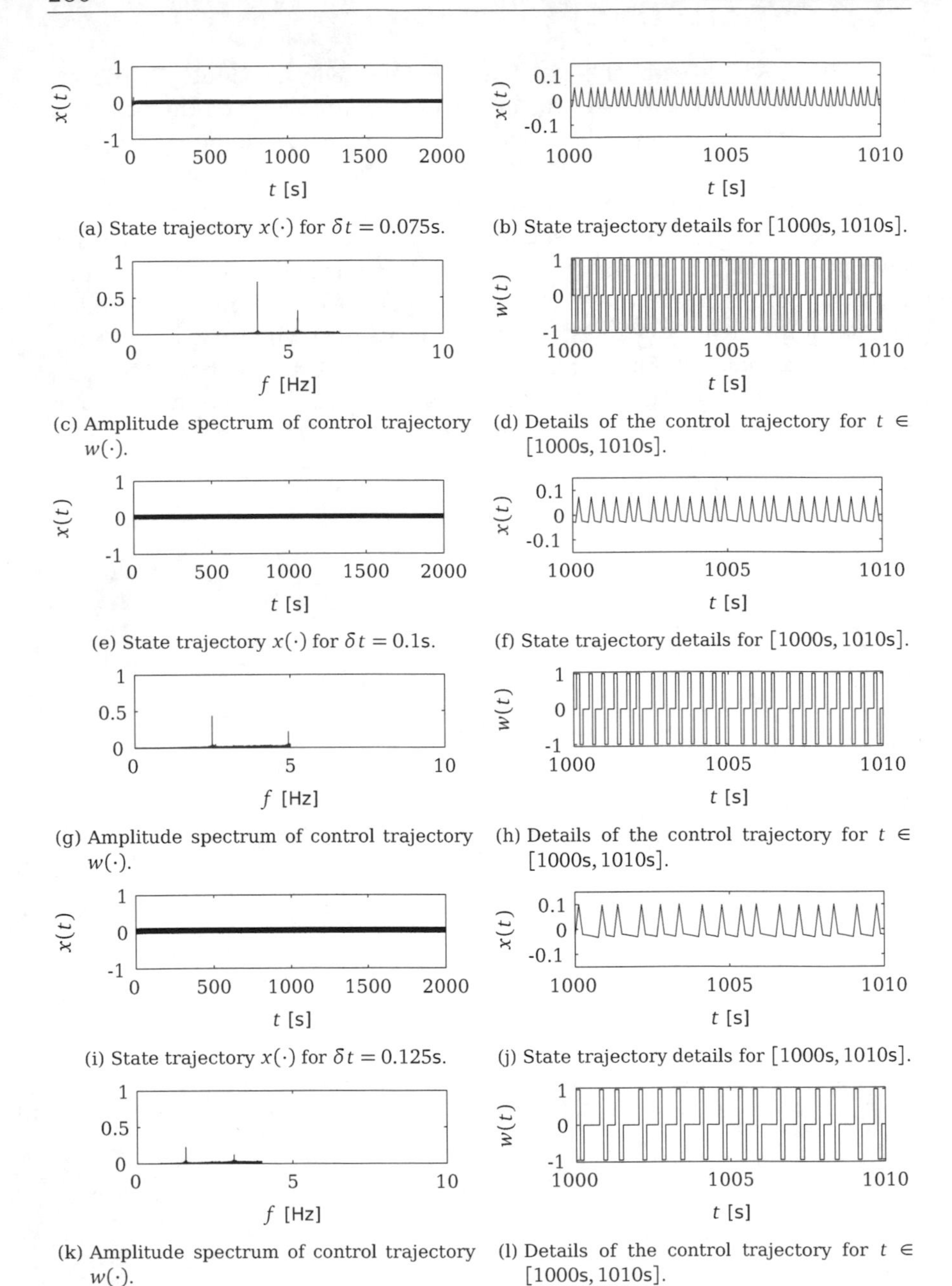


(a) State trajectory $x(\cdot)$ for $\delta t = 0.075$s.

(b) State trajectory details for $[1000\text{s}, 1010\text{s}]$.

(c) Amplitude spectrum of control trajectory $w(\cdot)$.

(d) Details of the control trajectory for $t \in [1000\text{s}, 1010\text{s}]$.

(e) State trajectory $x(\cdot)$ for $\delta t = 0.1$s.

(f) State trajectory details for $[1000\text{s}, 1010\text{s}]$.

(g) Amplitude spectrum of control trajectory $w(\cdot)$.

(h) Details of the control trajectory for $t \in [1000\text{s}, 1010\text{s}]$.

(i) State trajectory $x(\cdot)$ for $\delta t = 0.125$s.

(j) State trajectory details for $[1000\text{s}, 1010\text{s}]$.

(k) Amplitude spectrum of control trajectory $w(\cdot)$.

(l) Details of the control trajectory for $t \in [1000\text{s}, 1010\text{s}]$.

Figure 9.11: Contracting mixed–integer real–time iterations for example (9.6).

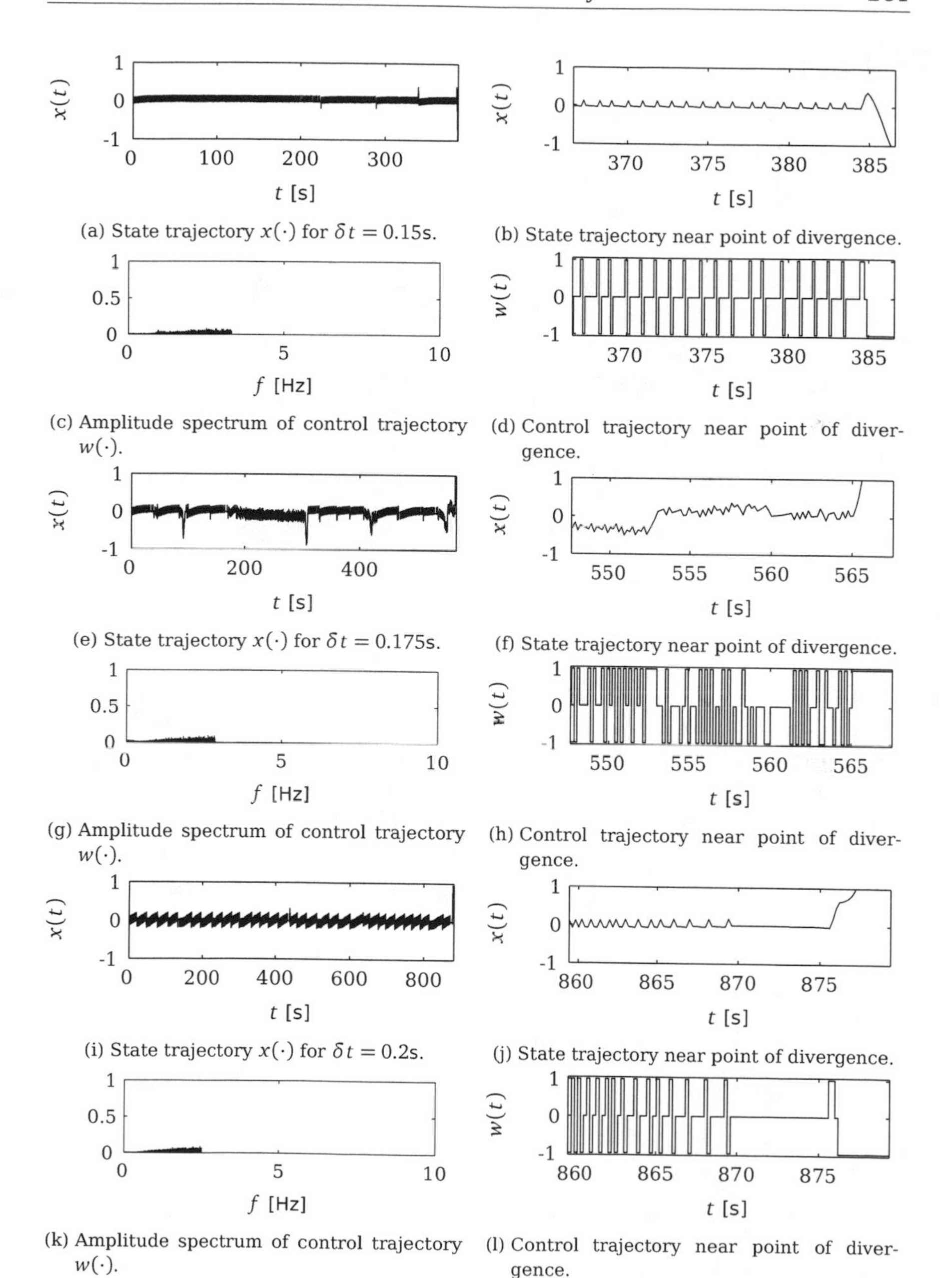


(a) State trajectory $x(\cdot)$ for $\delta t = 0.15$s.

(b) State trajectory near point of divergence.

(c) Amplitude spectrum of control trajectory $w(\cdot)$.

(d) Control trajectory near point of divergence.

(e) State trajectory $x(\cdot)$ for $\delta t = 0.175$s.

(f) State trajectory near point of divergence.

(g) Amplitude spectrum of control trajectory $w(\cdot)$.

(h) Control trajectory near point of divergence.

(i) State trajectory $x(\cdot)$ for $\delta t = 0.2$s.

(j) State trajectory near point of divergence.

(k) Amplitude spectrum of control trajectory $w(\cdot)$.

(l) Control trajectory near point of divergence.

Figure 9.12: Mixed–integer real–time iterations failing to control example (9.6).

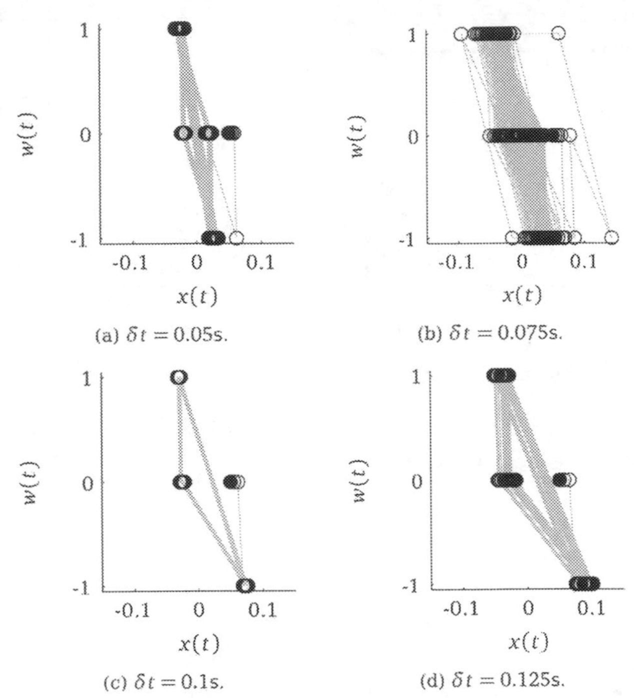


Figure 9.13: Phase space diagrams of the mixed–integer real–time iterates for gradually increasing sampling times. Choices $\delta t = 0.05$s, 0.075s, 0.1s, 0.125s for the sampling time succeed in controlling the system (9.6).

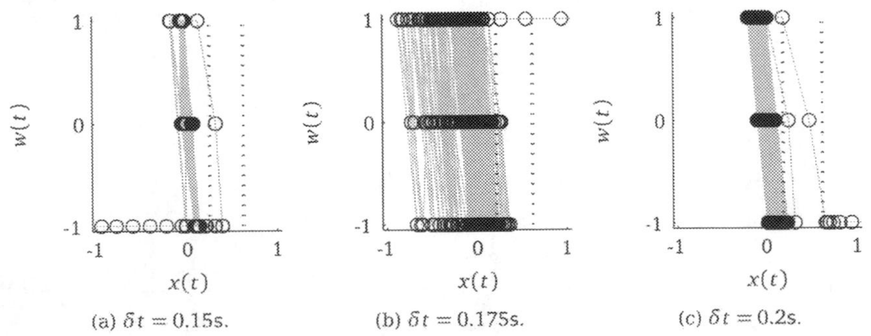


Figure 9.14: Phase space diagrams of the mixed–integer RTI for gradually increasing sampling times. Choices $\delta t = 0.15$s, 0.175s, 0.2s for the sampling time result in state trajectories that ultimately violate the estimated upper bound on $x(t)$ derived from the choice of $\delta t$ and diverge.

## 9.3 OCPs and NLPs with Vanishing Constraints

In this section we investigate an autonomous robot path–following and communication problem that has frequently been studied in the literature, see e.g. [2] and the references found therein. We give an Nonlinear Program (NLP) formulation of this problem that includes vanishing constraints that model communication restrictions. These constraints violate constraint qualifications and lead to severe ill–conditioning as detailed in chapter 5. We give a reformulation of this problem as an ODE dynamic optimal control problem with vanishing constraints. We show that the nonconvex active set SQP method developed in this thesis is indeed able solve this problem to optimality, even though it is unrelated to the outer convexification reformulation for which our method has been developed. We compared the obtained solutions to those obtained for the ill–posed NLP formulation using the popular interior point method IPOPT [214, 215].

### 9.3.1 A Multiple Robot Path Coordination Problem

The *multiple robot path coordination* problem arises in autonomous exploration and surveillance of areas by a swarm of autonomous vehicles referred to as *robots*. These robots shall traverse the designated area on predefined paths and can decide autonomously on acceleration or braking. The swarm

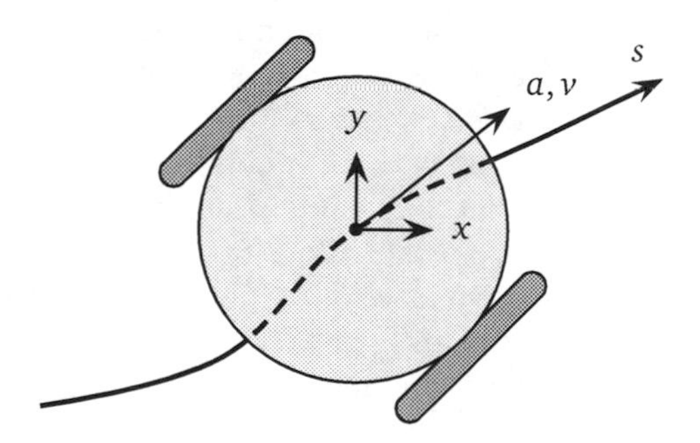


Figure 9.15: Schematic of a single robot with states and controls.

of $n > 1$ robots shall maintain a communication network which, for the purpose of our investigation, is modelled by a maximum transmission or communication radius $T > 0$, and a minimum connectivity $K > 0$. At all times, each of the $n$ robots may communicate with one or more of the other robots inside its communication radius. The swarm is expected to maintain connectivity by ensuring at all times that each robot can communicate with at least $K$ other robots. The goal is for all robots to reach the final point of their respective predefined paths under these constraints. This of course may mean for some of the robots to advance slower or to wait in appropriate positions in order to uphold connectivity of the entire swarm.

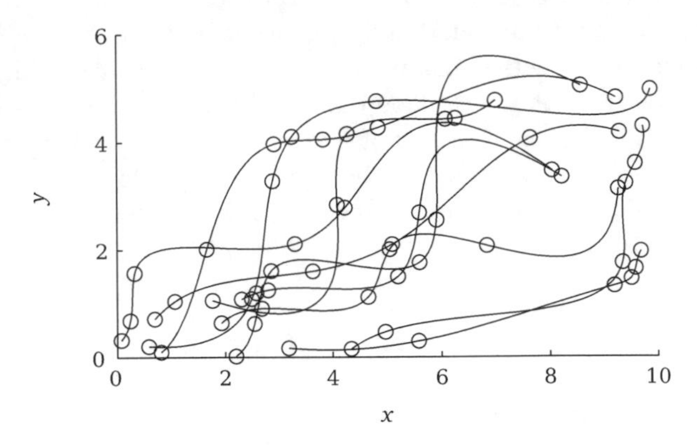


Figure 9.16: Predefined paths to be completed by the swarm of ten robots, cf. [2].

In this section, we present a variant of an NLP formulation for this problem due to [2] and develop an ODE dynamic optimal control problem formulation with vanishing constraints. Obviously, decreasing the communication radius $T$ or increasing the connectivity requirement $K$ increases the difficulty of this problem as the set of feasible solutions shrinks until the problem becomes infeasible for too small values of $T$ or too large values of $K$. A specific instance of this problem for $n = 10$ robots is given by the predefined paths whose cartesian coordinates $(x, y) \in \mathbb{R}^2$ in the plane are described by the piecewise cubic spline paths found in [124].

### 9.3.2 NLP Problem Formulation and Solution

Our NLP formulation of this multiple robot path coordination problem is given in (9.10).

$$\min_{\substack{a,c,s,\\v,x,y,h}} \quad hm \tag{9.10a}$$

$$\text{s.t.} \quad x_{i,j} = \sum_{l=0}^{3} \alpha_{x,j,k,l}(s_{i,j} - \tau_{j,k-1})^l \quad \forall i,j,\ s_{i,j} \in [\tau_{j,k-1}, \tau_{jk}], \tag{9.10b}$$

$$y_{i,j} = \sum_{l=0}^{3} \alpha_{y,j,k,l}(s_{i,j} - \tau_{j,k-1})^l \quad \forall i,j,\ s_{i,j} \in [\tau_{j,k-1}, \tau_{jk}], \tag{9.10c}$$

$$0 = x_{mj} - e_{x,j} \quad \forall j, \tag{9.10d}$$

$$0 = y_{mj} - e_{y,j} \quad \forall j, \tag{9.10e}$$

$$K \leqslant \sum_{j_2=1}^{n} c_{i,j_1,j_2} - 1 \quad \forall i, j_1, \tag{9.10f}$$

$$0 \leqslant (T - d_{i,j_1,j_2}) c_{i,j_1,j_2} \quad \forall i, j_1, j_2, \tag{9.10g}$$

$$s_{i,j} = s_{i-1,j} + \tfrac{h}{2}(v_{i,j-1} + v_{i,j}) \quad \forall i,j, \tag{9.10h}$$

$$v_{i,j} = v_{i-1,j} + h a_{i,j} \quad \forall i,j, \tag{9.10i}$$

$$0 \leqslant s_{i,j} \leqslant \tau_{j,5} \quad \forall i,j, \tag{9.10j}$$

$$0 \leqslant v_{i,j} \leqslant 0.5 \quad \forall i,j, \tag{9.10k}$$

$$-1 \leqslant a_{i,j} \leqslant 0.5 \quad \forall i,j, \tag{9.10l}$$

$$0 \leqslant c_{i,j} \leqslant 1 \quad \forall i,j. \tag{9.10m}$$

| Symbol | Dimension | Description |
|---|---|---|
| $\boldsymbol{a}$ | $m \times n$ | $a_{i,j}$ is the acceleration of robot $j$ in $[\tau_{i-1}, \tau_i]$ |
| $\boldsymbol{c}$ | $m \times n \times n$ | $c_{i,j_1,j_2} > 0$ if robots $j_1$ and $j_2$ can communicate at time $\tau_i$ |
| $h$ | 1 | The length in seconds of each of the $m$ time intervals |
| $\boldsymbol{s}$ | $m \times n$ | $s_{i,j}$ is the spline position parameter of robot $j$ at time $\tau_i$ |
| $\boldsymbol{v}$ | $m \times n$ | $v_{i,j}$ is the tangential velocity of robot $j$ at time $\tau_i$ |
| $(\boldsymbol{x}, \boldsymbol{y})$ | $m \times n$ | $(x_{i,j}, y_{i,j})$ are the cartesian coordinates of robot $j$ at $\tau_i$ |

Table 9.6: Variables of the NLP formulation for the robot path coordination problem.

An overview of the variables used in this formulation can be found in table 9.6. In all equations we assume $i \in \{1,\dots,m\}$, $j, j_1, j_2 \in \{1,\dots,n\}$. In addition, we define the initial positions, velocities, and start point of the spline parameterization as

$$s_{0,j} \stackrel{\text{def}}{=} 0, \quad v_{0,j} \stackrel{\text{def}}{=} 0, \quad \tau_{j,0} \stackrel{\text{def}}{=} 0. \tag{9.11}$$

We further require the definitions of the two distance functions $d_{i,j}$ of robot $j$ to the endpoint and $d_{i,j_1,j_2}$ of robot $j_1$ to robot $j_2$,

$$d_{i,j} \stackrel{\text{def}}{=} \sqrt{(x_{ij} - e_{x,j})^2 + (y_{ij} - e_{y,j})^2}, \tag{9.12a}$$

$$d_{i,j_1,j_2} \stackrel{\text{def}}{=} \sqrt{(x_{i,j_1} - x_{i,j_2})^2 + (y_{i,j_1} - y_{i,j_2})^2}, \tag{9.12b}$$

wherein $(e_{x,j}, e_{y,j}) \in \mathbb{R}^2$ are the terminal points of the $n$ spline paths.

In problem (9.10) we have discretized the time horizon required for completion of the scenario into $m > 0$ intervals of equidistant length $h$ subject to optimization The optimal objective then is $t_{\max} = hm$. Constraints (9.10b) and (9.10c) determine the cartesian coordinates of robot $j$ at time point $i$ required for distance computations, given the robot's advance $s_{ij}$ on the spline path. Constraints (9.10d) and (9.10e) ensure that all robots actually have arrived at their final destinations $(e_{x,j}, e_{y,j})$ at the end of the time horizon. The mentioned communication constraint is (9.10f) which requires each robot $j_1$ to be within the communication range of at least $K$ other robots. This communication range is computed in (9.10g) which forces the communication flag $c_{i,j_1,j_2}$ to zero if the pair $(j_1, j_2)$ of robots is farther apart than the transmission radius $T$. This constraint takes the shape of a *vanishing constraint* as presented in chapter 5. Note that in this case there is *no* outer convexification reformulation behind this constraint. Constraint (9.10h) and (9.10i) are discretizations of the integration of the acceleration and the tangential velocity $v(t)$ along the spline path. Constraints (9.10j), (9.10k), and (9.10l) impose simple upper and lower bounds on the position of the robot on the spline path, the tangential velocity of the robot, and the acceleration of the robot.

Snapshots of the optimal solution of this particular problem for $T = 2.5$ and $K = 2$ at different times $t \in [t_0, t_f]$ are shown in figure 9.17. Circles denote the communication radius of each robot. Note the active communication constraint for the lower right robot at $t \in [0.6h, 0.7h]$ and at $t = 0.9h$.

Problem (9.10) is a variation of the formulation presented in [2]. It dif-

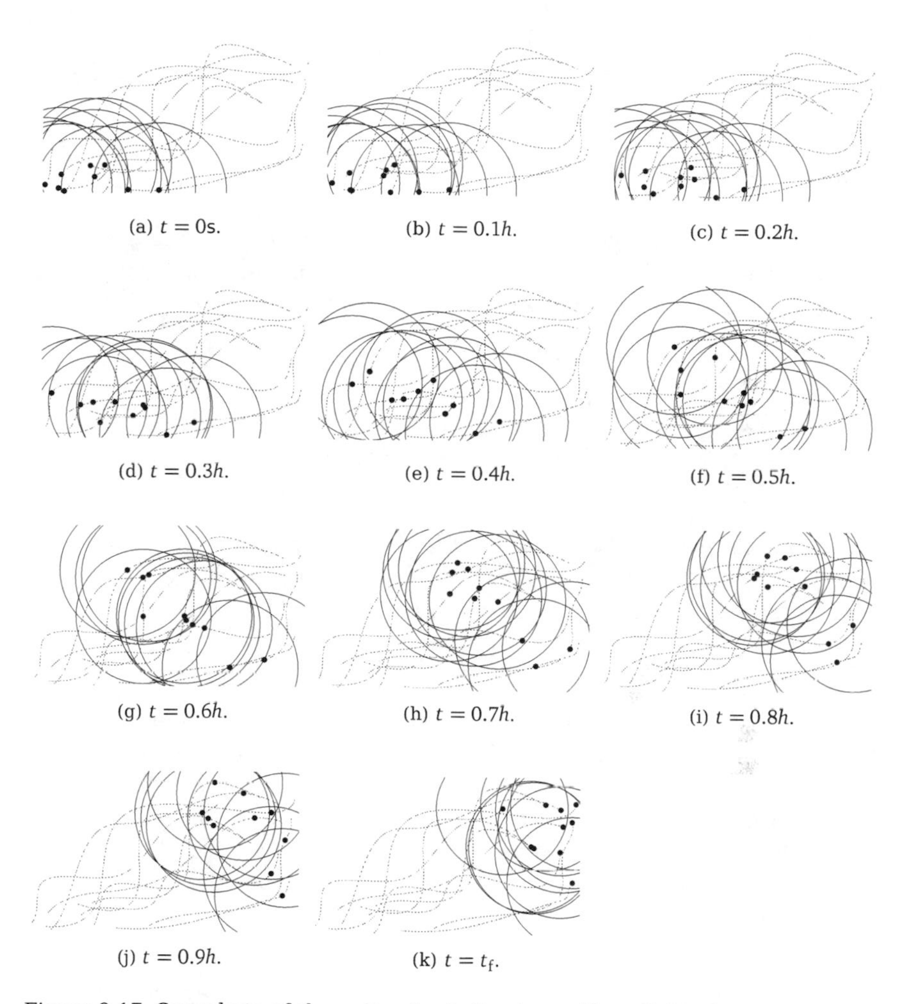

(a) $t = 0\text{s}$. (b) $t = 0.1h$. (c) $t = 0.2h$.

(d) $t = 0.3h$. (e) $t = 0.4h$. (f) $t = 0.5h$.

(g) $t = 0.6h$. (h) $t = 0.7h$. (i) $t = 0.8h$.

(j) $t = 0.9h$. (k) $t = t_\text{f}$.

Figure 9.17: Snapshots of the optimal solution to problem (9.10) for a communication radius $T = 2.5$ and constraint $K = 2$ at different times $t \in [0\text{s}, t_\text{f}]$, where $t_\text{f} = h = 7.995748\text{s}$.

fers in allowing for fractional objective function values $t_{\max} = hm$ whereas the reference employed a formulation that is restriced to integral ones. In order to achieve better comparability to our ODE dynamic formulation, we introduced acceleration of the robots as independent variables and modified

(9.10h) accordingly, whereas the reference imposed a secant constraint on the velocities.

### Choice of Initial Values

Problem (9.10) may for certain choices of the constraint parameters $T$ and $K$ have several local minima due to nonlinearity of the spline paths. The choice of initial values therefore is crucial to maintain comparability. We choose for all computations

$$a = 0, \quad \boldsymbol{c} = \mathbf{1}, \quad \boldsymbol{s} = \mathbf{0}, \quad \boldsymbol{v} = \mathbf{1}, \quad (\boldsymbol{x}, \boldsymbol{y}) = (\mathbf{0}, \mathbf{0}). \tag{9.13}$$

### NLP Solutions by IPOPT

Choosing a discretization of $m = 10$ intervals and a swarm of $n = 10$ robots, the NLP formulation of this problem has 1461 unknowns and 920 inequality constraints, 450 of which are vanishing constraints, and 380 equality constraints.

The time optimal solution for this formulation in absence of any communication constraint (i.e., $K = 0$), and hence in absence of any combinatorial structure in the problem, is 7.99575 seconds after 30 interior point iterations. Table 9.7 lists the solutions found by IPOPT (version 3.6) for choices of the maximal communication distance $T$ from 2 to 5 in steps of one half, and for choices of the communication constraint $K$ from 1 to 9. For all problem instances that could be solved successfully, the objective function $t_{\max}$ is shown together with the number of required interior point iterations.

Failure to converge to a locally optimal solution is indicated by (F). For these instances IPOPT without exception terminates with the message "Restoration phase converged to a feasible point that is unacceptable to the filter for the original problem. Restoration phase in the restoration phase failed.", indicating high degeneracy, wrong derivatives, or lack of constraint qualification as is the case for the problem at hand. Empty cells belong to instances that have been found infeasible by the AMPL presolve step.

From table 9.7 it can be clearly deduced that the violations of constraint qualification due to the combinatorial structure of the problem pose a significant challenge to this pure NLP solver. For low values of the communication constraint $K$ that do not cut off a significant part of the possible

| $T$ | Communication Constraint $K$ | | | | | | | | |
|---|---|---|---|---|---|---|---|---|---|
| | 1 | 2 | 3 | 4 | 5 | 6 | 7 | 8 | 9 |
| 2.0 | (F) | | | | | | | | |
| | 6512 | | | | | | | | |
| 2.5 | (F) | 24.1320 | **26.2376** | **33.2259** | | | | | |
| | 2935 | 2233 | 2026 | 1755 | | | | | |
| 3.0 | (F) | (F) | (F) | 20.2341 | 22.3159 | | | | |
| | 200 | 160 | 185 | 1774 | 1202 | | | | |
| 3.5 | (F) | (F) | (F) | (F) | 17.9407 | | | | |
| | 2415 | 1533 | 1691 | 1022 | 1045 | | | | |
| 4.0 | 7.99575 | 7.99575 | 7.99575 | (F) | (F) | 25.7857 | 25.8223 | 42.5870 | |
| | 1580 | 868 | 2412 | 866 | 1905 | 406 | 506 | 417 | |
| 4.5 | (F) | (F) | 7.99575 | (F) | 7.99575 | 7.99575 | 7.99575 | 30.9573 | **37.4242** |
| | 166 | 126 | 1197 | 164 | 409 | 993 | 1363 | 73 | 55 |
| 5.0 | 7.99575 | 7.99575 | 7.99575 | 7.99575 | (F) | 7.99575 | 7.99575 | 7.99575 | **20.4304** |
| | 339 | 397 | 363 | 379 | 91 | 474 | 373 | 567 | 41 |

Table 9.7: Objective function values and iteration counts for the locally optimal solutions found by AMPL and IPOPT (version 3.6) for NLP (9.10). Failure to converge to a locally optimal solution is indicated by (F). Solutions that are better than the ones found by our software package MuShROOM are printed in **boldface**, all others are inferior ones or are identical.

combinatorial choices, the iteration counts grow noticably and failures of convergence can be observed for 15 of the 41 instances. In addition, inconsistencies can be seen in table 9.7 for instances $(T,K) = (4.0,4)$, $(4.0,5)$, $(4.5,4)$ and $(5.0,4)$ which failed to converge to a locally optimal solution even though all adjacent instances did.

We close this investigation with the important note that the observed behavior of IPOPT *is expected* and can be understood from our analysis in chapter 5. We refer to e.g. [3, 105] for techniques for numerically efficient treatment of vanishing constraints in an interior point method; techniques that clearly cannot be expected to be present in the default configuration of IPOPT.

### 9.3.3 Dynamic Optimal Control Problem Formulation and Solution

We now give an ODE dynamic optimal control problem formulation for (9.10) that fits into the general problem class (1.15) of chapter 1.

$$
\begin{aligned}
\min_{\substack{\boldsymbol{a}(\cdot),\boldsymbol{s}(\cdot),\boldsymbol{v}(\cdot)\\ \boldsymbol{c}(\cdot),\boldsymbol{x}(\cdot),\boldsymbol{y}(\cdot),h}} \quad & h \\
\text{s. t.} \quad & \dot{s}_j(t) = h v_j(t), && t \in \mathcal{T},\ \forall j, \\
& \dot{v}_j(t) = h a_j(t), && t \in \mathcal{T},\ \forall j, \\
& x_j(t) = \sum_{l=0}^{3} \alpha_{\mathrm{x},j,k,l}(s_j(t) - \tau_{j,k-1})^l, && t \in \mathcal{T},\ \forall j, \\
& && s_j(t) \in [\tau_{j,k-1}, \tau_{jk}], \\
& y_j(t) = \sum_{l=0}^{3} \alpha_{\mathrm{y},j,k,l}(s_j(t) - \tau_{j,k-1})^l, && t \in \mathcal{T},\ \forall j, \\
& && s_j(t) \in [\tau_{j,k-1}, \tau_{jk}], \\
& 0 = x_j(t_{\mathrm{f}}) - e_{\mathrm{x},j} && \forall j, \\
& 0 = y_j(t_{\mathrm{f}}) - e_{\mathrm{y},j} && \forall j, \\
& K \leqslant \sum_{j_2=1}^{n} c_{j_1,j_2}(t) - 1, && t \in \mathcal{T},\ \forall j_1, \\
& 0 \leqslant \left(T - d_{j_1,j_2}(t)\right) c_{j_1,j_2}(t), && t \in \mathcal{T},\ \forall j_1, j_2, \\
& 0 \leqslant s_j(t) \leqslant \tau_{j,5}, && t \in \mathcal{T},\ \forall j, \\
& 0 \leqslant v_j(t) \leqslant 0.5, && t \in \mathcal{T},\ \forall j, \\
& -1 \leqslant a_j(t) \leqslant 0.5, && t \in \mathcal{T},\ \forall j, \\
& 0 = s_j(t_0) && \forall j, \\
& 0 = v_j(t_0) && \forall j, \\
& 0 \leqslant c_j(t) \leqslant 1, && t \in \mathcal{T},\ \forall j.
\end{aligned}
\tag{9.14}
$$

In all equations we again assume $i \in \{1, \ldots, m\}$, $j, j_1, j_2 \in \{1, \ldots, n\}$, and the normalized time horizon is $\mathcal{T} \stackrel{\text{def}}{=} [0,1] \subset \mathbb{R}$. Note again the vanishing constraint defining the communication function $c_{j_1,j_2}(t)$ for the pair $(j_1, j_2)$ of robots. After applying a direct multiple shooting discretization with $m = 10$

shooting intervals, all constraints are enforced in the shooting nodes only, as was the case for the time–discrete NLP (9.10). The ODE defining $s(\cdot)$ and $v(\cdot)$ is solved by a fixed–step 4th–order Runge–Kutta method, differing from (9.10) which used a single Euler step. As the exact solution to the ODEs is quadratic in $a_j(t)$ which is constant on each shooting interval, the obtained numerical solutions are identical up to machine precision.

### Optimal Control Problem Solutions by MuShROOM

The ODE dynamic optimal control problem formulation (9.14) with $n = 10$ robots has $n^{\mathrm{x}} = 21$ differential state trajectories (including the Lagrangian objective) and $n^{\mathrm{u}} = 55$ control trajectories. The multiple shooting discretization of the OCP with $m = 10$ multiple shooting intervals has 836 unknowns, 550 inequality constraints of which 450 are vanishing constraints, and 10 equality constraints.

The time optimal solution for this formulation in absence of any communication constraint (i.e., $K = 0$), and hence in absence of any combinatorial structure in the problem, is 7.995748 seconds after 14 SQP iterations and agrees up to the sixth digit with the one found by IPOPT for the NLP formulation. Table 9.8 lists the solutions found by our nonconvex SQP method of chapter 6 for choices of the maximal communication distance $T$ from 2 to 5 in steps of one half, and for choices of the communication constraint $K$ from 1 to 9.

For all problem instances that could be solved successfully, the objective function $t_{\max}$ is shown together with the number of SQP and nonconvex SQP iterations. Failure to converge to a locally optimal solution due to infeasibility of a Quadratic Program with Vanishing Constraints (QPVC) subproblem is indicated by (F). Again, empty cells belong to instances that have been found infeasible by the AMPL presolve step.

As can be seen from table 9.8 our nonconvex active set method succeeds in identifying an optimal solution of problem (9.14) for 38 of the 41 instances, and finds the globally optimal one resp. one that equals or beats the one identified by IPOPT in 36 of the 41 instances. Notably different from the previous table 9.7, we manage to find a solution for all cases with low values of $K$ that contain most of the combinatorial structure of the feasible set. Most QPVC iteration counts are smaller than the respective interior point iteration counts, which shows a performance advantage given that an active set exchange is significantly cheaper than an interior point iteration.

| T | Communication Constraint K | | | | | | | | |
|---|---|---|---|---|---|---|---|---|---|
| | 1 | 2 | 3 | 4 | 5 | 6 | 7 | 8 | 9 |
| 2.0 | **28.0376**<br>15/1254 | | | | | | | | |
| 2.5 | **7.99575**<br>14/1014 | **7.99575**<br>15/728 | 33.5581<br>10/914 | 52.44043<br>13/1004 | | | | | |
| 3.0 | **7.99575**<br>13/818 | **7.99575**<br>14/659 | **9.88516**<br>19/773 | **10.01863**<br>47/11257 | (F) | | | | |
| 3.5 | **7.99575**<br>13/751 | **7.99575**<br>13/760 | **7.99575**<br>14/792 | **8.97033**<br>23/1823 | 17.9407<br>15/1294 | | | | |
| 4.0 | 7.99575<br>13/644 | 7.99575<br>13/645 | 7.99575<br>13/658 | **7.99575**<br>13/667 | **7.99575**<br>13/725 | **12.1859**<br>13/649 | 25.8223<br>9/1096 | 42.5870<br>15/1459 | |
| 4.5 | **7.99575**<br>13/658 | **7.99575**<br>12/632 | 7.99575<br>13/636 | **7.99575**<br>13/630 | 7.99575<br>13/692 | 7.99575<br>18/951 | 7.99575<br>14/1035 | **16.7284**<br>14/1381 | (F) |
| 5.0 | 7.99575<br>13/624 | 7.99575<br>13/599 | 7.99575<br>13/602 | 7.99575<br>13/594 | **7.99575**<br>13/625 | 7.99575<br>13/652 | 7.99575<br>13/773 | 7.99575<br>12/891 | (F) |

Table 9.8: Objective function values and SQP and QPVC iteration counts for the locally optimal solutions found for problem (9.14) by the software package MuShROOM developed in this thesis. Instances found to be infeasible by the AMPL presolve step have not been evaluated. Termination due to infeasibility of a QPVC subproblem is indicated by (F). Solutions better than the ones found by IPOPT are printed in **boldface**, all others are identical or inferior ones.

We fail to identify a solution for the largest feasible choice of $K$ in three cases, and converge to a locally optimal solution that is inferior to the one found by IPOPT in two further cases. This may be attributed to the lack of an efficient globalization strategy in our nonconvex SQP algorithm, which has been omitted from the investigations in this thesis in view of the targeted model–predictive application.

### 9.3.4 Summary

In this section we investigated an NLP and an OCP formulation of a multi–robot pathfinding and communication problem frequently considered in the literature, cf. e.g. [2] and the references found therein. The problem features combinatorial constraints on the communication ability of the robot swarm and can in its NLP or direct multiple shooting discretized OCP variant be cast as a MPVC. By varying the communication radius and connectivity constraint, 41 feasible instances of this problem were created. We demonstrated the findings of our investigation in chapter 5 at the example of a popular interior point method without provision for MPVC and — as expected — found this method to fail on a significant number of the examined problem instances. It should be noted that appropriate relaxation schemes as e.g. briefly presented in chapter 5 can be employed to ensure convergence of interior point method on MPVC, cf. e.g. [3, 105] for details. We examined the numerical behavior of our developed nonconvex SQP and active set QP algorithms and could solve to optimality all but three problem instances. Future work includes a globalization method for our nonconvex SQP algorithm that promises to allow for the solution of the remaining three unsolved instances.

## 9.4 Block Structured Factorization and Updates

In this section we investigate the performance of our block structured active set QP solver qpHPSC at the example of a nonlinear time optimal mixed–integer control problem treated in the papers [119, 120, 122, 187] and in [80, 81] who also gives run times for a branch & bound based solution approach and a switching time optimization approach.

### 9.4.1 Vehicle Model

We give a brief description of the nonlinear vehicle dynamics that can be found in more detail in e.g. [80, 122]. We consider a single-track model, derived under the simplifying assumption that rolling and pitching of the car body can be neglected. Consequentially, only a single front and rear wheel are modeled, located in the virtual center of the original two wheels. Motion of the car body is considered on the horizontal plane only.

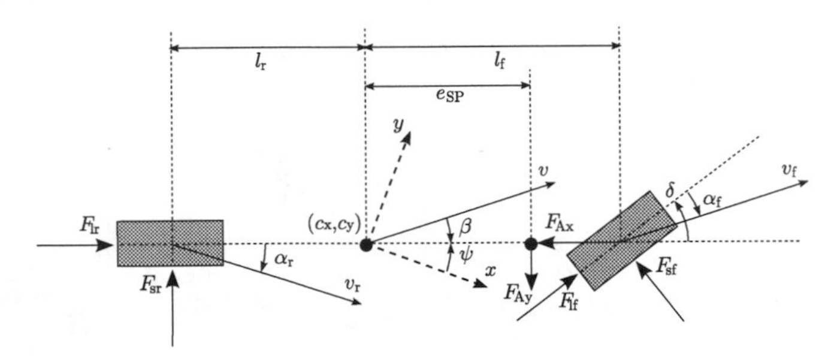


Figure 9.18: Coordinates and forces in the single-track vehicle model. The figure aligns with the vehicle's local coordinate system while dashed vectors denote the earth-fixed coordinate system chosen for computations.

Four controls represent the driver's choice on steering and velocity, and are listed in table 9.9. We denote with $w_\delta$ the steering wheel's angular velocity. The force $F_B$ controls the total braking force, while the accelerator pedal position $\varphi$ is translated into an accelerating force according to the torque model presented in (9.25). Finally, the selected gear $\mu$ influences the effective engine torque's transmission ratio. The single-track dynamics are described by a system of ordinary differential equations. The individual system states are listed in table 9.10. Figure 9.18 visualizes the choice of coordinates, angles, and forces. Equations of motion are derived as follows.

| Name | Description | Domain | Unit |
|---|---|---|---|
| $w_\delta$ | Steering wheel angular velocity | $[-0.5, 0.5]$ | $\frac{\text{rad}}{\text{s}}$ |
| $F_\text{B}$ | Total braking force | $[0, 1.5 \cdot 10^4]$ | N |
| $\varphi$ | Accelerator pedal position | $[0, 1]$ | – |
| $\mu$ | Selected gear | $\{1, \dots, n^\mu\}$ | – |

Table 9.9: Controls used in the single–track vehicle model.

The center of gravity is denoted by the coordinate pair $(c_\text{x}, c_\text{y})$ which is obtained by integration over the directional velocity,

$$\dot{c}_\text{x}(t) = v(t)\,\cos\big(\psi(t) - \beta(t)\big), \tag{9.15a}$$
$$\dot{c}_\text{y}(t) = v(t)\,\sin\big(\psi(t) - \beta(t)\big). \tag{9.15b}$$

Acceleration is obtained from the sum of forces attacking the car's mass $m$ in the direction of driving,

$$\dot{v}(t) = \frac{1}{m}\Big(\big(F^\mu_\text{lr} - F_\text{Ax}\big)\,\cos\beta(t) + F_\text{lf}\,\cos\big(\delta(t) + \beta(t)\big) \\ - \big(F_\text{sr} - F_\text{Ay}\big)\,\sin\beta(t) - F_\text{sf}\,\sin\big(\delta(t) + \beta(t)\big)\Big). \tag{9.16}$$

The steering wheel's angle is obtained from the corresponding controlled angular velocity,

$$\dot{\delta}(t) = w_\delta. \tag{9.17}$$

The slip angle's change is controlled by the steering wheel and counteracted by the sum of forces attacking perpendicular to the car's direction of driving. The forces' definitions are given in (9.21ff).

$$\dot{\beta}(t) = w_\text{z}(t) - \frac{1}{m\,v(t)}\Big(\big(F_\text{lr} - F_\text{Ax}\big)\,\sin\beta(t) + F_\text{lf}\,\sin\big(\delta(t) + \beta(t)\big) \\ + \big(F_\text{sr} - F_\text{Ay}\big)\,\cos\beta(t) + F_\text{sf}\,\cos\big(\delta(t) + \beta(t)\big)\Big). \tag{9.18}$$

The yaw angle is obtained by integrating over its change $w_\text{z}$,

$$\dot{\psi}(t) = w_\text{z}(t), \tag{9.19}$$

which in turn is the integral over the sum of forces attacking the front wheel

| Name | Description | Unit |
|---|---|---|
| $c_x$ | Horizontal position of the car | m |
| $c_y$ | Vertical position of the car | m |
| $v$ | Magnitude of directional velocity of the car | $\frac{m}{s}$ |
| $\delta$ | Steering wheel angle | rad |
| $\beta$ | Side slip angle | rad |
| $\psi$ | Yaw angle | rad |
| $w_z$ | Yaw angle velocity | $\frac{rad}{s}$ |

Table 9.10: Coordinates and states used in the single–track vehicle model.

in direction perpendicular to the car's longitudinal axis of orientation,

$$\dot{w}_z(t) = \frac{1}{I_{zz}} \left( F_{sf}\, l_f \cos\delta(t) - F_{sr}\, l_{sr} - F_{Ay}\, e_{SP} + F_{lf}\, l_f \sin\delta(t) \right). \tag{9.20}$$

We now list and explain the individual forces used in this ODE system. We first discuss lateral and longitudinal forces attacking at the front and rear wheels. In view of the convex reformulation we'll undertake later, we consider the gear $\mu$ to be fixed and denote dependencies on the selected gear by a superscript $\mu$. The side (lateral) forces on the front and rear wheels as functions of the slip angles $\alpha_f$ and $\alpha_r$ according to the so-called *magic formula* due to [164] are

$$F_{sf,sr}(\alpha_{f,r}) \stackrel{\text{def}}{=} D_{f,r} \sin\Big( C_{f,r} \arctan\big(B_{f,r}\, \alpha_{f,r} - E_{f,r}(B_{f,r}\, \alpha_{f,r} - \arctan(B_{f,r}\, \alpha_{f,r}))\big) \Big), \tag{9.21}$$

The front slip and rear slip angles are obtained from

$$\alpha_f \stackrel{\text{def}}{=} \delta(t) - \arctan\left( \frac{l_f\, \dot{\psi}(t) - v(t) \sin\beta(t)}{v(t) \cos\beta(t)} \right), \tag{9.22a}$$

$$\alpha_r \stackrel{\text{def}}{=} \arctan\left( \frac{l_r\, \dot{\psi}(t) + v(t) \sin\beta(t)}{v(t) \cos\beta(t)} \right). \tag{9.22b}$$

The longitudinal force at the front wheel is composed from braking force $F_{Bf}$ and resistance due to rolling friction $F_{Rf}$

$$F_{lf} \stackrel{\text{def}}{=} -F_{Bf} - F_{Rf}. \tag{9.23}$$

Assuming a rear wheel drive, the longitudinal force at the rear wheel is given by the transmitted engine torque $M_{\text{wheel}}$ and reduced by braking force $F_{\text{Br}}$ and rolling friction $F_{\text{Rr}}$. The effective engine torque $M^{\mu}_{\text{mot}}$ is transmitted twice. We denote by $i^{\mu}_{\text{g}}$ the gearbox transmission ratio corresponding to the selected gear $\mu$, and by $i_{\text{t}}$ the axle drive's fixed transmission ratio. $R$ is the rear wheel radius.

$$F^{\mu}_{\text{lr}} \stackrel{\text{def}}{=} \frac{i^{\mu}_{\text{g}}\, i_{\text{t}}}{R}\, M^{\mu}_{\text{mot}}(\varphi) - F_{\text{Br}} - F_{\text{Rr}}. \tag{9.24}$$

The engine's torque, depending on the acceleration pedal's position $\varphi$, is modeled as follows:

$$M^{\mu}_{\text{mot}}(\varphi) \stackrel{\text{def}}{=} f_1(\varphi)\, f_2(w^{\mu}_{\text{mot}}) + (1 - f_1(\varphi))\, f_3(w^{\mu}_{\text{mot}}), \tag{9.25a}$$

$$f_1(\varphi) \stackrel{\text{def}}{=} 1 - \exp(-3\,\varphi), \tag{9.25b}$$

$$f_2(w_{\text{mot}}) \stackrel{\text{def}}{=} -37.8 + 1.54\, w_{\text{mot}} - 0.0019\, w^2_{\text{mot}}, \tag{9.25c}$$

$$f_3(w_{\text{mot}}) \stackrel{\text{def}}{=} -34.9 - 0.04775\, w_{\text{mot}}. \tag{9.25d}$$

Here, $w^{\mu}_{\text{mot}}$ is the engines rotary frequency in Hertz. For a given gear $\mu$ it is computed from

$$w^{\mu}_{\text{mot}} \stackrel{\text{def}}{=} \frac{i^{\mu}_{\text{g}}\, i_{\text{t}}}{R}\, v(t). \tag{9.26}$$

The total braking force $F_{\text{B}}$ is controlled by the driver. The distribution to front and rear wheels is chosen as

$$F_{\text{Bf}} \stackrel{\text{def}}{=} \frac{2}{3}\, F_{\text{B}}, \qquad F_{\text{Br}} \stackrel{\text{def}}{=} \frac{1}{3}\, F_{\text{B}}. \tag{9.27}$$

The braking forces $F_{\text{Rf}}$ and $F_{\text{Rr}}$ due to rolling resistance are obtained from

$$F_{\text{Rf}}(v) \stackrel{\text{def}}{=} f_{\text{R}}(v)\, \frac{m\, l_{\text{r}}\, g}{l_{\text{f}} + l_{\text{r}}}, \qquad F_{\text{Rr}}(v) \stackrel{\text{def}}{=} f_{\text{R}}(v)\, \frac{m\, l_{\text{f}}\, g}{l_{\text{f}} + l_{\text{r}}}, \tag{9.28}$$

where the velocity-dependent amount of friction is modeled by

$$f_{\text{R}}(v) \stackrel{\text{def}}{=} 9 \cdot 10^{-3} + 7.2 \cdot 10^{-5}\, v + 5.038848 \cdot 10^{-10}\, v^4. \tag{9.29}$$

Finally, drag force due to air resistance is given by $F_{\text{Ax}}$, while we assume

| Name | Description | Value | Unit |
|---|---|---|---|
| $A$ | Effective flow surface | 1.437895 | $\text{m}^2$ |
| $B_\text{f}$ | Pacejka tyre model stiffness factor (front) | 10.96 | - |
| $B_\text{r}$ | Pacejka tyre model stiffness factor (rear) | 12.67 | - |
| $C_\text{f}, C_\text{r}$ | Pacejka tyre model shape factor (front, rear) | 1.3 | - |
| $D_\text{f}$ | Pacejka tyre model peak value (front) | 4560.40 | - |
| $D_\text{r}$ | Pacejka tyre model peak value (rear) | 3947.81 | - |
| $E_\text{f}, E_\text{r}$ | Pacejka tyre model curvature factor (front, rear) | $-0.5$ | - |
| $I_{\text{zz}}$ | Vehicle's moment of inertia | 1752 | $\text{kg m}^2$ |
| $R$ | Wheel radius | 0.302 | m |
| $c_\text{w}$ | Air drag coefficient | 0.3 | - |
| $e_{\text{SP}}$ | Distance of drag mount point to center of gravity | 0.5 | m |
| $g$ | Gravity force | 9.81 | $\frac{\text{kg}}{\text{m s}^2}$ |
| $i_\text{g}^\mu$ | Gearbox torque transmission ratio | various | - |
| $i_\text{t}$ | Engine torque transmission ratio | 3.91 | - |
| $l_\text{f}$ | Distance of front wheel to center of gravity | 1.19016 | m |
| $l_\text{r}$ | Distance of rear wheel to center of gravity | 1.37484 | m |
| $m$ | Mass of vehicle | 1239 | kg |
| $\varrho$ | Air density | 1.249512 | $\frac{\text{kg}}{\text{m}^3}$ |

Table 9.11: Parameters used in the single–track vehicle model.

that no sideward drag forces (e.g. side wind) are present.

$$F_{\text{Ax}} \stackrel{\text{def}}{=} \frac{1}{2} c_\text{w} \varrho A v^2(t), \qquad F_{\text{Ay}} \stackrel{\text{def}}{=} 0. \tag{9.30}$$

The values and units of all model parameters can be found in table 9.11.

**Track Model**

The double-lane change maneuver presented in [80, 81, 122] is realized by constraining the car's position onto a prescribed track at any time $t \in \mathcal{T} \stackrel{\text{def}}{=} [t_0, t_\text{f}]$, see Figure 9.19.

Starting in the left position with an initial prescribed velocity, the driver is asked to manage a change of lanes modeled by an offset of 3.5 meters in the track. Afterwards he is asked to return to the starting lane. This maneuver can be regarded as an overtaking move or as an evasive action taken

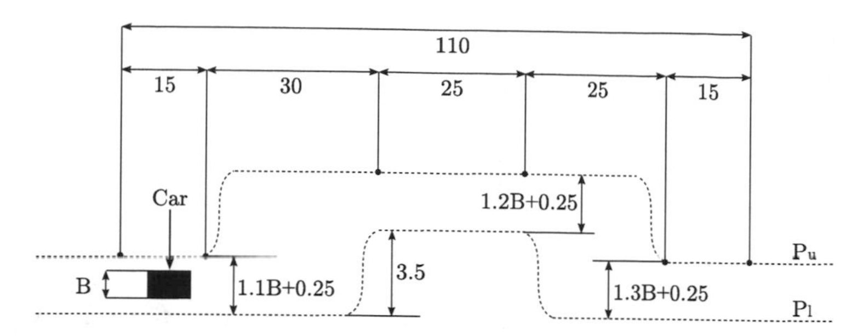


Figure 9.19: Track constraints of the double–lane change maneuvre for which a time–optimal solution is sought in problem (9.33).

to avoid hitting an obstacle suddenly appearing on the straight lane. The constraints $P_l(x)$ and $P_u(x)$ are given by the following piecewise splines assuming a width of $B = 1.5$ meters for the vehicle,

$$P_l(x) \stackrel{\text{def}}{=} \begin{cases} 0 & \text{if } x \in [0, 44], \\ 13.5(x-44)^3 & \text{if } x \in [44.0, 44.5], \\ 13.5(x-45)^3 + 3.5 & \text{if } x \in [44.5, 45.0], \\ 3.5 & \text{if } x \in [45.0, 70.0], \\ 13.5(70-x)^3 + 3.5 & \text{if } x \in [70.0, 70.5], \\ 13.5(71-x)^3 & \text{if } x \in [70.5, 71], \\ 0 & \text{if } x \in [71, 140], \end{cases} \tag{9.31a}$$

$$P_u(x) \stackrel{\text{def}}{=} \begin{cases} 1.9 & \text{if } x \in [0, 15], \\ 14.6(x-15)^3 + 1.9 & \text{if } x \in [15.0, 15.5], \\ 14.6(x-16)^3 + 5.55 & \text{if } x \in [15.5, 16.0], \\ 5.55 & \text{if } x \in [16.0, 94.0], \\ 13.4(94-x)^3 + 5.55 & \text{if } x \in [94.0, 94.5], \\ 13.4(95-x)^3 + 2.2 & \text{if } x \in [94.5, 95.0], \\ 2.2 & \text{if } x \in [95, 140]. \end{cases} \tag{9.31b}$$

### 9.4.2 Mixed–Integer Time–Optimal Control Problem

#### Problem Formulation

We denote with $\boldsymbol{x}$ the state vector of the ODE system and by $\boldsymbol{f}$ the corresponding right-hand side function as described in section 9.4.1. The vector

$\boldsymbol{u}$ shall be the vector of continuous controls, whereas the integer control $\mu(\cdot)$ will be written in a separate vector,

$$\boldsymbol{x}(t) \stackrel{\text{def}}{=} \begin{bmatrix} c_x(t) & c_y(t) & v(t) & \delta(t) & \beta(t) & \psi(t) & w_z(t) \end{bmatrix}, \tag{9.32a}$$

$$\boldsymbol{u}(t) \stackrel{\text{def}}{=} \begin{bmatrix} w_\delta(t) & F_B(t) & \varphi(t) \end{bmatrix}, \tag{9.32b}$$

$$\boldsymbol{w}(t) \stackrel{\text{def}}{=} \begin{bmatrix} \mu(t) \end{bmatrix}. \tag{9.32c}$$

With this notation, the resulting mixed-integer optimal control problem reads

$$\min_{\boldsymbol{x}(\cdot),\boldsymbol{u}(\cdot),\boldsymbol{w}(\cdot),t_f} \quad t_f + \int_0^{t_f} w_\delta^2(t)\,dt \tag{9.33a}$$

$$\text{s.t.} \quad \dot{\boldsymbol{x}}(t) = f(t, \boldsymbol{x}(t), \boldsymbol{u}(t), \boldsymbol{w}(t)) \qquad t \in [t_0, t_f], \tag{9.33b}$$

$$c_y(t) \in [P_l(c_x(t)) + 0.75, P_u(c_x(t)) - 0.75] \qquad t \in [t_0, t_f], \tag{9.33c}$$

$$w_\delta(t) \in [-0.5, 0.5] \qquad t \in [t_0, t_f], \tag{9.33d}$$

$$F_B(t) \in [0, 1.5 \cdot 10^4] \qquad t \in [t_0, t_f], \tag{9.33e}$$

$$\varphi(t) \in [0, 1] \qquad t \in [t_0, t_f], \tag{9.33f}$$

$$\mu(t) \in \{1, \dots, n^\mu\} \qquad t \in [t_0, t_f], \tag{9.33g}$$

$$\boldsymbol{x}(t_0) = (-30, \text{free}, 10, 0, 0, 0, 0), \tag{9.33h}$$

$$c_x(t_f) = 140, \tag{9.33i}$$

$$\psi(t_f) = 0. \tag{9.33j}$$

By employing the objective function (9.33a) we strive to minimize the total time $t_f$ required to traverse the test course, and to do so with minimal steering effort $w_\delta(t)$. At any time, the car must be positioned within the test course's boundaries; this requirement is formulated by the double inequality path constraint (9.33c). The system's initial values are fixed in (9.33h) with the exception of the car's initial vertical position on the track, which remains a free variable only constrained by the track's boundary. Finally, constraints (9.33i, 9.33j) guarantee that the car actually arrives at the end of the test course driving straight ahead.

### Solutions

The ODE system is solved using a fixed–step Runge–Kutta method with 20 steps per shooting interval. We use one–sided finite difference approximations to the derivatives of all model functions.

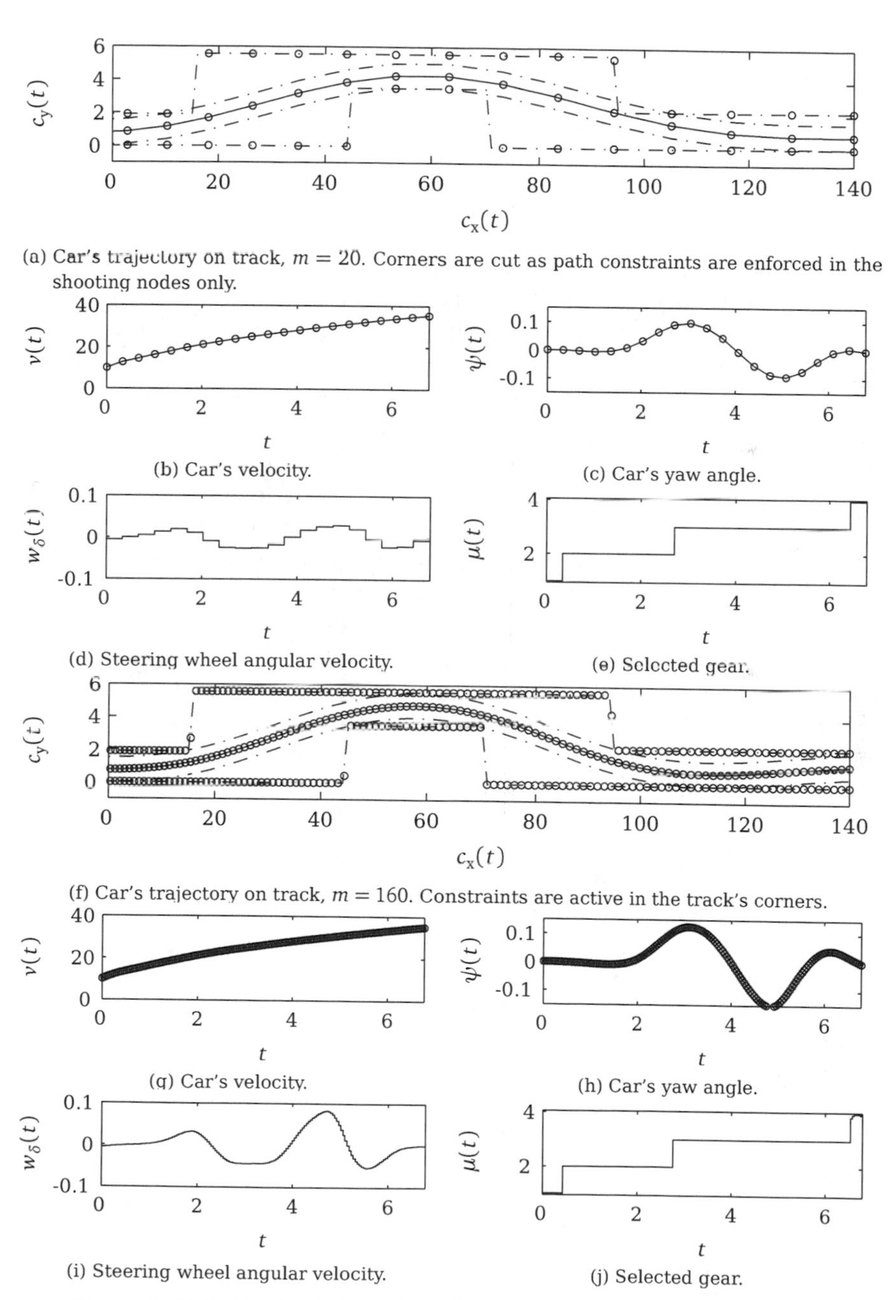


(a) Car's trajectory on track, $m = 20$. Corners are cut as path constraints are enforced in the shooting nodes only.

(b) Car's velocity.

(c) Car's yaw angle.

(d) Steering wheel angular velocity.

(e) Selected gear.

(f) Car's trajectory on track, $m = 160$. Constraints are active in the track's corners.

(g) Car's velocity.

(h) Car's yaw angle.

(i) Steering wheel angular velocity.

(j) Selected gear.

Figure 9.20: Optimal solution of problem (9.33) for $m = 20$ and $m = 160$.

The Hessian is approximated by a limited–memory BFGS scheme with a memory length of $l = 15$ in order to avoid accumulation of ill–conditioned secant information due to the rapidly changing curvature of the path constraint.

All QP subproblem are solved to an optimality tolerance of $10^{-8}$ by our block structured parametric active set method qpHPSC. The NLP problem is solved to a KKT tolerance of $10^{-10}$. The optimal solutions of problem (9.33) for $m = 20$ and $m = 160$ direct multiple shooting intervals are shown in figure 9.20. The discretization of the path constraint enforcing the track's boundaries has a significant influence on the minimal objective function value that can be attained, as corners can be cut for low values of $m$. Table 9.12 lists for $n^{\mu} = 4$ and various choices of $m$ the problem's dimensions, i.e., the number of unknowns $n_{\text{var}}$ and the number of equality and inequality constraints $n_{\text{con}}$, together with the obtained optimal solution $t_{\text{f}}$, the 2-norm of the remaining infeasibility, and the number $n_{\text{frac}}$ of fractional relaxed convex multipliers for the integer control. The number of SQP iterations and active set QP iterations is given along with the overall computation time. Total CPU time grows roughly quadratically with the number $m$ of multiple shooting nodes, as the runtime per SQP iteration is $\mathcal{O}(m)$ (in sharp contrast to classical condensing methods) and the number of SQP iterations to convergence appears to grow roughly linearly with $m$. Using our block structured active set QP method qpHPSC, over 90 percent of this time is spent in the solution of the ODE system and the computation of sensitivities.

| Dimensions | | | Solution | | | Computation | | |
|---|---|---|---|---|---|---|---|---|
| $m$ | $n_{\text{var}}$ | $n_{\text{con}}$ | Objective $t_{\text{f}}$ | Infeasibility | $n_{\text{frac}}$ | SQP | QP | CPU Time |
| 20 | 336 | 64 | 6.781176 | $1.83 \cdot 10^{-11}$ | 0 | 71 | 567 | 00m 04.3s |
| 40 | 656 | 124 | 6.791324 | $2.65 \cdot 10^{-11}$ | 0 | 142 | 1485 | 00m 16.7s |
| 80 | 1296 | 244 | 6.795562 | $3.03 \cdot 10^{-12}$ | 0 | 237 | 3434 | 00m 56.3s |
| 160 | 2576 | 484 | 6.804009 | $2.22 \cdot 10^{-12}$ | 3 | 334 | 8973 | 02m 47.7s |

Table 9.12: Optimal solutions of problem (9.33) for $n^{\mu} = 4$ as in [80, 81, 122]. With increasily fine discretization of the path constraint modelling the track, the time optimal objective increases as cutting of corners is reduced.

### 9.4.3 Comparison of Structure and Sparsity

In table 9.13 an account of the number of unknowns and constraints of the NLP obtained from the direct multiple shooting discretization, outer convexification, and relaxation of the MIOCP (9.33) is given. In addition, the amount of sparsity present in the Hessian and the constraints Jacobian of the resulting QP subproblem is listed. All compared algorithms work on this large structured QP subproblem. The classical condensing algorithm preprocesses it into a smaller but densely populated one. Its dimensions and sparsity can be compared to the original one by using table 9.14.

As can be seen, the condensed Hessian matrix is densely populated and does not contain any remaining structural zero elements. The condensed constraints Jacobian matrix is populated to almost 40%. With increasing length or granularity $m$ of the multiple shooting discretization, the number of nonzero elements grows twice as fast in the condensed constraints Jacobian compared to the block structured one. The dense active set method QPOPT however is not able to exploit the remaining sparsity but instead computes dense factors of this constraints Jacobian. Note in addition that the simple bounds on the additionally introduced shooting node values $\boldsymbol{s}_1, \dots, \boldsymbol{s}_m$ would become linear constraints after condensing as mentioned in chapter 7, and are thus *omitted* from the initial condensed problem.

In addition, as an be seen in table 9.15 the runtime complexity of $\mathcal{O}(m^2)$ leads to long condensing run times for larger values of $m$ that make this algorithmic approach at structure exploitation unattractive for long horizons or fine discretization of the control trajectory. With increasing number of $n^{\mathrm{q}} = 3 + n^{\mu}$ control parameters, the advantage of eliminating the additionally introduced shooting node variables is clearly surpassed by the disadvantage of loosing the structural properties of the problem. The number $n^{\mathrm{x}}$ of matching conditions used to eliminate the additionally introduced shooting node values $\boldsymbol{s}_1, \dots, \boldsymbol{s}_m$ is hidden by the dominant dimension $n^{\mathrm{q}}$ of the control parameters $\boldsymbol{q}_0, \dots, \boldsymbol{q}_{m-1}$ that remain in the QP's matrices.

### 9.4.4 Comparison of Runtimes

We study the run time of various combinations of numerical codes for the solution of the direct multiple shooting block structured QP, which include

1. The classical condensing algorithm for the block structured QP combined with the dense null–space active set QP solver QPOPT [86] with

default settings. This algorithmic approach is taken e.g. in the software package MUSCOD-II [132, 134].

2. Our parametric active set method qpHPSC with dense, sparse, and block structured solvers for the structured KKT system:
   - The block structured Hessian Projection Schur Complement (HPSC) factorization of chapter 7 with the matrix update techniques of chapter 8.
   - The block structured HPSC factorization of chapter 7 without matrix updates but otherwise identical settings.
   - The sparse multifrontal symmetric indefinite code MA57 [56], with standard settings.

| $m$ | $n^\mu$ | Hessian of the Lagrangian | | | | Jacobian of the Constraints | | | |
|---|---|---|---|---|---|---|---|---|---|
| | | Size | Elements | Nonzeros | | Rows | Elements | Nonzeros | |
| 20 | 4 | 336 | 112,896 | 5262 | 4.7% | 264 | 88,704 | 1906 | 2.1% |
| | 8 | 420 | 176,400 | 7878 | 4.5% | 264 | 110,880 | 2465 | 2.2% |
| | 12 | 504 | 254,016 | 11,918 | 4.7% | 264 | 133,056 | 3024 | 2.3% |
| | 16 | 588 | 345,744 | 16,254 | 4.7% | 264 | 155,232 | 3584 | 2.3% |
| 40 | 4 | 656 | 430,336 | 10,382 | 2.4% | 524 | 343,744 | 3806 | 1.1% |
| | 8 | 820 | 672,400 | 15,814 | 2.4% | 524 | 429,680 | 4924 | 1.1% |
| | 12 | 984 | 968,256 | 22,718 | 2.3% | 524 | 515,616 | 6044 | 1.2% |
| | 16 | 1148 | 1,317,904 | 31,934 | 2.4% | 524 | 601,552 | 7166 | 1.2% |
| 80 | 4 | 1296 | 1,679,616 | 20,622 | 1.2% | 1044 | 1,353,024 | 7607 | 0.6% |
| | 8 | 1620 | 2,624,400 | 30,950 | 1.2% | 1044 | 1,691,280 | 9845 | 0.6% |
| | 12 | 1944 | 3,779,136 | 46,478 | 1.2% | 1044 | 2,029,536 | 12,087 | 0.6% |
| | 16 | 2268 | 5,143,824 | 60,478 | 1.2% | 1044 | 2,367,792 | 14,325 | 0.6% |
| 160 | 4 | 2576 | 6,635,776 | 41,048 | 0.6% | 2084 | 5,368,384 | 15,208 | 0.3% |
| | 8 | 3220 | 10,368,400 | 62,316 | 0.6% | 2084 | 6,710,480 | 19,688 | 0.3% |
| | 12 | 3864 | 14,930,496 | 92,208 | 0.6% | 2084 | 8,052,576 | 24,169 | 0.3% |
| | 16 | 4508 | 20,322,064 | 121,924 | 0.6% | 2084 | 9,394,672 | 28,648 | 0.3% |

Table 9.13: Numbers of NLP (equiv. QP) unknowns and constraints, and percentage of nonzero elements in the Hessian and constraints Jacobian of the QP for problem (9.33), listed for various choices of the number $m$ of multiple shooting intervals and the number $n^q = 3 + n^\mu$ of control parameters.

  - The unsymmetric multifrontal LU code UMFPACK [50], with standard settings. Symmetry of the KKT system cannot be exploited here and two backsolves with the unsymmetric factors are required.

3. Our parametric active set method qpHPSC with the following *LAPACK* [9] routines as reference KKT solvers:

  - The banded unsymmetric LU code DGBTRF, with the same restrictions that apply to UMFPACK.
  - The dense symmetric indefinite code DSYTRF that does not exploit any structure and will therefore yield inferior performance.

| $m$ | $n^\mu$ | Hessian of the Lagrangian | | | | Jacobian of the Constraints | | | |
|---|---|---|---|---|---|---|---|---|---|
| | | Size | Elements | Nonzeros | | Rows | Elements | Nonzeros | |
| 20 | 4 | 130 | 16,900 | 16,900 | 100% | 64 (264) | 8,320 | 3117 | 37% |
| | 8 | 264 | 55,440 | 55,440 | 100% | 64 (264) | 13,440 | 5036 | 37% |
| | 12 | 290 | 84,100 | 84,100 | 100% | 64 (264) | 18,560 | 6955 | 37% |
| | 16 | 370 | 136,900 | 136,900 | 100% | 64 (264) | 23,680 | 8875 | 37% |
| 40 | 4 | 250 | 62,500 | 62,500 | 100% | 124 (524) | 31,000 | 11,017 | 36% |
| | 8 | 410 | 168,100 | 86,010 | 100% | 124 (524) | 50,840 | 18,055 | 36% |
| | 12 | 570 | 324,900 | 324,900 | 100% | 124 (524) | 70,680 | 24,095 | 36% |
| | 16 | 730 | 532,900 | 532,900 | 100% | 124 (524) | 90,520 | 32,137 | 36% |
| 80 | 4 | 490 | 240,100 | 240,100 | 100% | 244 (1044) | 119,560 | 41,218 | 34% |
| | 8 | 810 | 656,100 | 656,100 | 100% | 244 (1044) | 197,640 | 68,096 | 34% |
| | 12 | 1130 | 1,276,900 | 1,276,900 | 100% | 244 (1044) | 275,720 | 94,978 | 34% |
| | 16 | 1450 | 2,102,500 | 2,102,500 | 100% | 244 (1044) | 353,800 | 121,856 | 34% |
| 160 | 4 | 970 | 940,900 | 940,900 | 100% | 484 (2084) | 469,480 | 159,219 | 34% |
| | 8 | 1610 | 2,592,100 | 2,592,100 | 100% | 484 (2084) | 779,240 | 264,179 | 34% |
| | 12 | 2250 | 5,062,500 | 5,062,500 | 100% | 484 (2084) | 1,089,000 | 369,140 | 34% |
| | 16 | 2890 | 6,022,760 | 6,022,760 | 100% | 484 (2084) | 1,398,760 | 474,099 | 34% |

Table 9.14: Numbers of unknowns and constraints, and percentage of nonzero elements in the condensed Hessian and condensed constraints jacobian of the QP (cf. section 7.2.1) for problem (9.33), listed for various choices of the number $m$ of multiple shooting intervals and the number $n^q = 3 + n^\mu$ of control parameters. Structure and sparsity are lost as the number of control parameters increases.

Figure 9.21 and table 9.15 summarize the computational effort required by each of the investigated QP solving algorithms for increasingly fine discretizations $m$ of the control and an increasing number of available gear choices $n^\mu$ i.e., control parameters $n^q = 3 + n^\mu$ per shooting interval.

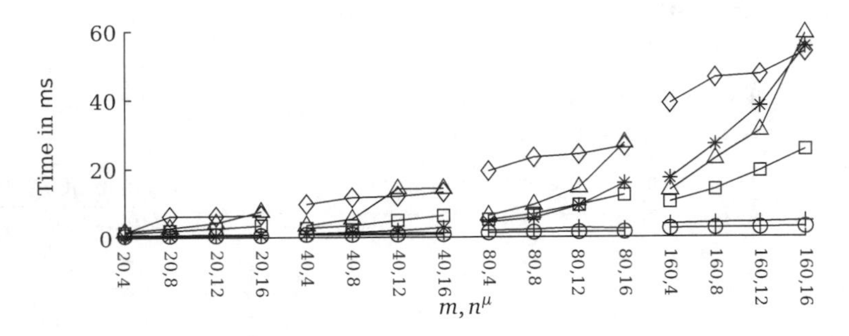


Figure 9.21: Average runtime in milliseconds (ms) per QP solver iteration for problem (9.33), depending on the numbers $(m, n^\mu)$ of multiple shooting intervals $m$ and control parameters $n^\mu$. (–⊖–) qpHPSC with matrix updates, (–+–) qpHPSC without updates, (–⊟–) MA57, (–△–) LAPACK DGBTRF, (–✳–) QPOPT on the condensed QP without runtime spent in condensing, (–◇–) UMFPACK. Runtime for LAPACK DSYTRF is not shown.

As expected, due to the runtime complexity of $\mathcal{O}(m^3 n^3)$ classical condensing and the dense active set QP solver quickly fall behind in performance as either of the problem dimensions increases. Quadratic growth of the condensing runtime in both $m$ and $n^q$ can be observed from the third column. For the smallest instances, the banded LU decomposition provided by LAPACK is sufficient for structure exploitation and even shows linear runtime growth in $m$. Its absolute performance falls behind as $m$ or $n^q$ get larger. All structure or sparsity exploiting approaches show linear growth of the runtime in $m$. The performance of MA57 and UMFPACK falls behind as the problem instances get larger. This can be explained by the densely populated Initial Value Problem (IVP) sensitivity matrices generated by the direct multiple shooting method. Generic sparse solvers are more appropriately employed in conjunction with collocation methods that yield much larger but sparsely populated sensitivity matrices. UMFPACK in addition cannot exploit the KKT system's symmetry and requires two backsolves with the unsymmetric factors. Matrix updates could be realized for the unsymmetric sparse LU factorization provided by UMFPACK, though, see e.g. [108]. Fill–in is however reported to happen.

The HPSC factorization with updates developed in this thesis may be regarded as a suitable substitute for structure exploitation in direct multiple shooting, and yields without exception the fastest run times among all investigated structure exploiting KKT solvers. This is true even without the application of our new update techniques, which reduce the runtime by a further factor of at most 2 for all investigated problem instances. A particular and unique feature of the HPSC factorization developed in this thesis is the very small and linear growth of the runtime with increasing number $n^q$

| $m$ | $n^\mu$ | Condensing (once) | QPOPT[1] | qpHPSC | qpHPSC +upd. | MA57 | UMFPACK | LAPACK DGBTRF | LAPACK DSYTRF |
|---|---|---|---|---|---|---|---|---|---|
| 20 | 4 | 4.49 | 0.247 | 0.737 | 0.457 | 1.366 | 1.396 | 1.734 | 35.89 |
| | 8 | 7.14 | 0.359 | 0.632 | 0.439 | 1.992 | 6.119 | 2.664 | 57.63 |
| | 12 | 10.3 | 0.365 | 0.765 | 0.426 | 2.550 | 6.129 | 4.116 | 108.6 |
| | 16 | 14.0 | 0.656 | 0.681 | 0.447 | 3.316 | 6.487 | 7.488 | 125.5 |
| 40 | 4 | 24.9 | 1.010 | 1.035 | 0.717 | 2.589 | 9.588 | 3.426 | 225.4 |
| | 8 | 42.6 | 1.392 | 1.130 | 0.750 | 3.429 | 11.58 | 5.304 | 380.3 |
| | 12 | 64.0 | 1.916 | 1.154 | 0.751 | 4.832 | 11.88 | 14.12 | 666.0 |
| | 16 | 90.6 | 2.763 | 1.234 | 0.797 | 6.299 | 13.09 | 14.15 | 1621 |
| 80 | 4 | 158 | 4.337 | 1.922 | 1.261 | 4.995 | 19.21 | 6.204 | —[2] |
| | 8 | 289 | 5.478 | 2.161 | 1.384 | 6.637 | 23.11 | 9.360 | |
| | 12 | 451 | 9.101 | 2.703 | 1.402 | 9.278 | 23.97 | 14.31 | |
| | 16 | 662 | 15.60 | 2.327 | 1.502 | 12.17 | 26.39 | 27.32 | |
| 160 | 4 | 1128 | 17.10 | 3.755 | 2.410 | 10.18 | 38.90 | 13.55 | —[2] |
| | 8 | 2224 | 26.92 | 4.115 | 2.601 | 13.88 | 46.35 | 22.67 | |
| | 12 | 3577 | 38.04 | 4.311 | 2.706 | 19.21 | 47.25 | 30.78 | |
| | 16 | 5322 | 55.30 | 4.586 | 2.907 | 25.39 | 53.79 | 59.06 | |

Table 9.15: Average runtime in milliseconds (ms) per QP solver iteration for problem (9.33), depending on the number $n^q$ of control parameters and the number $m$ of multiple shooting intervals. Averages have been taken over the first 30 SQP iterations.

[1]The QPOPT per iteration runtime applies to the condensed QP whose computation requires additional runtime as indicated by the third column.

[2]LAPACK DSYTRF performance has not been investigated for $m = 80$ and $m = 160$.

of control parameters. The largest investigated dimension $n^{\mathrm{q}} = 16$ makes the additional speedup gained by the matrix updates most apparent, but at the same time shows that much larger control dimension could be treated without significant loss in computational performance. The dense active set solver QPOPT yields faster per iteration run times only for the smallest problem instances, albeit at the cost of the condensing preprocessing step. Condensing costs the runtime equivalent of at least 20 and at most 100 iterations of QPOPT, or at least 10 and at most 2000 iterations of our code qpHPSC for the investigated problem instances. Our approach is easily competitive against recently emerging code generation approaches, e.g. [151]. Therein, an average iteration time of 425$\mu$s for an KKT system with 1740 nonzero elements is reported on a 1.7 GHz machine. For comparison, our problem instance $m = 20$, $n^{\mu} = 4$ has a KKT matrix of 9074 nonzero elements (well over 5 times more) and we achive a quite similar average iteration time of 457$\mu$s on a very similar machine about 1.5 times faster, running at 2.6 GHz.

### 9.4.5 Summary

In this section we investigated a MIOCP from automobile test driving and varied the problem's size by increasing the number $n^{\mu}$ of available gears and the granularity $m$ of the control discretization. At the example of this problem we examined the performance of our new active set QP code qpHPSC with block structured factorization and matrix updates and compared it to the classical condensing algorithm coupled with the dense active set QP code QPOPT as well as to various structure exploiting linear algebra codes used to solve the structured QP's KKT system. For optimal control problems, our new algorithm allows for the efficient treatment of problems with very long horizons or very fine discretizations of the control that could not be treated before. Consequentially for Model Predictive Control (MPC) problems it promises faster sampling times as the condensing preprocessing step is eliminated, and allows much longer prediction horizons to be treated under tight runtime constraints. Convex reformulations of MIOCPs that employ a large number of block local control parameters acting as convex multipliers pose a significant challenge to the classical condensing algorithm that effectively prevented larger problem instances from being solved efficiently. Such problems greatly benefit from the newly developed HPSC factorization with matrix updates.

## 9.5 Application: A Predictive Cruise Controller

In this section we apply the numerical theory and algorithms developed in this thesis to a challenging real–world industrial problem: real–time capable nonlinear model–predictive control of a heavy–duty truck including predictive choice of gears based on a 3D profile of the traveled roads.

### 9.5.1 Overview

Human drivers of heavy-duty trucks ideally control their vehicles in pursuit of maintaining a desired velocity, keeping the fuel consumption at a minimum. To this end, the driver repetitively chooses the truck's input controls, comprising engine torque, braking torque, and gear choice, according to human experience and anticipation of estimated road and traffic conditions. In this paper we present a novel numerical method for model-predictive control of heavy-duty trucks, acting as a cruise controller including fully automatic gear choice. The combination of nonlinear dynamics, constraints, and objective, with the hybrid nature of the gear choice makes this problem extremely difficult.

Coupled to a navigation system providing a 3D digital map of the road sections ahead of the truck, it is able to compute an optimal choice of controls not only with respect to the current system state but also with respect to anticipated future behavior of the truck on the prediction horizon. The presented algorithm is based on the direct multiple-shooting method for the efficient solution of optimal control problems constrained by ODEs or DAEs. Optimal control feedback to the truck is computed from the successive solution of a sequence of nonlinear programs resulting from the multiple-shooting discretization of the optimal control problem. Encouraging real-time capable feedback rates are achieved by exploitation of problem structures and similarities of the sequentially obtained solutions.

Today's heavy duty trucks feature a powertrain that is composed of several units. A diesel engine acts as driving unit, while several braking devices such as engine brakes, a retarder, and service brakes exist. Engine braking works by generating a retarding torque using the diesel engine. Unlike the service brakes, engine brakes and also the retarder don't suffer from wearout. Under normal circumstances, their usage is preferred over using the service brakes. The powertrain is also equipped with an automated manual gearbox with eight to sixteen gears.

In many cases an experienced truck driver chooses to accelerate, brake, or shift gears based on his ability to predict future load changes of the powertrain. In this, his chooses the operation point of the truck in a fashion suited to an oncoming period of time rather than for the current observable system state only. For example, the driver might shift down just right in time before entering a steep slope, knowing that initiating the relatively long lasting process of gear shifting later during the climb would cause too large a decrease in the truck's velocity. Also acceleration and braking of the truck can be adapted to the road conditions by an experienced driver. For instance, it may be desirable to gain speed while cruising through a valley in order to build up momentum for the next oncoming hill. Speed limit zones or slower drivers in front of the truck may require braking maneuvers or downshift of the gear as well.

It becomes clear that cruise controllers that operate solely on the knowledge of the truck's current system state inevitably will make control decisions inferior to those of an experienced heavy-duty truck driver [42, 209, 103]. The presented paper aims at the design and implementation of a cruise control system for heavy-duty trucks that predicts the behavior of the truck over a longer prediction horizon, taking into account information about the conditions of the road section ahead. To this end, we describe a novel numerical algorithm that repeatedly computes a set of controls for the truck that are optimal with respect to a performance criterion evaluated over the course of a prediction horizon. The algorithm thereby imitates the behavior of an experienced driver who repeatedly adapts his acceleration, brake, and gear choice to the desired velocity as well as the observed road and traffic conditions.

### 9.5.2 Dynamic Truck Model

This section holds a description of a 1D truck model with track slope and curvature information, introduced in different variants in [42, 103, 209] that has been used for all computations. More background information on modelling in automotive engineering can be found e.g. in [117].

We start the presentation of the truck model with the control inputs to be optimized later. The truck's acceleration is governed by the indicated engine torque, whose rate of change $R_{\text{ind}}$ can be controlled. The total braking torque's rate of change $R_{\text{brk}}$ can be controlled as well. The actual truck system uses three separate sources of brake torques: engine brake torque

$M_{\text{EB}}$, retarder torque $M_{\text{ret}}$, and service brakes torque $M_{\text{SB}}$, all with separate state-dependent upper bounds.

$$M_{\text{brk}}(s) \stackrel{\text{def}}{=} i_{\text{T}}(y)M_{\text{EB}}(s) + M_{\text{ret}}(s) + M_{\text{SB}}(s) \tag{9.34}$$

It is not necessary to separate these sources within the model used for optimization, though. We rather chose to perform an a-posteriori separation into three brake torques once the optimal sum $M_{\text{brk}}(s)$ has been decided upon. This opens up the additional possibility of modeling hierarchical brake systems e.g. to prefer using the retarder $M_{\text{ret}}$ over using the engine brakes $M_{\text{EB}}$, which again are preferred over using the service brakes $M_{\text{SB}}$. Finally, the gear $y$ enters the problem as an integer control variable that chooses from the available gearbox transmission ratios $i_{\text{T}}(y)$ and corresponding degrees of efficiency $\eta_{\text{T}}(y)$. The list of controls influencing the truck's behavior is given in table 9.16.

| Name | Description | Unit | Domain |
|---|---|---|---|
| $R_{\text{ind}}$ | Indicated engine torque rate | Nm/s | $[R_{\text{ind,min}}, R_{\text{ind,max}}]$ |
| $R_{\text{brk}}$ | Brake torque rate | Nm/s | $[R_{\text{brk,min}}, R_{\text{brk,max}}]$ |
| $y$ | Gear | – | $\{1, \ldots, y_{\max}\}$ |

Table 9.16: Controls of the truck model.

The ODE system of the truck model comprises four differential states. The location $s \in \mathcal{S} \stackrel{\text{def}}{=} [s_0, s_{\text{f}}]$ (in meters) on the map is chosen as the independent variable. This limits the model's applicability to the domain of strictly positive velocities. Time $t(s)$ depending on position $s$ and velocity $v(s)$ are recaptured using the differential equation

$$\dot{t}(s) = \frac{1}{v(s)}, \quad t(s_0) = 0. \tag{9.35}$$

The truck's velocity is computed from the summation of accelerating torques $M_{\text{acc}}$, braking torques $M_{\text{brk}}$, and resisting torques $M_{\text{air}}$ and $M_{\text{road}}$ due to turbulent and rolling friction. The parameter $m$ denotes the truck's mass. The rear axle's transmission ratio is denoted by $i_{\text{A}}$ while the static rear tire radius is $r_{\text{stat}}$. The acceleration is given by

$$\dot{v}(s) = \frac{1}{m\,v(s)} \left( \frac{i_A}{r_{\text{stat}}} \left(M_{\text{acc}} - M_{\text{brake}}\right) - M_{\text{air}} - M_{\text{road}} \right). \tag{9.36}$$

For rate-limited controls, we control the corresponding rates of change and recover the actual control values as follows.

$$\dot{M}_{\text{ind}}(s) = \frac{1}{v(s)} R_{\text{ind}}(s), \tag{9.37a}$$

$$\dot{M}_{\text{brk}}(s) = \frac{1}{v(s)} R_{\text{brk}}(s). \tag{9.37b}$$

The consumed amount of fuel is given by

$$\dot{Q}(s) = \frac{1}{v(s)} Q_{\text{fuel}}(M_{\text{ind}}(s), n_{\text{eng}}(s)) \tag{9.38}$$

where $Q_{\text{fuel}}$ gives the specific consumption rate depending on the indicated engine torque and engine speed. In table 9.17 the list of differential states of this vehicle model is given.

| Name | Description | Unit | Domain |
|---|---|---|---|
| $t$ | Time | s | $\mathbb{R}$ |
| $v$ | Velocity | m/s | $(0, v_{\text{max}}]$ |
| $M_{\text{ind}}$ | Indicated engine torque | Nm | $[0, M_{\text{ind,max}}]$ |
| $M_{\text{brk}}$ | Total brake torque | Nm | $[0, M_{\text{brk,max}}]$ |
| $Q$ | Fuel consumption | l | $[0, \infty)$ |

Table 9.17: Differential states of the truck model.

In the above system of differential equations, several terms are still missing and are computed from fixed algebraic formulas as follows. The accelerating torque $M_{\text{acc}}$ is computed from the corresponding control depending on the transmission ratio $i_T(y)$ and the degree of efficiency $\eta_T(y)$ of the selected gear $y$,

$$M_{\text{acc}}(s) \stackrel{\text{def}}{=} i_T(y)\,\eta_T(y)\,M_{\text{ind}}(s). \tag{9.39}$$

The sum of braking torques $M_{\text{brk}}$ is computed from $M_{\text{ret}}$ and $M_{\text{EB}}$, increased by resisting torques due to friction $M_{\text{fric}}$ in the engine. The value $n_{\text{eng}}$ denotes the engine's speed in revolutions per minute.

$$M_{\text{brake}}(s) \stackrel{\text{def}}{=} i_{\text{T}}(y) M_{\text{EB}}(s) + M_{\text{ret}}(s) + M_{\text{SB}}(s) + i_{\text{T}}(y)\, M_{\text{fric}}(n_{\text{eng}}(s)). \quad (9.40)$$

Additional braking torques, independent of the selected gear, due to turbulent friction $M_{\text{air}}$ and road conditions $M_{\text{road}}$ are taken into account. The parameter $A$ denotes the truck's effective flow surface, while $c_{\text{w}}$ is the aerodynamic shape coefficient and $\varrho_{\text{air}}$ the density of air,

$$M_{\text{air}}(s) \stackrel{\text{def}}{=} \tfrac{1}{2}\, c_{\text{w}}\, A\, \varrho_{\text{air}}\, v^2(s). \quad (9.41)$$

The road conditions term accounts for rolling friction with coefficient $f_{\text{r}}$ and downhill force depending on the slope $\gamma(s)$ available from the 3D map data of the road. The parameter $g$ is the gravity constant.

$$M_{\text{road}}(s) \stackrel{\text{def}}{=} m\, g\, (\sin\gamma(s) + f_{\text{r}} \cos\gamma(s)). \quad (9.42)$$

Finally, the engine's speed in revolutions per minute, depending on the selected gear $y$, can be recaptured from the truck's current velocity,

$$n_{\text{eng}}(s) \stackrel{\text{def}}{=} v(s)\, \frac{i_{\text{A}}\, i_{\text{T}}(y(s))}{r_{\text{stat}}}\, \frac{60\ [\text{s}]}{2\pi}. \quad (9.43)$$

Table 9.18 holds the list of fixed model parameters. The truck engine's characteristics are defined by the functions $M_{\text{ind,max}}$, $M_{\text{brk,max}}$, and $M_{\text{fric}}$ giving the maximum indicated torque, braking torque amd torque loss due to friction, all depending on the engine speed $n_{\text{eng}}$.

In addition, the specific fuel consumption rate $Q_{\text{fuel}}$ depends on both $M_{\text{ind}}$ and $n_{\text{eng}}$. Representative examples of those functions are shown in figure 9.22.

### 9.5.3 Environment Model

The truck system is subjected to various environmental conditions changing over time as the position $s$ of the truck advances. The unique feature of this predictive truck control problem is the changing slope $\gamma(s)$ and curvature $\kappa(s)$ of the road. This information is obtained from a 3D map available in electronic form on board the truck. Positioning information is made available with the help of the Global Positioning System (GPS) and allows to map a position $s$ on the track to cartesian coordinates $(x, y)$ and road conditions $(\gamma, \kappa)$. Figure 9.23 shows a representative section of 3D map data.

On the prediction horizon, the truck system needs to respect certain mechanical constraints, such as velocity and engine speed limits. Beside the bounds on the truck controls given in table 9.16 and on the truck system's differential states listed in table 9.17, the truck's velocity $v(s)$ is subject to several constraints, the most significant ones being the velocity limits imposed by law,

$$v(s) \leqslant v_{\text{law}}(s) \qquad \forall s \in \mathcal{S}. \tag{9.44}$$

From the available 3D map data, the predictive optimal control algorithm, the curvature $\kappa(s)$ of the road at position $s$ is extracted and converted to a maximum allowable velocity $v_{\text{curve}}(s)$,

$$v(s) \leqslant v_{\text{curve}}(\kappa(s)) \qquad \forall s \in \mathcal{S}. \tag{9.45}$$

The indicated and brake torques must respect state-dependent upper limits as specified by the engine characteristics

$$0 \leqslant M_{\text{ind}}(s) \leqslant M_{\text{ind,max}}\left(n_{\text{eng}}(s)\right) \qquad s \in \mathcal{S}, \tag{9.46a}$$

$$0 \leqslant M_{\text{brk}}(s) \leqslant M_{\text{brk,max}}\left(n_{\text{eng}}(s)\right) \qquad s \in \mathcal{S}. \tag{9.46b}$$

| Name | Description | Unit |
|---|---|---|
| $A$ | Front facing area | m$^2$ |
| $c_{\text{w}}$ | Aerodynamic shape coefficient | – |
| $f_{\text{r}}$ | Coefficient of rolling friction | – |
| $\gamma(s)$ | Road's slope | rad |
| $g$ | Gravity constant | m/s$^2$ |
| $i_{\text{A}}(y)$ | Rear axle transmission ratio | – |
| $i_{\text{T}}(y)$ | Gearbox transmission ratio | – |
| $\kappa(s)$ | Road's curvature | – |
| $m$ | Vehicle mass | kg |
| $n_{\text{eng,min}}$ | Minimum engine speed | 1/min |
| $n_{\text{eng,max}}$ | Maximum engine speed | 1/min |
| $\eta_{\text{T}}$ | Gearbox degree of efficiency | – |
| $\varrho_{\text{air}}$ | Air density | kg/m$^3$ |
| $r_{\text{stat}}$ | Static rear tire radius | m |

Table 9.18: Parameters of the truck model.

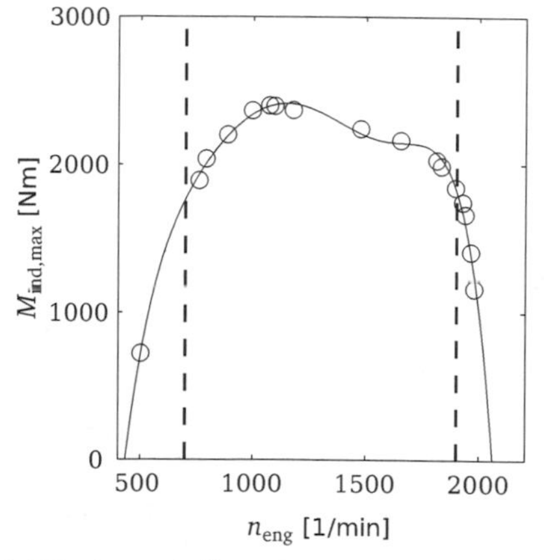


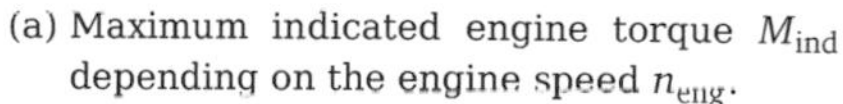

(a) Maximum indicated engine torque $M_{ind}$ depending on the engine speed $n_{eng}$.

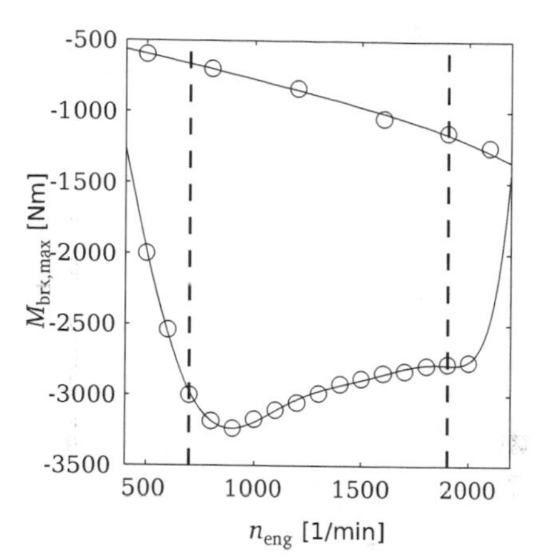


(b) Maximum engine brake torque $M_{EB}$ and retarder torque $M_{ret}$ depending on the engine speed $n_{eng}$.

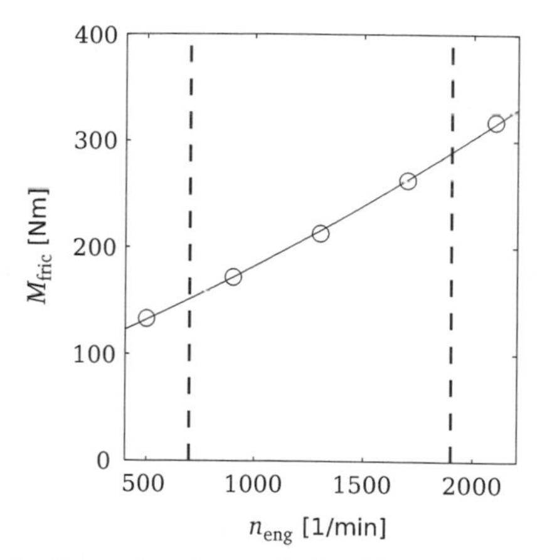


(c) Torque loss due to friction $M_{fric}$ depending on the engine speed $n_{eng}$.

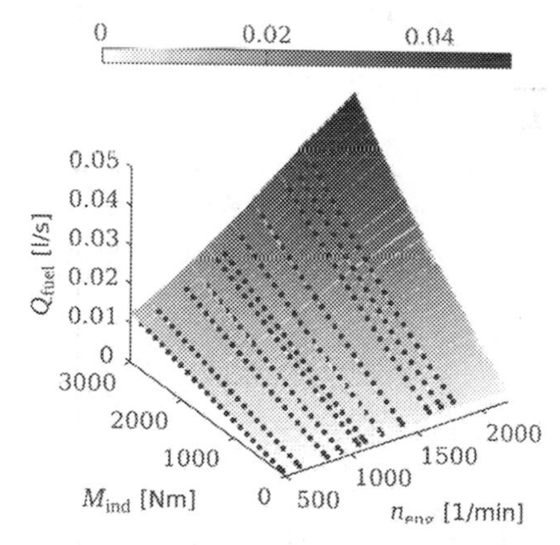


(d) Specific fuel consumption rate $Q_{fuel}$ depending on the indicated engine torque $M_{ind}$ and the engine speed $n_{eng}$.

Figure 9.22: Exemplary nonlinear truck engine characteristics.

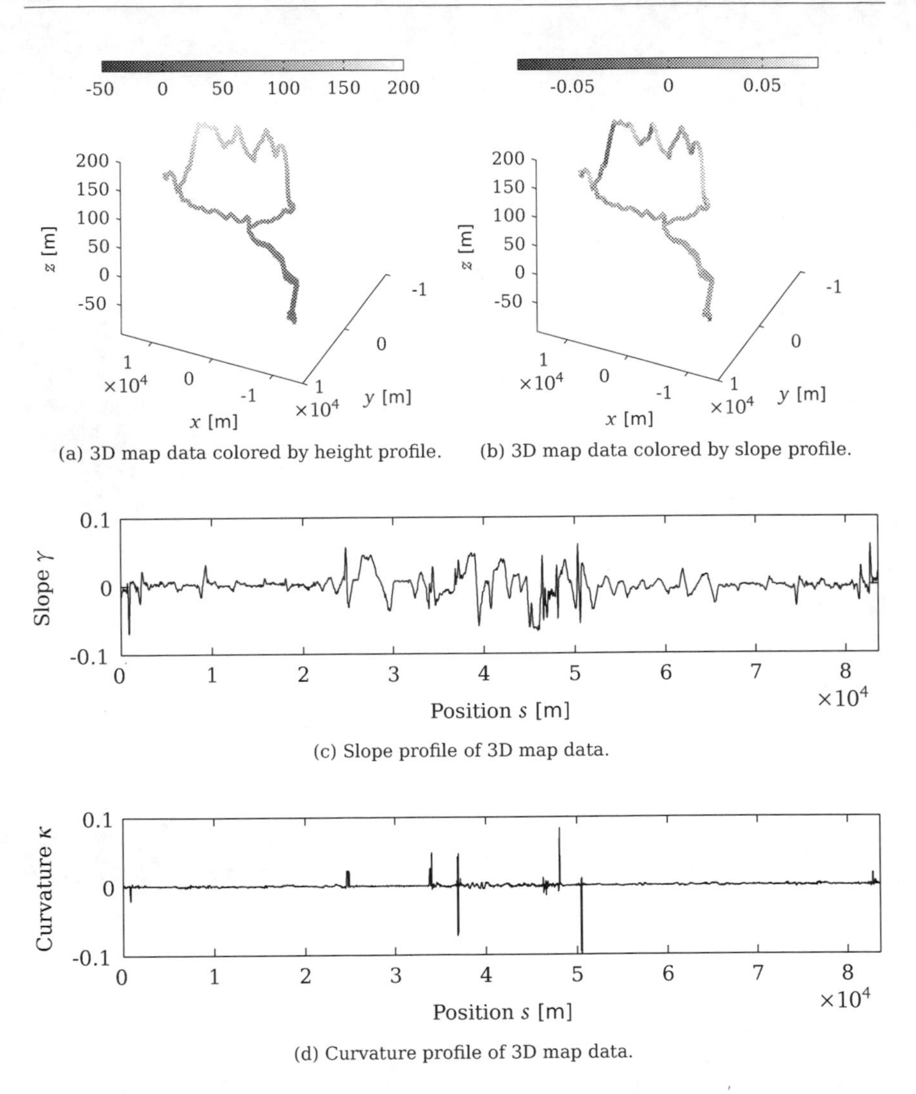


(a) 3D map data colored by height profile.

(b) 3D map data colored by slope profile.

(c) Slope profile of 3D map data.

(d) Curvature profile of 3D map data.

Figure 9.23: Exemplary real–world 3D map data describing road conditions for the truck predictive control problem on a track of over 80 kilometers length. Sharp spikes in the curvature profile 9.23d correspond to major changes in the general direction of travelling in figures 9.23a and 9.23b.

Finally, the engine's revolutionary speed $n_{\text{eng}}$, depending on the truck's velocity and the selected gear, must stay within prescribed limits according to the engine's specification,

$$n_{\text{eng,min}} \leqslant n_{\text{eng}}\left(v(s), y(s)\right) \leqslant n_{\text{eng,max}} \qquad \forall s \in \mathcal{S}. \tag{9.47}$$

### 9.5.4 Design of a Performance Index

We're interested in computing continuous optimal control trajectories $R_{\text{ind}}(\cdot)$, $R_{\text{brk}}(\cdot)$ and an integer optimal control trajectory $y(\cdot)$ on a section $\mathcal{S}$ of a track described by parameters $\gamma(\cdot)$ and $\kappa(\cdot)$ varying in space such that the truck completes this track satisfying a chosen compromise between minimal energy consumption and earliest arrival time. The integral cost criterion to be minimized in each of the successive steps of the moving horizon predictive control algorithm is composed of a weighted sum of three different objectives.

1. *Deviation from a desired velocity*:

   The deviation of the truck's observed velocity from the desired one is penalized in a least-squares sense over the length $H$ of the prediction horizon $\mathcal{S} \overset{\text{def}}{=} [s_0, s_0 + H]$ starting at the truck's current position $s_0$ on the road,

$$\Phi_{\text{dev}} \overset{\text{def}}{=} \int_{s_0}^{s_0+H} \left(v(s) - v_{\text{des}}(s)\right)^2 \,\mathrm{d}s. \tag{9.48}$$

   This computation of a profile of desired velocities to be tracked by the predictive controller is automated and will in general match the maximum velocity permitted by law. For sharp bends of the road, a heuristic is applied to adapt the velocity to be tracked to the truck's physical capabilities. This avoids large residuals of the objective that impede the numerical behavior of the Gauss–Newton least squares tracking. In addition, several parameters of this heuristic can be exposed to the truck driver to grant some additional freedom for adaptation to environmental conditions such as traffic or weather extending over the premises made by the predictive control algorithm.

2. *Fuel consumption*:

   The fuel consumption is identified from the specific fuel consumption rate map exemplarily shown in figure 9.22d, and is minimized by the objective contribution

$$\Phi_{\text{fuel}} \stackrel{\text{def}}{=} \int_{s_0}^{s_0+H} \frac{1}{v(s)} Q\left(n_{\text{eng}}(s), M_{\text{ind}}(s)\right) \, \mathrm{d}s. \tag{9.49}$$

3. *Driving comfort*:

   Rapid changes of the indicated engine torque degrade the driving comfort as experienced by the truck driver:

$$\Phi_{\text{comf}} \stackrel{\text{def}}{=} \int_{s_0}^{s_0+H} \frac{1}{v(s)} \left(R_{\text{ind}}(s) + R_{\text{brk}}(s)\right)^2 \, \mathrm{d}s. \tag{9.50}$$

   This objective part can be seen as tracking zero acceleration in absence of slope, or as a control regularization from a numerical point of view.

Weighting the objective function contributions and summing up, we obtain the combined objective function

$$\Phi \stackrel{\text{def}}{=} \lambda_1 \Phi_{\text{dev}} + \lambda_2 \Phi_{\text{fuel}} + \lambda_3 \Phi_{\text{comf}}. \tag{9.51}$$

The weighting factor $\lambda_3$ is chosen to be comparably small in our computations. The choice of $\lambda_1$ and $\lambda_2$ allows for a gradual selection of a compromise between meeting the desired velocity that can even be chosen on–line by the truck driver who may prefer to travel faster at the cost of increased fuel consumption or, being ahead of his schedule, may prefer to save on fuel by following an economic operating mode of the truck at the cost of longer travel times. For more details on the investigation of pareto–optimality of mixed–integer control problems we refer to our paper [141].

### 9.5.5 Mixed–Integer Optimal Control Problem

#### Problem Formulation

The MIOCP resulting from the presented vehicle and environment model is given in (9.54). We summarize the state vectors

$$\boldsymbol{x}(s) = \left[ v(s) \;\; M_{\text{ind}}(s) \;\; M_{\text{brk}} \;\; Q(s) \;\; t(s) \right] \tag{9.52}$$

and the continuous controls vectors

$$u(s) = \begin{bmatrix} R_{\text{ind}}(s) & R_{\text{brk}}(s) \end{bmatrix} \tag{9.53}$$

and denote by $w(s) = y(s)$ the gear choice control. In (9.54) the ODE system (9.54b) comprises the vehicle ODE model derived in section 9.5.2. The engine speed constraints (9.54e) depends on the integer control $y(s)$ in (9.54f).

$$\min_{x(\cdot),u(\cdot),w(\cdot)} \quad \lambda_1 \Phi_{\text{dev}} + \lambda_2 \Phi_{\text{fuel}} + \lambda_3 \Phi_{\text{comf}} \tag{9.54a}$$

$$\text{s.t.} \quad \dot{x}(s) = f(s, x(s), u(s), w(s), p) \quad s \in \mathcal{S}, \tag{9.54b}$$

$$v(s) \leqslant v_{\text{law}}(s) \quad s \in \mathcal{S},$$

$$v(s) \leqslant v_{\text{curve}}(s) \quad s \in \mathcal{S},$$

$$M_{\text{ind}}(s) \in [0, M_{\text{ind,max}}(v(s), y(s))] \quad s \in \mathcal{S}, \tag{9.54c}$$

$$M_{\text{brk}}(s) \in [0, M_{\text{ind,brk}}(v(s), y(s))] \quad s \in \mathcal{S}, \tag{9.54d}$$

$$R_{\text{ind}}(s) \in [R_{\text{ind,min}}, R_{\text{ind,max}}] \quad s \in \mathcal{S},$$

$$R_{\text{ind}}(s) \in [R_{\text{brk,min}}, R_{\text{brk,max}}] \quad s \in \mathcal{S},$$

$$n_{\text{eng}}(v(s), y(s)) \in [n_{\text{eng,min}}, n_{\text{eng,max}}] \quad s \in \mathcal{S}. \tag{9.54e}$$

$$y(s) \in \{1, \dots, n^{\text{y}}\} \quad s \in \mathcal{S}. \tag{9.54f}$$

## Outer Convexification Reformulation

We first address the reformulation of this MIOCP using outer convexification of the objective and ODE dynamics with respect to the integer gear choice, and relaxation of the introduced binary convex multipliers. We introduce $n^{\text{y}}$ binary control functions $\omega_j(\cdot) \in \{0, 1\}$, $1 \leqslant j \leqslant n^{\text{y}}$ each indicating whether the $j$-th gear is selected at location $s \in \mathcal{S}$ on the prediction horizon, together with their relaxed counterpart functions $\alpha_j(\cdot) \in [0, 1] \subset \mathbb{R}$. From these the selected gear may be computed as

$$y(s) = \sum_{j=1}^{n^{\text{y}}} j \alpha_j(s), \qquad s \in \mathcal{S}, \tag{9.55}$$

and is integral if the multipliers $\alpha_j(s)$ are binary. We convexify the contributing term $\Phi_{\text{fuel}}$ of the objective (9.51) of problem (9.54) with respect to the integer control $y(\cdot)$ as follows,

$$\Phi_{\text{fuel}}(s) \stackrel{\text{def}}{=} \int_{s_0}^{s_0+H} \frac{1}{v(s)} \sum_{j=1}^{n^y} \alpha_j(s) Q\left(n_{\text{eng}}(v(s), j), M_{\text{ind}}(s)\right) \, \mathrm{d}s, \tag{9.56}$$

while the contributing terms $\Phi_{\text{dev}}$ and $\Phi_{\text{comf}}$ are independent of $w(t)$ and hence remains unchanged. We further convexify the dynamics of problem (9.54) with respect to the integer control $y(\cdot)$ as follows,

$$\dot{x}(s) = \sum_{j=1}^{n^y} \alpha_j(s) \boldsymbol{f}(s, \boldsymbol{x}(s), \boldsymbol{u}(s), j, \boldsymbol{p}) \qquad \forall s \in \mathcal{S}, \tag{9.57}$$

where the gear choice constraint (9.54f) is replaced by

$$\alpha_j(s) \in [0,1], \quad 1 \leqslant j \leqslant n^y, \qquad \sum_{j=1}^{n^y} \alpha_j(s) = 1, \quad \forall s \in \mathcal{S}. \tag{9.58}$$

The torque constraints (9.54c) and (9.54d) are written as

$$0 \leqslant M_{\text{ind}}(s) \leqslant \sum_{j=1}^{n^y} \alpha_j(s) M_{\text{ind,max}}(v(s), j), \tag{9.59a}$$

$$0 \leqslant M_{\text{brk}}(s) \leqslant \sum_{j=1}^{n^y} \alpha_j(s) M_{\text{brk,max}}(v(s), j). \tag{9.59b}$$

**Reformulations of the Engine Speed Constraint**

In chapter 5 we have proposed several possible reformulations of path constraints directly depending on an integreal control. For the engine speed constraint (9.47) depending on the gear choice $y(s)$, we study again three of the proposed reformulations.

**Inner Convexification** We briefly look at the effect of treating $y(s)$ as a continuous variable, referred to as inner convexification of the gear choice in this works. This modelling approach results in the formulation

$$n_{\text{eng,min}} \leqslant n_{\text{eng}}(v(s), y(s)) \leqslant n_{\text{eng,max}}, \qquad y(s) \in [1, n^y] \subset \mathbb{R}. \tag{9.60}$$

From an engineering point of view, inner convexification amounts to assuming an idealized continuous transmission gearbox that is able to run on arbitrary ratios of the engine speed and the vehicle resp. wheel speed. An appropriate formulation might also introduce the engine speed $n_{\text{eng}}(s)$ as a free control trajectory on $\mathcal{S}$ subject to optimization, and impose a transmission ratio constraint,

$$i_{\text{T,min}} \leqslant \frac{n_{\text{eng}}(s)}{v(s)} \frac{r_{\text{stat}}}{i_{\text{A}}} \frac{2\pi}{60\ [\text{s}]} \leqslant i_{\text{T,max}}, \; n_{\text{eng}}(s) \in [n_{\text{eng,min}}, n_{\text{eng,max}}]. \quad (9.61)$$

This formulation is appropriate for vehicles with a built–in CVT (continuously variable transmission) drive, but for gearboxes with a finite number $n^{\text{y}}$ of available gears, several issues arise. Optimal solutions computed using this modelling approach need to to be "rounded" towards an actually available transmission ratio resp. engine speed. Bounds on the loss of optimality or feasibility of the rounded solution cannot be given. As an example, in figure 9.24 the constraint on $M_{\text{ind}}(s)$ is shown in its inner convexification reformulation i.e., with the gear choice $y(s)$ treated as continuous control, and in its outer convexification formulation. Constraint violations for fractional gear choices caused by the inner convexification reformulation are clearly visible in figure 9.24a and are avoided in figure 9.24b. Finally, engine and vehicle characteristics most often represented as tabulated and interpolated data need to be extended to physically unavailable transmission ratios in a sufficiently often continuously differentiable way in order to make the mathematical model of both engine and vehicle evaluatable.

**Standard Formulation after Outer Convexification** We next consider the formulation of (9.47) after outer convexification of the objective and dynamics with respect to the integer gear choice,

$$n_{\text{eng,min}} \leqslant \sum_{j=1}^{n^{\text{y}}} \alpha_j(s) n_{\text{eng}}(v(s), j) \leqslant n_{\text{eng,max}}. \quad (9.62)$$

Here, outer convexification with respect to $y(s)$ is applied to the engine speed $n_{\text{eng}}(v(s), y(s))$, and the constraint is imposed on the engine speed obtained by evaluation of the convex combination. Contrary to the inner convexification approach, this formulation relieves us from the need to evaluate the engine and vehicle model functions for fractional gear choices. Observe though that the engine speed $n_{\text{eng}}(v(s), j)$ resulting for an individual (partially) selected gear $j$ with $\alpha_j(s) > 0$ may violate either bound as long as there exists another gear $k$ *compensating* for this violation in the summed-

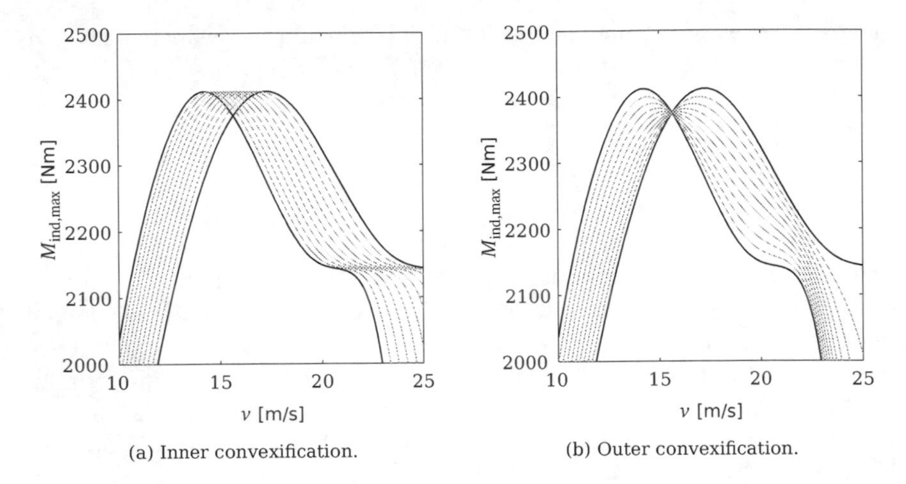


(a) Inner convexification. (b) Outer convexification.

Figure 9.24: Inner and outer convexification of the indicated engine torque constraint for two adjacent gears. Constraint violations for fractional gear choices caused by the inner convexification reformulation are clearly visible.

up residual (9.62). This effect is shown in figure 9.29 on page 308 for the mixed–integer optimal control scenario of figure 9.28, page 307.

**Outer Convexification of the Constraint** In this thesis we proposed for the first time to apply outer convexification also to the path constraint directly depending on the integer control,

$$0 \leqslant \alpha_j(s)\left(n_{\text{eng}}(v(s), j) - n_{\text{eng,min}}\right), \qquad 1 \leqslant j \leqslant n^{\text{y}}, \tag{9.63a}$$

$$0 \leqslant \alpha_j(s)\left(n_{\text{eng,max}} - n_{\text{eng}}(v(s), j)\right). \tag{9.63b}$$

Instead of a single constraint on the engine speed resulting from a convex combination, we now impose a separate constraint on the engine speed for each available gear. Clearly, if $\alpha_j(s) = 0$ and gear $j$ is not chosen at point $s \in \mathcal{S}$, the two constraints (9.63) are feasible regardless of the actual velocity $v(s)$. If $\alpha_j(s) > 0$ and gear $j$ enters the convex combination, feasibility of the associated engine speed $n_{\text{eng}}(v(s), j)$ is required.

Different from the previous formulation, this requirement must be sat-

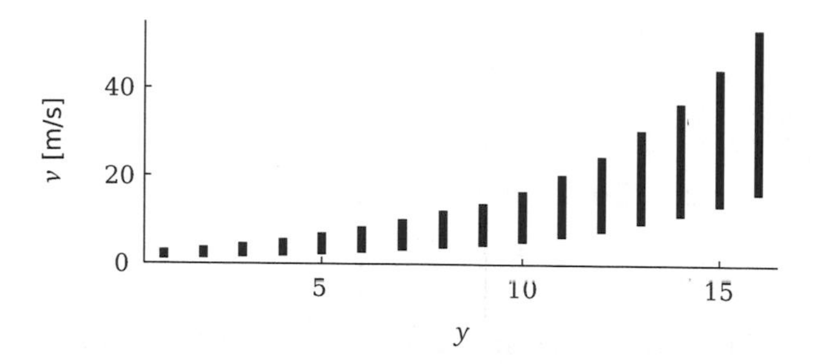


Figure 9.25: Feasible vehicle speeds $v(s)$ for a given choice $y(s) = j$ of the gear. The figure reveals the combinatorial structure of the feasible set created by the constraints $n_{\text{eng,min}} \leqslant n_{\text{eng}}(v(s), j) \leqslant n_{\text{eng,max}}$ that vanish if the associated convex multiplier $\alpha_j(s)$ is zero.

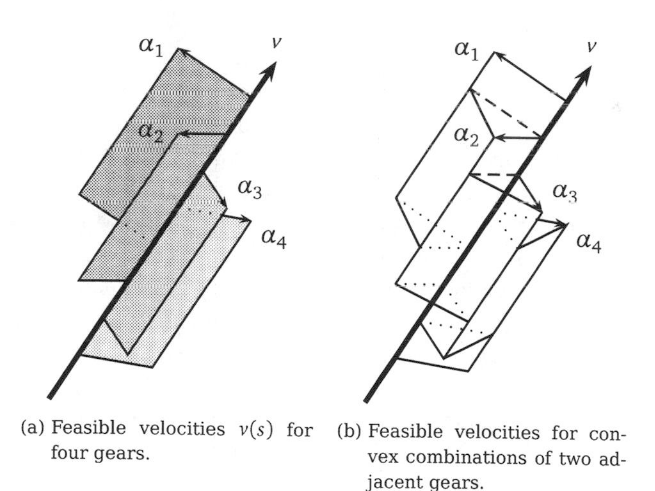


(a) Feasible velocities $v(s)$ for four gears.

(b) Feasible velocities for convex combinations of two adjacent gears.

Figure 9.26: Schematic of the nonconvex feasible set for the vehicle velocity $v$ described by the engine control reformulation (9.63) with four gears ($n^{\text{y}} = 4$). Arrows indicate axes of the space $(v, \boldsymbol{\alpha}) \in \mathbb{R}^+ \times [0,1]^{n^{\text{y}}}$. The feasible set is depicted in 9.26a for integral choices of the gear ($\alpha_j = 1$ for some $1 \leqslant j \leqslant n^{\text{y}}$, $\alpha_k = 0$ for all $k \neq j$), and in 9.26b for convex combinations formed between two adjacent gears, i.e., an SOS 2 constraint imposed on the convex multipliers $\alpha_j$, $1 \leqslant j \leqslant n^{\text{y}}$.

isfied even though the convex multiplier $\alpha_j(s)$ might be less than 1. Consequentially, the individual engine speeds of all gears entering the convex combination are indeed feasible on their own. Rounding of a fractional relaxed optimal solution to the partially convexified problem (9.54) hence does not violate the critical engine speed constraint.

### 9.5.6 Exemplary Mixed–Integer Optimal Control Scenarios

In this section we present MIOCP solutions to problem (9.54) that demonstrate the predictive nature of the modelled cruise controller at the example of two short road sections with a steep slope and a speed limit imposed. Compensation effects of the standard constraint formulation after outer convexification are investigated.

#### Example: Steep Slope Scenario

Figure 9.27 on page 306 shows the optimal solution to a mixed–integer optimal control scenario on a road section of two kilometers length with a steep slope of 8% for 500 meters, starting at 500 meters into the section. No curvature is present. The desired velocity is set at $v_{\text{des}} = 80$ km/h $= 22.\overline{2}$ m/s as indicated in figure 9.27a, while the initial velocity at the start of the scenario is set to 19 m/s. Objective function weights are chosen as $\lambda_{\text{dev}} = 10^{-2}$, $\lambda_{\text{fuel}} = 10^{-2}$, $\lambda_{\text{comf}} = 10^{-4}$ such that contribution tracking the desired velocity dominates the objective. The truck enters the slope with a velocity exceeding the desired one, as can be seen in figure 9.27a. This happens in order to maximize the remaining exit velocity after the slope has been tackled. Figure 9.27b shows the accumulated fuel consumption. In figure 9.27c the effective torque $M_{\text{ind}}(s) - M_{\text{brk}}(s)$ can be seen together with its engine speed dependent upper bound. Figure 9.27d shows the associated torque rate. In figure 9.27e downshifting of the gear from 13 down to 10 can be seen in order to maintain an engine speed of over 1500 1/min seen in figure 9.27f, as the velocity drops from 25 m/s down to 13 m/s at the exit of the slope. The fast acceleration sequence starting at 1000 meters into the section is accompanied by rapid upshifting to the highest gear number 16 as the desired velocity has been reached again.

### Example: Speed Limit Scenario

In figure 9.28 on page 307, the slope has been replaced by a speed limit of 40 km/h and the truck's initial speed has been set to 40 km/h as well. Acceleration and braking can be seen in figure 9.28a and 9.28c to minimize the time until the entry of the speed limit zone. This zone is crossed at the lowest possible engine speed in order to save fuel. The acceleration back to the desired velocity is accompanied by upshift of the gear to keep the engine speed above 1500 1/min again, while the remainder of the road section is completed at around 800 1/min in gear 16 as the desired velocity has been reached.

### Example: Compensatory Effects

In figure 9.29 on page 308 the choice convex multipliers $\boldsymbol{\alpha}(s)$ for the gear $y(s)$ can be seen if the first example of figure 9.28 is solved with the standard formulation (9.62) of the engine speed constraint (9.47) after outer convexification, instead of using the vanishing constraint formulation proposed in this thesis. Figure 9.29 shows for each of the 16 available gears the resulting engine speed if a (partial) selection of the respective gear is indicated by the convex relaxed multipliers $\boldsymbol{\alpha}(s)$. Compensatory effects are clearly revealed. They allow for a selection of infeasible gears violating the engine speed constraints (e.g. gears 5 to 9 in figure 9.29), if this violation is compensated for by (partial) selection of other gears that either do not hit the engine speed constraint or violate the opposite constraint (e.g. gears 14 to 16) such that violations cancel out in (9.62)

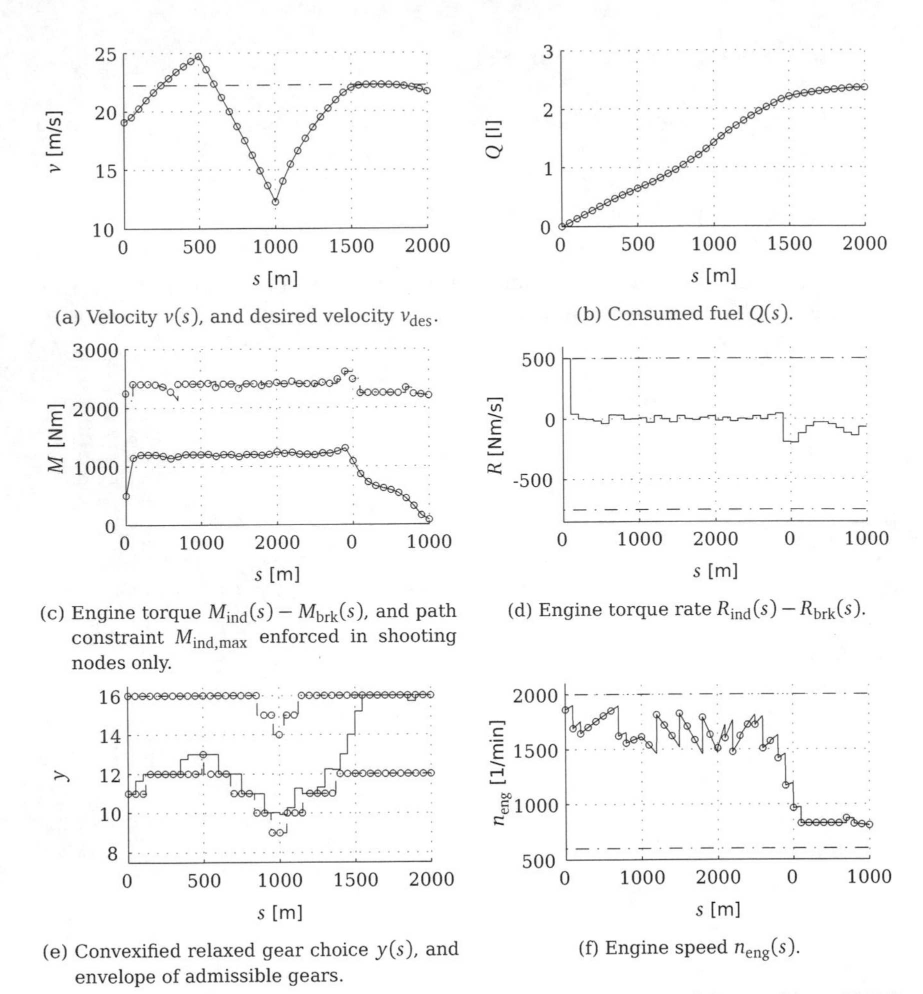


(a) Velocity $v(s)$, and desired velocity $v_{des}$.

(b) Consumed fuel $Q(s)$.

(c) Engine torque $M_{ind}(s) - M_{brk}(s)$, and path constraint $M_{ind,max}$ enforced in shooting nodes only.

(d) Engine torque rate $R_{ind}(s) - R_{brk}(s)$.

(e) Convexified relaxed gear choice $y(s)$, and envelope of admissible gears.

(f) Engine speed $n_{eng}(s)$.

Figure 9.27: Optimal engine torque and gear choice computed for problem (9.54) with outer convexification and relaxation applied to dynamics and path constraints. On a road section of two kilometers length with a slope of 8% for one kilometer, the controller aims at a compromise between maintaining a velocity of 80 km/h and burning as little fuel as possible. The entry velocity exceeds the desired one in figure 9.28a in order to maximize the exit velocity that drops due to the steep slope. Downshifts from gear 13 to gear 10 in figure 9.28e keep the engine speed above 1500 1/min in figure 9.28f. Once the slope has been tackled, the highest available gear 16 is selected and the engine speed drops to 800 1/min in order to save fuel while maintaining the desired velocity.

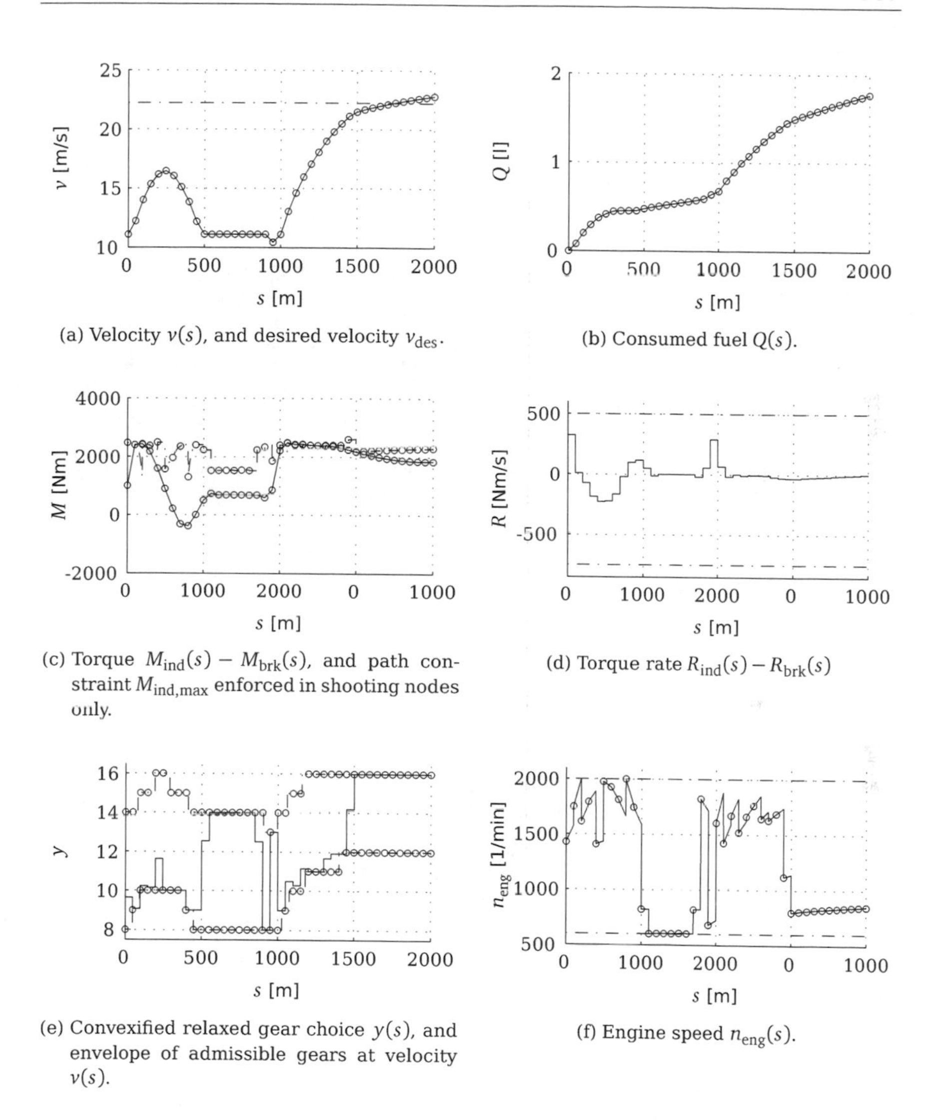


(a) Velocity $v(s)$, and desired velocity $v_{\text{des}}$.

(b) Consumed fuel $Q(s)$.

(c) Torque $M_{\text{ind}}(s) - M_{\text{brk}}(s)$, and path constraint $M_{\text{ind,max}}$ enforced in shooting nodes only.

(d) Torque rate $R_{\text{ind}}(s) - R_{\text{brk}}(s)$

(e) Convexified relaxed gear choice $y(s)$, and envelope of admissible gears at velocity $v(s)$.

(f) Engine speed $n_{\text{eng}}(s)$.

Figure 9.28: Optimal engine torque and gear choice computed for problem (9.54) with outer convexification and relaxation applied to dynamics and path constraints. On a road section of two kilometers length with a speed limit of 40 km/h imposed for a section of 500 meters, the controller aims at maintaining a velocity of 80 km/h.

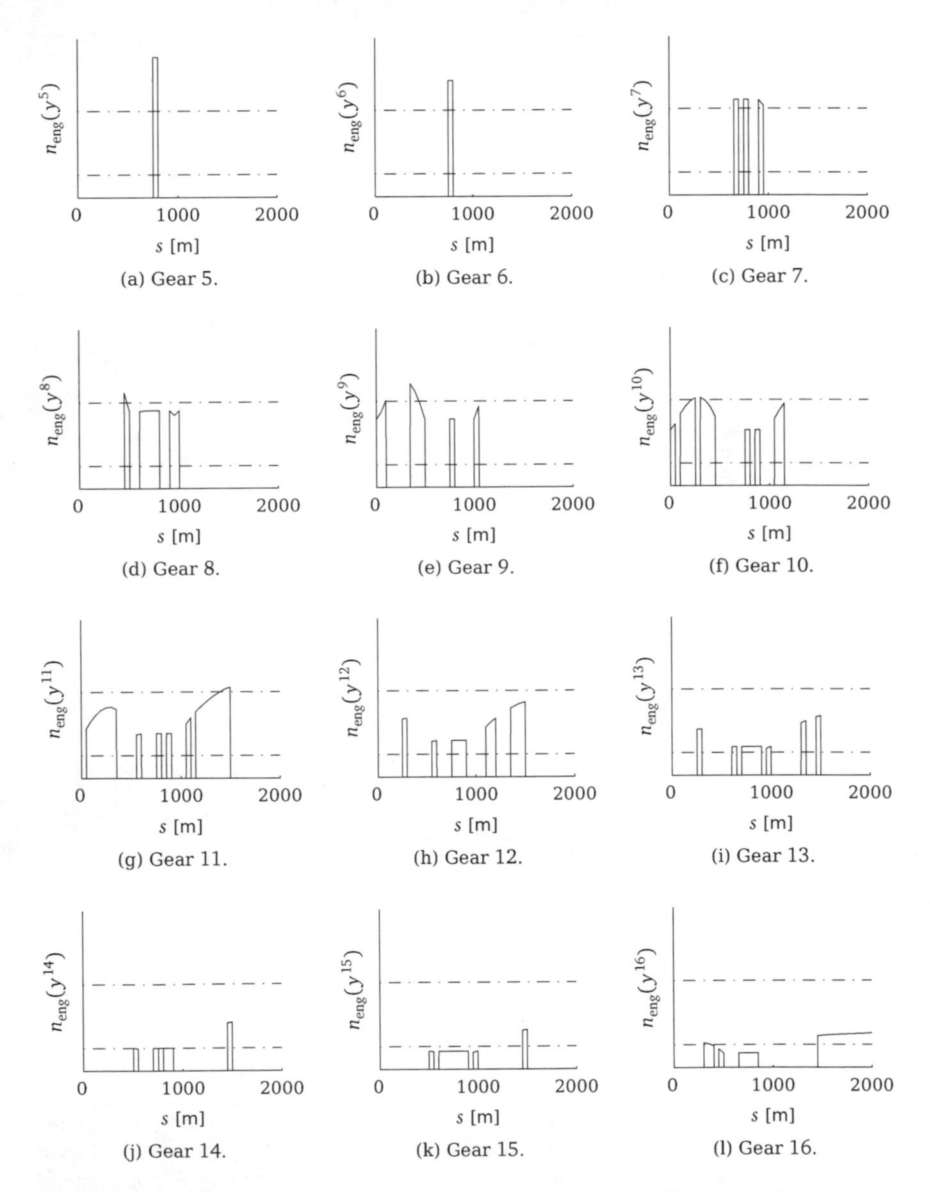


Figure 9.29: Compensatory effects arising for inner convexification of the engine speed constraint at the example of figure 9.28. Engine speeds associated with each partially selected gear are indicated and can be seen to violate phyical limits.

### 9.5.7 Mixed–Integer Predictive Control Results

We finally present mixed–integer model–predictive control results for their cruise controller using the above MIOCP formulation. Using a prediction horizon of 2000 meters length, discretized into $m = 40$ direct multiple shooting intervals of 50 meters length each, we compute mixed–integer feedback control trajectories for a south german highway of 150 kilometers length. We give feedback every 10 meters and use the warm starting strategy of section 4.2.3 for initialization.

In figure 9.30 spread across pages 311 to 313 the highway's slope profile, the realized feedback controls, and the obtained system states are shown for the entire distance of 150 kilometers. For this solution, we chose $\lambda_{\text{dev}} = 10^{-2}$, $\lambda_{\text{fuel}} = 10^{-2}$, and $\lambda_{\text{comf}} = 10^{-4}$ to obtain a speed–oriented behavior of the cruise controller.

We can see in figure 9.30d that the desired velocity $v_{\text{des}}$ of 80 km/h is kept during the majority of the truck's journey. Minor deviations are accepted to prevent unnecessary acceleration or braking maneuvres that would impact the driver's comfort and burn additional fuel. Major deviations have to be accepted on steep sloped parts of the highway that make it impossible for the truck to sustain the desired velocity at a load of $m = 40$ metric tonnes. Gear 12 is selected most of as seen in figure 9.30g, keeping the engine speed $n_{\text{eng}}$ well above 1500 rpm to ensure maximum responsiveness of the truck according to its engine characteristics. This of course comes at the cost of increased fuel consumption. Sloped parts of the highway see very regular downshift and upshift sequences of the gear as the truck enters and exits the sloped sections. The highest available gear 16 is occasionally chosen on flat sections to save on fuel if the desired velocity has already been reached.

In figure 9.31 on page 314, details of this feedback solution can be studied for a highway section of 6 kilometers length starting after a traveled distance of 54 kilometers. Here, a steep slope of two kilometers length is present with the relative altitude of the highway rising to 400 meters, up from around 230 meters above the truck's initial position, see figure 9.31b. The predictive nature of the controller's behavior can be seen clearly. In figure 9.31c the velocity can be seen to rise above the desired velocity well before the truck actually enters the steep sloped section. This ameliorates the exit velocity that nonetheless drops to about 40 km/h down from over 90 km/h. The engine runs at maximum indicated torque (figure 9.31e) and the total amount of consumed fuel rises (figure 9.31d) accordingly. Downshifts of the gear seen in figure 9.31g keep the engine's speed around 1500 rpm as seen in figure 9.31h, the engine's most efficient mode of operation. An upshift sequence ending at the highest available gear 16 completes the studied excerpt once the steep slope has been crossed successfully. As already noted above, the predictive controller returns to gear 12 as the desired velocity has been reached again.

Figure 9.32 shows mixed–integer feedback control trajectories for the same scenario with modified objective function weights $\lambda_{\text{dev}} = 10^{-3}$, $\lambda_{\text{fuel}} = 10^{-1}$, and $\lambda_{\text{comf}} = 10^{-4}$ to achieve a more fuel consumption aware behavior of the predictive cruise controller. Consequentially, the highest available gear 16 is chosen significantly more frequently, leading to an overall engine speed that is much lower than before. Downshift and upshift sequences during sloped parts of the highway are more prominent, and less chattering of the gear choice is observed in figure 9.32g. The fuel consumption over 150 kilometers has been noticeably reduced (figure 9.32c). This of course comes at the cost of larger deviations from the desired velocity (figure 9.32d).

### 9.5.8 Computational Demand

We finally investigate the computational demand of our mixed–integer model–predictive cruise controller at the example of the results presented in figure 9.30 on page 313. The number of iterations of our QPVC active set method is shown in figure 9.33a. The problem's dimensions are $m = 40$, $n^{\text{x}} = 3$, and $n^{\text{q}} = 17$. Clearly, at least one iteration is performed per feedback phase of the mixed–integer real–time iteration scheme. Most iteration counts remain below 20, with few exceptions attributed to searching adjacent convex feasible subsets for improvements towards global optimality as described in chapter 6.

The computational effort of the feedback phase of our mixed–integer real–time iteration scheme is shown in figure 9.33b and is closely correlated with the number of QPVC iterations. This indicates that the runtime of the HPSC factorization and updates is dominated by the length $m = 40$ of the direct multiple shooting grid. Approximately 500 $\mu$s are spent per active set iteration, such that the total feedback delay remains below 10 milliseconds for most mixed–integer feedback controls computed.

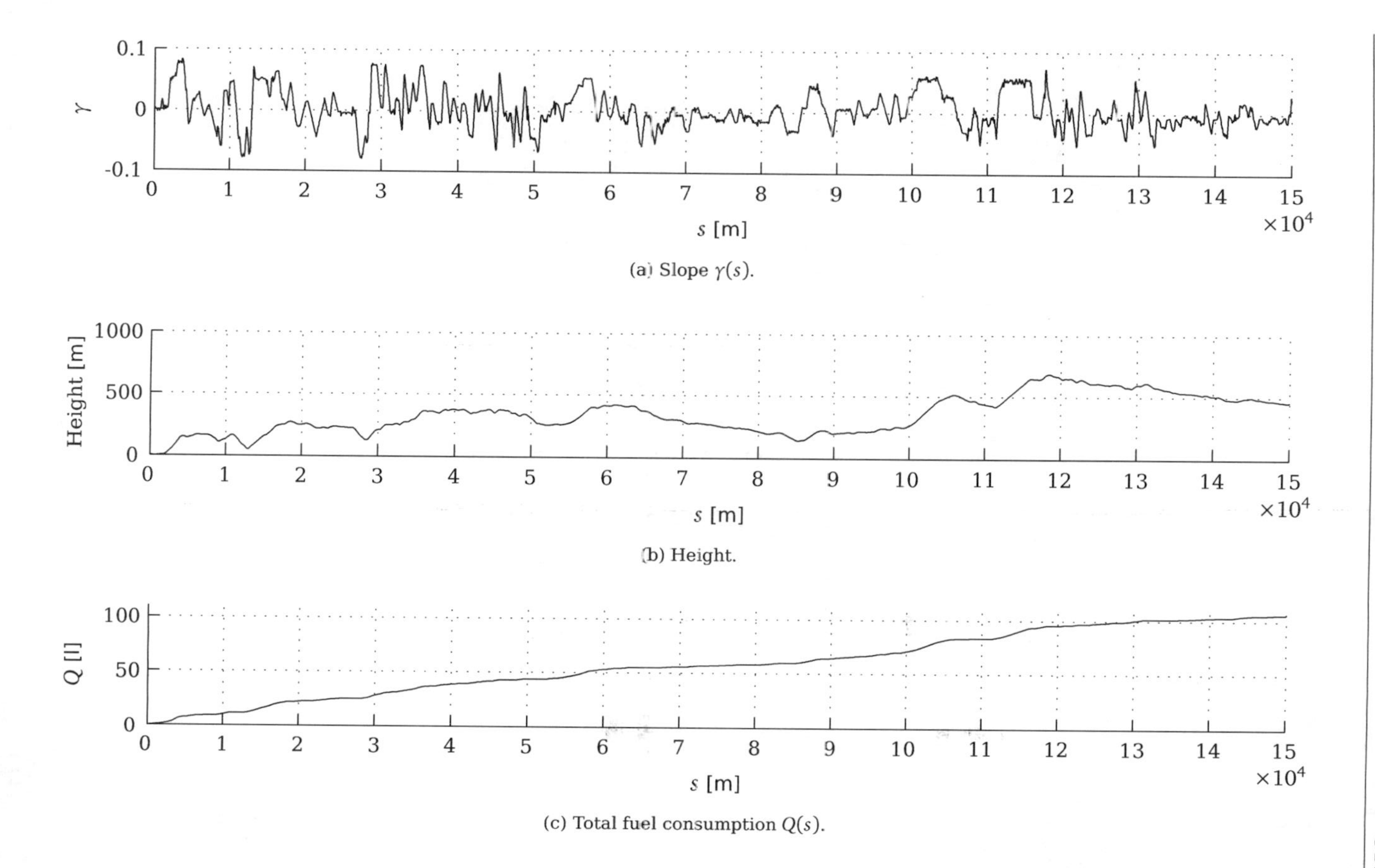


(a) Slope $\gamma(s)$.

(b) Height.

(c) Total fuel consumption $Q(s)$.

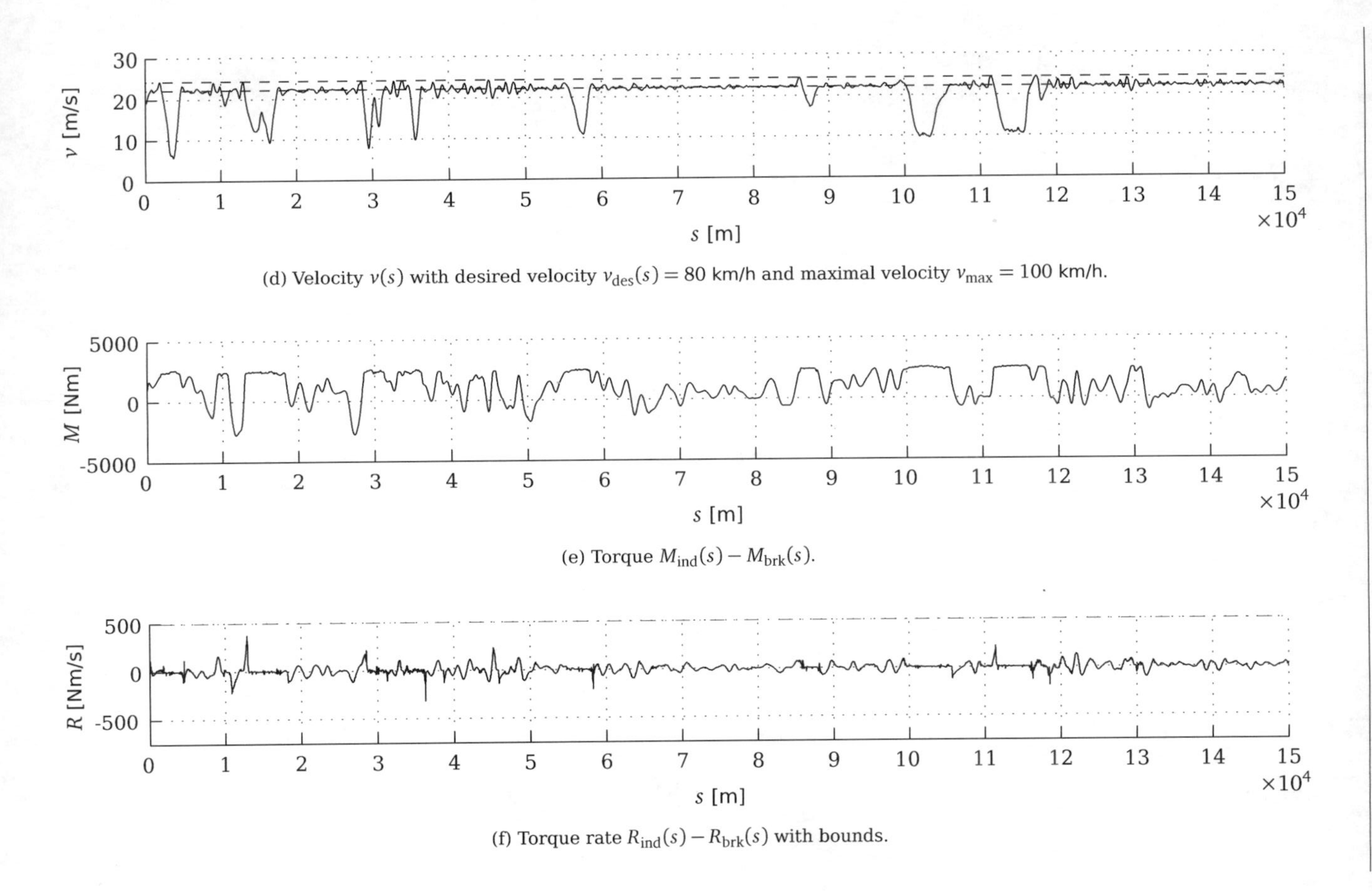


(d) Velocity $v(s)$ with desired velocity $v_{\mathrm{des}}(s) = 80$ km/h and maximal velocity $v_{\max} = 100$ km/h.

(e) Torque $M_{\mathrm{ind}}(s) - M_{\mathrm{brk}}(s)$.

(f) Torque rate $R_{\mathrm{ind}}(s) - R_{\mathrm{brk}}(s)$ with bounds.

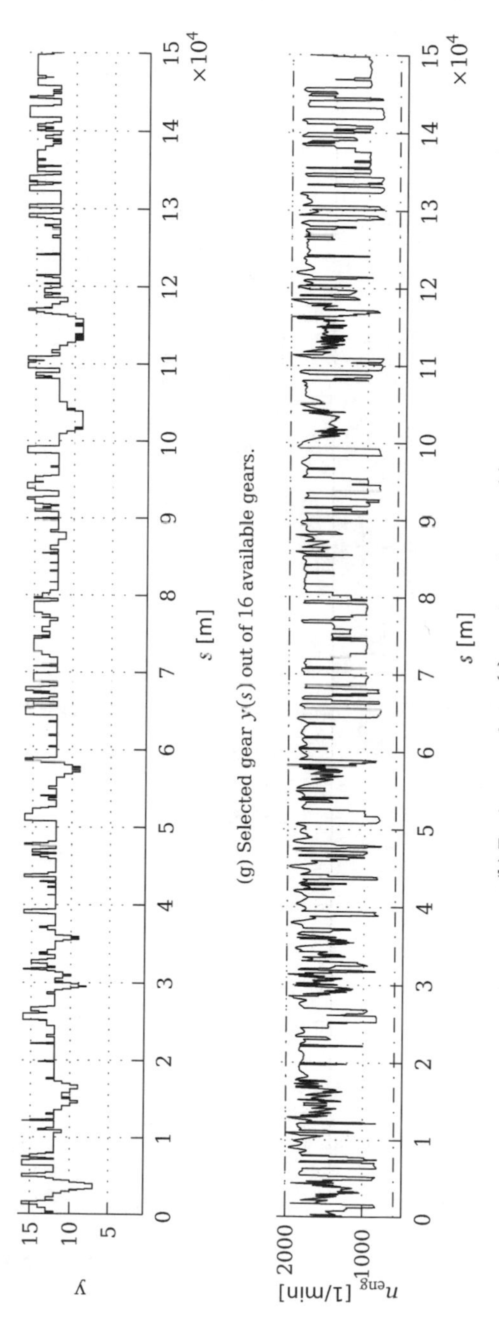


(g) Selected gear $y(s)$ out of 16 available gears.

(h) Engine speed $n_{\text{eng}}(s)$ at selected gear $y(s)$.

Figure 9.30: Simulated mixed–integer model–predictive control results for the predictive cruise controller problem on a southern German highway section of 150 kilometers length.

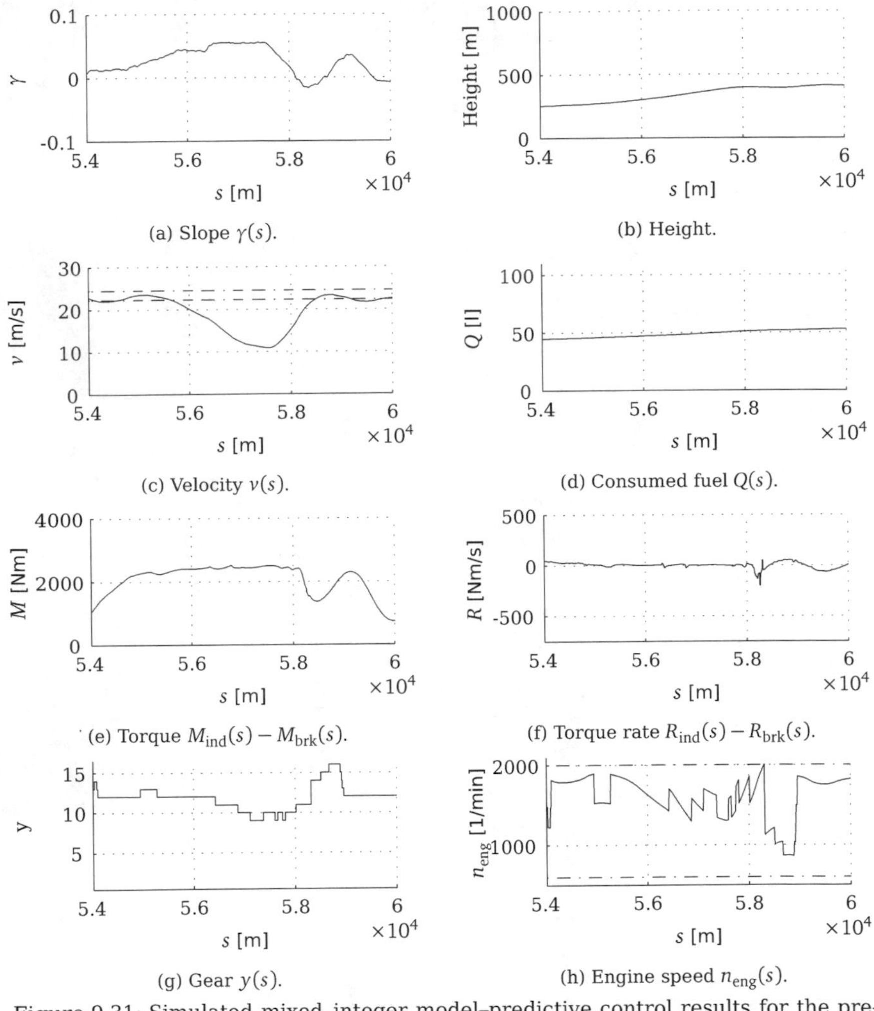


(a) Slope $\gamma(s)$.

(b) Height.

(c) Velocity $v(s)$.

(d) Consumed fuel $Q(s)$.

(e) Torque $M_{\text{ind}}(s) - M_{\text{brk}}(s)$.

(f) Torque rate $R_{\text{ind}}(s) - R_{\text{brk}}(s)$.

(g) Gear $y(s)$.

(h) Engine speed $n_{\text{eng}}(s)$.

Figure 9.31: Simulated mixed–integer model–predictive control results for the predictive cruise controller problem on a highway in southern Germany presented in figure 9.30. Details are shown for a sloped section of 6 km length at 54 km into the track. Climbing the steep slope causes the truck with a mass of 40,000 kg to slow down below 40 km/h. The predictive cruise controller anticipates this and accelerates to the maximum permitted velocity just before the steep slope starts. Downshifts of the gear $y$ keep the engine speed $n_{\text{eng}}$ up, maintaining the maximum possible torque $M_{\text{ind}}$. More economic low engine speeds are chosen by an appropriate upshift as the slope decreases, saving on fuel.

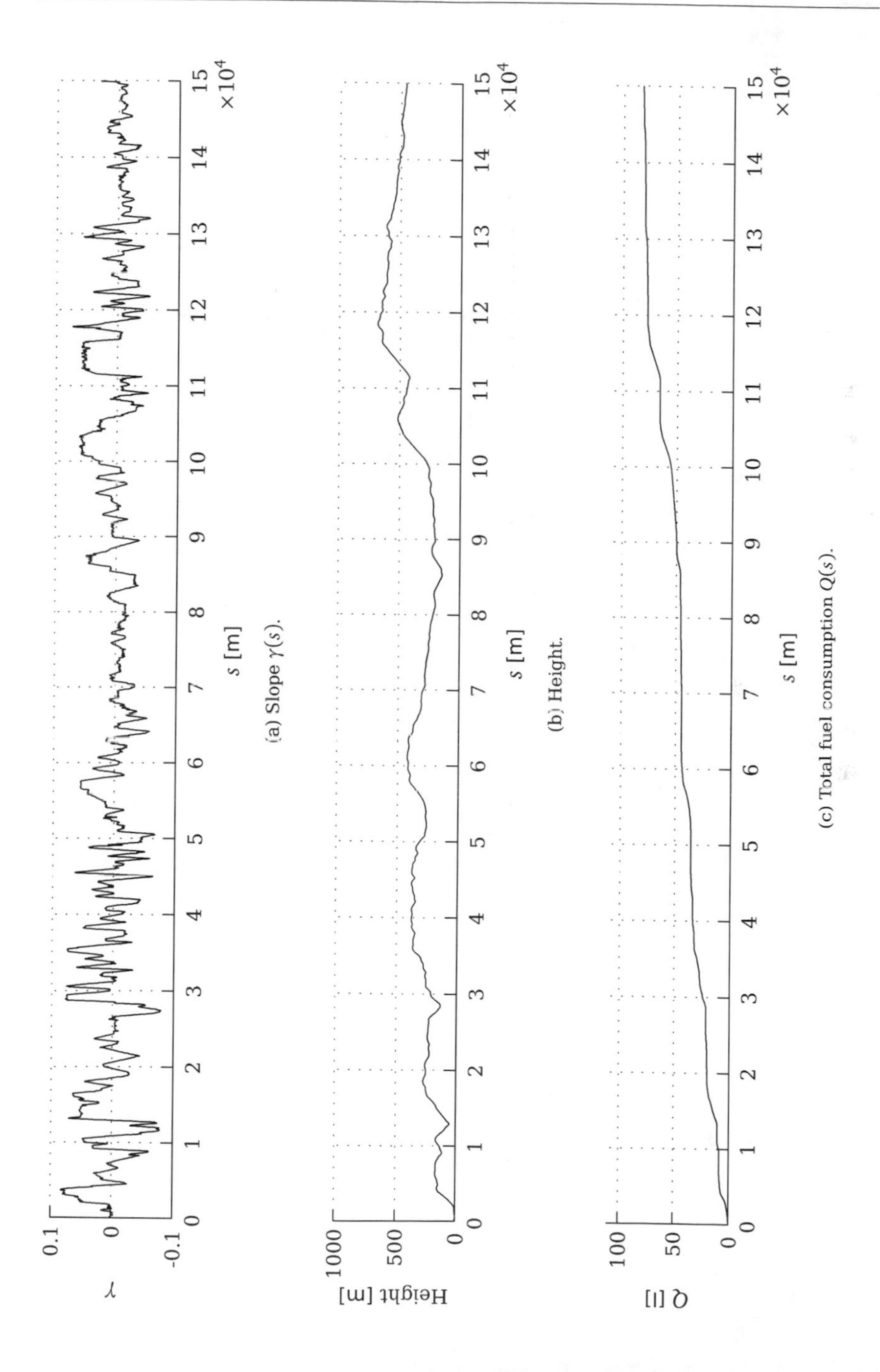


(a) Slope $\gamma(s)$.

(b) Height.

(c) Total fuel consumption $Q(s)$.

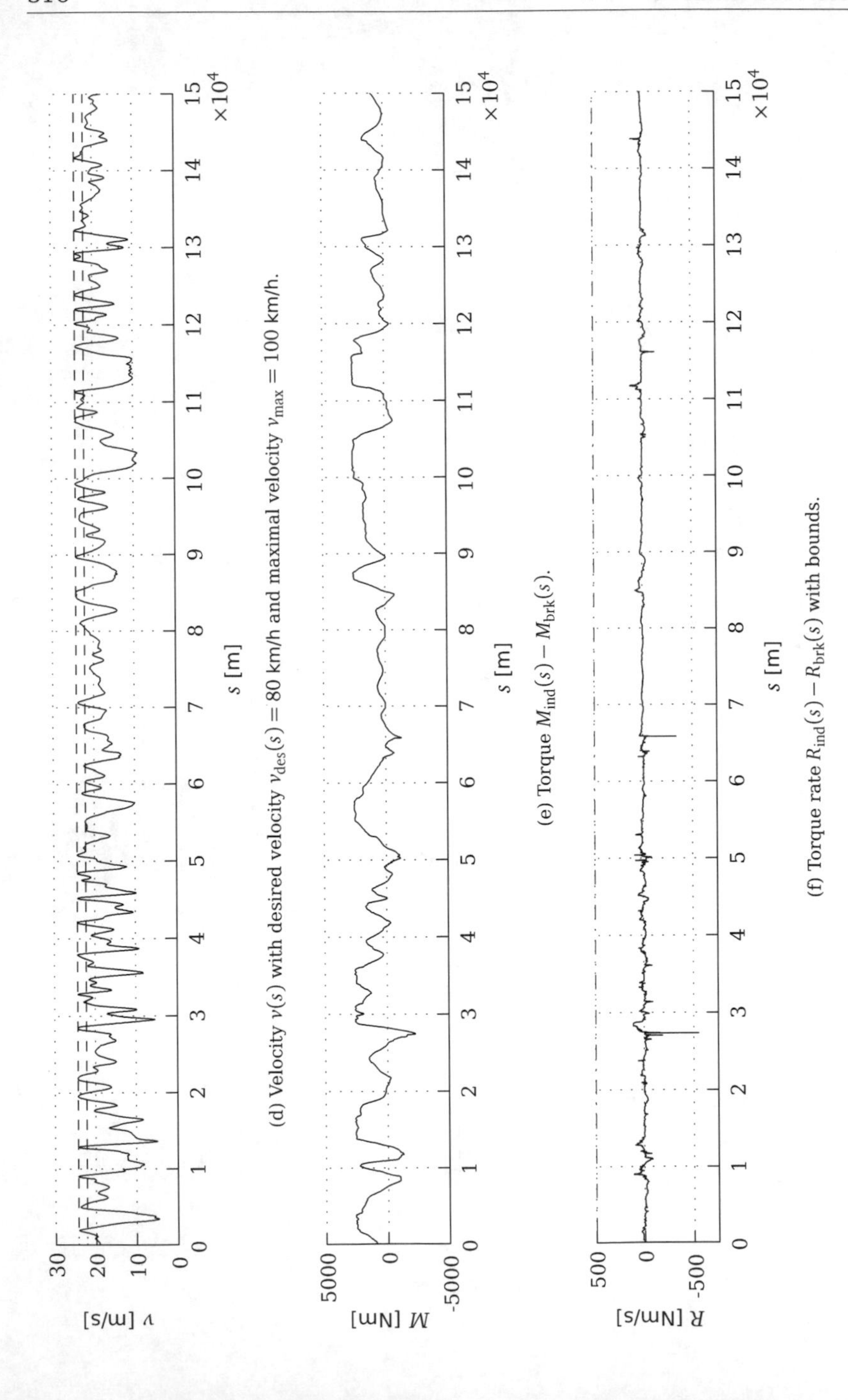


(d) Velocity $v(s)$ with desired velocity $v_{\text{des}}(s) = 80$ km/h and maximal velocity $v_{\text{max}} = 100$ km/h.

(e) Torque $M_{\text{ind}}(s) - M_{\text{brk}}(s)$.

(f) Torque rate $R_{\text{ind}}(s) - R_{\text{brk}}(s)$ with bounds.

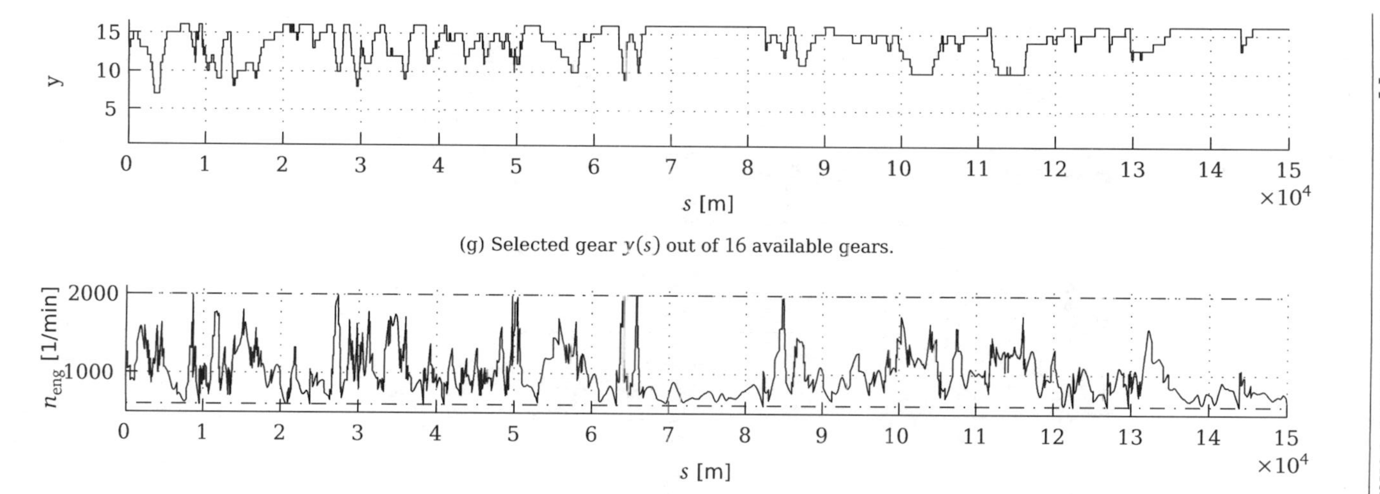


(g) Selected gear $y(s)$ out of 16 available gears.

(h) Engine speed $n_{\text{eng}}(s)$ at selected gear $y(s)$.

Figure 9.32: Simulated mixed–integer model–predictive control results for the predictive cruise controller problem on a southern German highway section of 150 kilometers length. The desired and permitted velocities are $v_{\text{des}} =$ 80 km/h and $v_{\max} = 100$ km/h. A prediction horizon of $H = 2000$ m length was used, discretized into $m = 40$ direct multiple shooting intervals. Control feedback was given every 10 m. Objective function weights $\lambda_{\text{dev}} = 10^{-3}$, $\lambda_{\text{fuel}} = 10^{-1}$, and $\lambda_{\text{comf}} = 10^{-4}$ were chosen and yield a compromise between speed–oriented and fuel–saving behavior. In figure 9.32h it can be seen that truck is operated at much lower engine speeds than in the previous set of results. The highest available gear 16 is selected in figure 9.32g unless slopes require acceleration or braking maneuvers. Consequentially, the total amount of fuel consumed is reduced in figure 9.32c. This comes at the cost of much larger deviations from the desired velocity in figure 9.32d.

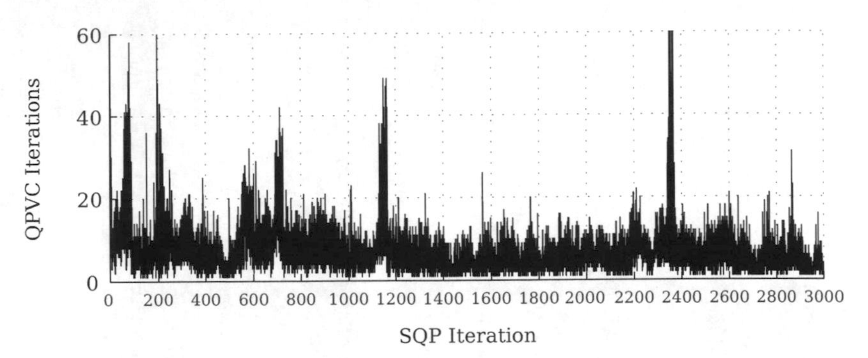


(a) Number of QPVC iterations for the results of figure 9.30.

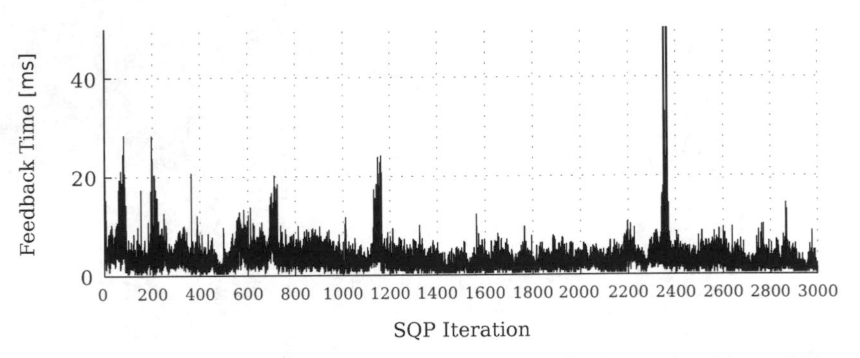


(b) Overall per–SQP iteration runtime of the QPVC solver for the results of figure 9.30.

Figure 9.33: Number of QPVC iterations required to compute the mixed–integer control feedback of the predictive cruise controller ($m = 40$ shooting intervals, horizon length $H = 2000$ m, feedback given every 10 m, 16 available gears). Major spikes in the number of QPVC iterations are caused by searching adjacent convex feasible sets in order to improve an identified stationary point.

Concerning real–time capability of the presented algorithm, consider that for a feedback granularity of 10 meters as employed for the presented results, the feedback has to be ready after at most $10[\mathrm{m}]/v(s)$ seconds, e.g. after 360 ms at $v(s) = 100$ km/h. Reserving half of this time for the preparation phase, our QPVC method is real–time capable below $180\ \mathrm{ms} \cdot 500\ \mu\mathrm{s}/\mathrm{iteration}$, i.e., 360 iterations on the desktop machine used for the presented computations. Considering that most mixed–integer feedback controls could be computed in 20 QPVC iterations, and given the additional possibility to constrain our primal–dual parametric algorithm to 20 iterations while still obtaining a physically meaningful iterate, we come up with a rough estimate of a factor 15 that an industrial target device could be slower than our desktop machine. We finally note that a condensing method on this problem would require a computation time in excess of ten seconds, though attributed to the preparation phase of the real–time iteration scheme, before the computation of a feedback control could be started.

### 9.5.9 Summary

In this section we have applied the new algorithmic techniques developed in this thesis, including the mixed–integer real–time iteration scheme, the nonconvex active set method for QPVCs, and the block structured HPSC factorization, to a challenging predictive cruise control problem of high relevance to current industrial practice. We have formulated a vehicle model including nonlinear dynamics, and various nonlinear physical constraints e.g. on engine torque and velocity. On a prediction horizon supplied with look–ahead information on the road conditions, an optimal control problem has been formulated to minimize a nonlinear objective function composed from several contributing terms, motivated by the predictive cruise controller's goals such as tracking of a prescribed velocity or minimizing fuel consumption. The described problem becomes a combinatorial one by including the optimal choice of transmission ratio and associated gearbox degree of efficiency, assuming, in contrast to a CVT drive, a gearbox with a fixed number of available gears. We have considered the gear choice in each discretization point on the whole of the prediction horizon, which led to a large search space prohibiting the use of exhaustive search techniques.

Previous work has attacked this predictive control problem using dynamic programming techniques, or by considering only a single gear shift on the prediction horizon. We have demonstrated that the techniques developed in this thesis allow for a computationally very efficient solution of this mixed–integer model–predictive control problem. We achieved mixed–integer control feedback times as low as one millisecond, staying below 10 milliseconds for most feedback computations and below 50 milliseconds for all scenarios investigated.

# A Implementation

## A.1 The Multiple–Shooting Real–Time Online Optimization Method MuShROOM

The mixed–integer optimal control algorithm developed and presented in this thesis has been implemented in C as the "multiple shooting real–time online optimization method", short MuShROOM. This section contains a brief discussion of the modular architecture, data structures, model functions and problem description, and the command line based user interface of this software.

### A.1.1 Software Architecture

#### Modular Architecture

We have adopted the paradigm of two central data structures containing the static description of the Mixed–Integer Optimal Control Problem (MIOCP) model on the one hand, and the iterative data for the mixed–integer real–time iteration scheme on the other hand. The algorithmic components of this scheme can be naturally separated into five modules as depicted in figure A.1 that have read access to the static model description, and read/write access to the iterative data. While this paradigm does not provide per–module protection of data, it gives the freedom to change data access patterns as algorithms continue to be developed.

#### Algorithmic Variants

The five modules depicted in figure A.1 are abstract placeholders for one of several concrete implementations of the respective module. Currently available implementations are listed in table A.1. For the evaluation module, the integrator module, and the SQP module, only one standard implementation is available. The Hessian approximation can be chosen as appropriate for

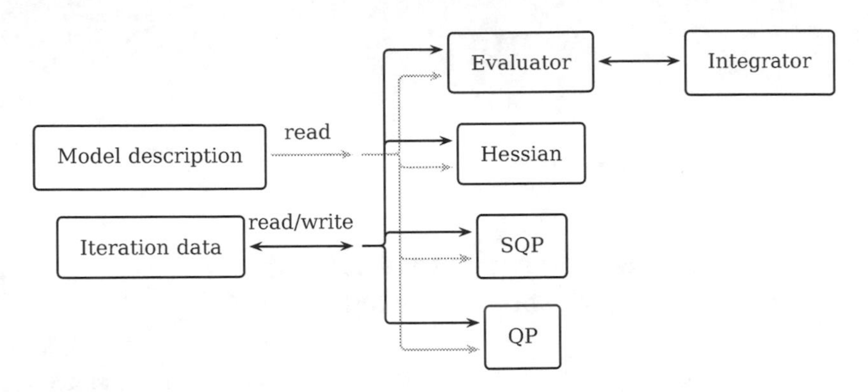


Figure A.1: Modular architecture of the software MuShROOM.

the problem instance under investigation. The Gauß–Newton Hessian requires a least–squares objective. Two QP solver modules are available that provide the classical condensing algorithm and the parametric active set strategy for QPVCs. Details on the latter can be found in section A.2.

| Module | Variants | Description |
|---|---|---|
| Evaluator | `EvaluatorStd` | Standard module for evaluation of multiple shooting model functions and derivatives |
| Hessian | `HessianDiagonal` | Diagonal approximation of the Hessian |
| | `HessianGaussNewton` | Gauß–Newton approximation |
| | `HessianBFGS` | BFGS approximation |
| | `HessianLBFGS` | Limited–memory BFGS approximation |
| Integrator | `IntegratorRK` | Explicit fixed–step Runge–Kutta method |
| SQP | `SqpStdSolver` | Standard SQP method |
| QP | `QpCondenseSolver` | Condensing QP solver; the condensed QP is solved by QPOPT [86] |
| | `QpHPSCSolver` | Parametric active–set QP solver for QPVCs; see section A.2 for details |

Table A.1: Available algorithmic variants for modules of MuShROOM.

## A.1.2 Description, Implementation, and Solution of a Problem

This section describes the interface of the MuShROOM software exposed to the user implementing a MIOCP problem instance. We present the layout and creation of all input files required to describe and implement an optimal control problem. We further address to different ways of using the MuShROOM software in order to solve the problem, namely using the command line interface or using a set of C functions provided for this purpose.

### Input and Output

An Optimal Control Problem (OCP) is described by a MATLAB file holding the static model description, and a shared object file holding all compiled C model functions, i.e., the ODE system's right hand side function, the objective function terms, and all constraint functions. Furthermore, concrete implementations of the five algorithmic modules are made available as shared objects as well and can be selected via the MATLAB model description. The solution of an OCP is made available both in MATLAB and in plain text format. If the command line interface is used, the iterative solution process can be followed on the screen. This setup is depicted in figure A.2 and will be explained in more detail in the ensuing sections.

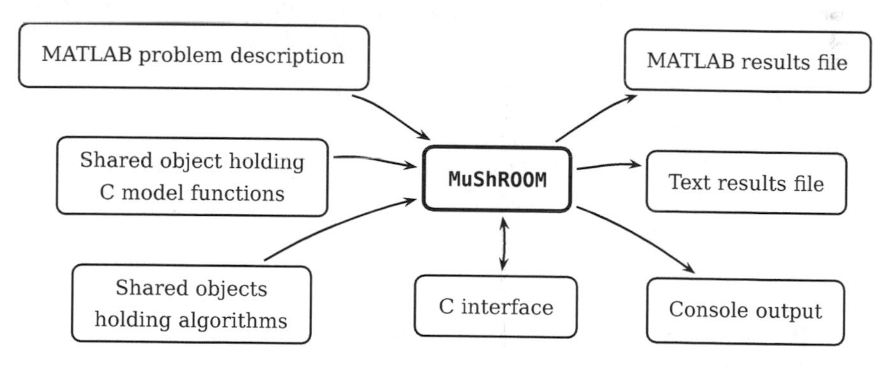


Figure A.2: Input and output files of the MuShROOM software.

### MATLAB Problem Description

The problem description is read from a MATLAB file which is expected to contain a structured variable named model with structure fields as listed in table A.3. The dimensions notation indicates the MATLAB array type expected; {·} denotes a cell array and (·) denotes a vector or matrix. The `flags` structure mandates some more detailed explanation. The software supports two–sided point constraints (`rtype` 3) only if the two residuals are affine linear, i.e., their jacobians coincide. A point constraint can be declared as a vanishing constraint by associating it with the lower bound of an unknown on the same shooting node via the flag `vanish`.

Possible values for the algorithm fields are listed in table A.1. The individual algorithms accept several options and settings as listed in table A.2. Options inappropriate for the selected algorithmic module are silently ignored.

### C Model Functions

All model functions are expected to be implemented as callable C functions residing in the shared object file specified in the MATLAB problem description. The shared object is expected to export a single function `int initialize (model)` that is called by the `MuShROOM` software to obtain function pointers to all model functions. The shared object is expected to make these function pointers known by calling for each required function and for each shooting node or interval the callback function

```
modelSetFunction (model, function, range, kind, level,
                  pointer);
```

with arguments as described in table A.4. The ODE nominal right hand side and one nominal objective function term are mandatory. A Mayer type objective can be set on the final shooting node only. For all functions, derivatives in the form of dense Jacobians may optionally be provided. Unavailable dense Jacobians are approximated by one–sided finite differences. For the ODE system's right hand side, sparse Jacobians or dense directional forward derivatives may be provided in addition. Unavailable directional forward derivatives of the ODE system's right hand side are computed by multiplication with the sparse (preferred) or dense (discouraged) Jacobian, or are approximated by directional one–sided finite differences.

All model functions are expected to have the C signature

```
int function (args)
```

regardless of their type and level. They all accept the same call arguments structure `args` with fields listed in table A.5 as single argument, and to return a success or error code according to table A.7. The valid input fields and expected output fields of this arguments structure depending on the model function and type are listed in table A.6. Jacobians of the function's value are stored in either dense column–major format or in sparse triplets format. Directional derivatives are always stored in dense format and are used for the ODE system's right hand side function only. The point constraint functions evaluation two–sided point constraint (`rtype` 3) compute two residuals. The single Jacobian is computes in the lower residual, such that two–sided point constraints must always be affine linear. The extended matching condition functions couple $(\boldsymbol{x}(t_{i+1}), \boldsymbol{u}_i)$ and $(\boldsymbol{s}_{i+1}, \boldsymbol{i}_{i+1})$ values belonging to shooting node $i+1$ and the two adjacent shooting intervals $[t_i, t_{i+1}]$ and $[t_{i+1}, t_{i+2}]$. Consequentially, Jacobians with respect to the four inputs must be computed.

### An Exemplary Model

An exemplary MATLAB file for the very simple NMPC problem

$$
\begin{aligned}
\min_{x(\cdot),u(\cdot)} \quad & \int_0^3 x^2(t) + u^2(t)\,\mathrm{d}t && && \text{(A.1)} \\
\text{s.t.} \quad & \dot{x}(t) = (1+x(t))x(t) + u(t), && t \in [0,3], \\
& x(0) = x_{\text{meas}}, \\
& x(3) = 0, \\
& x(t) \in [-1,1], && t \in [0,3], \\
& u(t) \in [-1,1], && t \in [0,3].
\end{aligned}
$$

that properly sets up a model description structure is given in figure A.3. An exemplary C file for problem (A.1) that makes use of the documented functions is given in figure A.5. As can be seen, nominal functions and dense Jacobian functions for the ODE system's right hand side, a least–squares objective, and an end–point constraint are implemented, and are registered with `MuShROOM` during initialization of the shared object containing this model.

| Field | Type | Def. | Description |
|---|---|---|---|
| `integrator` | | | |
| `.steps` | double | 20 | Number of integrator steps |
| `hessian` | | | |
| `.limit` | int | 30 | Memory length of the L–BFGS Hessian |
| `.levmar` | double | 0.0 | Levenberg–Marquardt regularization of the Gauß–Newton Hessian |
| `sqpsolver` | | | |
| `.kktacc` | double | $10^{-6}$ | Acceptable KKT tolerance for termination |
| `.itmax` | int | 100 | Maximum number of SQP iterations |
| `qpsolver` | | | |
| `.featol` | double | $10^{-8}$ | Tolerance for violation of QP constraints |
| `.opttol` | double | $10^{-8}$ | Optimality tolerance for QP solution |
| `.ranktol` | double | $10^{-14}$ | Active set degeneracy tolerance |
| `.itmax` | int | $10^4$ | Maximum number of QP solver iterations |
| `.subsolver` | string | `QpHPSCSolver`: KKT solver for the block structured KKT system; see section A.2 for possible values `QpCondenseSolver`: QP solver for the condensed QP; must be `QpOptSolver` to use `QPOPT` [86] | |
| `.expand` | int | 50 | `QPOPT` anti–cycling strategy setting |
| `.feaitmax` | int | $10^4$ | `QPOPT` max. number of phase one iterations |
| `.relax` | double | 1.1 | `QPOPT` infeasible constraints relaxation factor |
| `.print` | int | 0 | `QPOPT` print level for iterations |
| `.updates` | int | 1 | `qpHPSC` flag to enable/disable matrix updates; effective only for `KKTSolverHPSC`. |
| `.kktcheck` | int | 0 | `qpHPSC` debug consistency checks |
| `.refine` | int | 0 | `qpHPSC` number of iterative refinement steps |
| `.print_iter` | int | 0 | `qpHPSC` print level for iterations |
| `.print_primal` | int | 0 | `qpHPSC` print level for primal blocking tests |
| `.print_dual` | int | 0 | `qpHPSC` print level for dual blocking tests |
| `.print_degen` | int | 0 | `qpHPSC` print level for degeneracy resolution |

Table A.2: Algorithmic settings in the MATLAB model description file.

| Field | Dimen. | Type | Description |
|---|---|---|---|
| `library` | 1 | string | Name of shared object file |
| `algorithm.?.name` | | | |
| `evaluator` | 1 | string | Name of the Evaluate module to use |
| `hessian` | 1 | string | Name of the Hessian module to use |
| `integrator` | 1 | string | Name of the Integrator module to use |
| `sqpsolver` | 1 | string | Name of the SQP module to use |
| `qpsolver` | 1 | string | Name of the QP module to use |
| `dim.lsq` | 1 | int | Dim. $l$ of the least–squares obj. $\boldsymbol{l}(\cdot) \in \mathbb{R}^l$ |
| `dim.nodes` | 1 | int | Number $m$ of multiple shooting nodes |
| `dim.p` | 1 | int | Number $n^{\mathrm{p}}$ of global parameters $\boldsymbol{p} \in \mathbb{R}^{n^{\mathrm{p}}}$ |
| `dim.q` | 1 | int | Number $n^{\mathrm{q}}$ of control parameters $\boldsymbol{q} \in \mathbb{R}^{n^{\mathrm{q}}}$ |
| `dim.r` | $(m)$ | int | Numbers $n_i^{\mathrm{r}}$ of point constraints $\boldsymbol{r}_i(\cdot) \in \mathbb{R}^{n_i^{\mathrm{r}}}$ |
| `dim.x` | 1 | int | Number $n^{\mathrm{x}}$ of differential states $\boldsymbol{x}(\cdot) \in \mathbb{R}^{n^{\mathrm{x}}}$ |
| `flags.rtype` | $\{m\}(n_i^{\mathrm{r}})$ | int | Types of the point constraints $\boldsymbol{r}_i(\cdot)$; 0 : lower, 1 : upper, 2 : both, 3 : equality |
| `flags.vanish` | $\{m\}(n_i^{\mathrm{r}})$ | int | Index of bound controlling $r_{ij}(\cdot)$ as a vanishing constraint; $-1$ for a normal one |
| `flags.xspec` | 1 | int | Initialization of the shooting node states; 0 : given, 1 : interpolation, 2 : integration |
| `flags.ive` | 1 | int | Enable or disable initial–value embedding |
| `min, max.p` | $(n^{\mathrm{p}})$ | double | Lower and upper bound for parameters $\boldsymbol{p}$ |
| `min, max.q` | $\{m\}(n^{\mathrm{q}})$ | double | L. and u. bound for control parameters $\boldsymbol{q}_i$ |
| `min, max.t` | 1 | double | Start and end of time horizon |
| `min, max.x` | $\{m\}(n^{\mathrm{x}})$ | double | L. and u. bound for shooting node states $\boldsymbol{s}_i$ |
| `val.p` | $(n^{\mathrm{p}})$ | double | Initial values for parameters $\boldsymbol{p}$ |
| `val.q` | $\{m\}(n^{\mathrm{q}})$ | double | Initial value for control parameters $\boldsymbol{q}_i$ |
| `val.x` | $\{m\}(n^{\mathrm{x}})$ | double | Initial value for node states $\boldsymbol{s}_i$ |
| `sca.p` | $(n^{\mathrm{p}})$ | double | Scale factors for parameters $\boldsymbol{p}$ |
| `sca.q` | $\{m\}(n^{\mathrm{q}})$ | double | Scale factors for control parameters $\boldsymbol{q}_i$ |
| `sca.r` | $\{m\}(n_i^{\mathrm{r}})$ | double | Scale factors for point constraint residuals $\boldsymbol{r}_i$ |
| `sca.x` | $\{m\}(n^{\mathrm{x}})$ | double | Scale factors for node states $\boldsymbol{s}_i$ |
| `fix.q` | $\{m\}(n^{\mathrm{q}})$ | int | Fixation flags for control parameters $\boldsymbol{q}_i$ |
| `fix.x` | $\{m\}(n^{\mathrm{x}})$ | int | Fixation flags for shooting node states $\boldsymbol{s}_i$ |

Table A.3: Data fields of the MATLAB model description file.

| Argument | Possible Values | Description |
|---|---|---|
| `model` | | Pointer to the static model description |
| `function` | `rightHandSide` | Sets the ODE right hand side function |
| | `leastSquaresObjective` | Sets the least–squares objective function |
| | `mayerObjective` | Sets the Mayer type objective function |
| | `pointConstraint` | Sets the decouple point constraint func. |
| | `continuityCondition` | Sets the ext. matching condition function |
| `range` | `allIntervals` | Selects all intervals |
| | `allNodes` | Select all nodes |
| | `endNode` | Select the last node |
| | `interiorNode` | Select all nodes except boundary nodes |
| | `startNode` | Select the first node or interval |
| | any value $0 \leqslant i \leqslant m$ | Selects the single node $0 \leqslant i \leqslant m$ or the single interval $0 \leqslant i \leqslant m-1$ |
| `kind` | `C` | The function is a C function |
| | `finiteDifference` | The function should be approximated by finite differences, available only if `level` is not `nominal`. |
| | `sparseConvert` | The dense Jacobian should be computed by converting the sparse one, available only if `level` is `denseJac`. |
| | `sparseMultiply` | The directional derivative should be computed by multiplication with the sparse jacobian, available only if `level` is `fwdDirDer` or `adjDirDer`. |
| | `denseMultiply` | The directional derivative should be computed by multiplication with the dense Jacobian, available only if `level` is `fwdDirDer` or `adjDirDer`. |
| `level` | `nominal` | Nominal model function |
| | `denseJac` | Dense Jacobian function |
| | `sparseJac` | Sparse Jacobian function (ODE) |
| | `fwdDirDer` | Directional derivatives function (ODE) |
| `pointer` | | Pointer to the model function (`kind=C`) |

Table A.4: Call arguments of the routine `setModelFunction`.

| Field | Dimen. | Type | Description |
|---|---|---|---|
| `t` | 1 | double | Model time $t \in [t_0, t_f]$ |
| `x` | $n^x$ | double | Differential states $\boldsymbol{x}(t)$ |
| `u` | $n^u$ | double | Controls $\boldsymbol{u}(t)$ |
| `p` | $n^p$ | double | Global model parameters $\boldsymbol{p}$ |
| `res0` | $n^f$ | double | Function's primary return value |
| `res1` | $n^f$ | double | Function's secondary return value |
| `dt` | $n^f \times 1$ | depends | Derivative of `res0` w.r.t. the time $t$ |
| `dx` | $n^f \times n^x$ | depends | Derivative of `res0` w.r.t. the differential states $\boldsymbol{x}(t)$ |
| `du` | $n^f \times n^u$ | depends | Derivative of `res0` w.r.t. the control $\boldsymbol{u}(t)$ |
| `dir[3]` | $(1, n^x, n^u) \times n^d$ | double | $(t, \boldsymbol{x}, \boldsymbol{u})$ parts of the derivative directions |
| `der` | $n^f \times n^d$ | double | Directional derivatives into directions `dir` |

Table A.5: Fields of the model function call arguments structure `args`. The dimension $n^f$ is a placeholder for the function value's actual dimension that depends on the function type.

| Function | Level | Arguments structure fields | | | | | | | | | | | | | | |
|---|---|---|---|---|---|---|---|---|---|---|---|---|---|---|---|---|
| | | `t` | `x` | `x2` | `u` | `u2` | `p` | `res0` | `res1` | `dt` | `dx` | `dx2` | `du` | `du2` | `dir` | `der` |
| `rightHandSide` | `nominal` | I | I | | I | | I | O | | | | | | | | |
| | `denseJac` | I | I | | I | | I | I | | O | O | | O | | | |
| | `sparseJac` | I | I | | I | | I | I | | O | O | | O | | | |
| | `fwdDirDer` | I | I | | I | | I | I | | | | | | | I | O |
| `leastSquares-Objective` | `nominal` | I | I | | I | | I | O | | | | | | | | |
| | `denseJac` | I | I | | I | | I | I | | O | O | | O | | | |
| `mayer-Objective` | `nominal` | I | I | | I | | I | O | | | | | | | | |
| | `denseJac` | I | I | | I | | I | I | | O | O | | O | | | |
| `point-Constraint` | `nominal` | I | I | | I | | I | O | O | | | | | | | |
| | `denseJac` | I | I | | I | | I | I | | O | O | | O | | | |
| `continuity-Condition` | `nominal` | I | I | I | I | I | I | O | | | | | | | | |
| | `denseJac` | I | I | I | I | I | I | I | | O | O | O | O | O | | |

Table A.6: Valid input and output fields in the arguments structure for all functions and levels. "I" denotes inputs and "O" denotes outputs. Empty fields are unused and should not be read or written to by the called function.

```
model.library = 'libnmpc1';

model.algorithm.condenser.name   = 'CondenserDense';
model.algorithm.evaluator.name   = 'Evaluator';
model.algorithm.hessian.name     = 'HessianGaussNewton';
model.algorithm.integrator.name  = 'IntegratorRK';
model.algorithm.sqpsolver.name   = 'SqpStdSolver';
model.algorithm.qpsolver.name    = 'QpCondenseSolver';

model.algorithm.hessian.levmar   = 0.0;
model.algorithm.integrator.steps = int32(5);
model.algorithm.sqpsolver.itmax  = int32(1000);
model.algorithm.sqpsolver.kktacc = 1e-8;

model.dim.nodes = int32(31);
model.dim.lsq   = int32(2);
model.dim.p     = int32(1);
model.dim.q     = int32(1);
model.dim.x     = int32(1);
model.dim.sos   = int32(0);
model.dim.hf    = int32(0);
model.dim.r(model.dim.nnodes)  = int32(1);

model.min.t = 0.0;
model.max.t = 3.0;

model.min.of       = 0.0;
model.max.of       = 0.2;
model.sca.of       = 1.0;
model.sca.may      = 1.0;
model.sca.lsq(1:2) = 1.0;
model.sca.r{model.dim.nnodes}(1) = 1.0;
model.flags.rtype(model.dim.nnodes) = int32(3);
model.flags.xspec = 2;
model.flags.ive   = 1;

for ii = 1:model.dim.nnodes
    model.val.x(ii,1) =  0.05;
    model.min.x(ii,1) = -1.0;
```

Figure A.3: An exemplary MuShROOM description of problem (A.1) in MATLAB (continued on page 331).

```
    model.max.x(ii,1) =  1.0;
    model.sca.x(ii,1) =  1.0;
    model.fix.x(ii,1) =  0;
end

for ii = 1:model.dim.nnodes-1
    model.val.q(ii,1) =  0.0;
    model.min.q(ii,1) = -1.0;
    model.max.q(ii,1) =  1.0;
    model.sca.q(ii,1) =  1.0;
    model.fix.q(ii,1) =    0;
end
```

Figure A.4: An exemplary MuShROOM description of problem (A.1) in MATLAB (continued from page 330).

```
int ffcn (struct model_TNominalArgs *args) {
    res(0) = (1.0 + x(0)) * x(0) + u(0);
    return 0;
}

int ffcn_d (struct model_TDenseJacArgs *args) {
    dx(0,0) = 2.0 * x(0) + 1.0;
    du(0,0) = 1.0;
    return 0;
}

int lsqfcn (struct model_TNominalArgs *args) {
    res(0) = x(0);
    res(1) = u(0);
    return 0;
}

int lsqfcn_d (struct model_TDenseJacArgs *args) {
    dx(0,0) = 1.0;
    du(1,0) = 1.0;
    return 0;
}
```

Figure A.5: An exemplary MuShROOM C model file for problem (A.1) (continued on page 332).

```
int rdfcn_e (struct model_TNominalArgs *args) {
    res(0) = x(0);
    return 0;
}

int rdfcn_e_d (struct model_TDenseJacArgs *args) {
    dx(0,0) = 1.0;
    return 0;
}

int initialize (struct model_TModel *pModel) {
    modelSetFunction (pModel, modelRightHandSide, modelAllIntervals,
                      modelC, modelNominal,  ffcn);
    modelSetFunction (pModel, modelRightHandSide, modelAllIntervals,
                      modelC, modelDenseJac, ffcn_d);
    modelSetFunction (pModel, modelLeastSquaresObjective,
                      modelAllNodes, modelC, modelNominal,  lsqfcn);
    modelSetFunction (pModel, modelLeastSquaresObjective,
                      modelAllNodes, modelC, modelDenseJac, lsqfcn_d);
    modelSetFunction (pModel, modelPointConstraint, modelEndNode,
                      modelC, modelNominal,  rdfcn_e);
    modelSetFunction (pModel, modelPointConstraint, modelEndNode,
                      modelC, modelDenseJac, rdfcn_e_d);
    return 0;
}
```

Figure A.6: An exemplary MuShROOM C model file (continued from page 331).

| Error code | Description |
|---|---|
| $< 0$ | Indicates an error during the evaluation of the function. |
| $= 0$ | Indicates successful evaluation of the function. |
| growSparseMatrix | Indicates insufficient storage space for a sparse Jacobian. The caller should grow the storage space and retry evaluation. |

Table A.7: Possible return codes of a model function.

### A.1.3 User Interface

Two user interfaces are provided for this software package. The command line interface allows to solve off–line optimal control problems. The C interface provides a facility for embedding measured or estimated system states using the initial value embedding techniques, thus solving a simulated or real on–line optimal control problem.

#### Command Line Interface

The MuShROOM software can be used through a command line interface invoked by the shell command

```
./mushroom problem [--switch[=value] ...]
```

where *problem* must be substituted by the full file name of a MATLAB problem description file. All algorithmic settings, tolerances, etc. are specified in this MATLAB file as detailed in the previous section. One or more command line switches from table A.8 may be specified in addition to the problem's name.

| Switch | Description |
|---|---|
| `--[no]aset` | (Do not) print the active set after each SQP iteration |
| `--continue` | Continue solution in iterate read from results file |
| `--info` | Print information about the MuShROOM build, and quit |
| `--plugin=`*filename* | Load an additional shared object (library) and initialize it |
| `--verbose=`*p* | Set verbosity level of text output to $p \geqslant 0$ |
| `--[no]writemodel` | (Do not) write a textual protocol of the model description |

Table A.8: MuShROOM command line switches.

The output of MuShROOM if called using the command line interface for problem (A.1) with embedded initial value $x_{\text{meas}} = 0$ and the settings from figure A.3 is shown in figure A.7. The individual columns of the textual output are described in table A.9.

#### C Interface

The MuShROOM software can also be invoked from inside another program. To this end, a selection of function calls are provided that deal with proper

```
it  qp    kkttol          sobj    infcon infmatch       laggrd  varstep  mulstep
 0              4.763137e-01 0.00e+00 5.85e-01
 1   1  2.10e-01 4.630546e-02 1.33e-15 2.32e-01 4.318604e-02 2.42e+00 2.95e-01
 2   1  8.25e-03 2.868694e-03 1.25e-16 2.34e-02 2.322964e-03 8.51e-01 2.47e-01
 3   1  3.37e-04 2.535345e-03 1.57e-16 3.05e-04 2.219919e-04 1.15e-01 6.07e-02
 4   1  8.66e-06 2.543811e-03 7.16e-19 1.59e-08 1.707440e-06 1.98e-03 1.89e-03
 5   1  5.15e-10 2.543811e-03 4.05e-22 5.27e-13 1.895692e-08 8.35e-06 7.19e-06

it  qp    kkttol          sobj  infcon infmatch       laggrd  varstep  mulstep
 6   1  3.33e-14 2.543811e-03 6.31e-23 1.50e-16 3.278265e-10 9.77e-08 8.03e-08
```

Figure A.7: Output of the command line interface to MuShROOM for the exemplary problem (A.1).

| Column | Description |
|---|---|
| it | Running count of SQP iterations |
| qp | Number of QP or QPVC solver iterations spent |
| kkttol | Convergence criterion, satisfied if KKTTol < kktacc |
| sobj | Scaled objective function to be minimized |
| infcon | Infeasibility of the point and vanishing constraints |
| infmatch | Infeasibility of the matching conditions |
| laggrd | Euclidean norm of the gradient of the (MPVC–)Lagrangian |
| step | Euclidean norm of the primal step |
| mulstep | Euclidean norm of the dual step |

Table A.9: Columns of the textual output provided by MuShROOM.

initialization, loading of the problem description, and execution of the SQP iterations providing control feedback.

The C interface mirrors the software architecture described in this section. All algorithmic modules as well as all model functions are contained in shared object files, referred to as *libraries*, and the C interface provides facilities to load one or more libraries and maintains a list of loaded libraries. The static model description is contained in a MATLAB file, and the C interface can create and fill a model description structure given the name of a MATLAB file. One or more sets of iterative data can be created and initialized given a model description. To these data sets, the real–time iteration scheme can be applied by way of three C interface functions mirroring the three phases of that scheme. The most important functions are listed in table A.10 together with a brief explanation of their arguments. For a more extensive description we refer to the program's documentation.

| Function | Description |
|---|---|
| `modelReadModel` | Load a model description from a MATLAB file |
| `model` | Static model description structure to be initialized |
| `file` | Open MATLAB file to read from |
| `name` | Variable containing the model description |
| `modelDelModel` | Delete a model description |
| `model` | Initialized model description structure to be deleted |
| `interfaceLoadLibraries` | Load all algorithm libraries found in a directory |
| `libs` | Library structure to be initialized or appended |
| `path` | Directory to be searched for algorithm modules |
| `interfaceLoadLibrary` | Load a specific algorithm or model library |
| `libs` | Library structure to be initialized or appended |
| `name` | Full path and name of the shared object to be loaded |
| `interfaceUnload-Libraries` | Unload all loaded algorithm or model libraries |
| `libs` | Library structure containing all loaded libraries |
| `interfaceInitialize-Libraries` | Initialize all loaded algorithm or model libraries |
| `libs` | Library structure containing all loaded libraries |
| `model` | Initialized static model description structure |
| `interfaceFinalize-Libraries` | Finalize all loaded algorithm or model libraries. |
| `libs` | Library structure containing all loaded libraries |
| `interfaceNewData` | Create an iterative data structure for a model |
| `model` | Initialized static model description structure |
| `data` | Iterative data structure to be initialized |
| `interfaceDeleteData` | Delete an iterative data structure |
| `model` | Initialized static model description structure |
| `data` | Initialized iterative data structure to be deleted |

Table A.10: Functions of the C interface to the MuShROOM software (continued on page 336).

| | |
|---|---|
| sqpStartup | Initialize the SQP for the first iteration |
|     sqpsolver | SQP solver module instance |
|     sqpdata | SQP subset of the iterative data structure to use |
|     start | sqpColdStart/sqpWarmStart for a cold/warm start |
| sqpPrepareStep | Preparation phase of an SQP iteration |
|     sqpsolver | SQP solver module instance |
|     sqpdata | SQP subset of the iterative data structure to use |
| sqpFeedbackStep | Feedback phase of an SQP iteration |
|     sqpsolver | SQP solver module instance |
|     sqpdata | SQP subset of the iterative data structure to use |
|     state | Measured or estimated state to be embedded |
|     control | Control to be fed back to the process |
| sqpTransitionStep | Transition phase of an SQP iteration |
|     sqpsolver | SQP solver module instance |
|     sqpdata | SQP subset of the iterative data structure to use |

Table A.11: Functions of the C interface to the MuShROOM software (continued from page 335).

#### Model Description Structure

The static model description data structure shown in table A.12 reflects that of a MIOCP problem. It is initialized from a user provided problem description and is not modified during the program's execution.

#### Iteration Data Structure

The iteration data structure holds all values computed for one iteration of the mixed–integer real–time iteration scheme. The evaluation modules computes function values and derivatives in the current iterate and stores the in the `eval.fun` and `eval.der` structures. The Hessian module computes an node–wise approximation of the Hessian of the Lagrangian based on these values, and stores the Hessian blocks in the structure `hessian`. The sparse QP structure `qp.sparse` is populated with references to these values. Depending on the selected QP solver, this QP is either solved directly by the `QpHPSC` module, or condensed to a smaller dense QP by the `QpCondense` module, which can then be found in the `qp.dense` structure.

The SQP module uses the solution of this Quadratic Program (QP) to populate the `sqp.iterStep` and `sqp.mulStep` or `sqp.iter` and `sqp.mul`, depending on the setup of the QP.

| Field | Subfields | Description |
|---|---|---|
| `bndl` | `of, q, x` | Lower bounds of the objective, control parameters, and node states |
| `bndu` | `of, q, x` | Upper bounds of the objective, control parameters, and node states |
| `dim` | | Problem dimensions |
| `fix` | `q, x` | Fixation flags for control parameters and node states |
| `flag` | | Miscellaneous flags, see text |
| `func` | | Pointers to the ANSI C model functions |
| `init` | `p, q, x` | Initial values of the model parameters, control parameters, and node states |
| `scal` | `of, p, q, x` | Scale factors of the objective, the model parameters, control parameters, and node states |
| `time` | `tmin, tmax` | Start and end time of the time horizon |

Table A.12: Static model description structure.

| Field | Subfields | Description |
|---|---|---|
| `eval.fun` | `bndl` | Residuals of the lower bounds |
| | `bndu` | Residuals of the upper bounds |
| | `may` | Mayer type objective function value |
| | `lsq` | Least–squares objective function value |
| | `match` | Residuals of the matching conditions |
| | `resl` | Residuals of the lower constraints |
| | `resu` | Residuals of the upper constraints |
| `eval.der` | `may` | Gradient of the Mayer type objective |
| | `lsq` | Jacobian of the least–squares objective |
| | `sens` | ODE sensitivity w.r.t. the initial values |
| | `coup` | Coupling matrix |
| | `res` | Jacobian of the constraints residuals |

Table A.13: Iteration data structure (continued on page 338).

| | | |
|---|---|---|
| hessian | qq | Hessian of the Lagrangian w.r.t. the controls |
| | qx, xq | Mixed Hessian of the Lagrangian |
| | xx | Hessian of the Lagrangian w.r.t. the node states |
| qp.dense | constraints | Condensed constraints matrix |
| | gradient | Condensed gradient vector |
| | hessian | Condensed Hessian |
| | resBndl | Residuals of the lower bounds |
| | resBndu | Residuals of the upper bounds |
| | resl | Condensed residuals of the lower constraints |
| | resu | Condensed residuals of the upper constraints |
| | activeSet | Active set of the QP solution |
| | dualPoint | Dual solution of the condensed QP |
| | primalPoint | Primal solution of the condensed QP |
| qp.sparse | conCoup | Matching condition coupling Jacobians |
| | conPoint | Point constraint Jacobian |
| | conSens | Matching condition sensitivity Jacobians |
| | gradient | Node gradients of the objective |
| | hessian | Node Hessians of the Lagrangian |
| | resBndl | Node residuals of the lower bounds |
| | resBndu | Node residuals of the upper bounds |
| | resMatch | Node residuals of the matching conditions |
| | resPointl | Node residuals of the lower constraints |
| | resPointu | Node residuals of the upper constraints |
| | activeSet | Active set of the QP solution |
| | dualPoint | Dual solution of the QP |
| | primalPoint | Primal solution of the QP |
| sqp | aset | Active set |
| | iter | Primal iterate |
| | iterStep | Full primal step to next iterate |
| | mul | Dual iterate |
| | mulStep | Full dual step to next iterate |
| | obj | Objective function value |
| | stepInfo | Information about the current step |

Table A.14: Iteration data structure (continued from page 337).

## An Exemplary NMPC Loop

An exemplary NMPC loop realizing the real–time iteration scheme using this C interface to MuShROOM is given in figure A.8.

```
file = ioFileNew (MatlabFile, "model.mat", 0ul,
                  ioOpenRead|ioOpenExisting);
modelReadModel (&model, file, "model");
delete(file);

interfaceLoadLibrariesInPath (&libs, "");
interfaceLoadLibrary (&libs, "model.so");
interfaceInitializeLibraries (&libs, &model);
interfaceCreateInstances (&model.inst, &model);
interfaceNewData (&model, &data);

sqpStartup (model.inst.sqpsolver, &data.sqp, sqpColdStart);

while (data.sqp.stepInfo.uStatus != sqpStopped) {

    sqpPrepareStep  (model.inst.sqpsolver, &data.sqp);

    // To do: obtain or estimate system state 'system_state'

    sqpFeedbackStep (model.inst.sqpsolver, &data.sqp,
                     system_state, &control_feedback);

    // To do: feed 'control_feedback' back to process

    sqpTransitionStep (model.inst.sqpsolver, &data.sqp);
}

interfaceDeleteData (&model, &data);
interfaceFinalizeLibraries (&libs);
modelDelModel (&model);
interfaceUnloadLibraries (&libs);
```

Figure A.8: An exemplary NMPC loop using the C interface to MuShROOM.

## A.2 The Block Structured PQP Code qpHPSC

The parametric active set code qpHPSC is part of the MuShROOM software. Its purpose is to efficiently compute optimal solutions of a sequence of parametric QPs with vanishing constraints and with block structure as induced by a direct multiple shooting discretization. It can be used as a QP solver module replacing the condensing and QPOPT modules. In addition, it can be used as a stand–alone tool for Model Predictive Control (MPC) of discrete–time linear systems with vanishing constraints. qpHPSC implements the parametric active set strategy of chapter 6 and its extension to the special class of Quadratic Programs with Vanishing Constraints (QPVCs) obtained from applying Sequential Quadratic Programming (SQP) methods to a discretized MIOCP with constraints treated by outer convexification. For the solution of the KKT system, the HPSC factorization with matrix updates as developed in chapters 7 and 8 is implemented along with interfaces to several dense and sparse linear algebra packages for comparison purposes.

### A.2.1 Software Architecture

The software architecture of qpHPSC is depicted in figure A.9. The main active set loop is implemented and executed as detailed in chapter 6, with the possibility to either update or recompute the factorization of the Karush–Kuhn–Tucker (KKT) system using one of several KKT solvers. The main internal data structures of our implementation are listed in table A.15.

#### KKT Solvers

For the factorization of the block structured KKT system, one of several available decomposition codes can be chosen by the user. Choices are two applicable *LAPACK* routines for dense symmetric and banded unsymmetric factorization [9], the sparse symmetric indefinite factorization code *MA57* [56] of the *Harwell Subroutine Library (HSL)*, and the unsymmetric multifrontal factorization code *UMFPACK* [50] may be used. The newly developed Hessian Projection Schur Complement (HPSC) factorization of chapter 7 is tailored to the direct multiple shooting block structure. Matrix updates for the KKT system factorization are provided for this last factorization only, as detailed in chapter 8. Note that matrix updates could in principle be realized for a dense or sparse *LU* factorization of the structured KKT system as well, as detailed in chapter 7 and done e.g. by [108].

| Field | Subfield | Dim. | Description |
|---|---|---|---|
| `dim` | `nnodes` | 1 | Number $m$ of multiple shooting nodes |
| | `nx` | $m$ | Numbers $n$ of states+controls per node |
| | `nF` | $m$ | Numbers $n_i^{\mathrm{r}}$ of point constraints |
| | `nG` | $m$ | Number of matching conditions |
| | `nP` | $m$ | Number of coupling conditions |
| `aset.bidx`, | `states` | $m \times n$, $n_i^{\mathrm{r}}$ | Active set states of the simple bounds and point constraints |
| `aset.Fidx` | `active` | $m \times \lvert\mathcal{A}\rvert$, $\lvert\mathcal{X}\rvert$ | Indices of the active simple bounds and point constraints |
| | `inactive` | $m \times \lvert\mathcal{A}^{\mathrm{C}}\rvert$, $\lvert\mathcal{F}\rvert$ | Indices of the inactive/vanished simple bounds and point constraints |
| | `vanish` | $m \times n$, $n_i^{\mathrm{r}}$ | Mapping $\xi$ of simple lower bounds to vanishing constraints and vv. |
| `mat` | `F` | $m \times n_i^{\mathrm{r}} \times n$ | Decoupled point constraints Jac. $\boldsymbol{R}_i$ |
| | `G` | $m \times n^{\mathrm{x}} \times n$ | Matching condition sensitivities $\boldsymbol{G}_i$ |
| | `P` | $m \times n^{\mathrm{x}} \times n$ | Matching condition coupling mat. $\boldsymbol{P}_i$ |
| | `H` | $m \times n \times n$ | Hessian matrices $\boldsymbol{H}_i$ |
| `current`, | `f` | $m \times n$ | Current, step, and final gradient on the homotopy path |
| `delta`, | `bl`, `bu` | $m \times n$ | Simple bounds on homotopy path |
| `end` | `el`, `eu` | $m \times n_i^{\mathrm{r}}$ | Constraint residuals on homotopy path |
| | `h` | $m \times n^{\mathrm{x}}$ | Matching condition residuals on path |
| `iter` | `x` | $m \times n$ | Optimal primal iterate $(\boldsymbol{s}, \boldsymbol{q})$ |
| | `l` | $m \times n^{\mathrm{x}}$ | Optimal matching condition mult. $\boldsymbol{\lambda}$ |
| | `m` | $m \times n_i^{\mathrm{r}}$ | Optimal point constraints mult. $\boldsymbol{\mu}$ |
| | `n` | $m \times n$ | Optimal simple bounds mult. $\boldsymbol{\nu}$ |
| `kktmat` | `F`, `Ff`, `Fx`, `Fi` | $m \times \ldots$ | Point constraint jacbobian blocks $\boldsymbol{F}$, $\boldsymbol{F}^{\mathcal{AF}}$, $\boldsymbol{F}^{\mathcal{AX}}$, $\boldsymbol{F}^{\mathcal{A}^{\mathrm{C}}}$ |
| | `G`, `Gf`, `Gx` | $m \times \ldots$ | Matching condition sensitivity blocks $\boldsymbol{G}$, $\boldsymbol{G}^{\mathcal{F}}$, $\boldsymbol{G}^{\mathcal{X}}$ |
| | `P`, `Pf`, `Px` | $m \times \ldots$ | Matching condition coupling blocks $\boldsymbol{P}$, $\boldsymbol{P}^{\mathcal{F}}$, $\boldsymbol{P}^{\mathcal{X}}$ |
| | `H`, `Hff`, `Hxf`, `Hxx` | $m \times \ldots$ | Hessian matrix blocks $\boldsymbol{H}$, $\boldsymbol{H}^{\mathcal{FF}}$, $\boldsymbol{H}^{\mathcal{XF}}$, $\boldsymbol{H}^{\mathcal{XX}}$ |

Table A.15: Main data structures of the block structured QPVC solver code `qpHPSC` (continued on page 342).

| | | | |
|---|---|---|---|
| kktrhs | f,ff,fx | $m \times n$, $n_i^{\mathcal{F}}$, $n_i^{\mathcal{X}}$ | Gradient vectors $\boldsymbol{g}$, $\boldsymbol{g}^{\mathcal{F}}$, $\boldsymbol{g}^{\mathcal{X}}$ |
| | ba | $m \times n_i^{\mathcal{X}}$ | Active simple bounds $\boldsymbol{b}^{\mathcal{X}}$ |
| | ea | $m \times n_i^{\mathcal{A}}$ | Active point constraint residuals $\boldsymbol{e}^{\mathcal{A}}$ |
| | ha | $m \times n^{\mathrm{x}}$ | Matching condition residuals $\boldsymbol{h}$ |
| kktsol | x, xf, xx | $m \times n$, $n_i^{\mathcal{F}}$, $n_i^{\mathcal{X}}$ | Primal step $\boldsymbol{\delta x}$ |
| | la | $m \times n^{\mathrm{x}}$ | Matching condition multipliers step $\boldsymbol{\delta \lambda}$ |
| | ma | $m \times n_i^{\mathcal{A}}$ | Active point constraints mult. step $\boldsymbol{\delta \mu}^{\mathcal{A}}$ |
| | na | $m \times n_i^{\mathcal{X}}$ | Active simple bounds mult. step $\boldsymbol{\delta \nu}^{\mathcal{X}}$ |

Table A.16: Main data structures of the block structured QPVC solver code qpHPSC (continued from page 341).

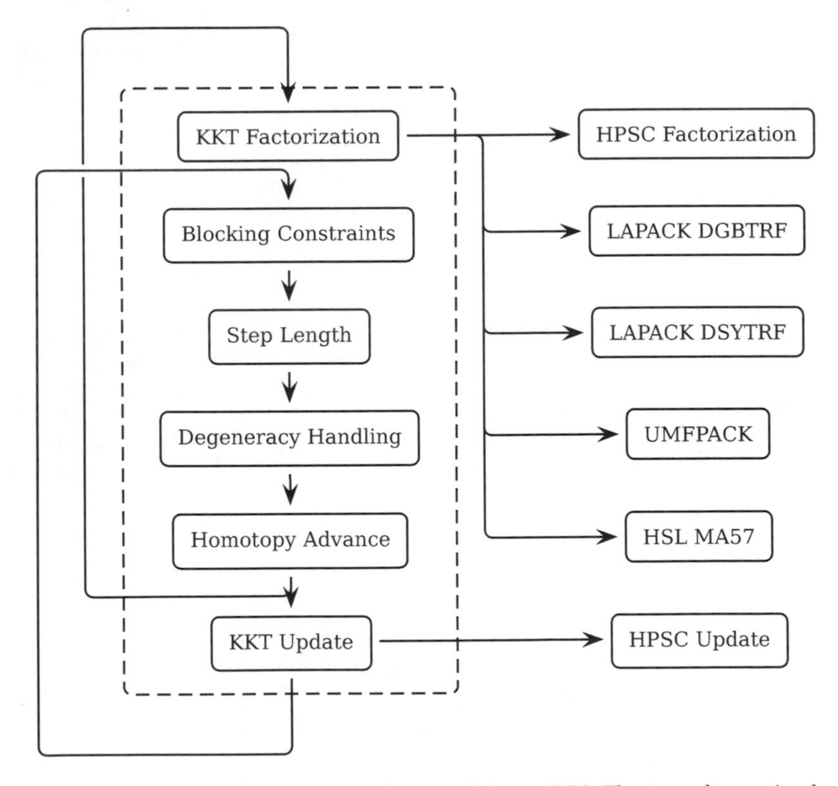


Figure A.9: Control flow of the QP solver module qpHPSC. The two alternative loops show recomputation or updating of the KKT system's factorization after an active set change.

### A.2.2 C Interface

#### C Interface Functions

The C interface to the qpHPSC code is straightforward. We provide two functions that start and continue the solution of a sequence of QPVCs, see table A.18. The problem description is expected as detailed for the field `qp.sparse` of the `MuShROOM` iterative data structure in table A.13. Upon successful solution, the fields `primalPoint`, `dualPoint` and `activeSet` hold the primal–dual optimal solution of the QPVC at the end of the homotopy path together with the associated active set.

#### Output

At the user's choice, the qpHPSC code provides an account of the steps and active set exchanges performed. Some exemplary iterations are shown in figure A.10 and the columns are further explained in table A.17. In iteration 1232 a continuation of the QPVC solution process in an adjacent convex subset can be seen, indicated by a reset of the homotopy parameter.

```
  it    distance      length     ki ty  no  id  vc sides  case act  regularity
1225  8.7174e-01  1.2826e-01  pr bd   9   1   . fr>up      .   .  8.9796e+08
1226  8.0012e-01  8.2157e-02  du bd  19  16  16 lo>fr      .   .          na
1227  7.8728e-01  1.6045e-02  pr bd  19  15  15 fr>lo      .   .  2.9517e+07
1228  7.2650e-01  7.7206e-02  du bd  19  17  17 lo>fr      .   .          na
1229  7.1765e-01  1.2181e-02  pr bd  19  16  16 fr>lo      .   .  5.1027e+07
1230  5.2708e-01  2.6555e-01  du bd  18  15  15 lo>fr      .   .          na
1231  5.1628e-01  2.0490e-02  pr bd  18  14  14 fr>lo      .   .  1.7937e+07
1232  2.5533e-01  5.0544e-01  pr bd  17  14  14 fr>lo +0>0- deg  0.0000e+00
                  1.2822e-06  du bd  17  13  13 lo>fr +0>0-   .          na
1233  1.0000e+00  0.0000e+00  pr bd  17  13  13 fr>lo +0>0- deg  0.0000e+00
                  3.2162e-06  du bd  17  15  15 lo>fr +0>0-   .          na
1234  5.9516e-01  4.0484e-01  du bd  18  16  16 lo>fr      .   .          na
1235  5.7512e-01  3.3666e-02  pr bd  18  15  15 fr>lo      .   .  2.0223e+07
1236  4.4086e-01  2.3344e-01  du bd  18  17  17 lo>fr      .   .          na
1237  4.2675e-01  3.2023e-02  pr bd  18  16  16 fr>lo      .   .  3.5430e+07
1238  2.9008e-01  3.2025e-01  pr bd  10   1   . fr>up      .   .  9.2597e+08
1239  0.0000e+00  1.0000e+00  ub  .   .   .   .  .   .    . opt
```

Figure A.10: Output of some exemplary active set iterations as provided by the qpHPSC code.

| Column | Description |
|---|---|
| `it` | Overall running count of QP iterations completed |
| `distance` | Remaining distance $1 - \tau^k$ to the end $\tau_1 = 1$ of the homotopy path |
| `length` | Length of the step onto the next blocking constraint on the remainder $[\tau^k, 1]$ of the homotopy path |
| `ki` | The *kind* of a blocking; `pr` for primal and `du` for dual |
| `ty` | The *type* of a blocking; `bd`: simple bound, `pc`: point constraint |
| `no` | The shooting node of the blocking bound or constraint |
| `id` | The index of the blocking bound or constraint on the shooting node |
| `vc` | The associated vanishing constraint, if any |
| `sides` | The active set exchange taking place |
| `case` | The MPVC index set exchange taking place, if any |
| `act` | `deg` for degeneracy resolution, `inf` for an infeasible problem, and `opt` for the optimal solution at the end of the homotopy path |
| `regularity` | The linear independence measure |

Table A.17: Columns of the per–iteration textual output of the `qpHPSC` code.

| Function | Description | |
|---|---|---|
| `hpscHomotopyStart` | | Start the solution of a new parametric QPVC |
| | `this` | Instance of the `qpHPSC` solver to use |
| | `data` | Pointer to a sparse QP data input/output structure according to the field `qp.sparse` in table A.13 |
| `hpscHomotopyContinue` | | Continue the solution of a started parametric QPVC |
| | `this` | Instance of the `qpHPSC` solver to use |
| | `data` | Pointer to a sparse QP data input/output structure according to the field `qp.sparse` in table A.13 |

Table A.18: Functions of the C interface to the code `qpHPSC`.

| KKT solver | Software | Ref. | Str. | Upd. | Description |
|---|---|---|---|---|---|
| `KKTSolverHPSC` | This thesis | | yes | yes | See chapters 7 and 8 |
| `KKTSolverBandLU` | *DGBTRF* | [9] | yes | no | Banded unsymm. $LU$ |
| `KKTSolverDenseLBL` | *DSYTRF* | [9] | no | no | Dense symmetric $LBL^T$ |
| `KKTSolverMA57` | *HSL MA57* | [56] | yes | no | Sparse symmetric $LBL^T$ |
| `KKTSolverUMFPACK` | *UMFPACK* | [50] | yes | no | Sparse unsymmetric LU |

Table A.19: Available KKT solvers for the block structured KKT system. Note that the dense $LBL^T$ decomposition yields inferior performance, and that dense updates for a dense or sparse $LU$ decomposition could be realized [108].

# Bibliography

[1] J. Abadie. On the Kuhn–Tucker theorem. In J. Abadie and S. Vajda, editors, *Nonlinear Programming*, pages 21–36. John Wiley & Sons, Inc., New York, NY, 1967.

[2] P. Abichandani, H. Benson, and M. Kam. Multi-vehicle path coordination under communication constraints. In *American Control Conference*, pages 650–656, 2008.

[3] W. Achtziger and C. Kanzow. Mathematical programs with vanishing constraints: optimality conditions and constraint qualifications. *Mathematical Programming Series A*, 114:69–99, 2008.

[4] J. Albersmeyer. *Adjoint based algorithms and numerical methods for sensitivity generation and optimization of large scale dynamic systems*. PhD thesis, Ruprecht–Karls–Universität Heidelberg, 2010.

[5] J. Albersmeyer and H. Bock. Sensitivity Generation in an Adaptive BDF-Method. In H. G. Bock, E. Kostina, X. Phu, and R. Rannacher, editors, *Modeling, Simulation and Optimization of Complex Processes: Proceedings of the International Conference on High Performance Scientific Computing, March 6–10, 2006, Hanoi, Vietnam*, pages 15–24. Springer Verlag Berlin Heidelberg New York, 2008.

[6] J. Albersmeyer and M. Diehl. The Lifted Newton Method and its Application in Optimization. *SIAM Journal on Optimization*, 20(3):1655–1684, 2010.

[7] F. Allgöwer and A. Zheng. *Nonlinear Predictive Control*, volume 26 of *Progress in Systems Theory*. Birkhäuser, Basel Boston Berlin, 2000.

[8] F. Allgöwer, T. Badgwell, J. Qin, J. Rawlings, and S. Wright. Nonlinear predictive control and moving horizon estimation – An introductory overview. In P. Frank, editor, *Advances in Control, Highlights of ECC'99*, pages 391–449. Springer Verlag, 1999.

[9] E. Anderson, Z. Bai, C. Bischof, S. Blackford, J. Demmel, J. Dongarra, J. Du Croz, A. Greenbaum, S. Hammarling, A. McKenney, and D. Sorensen. *LAPACK Users' Guide*. Society for Industrial and Ap-

plied Mathematics, Philadelphia, PA, 3rd edition, 1999. ISBN 0-89871-447-8 (paperback).

[10] V. Bär. Ein Kollokationsverfahren zur numerischen Lösung allgemeiner Mehrpunktrandwertaufgaben mit Schalt- und Sprungbedingungen mit Anwendungen in der optimalen Steuerung und der Parameteridentifizierung. Diploma thesis, Rheinische Friedrich-Wilhelms-Universität Bonn, 1983.

[11] M. Bartlett. An inverse matrix adjustment arising in discriminant analysis. *Annals of Mathematical Statistics*, 22(1):107–111, 1951.

[12] R. Bartlett and L. Biegler. QPSchur: A dual, active set, schur complement method for large-scale and structured convex quadratic programming algorithm. *Optimization and Engineering*, 7:5–32, 2006.

[13] R. Bartlett, A. Wächter, and L. Biegler. Active set vs. interior point strategies for model predicitve control. In *Proceedings of the American Control Conference*, pages 4229–4233, Chicago, IL, 2000.

[14] R. Bartlett, L. Biegler, J. Backstrom, and V. Gopal. Quadratic programming algorithms for large-scale model predictive control. *Journal of Process Control*, 12:775–795, 2002.

[15] I. Bauer. *Numerische Verfahren zur Lösung von Anfangswertaufgaben und zur Generierung von ersten und zweiten Ableitungen mit Anwendungen bei Optimierungsaufgaben in Chemie und Verfahrenstechnik*. PhD thesis, Ruprecht-Karls-Universität Heidelberg, 1999.

[16] B. Baumrucker and L. Biegler. MPEC strategies for optimization of a class of hybrid dynamic systems. *Journal of Process Control*, 19 (8):1248 - 1256, 2009. ISSN 0959-1524. Special Section on Hybrid Systems: Modeling, Simulation and Optimization.

[17] B. Baumrucker, J. Renfro, and L. Biegler. MPEC problem formulations and solution strategies with chemical engineering applications. *Computers and Chemical Engineering*, 32:2903–2913, 2008.

[18] R. Bellman. *Dynamic Programming*. University Press, Princeton, N.J., 6th edition, 1957. ISBN 0-486-42809-5 (paperback).

[19] P. Belotti. Couenne: a user's manual. Technical report, Lehigh University, 2009.

[20] A. Bemporad, F. Borrelli, and M. Morari. Model Predictive Control Based on Linear Programming — The Explicit Solution. *IEEE Trans-*

*actions on Automatic Control*, 47(12):1974–1985, 2002.

[21] A. Bemporad, S. di Cairano, E. Henriksson, and K. Johansson. Hybrid model predictive control based on wireless sensor feedback: An experimental study. *International Journal of Robust and Nonlinear Control*, 20:209–225, 2010.

[22] A. Berkelaar, K. Roos, and T. Terlaky. *Recent Advances in Sensitivity Analysis and Parametric Programming*, chapter 6: The Optimal Set and Optimal Partition Approach to Linear and Quadratic Programming, pages 6–1–6–45. Kluwer Publishers, Dordrecht, 1997.

[23] D. Bertsekas. *Dynamic Programming and Optimal Control*, volume 1 and 2. Athena Scientific, Belmont, MA, 1995.

[24] D. Bertsekas. *Nonlinear Programming*. Athena Scientific, Belmont, MA, 2003.

[25] M. Best. *An Algorithm for the Solution of the Parametric Quadratic Programming Problem*, chapter 3, pages 57–76. Applied Mathematics and Parallel Computing. Physica-Verlag, Heidelberg, 1996.

[26] L. Biegler. Solution of dynamic optimization problems by successive quadratic programming and orthogonal collocation. *Computers and Chemical Engineering*, 8:243–248, 1984.

[27] L. Biegler. An overview of simultaneous strategies for dynamic optimization. *Chemical Engineering and Processing*, 46:1043–1053, 2007.

[28] L. Biegler, O. Ghattas, M. Heinkenschloss, and B. Bloemen Waanders. *Large-Scale PDE-Constrained Optimization*, volume 30 of *Lecture Notes in Computational Science and Engineering*. Springer Verlag, Heidelberg, 2003.

[29] H. Bock. Numerische Optimierung zustandsbeschränkter parameterabhängiger Prozesse mit linear auftretender Steuerung unter Anwendung der Mehrzielmethode. Diploma thesis, Universität zu Köln, 1974.

[30] H. Bock. Zur numerischen Behandlung zustandsbeschränkter Steuerungsprobleme mit Mehrzielmethode und Homotopieverfahren. *Zeitschrift für Angewandte Mathematik und Mechanik*, 57(4):T266–T268, 1977.

[31] H. Bock. Numerical solution of nonlinear multipoint boundary value

problems with applications to optimal control. *Zeitschrift für Angewandte Mathematik und Mechanik*, 58:407, 1978.

[32] H. Bock. Numerical treatment of inverse problems in chemical reaction kinetics. In K. Ebert, P. Deuflhard, and W. Jäger, editors, *Modelling of Chemical Reaction Systems*, volume 18 of *Springer Series in Chemical Physics*, pages 102–125. Springer Verlag, Heidelberg, 1981. URL `http://www.iwr.uni-heidelberg.de/groups/agbock/FILES/Bock1981.pdf`.

[33] H. Bock. Recent advances in parameter identification techniques for ODE. In P. Deuflhard and E. Hairer, editors, *Numerical Treatment of Inverse Problems in Differential and Integral Equations*, pages 95–121. Birkhäuser, Boston, 1983. URL `http://www.iwr.uni-heidelberg.de/groups/agbock/FILES/Bock1983.pdf`.

[34] H. Bock. *Randwertproblemmethoden zur Parameteridentifizierung in Systemen nichtlinearer Differentialgleichungen*, volume 183 of *Bonner Mathematische Schriften*. Rheinische Friedrich–Wilhelms–Universität Bonn, 1987. URL `http://www.iwr.uni-heidelberg.de/groups/agbock/FILES/Bock1987.pdf`.

[35] H. Bock and R. Longman. Computation of optimal controls on disjoint control sets for minimum energy subway operation. In *Proceedings of the American Astronomical Society. Symposium on Engineering Science and Mechanics*, Taiwan, 1982.

[36] H. Bock and K. Plitt. A Multiple Shooting algorithm for direct solution of optimal control problems. In *Proceedings of the 9th IFAC World Congress*, pages 242–247, Budapest, 1984. Pergamon Press. URL `http://www.iwr.uni-heidelberg.de/groups/agbock/FILES/Bock1984.pdf`.

[37] H. Bock, M. Diehl, E. Kostina, and J. Schlöder. Constrained Optimal Feedback Control for DAE. In L. Biegler, O. Ghattas, M. Heinkenschloss, D. Keyes, and B. van Bloemen Waanders, editors, *Real-Time PDE-Constrained Optimization*, chapter 1, pages 3–24. SIAM, 2007.

[38] P. Bonami, L. Biegler, A. Conn, G. Cornuéjols, I. Grossmann, C. Laird, J. Lee, A. Lodi, F. Margot, N. Sawaya, and A. Wächter. An algorithmic framework for convex mixed integer nonlinear programs. *Discrete Optimization*, 5(2):186–204, 2009.

[39] B. Borchers and J. Mitchell. An improved Branch and Bound algo-

rithm for Mixed Integer Nonlinear Programming. *Computers and Operations Research*, 21(4):359–367, 1994.

[40] U. Brandt-Pollmann. *Numerical solution of optimal control problems with implicitly defined discontinuities with applications in engineering*. PhD thesis, Ruprecht–Karls–Universität Heidelberg, 2004.

[41] C. G. Broyden. The convergence of a class of double–rank minimization algorithms. *Journal of the Institute of Mathematics and its Applications*, 6:76–90, 1970.

[42] A. Buchner. Auf Dynamischer Programmierung basierende nichtlineare modellprädiktive Regelung für LKW. Diploma thesis, Ruprecht–Karls–Universität Heidelberg, January 2010. URL `http://mathopt.de/PUBLICATIONS/Buchner2010.pdf`.

[43] R. Bulirsch. Die Mehrzielmethode zur numerischen Lösung von nichtlinearen Randwertproblemen und Aufgaben der optimalen Steuerung. Technical report, Carl–Cranz–Gesellschaft, Oberpfaffenhofen, 1971.

[44] J. Bunch and B. Parlett. Direct methods for solving symmetric indefinite systems of linear equations. *SIAM Journal of Numerical Analysis*, 8(4):639–655, December 1971.

[45] J. Butcher. *The Numerical Analysis of Ordinary Differential Equations: Runge–Kutta and General Linear Methods*. Wiley, 1987. ISBN 0-471-91046-5 (paperback).

[46] E. Camacho and C. Bordons. *Model Predictive Control*. Springer Verlag, London, 2004.

[47] Y. Chen and M. Florian. The nonlinear bilevel programming problem: Formulations, regularity, and optimality conditions. *Optimization*, 32: 193–209, 1995.

[48] V. Chvatal. Edmonds polytopes and weakly Hamiltonian graphs. *Mathematical Programming*, 5:29–40, 1973.

[49] R. Dakin. A tree-search algorithm for mixed integer programming problems. *The Computer Journal*, 8:250–255, 1965.

[50] T. Davis. Algorithm 832: UMFPACK - an unsymmetric-pattern multifrontal method with a column pre-ordering strategy. *ACM Trans. Math. Software*, 30:196–199, 2004.

[51] M. Diehl. *Real-Time Optimization for Large Scale Nonlinear Pro-*

*cesses*. PhD thesis, Ruprecht–Karls–Universität Heidelberg, 2001. URL http://www.ub.uni-heidelberg.de/archiv/1659/.

[52] M. Diehl, H. Bock, J. Schlöder, R. Findeisen, Z. Nagy, and F. Allgöwer. Real-time optimization and nonlinear model predictive control of processes governed by differential-algebraic equations. *J. Proc. Contr.*, 12(4):577–585, 2002. URL http://www.iwr.uni-heidelberg.de/groups/agbock/FILES/Diehl2002b.pdf.

[53] M. Diehl, H. Bock, and J. Schlöder. A real-time iteration scheme for nonlinear optimization in optimal feedback control. *SIAM Journal on Control and Optimization*, 43(5):1714–1736, 2005.

[54] M. Diehl, H. Ferreau, and N. Haverbeke. Efficient numerical methods for nonlinear mpc and moving horizon estimation. In L. Magni, D. Raimondo, and F. Allgöwer, editors, *Nonlinear Model Predictive Control*, volume 384 of *Springer Lecture Notes in Control and Information Sciences*, pages 391–417. Springer Verlag, Berlin, Heidelberg, New York, 2009.

[55] J. Dormand and P. Prince. A reconsideration of some embedded Runge–Kutta formulae. *Journal of Computational and Applied Mathematics*, 15(2):203–211, 1986.

[56] I. Duff. MA57 — a code for the solution of sparse symmetric definite and indefinite systems. *ACM Transactions on Mathematical Software*, 30(2):118–144, 2004.

[57] I. Duff and J. Reid. The multifrontal solution of indefinite sparse symmetric linear equations. *ACM Transactions on Mathematical Software*, 9(3):302–325, 1983.

[58] M. Dür and V. Stix. Probabilistic subproblem selection in branch-and-bound algorithms. *Journal of Computational and Applied Mathematics*, 182(1):67–80, 2005.

[59] M. Duran and I. Grossmann. An outer-approximation algorithm for a class of mixed-integer nonlinear programs. *Mathematical Programming*, 36(3):307–339, 1986.

[60] M. Egerstedt, Y. Wardi, and H. Axelsson. Transition-time optimization for switched-mode dynamical systems. *IEEE Transactions on Automatic Control*, 51:110–115, 2006.

[61] E. Eich. *Projizierende Mehrschrittverfahren zur numerischen Lösung von Bewegungsgleichungen technischer Mehrkörpersysteme*

*mit Zwangsbedingungen und Unstetigkeiten.* PhD thesis, Universität Augsburg 1991, erschienen als Fortschr.-Ber. VDI Reihe 18 Nr. 109. VDI-Verlag, Düsseldorf, 1992.

[62] S. Eldersveld and M. Saunders. A block-LU update for large scale linear programming. *SIAM Journal of Matrix Analysis and Applications*, 13:191–201, 1992.

[63] M. Engelhart. Modeling, simulation, and optimization of cancer chemotherapies. Diploma thesis, Ruprecht–Karls–Universität Heidelberg, 2009. URL `http://mathopt.uni-hd.de/PUBLICATIONS/Engelhart2009.pdf`.

[64] W. Enright. Continuous numerical methods for ODEs with defect control. *Journal of Computational and Applied Mathematics*, 125(1):159–170, December 2001.

[65] E. Fehlberg. Klassische Runge-Kutta-Formeln fünfter und siebenter Ordnung mit Schrittweiten-Kontrolle. *Computing*, 4:93–106, 1969.

[66] E. Fehlberg. Klassische Runge-Kutta-Formeln vierter und niedrigerer Ordnung mit Schrittweiten-Kontrolle und ihre Anwendung auf Wärmeleitungsprobleme. *Computing*, 6:61–71, 1970.

[67] H. Ferreau. An online active set strategy for fast solution of parametric quadratic programs with applications to predictive engine control. Diploma thesis, Ruprecht–Karls–Universität Heidelberg, 2006. URL `http://homes.esat.kuleuven.be/~jferreau/pdf/thesisONLINE.pdf`.

[68] H. Ferreau, G. Lorini, and M. Diehl. Fast nonlinear model predictive control of gasoline engines. In *Proceedings of the IEEE International Conference on Control Applications, Munich*, pages 2754–2759, 2006.

[69] H. Ferreau, H. Bock, and M. Diehl. An online active set strategy to overcome the limitations of explicit MPC. *International Journal of Robust and Nonlinear Control*, 18(8):816–830, 2008.

[70] A. Fiacco. *Introduction to sensitivity and stability analysis in nonlinear programming.* Academic Press, New York, 1983.

[71] R. Fletcher. A new approach to variable metric algorithms. *Computer Journal*, 13:317–322, 1970.

[72] R. Fletcher. *Practical Methods of Optimization.* Wiley, Chichester,

2nd edition, 1987. ISBN 0-471-49463-1 (paperback).

[73] R. Fletcher. Resolving degeneracy in quadratic programming. Numerical Analysis Report NA/135, University of Dundee, Dundee, Scotland, 1991.

[74] C. Floudas. *Nonlinear and Mixed-Integer Optimization - Fundamentals and Applications.* Topics in Chemical Engineering. University Press, Oxford, 1995.

[75] M. Fukushima and P. Tseng. An implementable active–set algorithm for computing a B–stationary point of a mathematical program with linear complementarity constraints. *SIAM Journal on Optimization,* 12:724–739, 1999.

[76] A. Fuller. Study of an optimum nonlinear control system. *Journal of Electronics and Control,* 15:63–71, 1963.

[77] C. Garcia and M. Morari. Internal model control. 1. a unifying review and some new results. *Ind. Eng. Chem. Process Des. Dev.,* 24:472–484, 1985.

[78] M. Garey and D. Johnson. *Computers and Intractability: A Guide to the Theory of NP-Completeness.* W.H. Freeman, New York, 1979.

[79] A. Geoffrion. Generalized Benders decomposition. *Journal of Optimization Theory and Applications,* 10:237–260, 1972.

[80] M. Gerdts. Solving mixed-integer optimal control problems by Branch&Bound: A case study from automobile test-driving with gear shift. *Optimal Control Applications and Methods,* 26:1–18, 2005.

[81] M. Gerdts. A variable time transformation method for mixed-integer optimal control problems. *Optimal Control Applications and Methods,* 27(3):169–182, 2006.

[82] E. Gertz and S. Wright. Object-oriented software for quadratic programming. *ACM Transactions on Mathematical Software,* 29:58–81, 2003.

[83] P. Gill, G. Golub, W. Murray, and M. A. Saunders. Methods for modifying matrix factorizations. *Mathematics of Computation,* 28(126): 505–535, 1974.

[84] P. Gill, W. Murray, M. Saunders, and M. Wright. Procedures for optimization problems with a mixture of bounds and general linear constraints. *ACM Transactions on Mathematical Software,* 10(3):282–

298, 1984.

[85] P. Gill, W. Murray, M. Saunders, and M. Wright. A practical anti-cycling procedure for linearly constrained optimization. *Mathematical Programming*, 45(1–3):437–474, 1989.

[86] P. Gill, W. Murray, and M. Saunders. *User's Guide For QPOPT 1.0: A Fortran Package For Quadratic Programming*, 1995.

[87] D. Goldfarb. A family of variable metric updates derived by variational means. *Mathematics of Computation*, 24:23–26, 1970.

[88] D. Goldfarb and A. Idnani. A numerically stable dual method for solving strictly convex quadratic programs. *Mathematical Programming*, 27:1–33, 1983.

[89] G. Golub and C. van Loan. *Matrix Computations*. Johns Hopkins University Press, Baltimore, 3rd edition, 1996.

[90] R. Gomory. Outline of an algorithm for integer solutions to linar programs. *Bulletin of the American Mathematical Society*, 64:275–278, 1958.

[91] J. Gondzio. Multiple centrality corrections in a primal–dual interior point method for linear programming. *Computational Optimization and Applications*, 6:137–156, 1996.

[92] A. Griewank. *Evaluating Derivatives, Principles and Techniques of Algorithmic Differentiation*. Number 19 in Frontiers in Applied Mathematics. SIAM, Philadelphia, 2000.

[93] I. Grossmann. Review of nonlinear mixed-integer and disjunctive programming techniques. *Optimization and Engineering*, 3:227–252, 2002.

[94] I. Grossmann, P. Aguirre, and M. Barttfeld. Optimal synthesis of complex distillation columns using rigorous models. *Computers and Chemical Engineering*, 29:1203–1215, 2005.

[95] M. Grötschel, L. Lovász, and A. Schrijver. *Geometric Algorithms and Combinatorial Optimization*, volume 2 of *Algorithms and Combinatorics*. Springer Verlag, 1988. ISBN 3-540-13624-X, 0-387-13624-X (U.S.).

[96] J. Guddat, F. G. Vasquez, and H. Jongen. *Parametric Optimization: Singularities, Pathfollowing and Jumps*. Teubner, Stuttgart, 1990.

[97] M. Guignard. Generalized Kuhn–Tucker conditions for mathematical

programming problems in a Banach space. *SIAM Journal on Control*, 7(2):232–241, 1969.

[98] J. Hall and K. McKinnon. The simplest examples where the simplex method cycles and conditions where the EXPAND method fails to prevent cycling. *Mathematical Programming: Series A and B*, 100(1): 133–150, 2004.

[99] S. Hammarling. A note on modifications to the givens plane rotation. *J. Inst. Maths Applics*, 13:215–218, 1974.

[100] S. Han. Superlinearly convergent variable-metric algorithms for general nonlinear programming problems. *Mathematical Programming*, 11:263–282, 1976.

[101] C. Hargraves and S. Paris. Direct trajectory optimization using nonlinear programming and collocation. *AIAA J. Guidance*, 10(4):338–342, 1987.

[102] N. Haverbeke, M. Diehl, and B. de Moor. A structure exploiting interior-point method for moving horizon estimation. In *Proceedings of the 48th IEEE Conference on Decision and Control (CDC09)*, pages 1–6, 2009.

[103] E. Hellström, M. Ivarsson, J. Aslund, and L. Nielsen. Look-ahead control for heavy trucks to minimize trip time and fuel consumption. *Control Engineering Practice*, 17:245–254, 2009.

[104] H. Hermes and J. Lasalle. *Functional analysis and time optimal control*, volume 56 of *Mathematics in science and engineering*. Academic Press, New York and London, 1969.

[105] T. Hoheisel. *Mathematical Programs with Vanishing Constraints*. PhD thesis, Julius–Maximilians–Universität Würzburg, July 2009.

[106] T. Hoheisel and C. Kanzow. First- and second-order optimality conditions for mathematical programs with vanishing constraints. *Applications of Mathematics*, 52(6):459–514, 2007.

[107] T. Hoheisel and C. Kanzow. Stationary conditions for mathematical programs with vanishing constraints using weak constraint qualifications. *J. Math. Anal. Appl.*, 337:292–310, 2008.

[108] H. Huynh. *A Large-Scale Quadratic Programming Solver Based On Block-LU Updates of the KKT System*. PhD thesis, Stanford University, 2008.

[109] A. Izmailov and M. Solodov. Mathematical programs with vanishing constraints: Optimality conditions, sensitivity, and a relaxation method. *Journal of Optimization Theory and Applications*, 142:501–532, 2009.

[110] E. Johnson, G. Nemhauser, and M. Savelsbergh. Progress in linear programming-based algorithms for integer programming: An exposition. *INFORMS Journal on Computing*, 12(1):2–23, 2000.

[111] J. Júdice, H. Sherali, I. Ribeiro, and A. Faustino. Complementarity active-set algorithm for mathematical programming problems with equilibrium constraints. *Journal of Optimization Theory and Applications*, 134:467–481, 2007.

[112] S. Kameswaran and L. Biegler. Simultaneous dynamic optimization strategies: Recent advances and challenges. *Computers and Chemical Engineering*, 30:1560–1575, 2006.

[113] N. Karmarkar. A new polynomial time algorithm for linear programming. *Combinatorica*, 4(4):373–395, 1984.

[114] W. Karush. Minima of functions of several variables with inequalities as side conditions. Master's thesis, Department of Mathematics, University of Chicago, 1939.

[115] G. Kedem. Automatic differentiation of computer programs. *ACM Trans. on Math. Soft.*, 6:150–165, 1980.

[116] F. Kehrle. Optimal control of vehicles in driving simulators. Diploma thesis, Ruprecht–Karls–Unversität Heidelberg, March 2010. URL `http://mathopt.de/PUBLICATIONS/Kehrle2010.pdf`.

[117] U. Kiencke and L. Nielsen. *Automotive Control Systems*. Springer Verlag, 2000.

[118] C. Kirches. A numerical method for nonlinear robust optimal control with implicit discontinuities and an application to powertrain oscillations. Diploma thesis, Ruprecht–Karls–Universität Heidelberg, October 2006. URL `http://mathopt.de/PUBLICATIONS/Kirches2006.pdf`.

[119] C. Kirches, H. Bock, J. Schlöder, and S. Sager. Complementary condensing for the direct multiple shooting method. To appear in H. Bock, E. Kostina, H. Phu, and R. Rannacher, editors, *Proceedings of the Fourth International Conference on High Performance Scientific Computing: Modeling, Simulation, and Optimization of Complex*

*Processes, Hanoi, Vietnam, March 2–6, 2009*, Springer Verlag, 2011. URL http://mathopt.de/PUBLICATIONS/Kirches2010a.pdf.

[120] C. Kirches, H. Bock, J. Schlöder, and S. Sager. Block structured quadratic programming for the direct multiple shooting method for optimal control. *Optimization Methods and Software*, 26(2):239–257, April 2011.

[121] C. Kirches, H. Bock, J. Schlöder, and S. Sager. A factorization with update procedures for a KKT matrix arising in direct optimal control. *Mathematical Programming Computation*, 3(4), Fall 2011 (to appear).

[122] C. Kirches, S. Sager, H. Bock, and J. Schlöder. Time-optimal control of automobile test drives with gear shifts. *Optimal Control Applications and Methods*, 31(2):137–153, March/April 2010. URL http://mathopt.de/PUBLICATIONS/Kirches2010.pdf.

[123] C. Kirches, L. Wirsching, S. Sager, and H. Bock. Efficient numerics for nonlinear model predictive control. In M. Diehl, F. Glineur, E. Jarlebring, and W. Michiels, editors, *Recent Advances in Optimization and its Applications in Engineering*, pages 449–459, Springer Verlag, 2010. URL http://mathopt.de/PUBLICATIONS/Kirches2010c.pdf.

[124] C. Kirches. Fast numerical methods for mixed–integer nonlinear model–predictive control. PhD thesis, Ruprecht–Karls–Universität Heidelberg, 2010. URL http://mathopt.de/PUBLICATIONS/Kirches2010e.pdf.

[125] B. Korte and J. Vygen. *Combinatorial Optimization*. Springer Verlag, Berlin Heidelberg New York, 3rd edition, 2006. ISBN 3-540-25684-9.

[126] H. Kuhn and A. Tucker. Nonlinear programming. In J. Neyman, editor, *Proceedings of the Second Berkeley Symposium on Mathematical Statistics and Probability*, Berkeley, 1951. University of California Press.

[127] M. Kutta. Beitrag zur näherungsweisen Integration totaler Differentialgleichungen. *Zeitschrift für Mathematik und Physik*, 46:435–453, 1901.

[128] A. Land and A. Doig. An automatic method of solving discrete programming problems. *Econometrica*, 28:497–520, 1960.

[129] S. Lauer. SQP–Methoden zur Behandlung von Problemen mit indefiniter reduzierter Hesse–Matrix. Diploma thesis, Ruprecht–Karls–Universität Heidelberg, February 2010.

[130] D. Lebiedz, S. Sager, H. Bock, and P. Lebiedz. Annihilation of limit cycle oscillations by identification of critical phase resetting stimuli via mixed-integer optimal control methods. *Physical Review Letters*, 95:108303, 2005.

[131] D. Lebiedz, S. Sager, O. Shaik, and O. Slaby. Optimal control of self-organized dynamics in cellular signal transduction. In *Proceedings of the 5th MATHMOD conference*, ARGESIM-Reports, ISBN 3-901608-25-7, Vienna, 2006.

[132] D. Leineweber. Analyse und Restrukturierung eines Verfahrens zur direkten Lösung von Optimal-Steuerungsproblemen. Diploma thesis, Ruprecht–Karls–Universität Heidelberg, 1995.

[133] D. Leineweber. *Efficient reduced SQP methods for the optimization of chemical processes described by large sparse DAE models*, volume 613 of *Fortschritt-Berichte VDI Reihe 3, Verfahrenstechnik*. VDI Verlag, Düsseldorf, 1999.

[134] D. Leineweber, I. Bauer, A. Schäfer, H. Bock, and J. Schlöder. An efficient multiple shooting based reduced SQP strategy for large-scale dynamic process optimization (Parts I and II). *Computers and Chemical Engineering*, 27:157–174, 2003.

[135] S. Leyffer. *Deterministic methods for mixed-integer nonlinear programming*. PhD thesis, University of Dundee, 1993.

[136] S. Leyffer. Integrating SQP and branch-and-bound for mixed integer nonlinear programming. *Computational Optimization and Applications*, 18(3):295–309, 2001.

[137] S. Leyffer. Complementarity constraints as nonlinear equations: Theory and numerical experiences. Technical report, Argonne National Laboratory, June 2003.

[138] S. Leyffer. The return of the active–set method. preprint ANL/MCS-P1277-0805, Argonne National Laboratory, 9700 South Cass Avenue, Argonne, IL 60439, USA, February 2005.

[139] S. Leyffer, G. López-Calva, and J. Nocedal. Interior methods for mathematical programs with complementarity constraints. *SIAM Journal on Optimization*, 17(1):52–77, 2006.

[140] J. Linderoth and M. Savelsbergh. A computational study of branch and bound search strategies for mixed integer programming. *INFORMS Journal on Computing*, 11:173–187, 1999.

[141] F. Logist, S. Sager, C. Kirches, and J. van Impe. Efficient multiple objective optimal control of dynamic systems with integer controls. *Journal of Process Control*, 20(7):810–822, August 2010.

[142] C. Long and E. Gatzke. Model predictive control algorithm for prioritized objective inferential control of unmeasured states using propositional logic. *Ind. Eng. Chem. Res.*, 44:3575–3584, 2005.

[143] D. Luenberger. *Optimization by vector space methods.* Wiley Professional Paperback Series. John Wiley & Sons, Inc., New York, NY, 1969. ISBN 0471-18117-X (paperback).

[144] Z. Luo, J. Pang, and D. Ralph. *Mathematical Programs with Equlibrium Constraints.* Cambridge University Press, Cambridge, 1996.

[145] J. Lyness and C. Moler. Numerical differentiation of analytic functions. *SIAM Journal on Numerical Analysis*, 4:202–210, 1967.

[146] J. Macki and A. Strauss. *Introduction to optimal control theory.* Springer Verlag, Heidelberg, 1995.

[147] L. Magni, D. Raimondo, and F. Allgöwer, editors. *Nonlinear Model Predictive Control: Towards New Challenging Applications*, volume 384 of *Lecture Notes in Control and Information Sciences.* Springer Verlag, 2009.

[148] L. Magni, D. Raimondo, and F. Allgöwer, editors. *Proceedings of the international workshop on assessment and future directions of nonlinear model predictive control (NMPC'08), Pavia, Italy, September 5–9, 2008*, volume 384 of *Lecture Notes in Control and Information Sciences*, Berlin Heidelberg New York, 2009. Springer Verlag.

[149] O. Mangasarian and S. Fromovitz. Fritz John necessary optimality conditions in the presence of equality and inequality constraints. *Journal of Mathematical Analysis and Applications*, 17:37–47, 1967.

[150] Maplesoft. *Maple 13.* Maplesoft, Inc., 2009.

[151] J. Mattingley and S. Boyd. Automatic code generation for real-time convex optimization. In Y. Eldar and D. Palomar, editors, *Convex Optimization in Signal Processing and Communications.* Cambridge University Press, 2010.

[152] D. Q. Mayne. Nonlinear model predictive control: Challenges and opportunities. In F. Allgöwer and A. Zheng, editors, *Nonlinear Predictive Control*, volume 26 of *Progress in Systems Theory*, pages 23–44,

Basel Boston Berlin, 2000. Birkhäuser.

[153] D. Q. Mayne and H. Michalska. Receding horizon control of nonlinear systems. *IEEE Transactions on Automatic Control*, 35(7):814–824, 1990.

[154] D. Q. Mayne and S. Rakovic. Optimal control of constrained piecewise affine discrete-time systems. *Computational Optimization and Applications*, 25:167–191, 2003.

[155] S. Mehrotra. On the implementation of a primal-dual interior point method. *SIAM Journal on Optimization*, 2(4):575–601, 1992.

[156] K. Mombaur. *Stability Optimization of Open-loop Controlled Walking Robots*. PhD thesis, Ruprecht–Karls–Universität Heidelberg, 2001. URL `http://www.ub.uni-heidelberg.de/archiv/1796`.

[157] K. Murty. Some NP-complete problems in quadratic and nonlinear programming. *Mathematical Programming*, 39:117–129, 1987.

[158] J. Nocedal and S. Wright. *Numerical Optimization*. Springer Verlag, Berlin Heidelberg New York, 2nd edition, 2006. ISBN 0-387-30303-0 (hardcover).

[159] I. Nowak. *Relaxation and Decomposition Methods for Mixed Integer Nonlinear Programming*. Birkhäuser, Basel Boston Berlin, 2005.

[160] J. Oldenburg. *Logic–based modeling and optimization of discrete–continuous dynamic systems*, volume 830 of *Fortschritt-Berichte VDI Reihe 3, Verfahrenstechnik*. VDI Verlag, Düsseldorf, 2005.

[161] J. Oldenburg, W. Marquardt, D. Heinz, and D. Leineweber. Mixed logic dynamic optimization applied to batch distillation process design. *AIChE Journal*, 49(11):2900–2917, 2003.

[162] M. Osborne. On shooting methods for boundary value problems. *Journal of Mathematical Analysis and Applications*, 27:417–433, 1969.

[163] B. Owren and M. Zennaro. Derivation of efficient continuous explicit Runge–Kutta methods. *SIAM Journal on Scientific and Statistical Computing*, 13:1488–1501, 1992.

[164] H. Pacejka and E. Bakker. The magic formula tyre model. *Vehicle System Dynamics*, 21:1–18, 1993.

[165] D. Peterson. A review of constraint qualifications in finite dimensional spaces. *SIAM Review*, 15(3):639–654, July 1973.

[166] L. Petzold, S. Li, Y. Cao, and R. Serban. Sensitivity analysis of differential-algebraic equations and partial differential equations. *Computers and Chemical Engineering*, 30:1553–1559, 2006.

[167] K. Plitt. Ein superlinear konvergentes Mehrzielverfahren zur direkten Berechnung beschränkter optimaler Steuerungen. Diplomarbeit, Rheinische Friedrich–Wilhelms–Universität Bonn, 1981.

[168] A. Potschka. Handling path constraints in a direct multiple shooting method for optimal control problems. Diploma thesis, Ruprecht–Karls–Universität Heidelberg, 2006. URL `http://apotschka.googlepages.com/APotschka2006.pdf`.

[169] A. Potschka, H. Bock, and J. Schlöder. A minima tracking variant of semi-infinite programming for the treatment of path constraints within direct solution of optimal control problems. *Optimization Methods and Software*, 24(2):237–252, 2009.

[170] M. Powell. Algorithms for nonlinear constraints that use Lagrangian functions. *Mathematical Programming*, 14(3):224–248, 1978.

[171] S. Qin and T. Badgwell. Review of nonlinear model predictive control applications. In B. Kouvaritakis and M. Cannon, editors, *Nonlinear model predictive control: theory and application*, pages 3–32, London, 2001. The Institute of Electrical Engineers.

[172] I. Quesada and I. Grossmann. An LP/NLP based branch and bound algorithm for convex MINLP optimization problems. *Computers and Chemical Engineering*, 16:937–947, 1992.

[173] A. Raghunathan and L. Biegler. Mathematical programs with equilibrium constraints (MPECs) in process engineering. *Computers and Chemical Engineering*, 27:1381–1392, 2003.

[174] A. Raghunathan, M. Diaz, and L. Biegler. An mpec formulation for dynamic optimization of distillation operations. *Computers and Chemical Engineering*, 28:2037–2052, 2004.

[175] D. Ralph and S. J. Wright. Some properties of regularization and penalization schemes for mpecs. *Optimization Methods and Software*, 19:527–556, 2004.

[176] C. Rao, S. Wright, and J. Rawlings. Application of interior-point methods to model predictive control. *Journal of Optimization Theory and Applications*, 99:723–757, 1998.

[177] J. Rawlings and D. Mayne. *Model Predictive Control: Theory and Design*. Nob Hill Publishing, LLC, 2009.

[178] J. Rawlings, E. Meadows, and K. Muske. Nonlinear model predictive control: A tutorial and survey. In *Proc. Int. Symp. Adv. Control of Chemical Processes, ADCHEM*, Kyoto, Japan, 1994.

[179] A. Richards and J. How. Model predictive control of vehicle maneuvers with guaranteed completion time and robust feasibility. In *Proceedings of the IEEE American Control Conference (ACC 2003), Denver, CO, USA, June 4–6, 2003*, volume 5, pages 4034–4040, 2003.

[180] S. Robinson. Perturbed kuhn-tucker points and rates of convergence for a class of nonlinear programming algorithms. *Mathematical Programming*, 7:1–16, 1974.

[181] C. D. T. Runge. Über die numerische Auflösung von Differentialgleichungen. *Mathematische Annalen*, 46(2):167–178, 1895.

[182] R. Russell and L. Shampine. A collocation method for boundary value problems. *Numerische Mathematik*, 19:1–28, 1972.

[183] S. Sager. *Numerical methods for mixed–integer optimal control problems*. Der andere Verlag, Tönning, Lübeck, Marburg, 2005. URL `http://sager1.de/sebastian/downloads/Sager2005.pdf`. ISBN 3-89959-416-9.

[184] S. Sager, H. Bock, M. Diehl, G. Reinelt, and J. Schlöder. Numerical methods for optimal control with binary control functions applied to a Lotka-Volterra type fishing problem. In A. Seeger, editor, *Recent Advances in Optimization (Proceedings of the 12th French-German-Spanish Conference on Optimization)*, volume 563 of *Lectures Notes in Economics and Mathematical Systems*, pages 269–289, Heidelberg, 2006. Springer Verlag.

[185] S. Sager, M. Diehl, G. Singh, A. Küpper, and S. Engell. Determining SMB superstructures by mixed-integer control. In K.-H. Waldmann and U. Stocker, editors, *Proceedings OR2006*, pages 37–44, Karlsruhe, 2007. Springer Verlag.

[186] S. Sager, H. Bock, and M. Diehl. Solving mixed-integer control problems by sum up rounding with guaranteed integer gap. Preprint, IWR, University of Heidelberg, 2008. URL `http://www.ub.uni-heidelberg.de/archiv/8384`.

[187] S. Sager, C. Kirches, and H. Bock. Fast solution of periodic optimal

control problems in automobile test-driving with gear shifts. In *Proceedings of the 47th IEEE Conference on Decision and Control (CDC 2008), Cancun, Mexico*, pages 1563–1568, 2008. ISBN: 978-1-4244-3124-3.

[188] S. Sager, H. Bock, and M. Diehl. The integer approximation error in mixed-integer optimal control. To appear in *Mathematical Programming*, 2011. URL `http://mathopt.de/PUBLICATIONS/Sager2011.pdf`.

[189] S. Sager, G. Reinelt, and H. Bock. Direct methods with maximal lower bound for mixed-integer optimal control problems. *Mathematical Programming*, 118(1):109–149, 2009. URL `http://mathopt.de/PUBLICATIONS/Sager2009.pdf`.

[190] S. Sager, M. Jung, and C. Kirches. Combinatorial Integral Approximation. *Mathematical Methods of Operations Research*, 73(3):363–380, 2011. URL `http://mathopt.de/PUBLICATIONS/Sager2011a.pdf`.

[191] A. Schäfer. *Efficient reduced Newton-type methods for solution of large-scale structured optimization problems with application to biological and chemical processes*. PhD thesis, Ruprecht–Karls–Universität Heidelberg, 2005. URL `http://archiv.ub.uni-heidelberg.de/volltextserver/volltexte/2005/5264/`.

[192] S. Scholtes. Convergence properties of a regularization scheme for mathematical programs with complementarity constraints. *SIAM Journal on Optimization*, 11:918–936, 2001.

[193] S. Scholtes. Nonconvex structures in nonlinear programming. *Operations Research*, 52(3):368–383, May–June 2004.

[194] V. Schulz, H. Bock, and M. Steinbach. Exploiting invariants in the numerical solution of multipoint boundary value problems for DAEs. *SIAM Journal on Scientific Computing*, 19:440–467, 1998.

[195] C. Schweiger and C. Floudas. Interaction of design and control: Optimization with dynamic models. In W. Hager and P. Pardalos, editors, *Optimal Control: Theory, Algorithms, and Applications*, pages 388–435. Kluwer Academic Publishers, 1997.

[196] O. Shaik, S. Sager, O. Slaby, and D. Lebiedz. Phase tracking and restoration of circadian rhythms by model-based optimal control. *IET Systems Biology*, 2:16–23, 2008.

[197] L. F. Shampine. Interpolation for Runge–Kutta formulas. *SIAM Jour-*

nal on Numerical Analysis, 22(5):1014–1027, October 1985.

[198] D. F. Shanno. Conditioning of Quasi–Newton methods for function minimization. *Mathematics of Computation*, 24(111):647–656, July 1970.

[199] M. Soliman, C. Swartz, and R. Baker. A mixed-integer formulation for back–off under constrained predictive control. *Computers and Chemical Engineering*, 32:2409–2419, 2008.

[200] B. Speelpenning. *Compiling fast partial derivatives of functions given by algorithms.* PhD thesis, University of Illinois at Urbana-Champaign, 1980.

[201] M. Steinbach. A structured interior point SQP method for nonlinear optimal control problems. In R. Bulirsch and D. Kraft, editors, *Computational Optimal Control*, volume 115 of *International Series of Numerical Mathematics*, pages 213–222. Birkhäuser, Basel Boston Berlin, 1994. ISBN 0-8176-5015-6.

[202] M. Steinbach. *Fast recursive SQP methods for large-scale optimal control problems.* PhD thesis, Ruprecht–Karls–Universität Heidelberg, 1995.

[203] M. Steinbach. Structured interior point SQP methods in optimal control. *Zeitschrift für Angewandte Mathematik und Mechanik*, 76(S3): 59–62, 1996.

[204] M. Steinbach. Tree–sparse convex programs. *Mathematical Methods of Operations Research*, 56(3):347–376, 2002.

[205] J. Stoer and R. Bulirsch. *Introduction to Numerical Analysis.* Springer Verlag, 1992.

[206] O. Stryk. Numerical solution of optimal control problems by direct collocation. In *Optimal Control: Calculus of Variations, Optimal Control Theory and Numerical Methods*, volume 111, pages 129–143. Bulirsch et al., 1993.

[207] R. Stubbs and S. Mehrotra. Generating convex quadratic inequalities for mixed 0–1 programs. *Journal of Global Optimization*, 24:311–332, 2002.

[208] O. Stursberg and S. Engell. Optimal control of switched continuous systems using mixed-integer programming. In *15th IFAC World Congress*, Barcelona, 2002. Paper Th-A06-4.

[209] S. Terwen, M. Back, and V. Krebs. Predictive powertrain control for heavy duty trucks. In *Proceedings of IFAC Symposium in Advances in Automotive Control*, pages 451–457, Salerno, Italy, 2004.

[210] T. Tsang, D. Himmelblau, and T. Edgar. Optimal control via collocation and non-linear programming. *International Journal on Control*, 21: 763–768, 1975.

[211] G. Vainikko. On the stability and convergence of the collocation method. *Differentsial'nye Uravneniya*, 1:244–254, 1965. (In Russian. Translated in Differential Equations, 1 (1965), pp. 186–194).

[212] R. Vanderbei. LOQO: An interior point code for quadratic programming. *Optimization Methods and Software*, 11(1–4):451–484, 1999.

[213] J. H. Verner. Explicit Runge–Kutta methods with estimates of the local truncation error. *SIAM Journal on Numerical Analysis*, 15(4):772–790, August 1978.

[214] A. Wächter. *An Interior Point Algorithm for Large-Scale Nonlinear Optimization with Applications in Process Engineering*. PhD thesis, Carnegie Mellon University, 2002.

[215] A. Wächter and L. Biegler. On the implementation of an interior-point filter line-search algorithm for large-scale nonlinear programming. *Mathematical Programming*, 106(1):25–57, 2006.

[216] X. Wang. Resolution of ties in parametric quadratic programming. Master's thesis, University of Waterloo, Ontario, Canada, 2004.

[217] Y. Wang and S. Boyd. *Fast Model Predictive Control Using Online Optimization*. 2008.

[218] R. Wengert. A simple automatic derivative evaluation program. *Commun. ACM*, 7(8):463–464, 1964.

[219] R. Whaley and A. Petitet. Minimizing development and maintenance costs in supporting persistently optimized BLAS. *Software: Practice and Experience*, 35(2):101–121, February 2005.

[220] J. Wilkinson. *The Algebraic Eigenvalue Problem*. Clarendon Press, Oxford, 1965.

[221] R. Wilson. *A simplicial algorithm for concave programming*. PhD thesis, Harvard University, 1963.

[222] Wolfram Research, Inc. *Mathematica*. Wolfram Research, Inc., Champaign, Illinois, 5.0 edition, 2003.

[223] E. Zafiriou. Robust model predictive control of processes with hard constraints. *Computers & Chemical Engineering*, 14(4–5):359–371, 1990.

[224] V. Zavala and L. Biegler. The advanced–step NMPC controller: optimality, stability and robustness. *Automatica*, 45(1):86–93, January 2009. submitted.

[225] V. Zavala, C. Laird, and L. Biegler. A fast computational framework for large-scale moving-horizon estimation. In *Proceedings of the 8th International Symposium on Dynamics and Control of Process Systems (DYCOPS)*, Cancun, Mexico, 2007.